Understanding Vision

Understanding Vision
Theory, Models, and Data

Li Zhaoping
University College London
London, UK

OXFORD
UNIVERSITY PRESS

OXFORD
UNIVERSITY PRESS

Great Clarendon Street, Oxford, OX2 6DP,
United Kingdom

Oxford University Press is a department of the University of Oxford.
It furthers the University's objective of excellence in research, scholarship,
and education by publishing worldwide. Oxford is a registered trade mark of
Oxford University Press in the UK and in certain other countries

Published in the United States of America by Oxford University Press
198 Madison Avenue, New York, NY 10016, United States of America

British Library Cataloguing in Publication Data
Data available

Library of Congress Control Number: 2013957434

ISBN 978–0–19–956466–8

As printed and bound by
CPI Group (UK) Ltd, Croydon, CR0 4YY

Oxford University Press makes no representation, express or implied, that the
drug dosages in this book are correct. Readers must therefore always check
the product information and clinical procedures with the most up-to-date
published product information and data sheets provided by the manufacturers
and the most recent codes of conduct and safety regulations. The authors and
the publishers do not accept responsibility or legal liability for any errors in the
text or for the misuse or misapplication of material in this work. Except where
otherwise stated, drug dosages and recommendations are for the non-pregnant
adult who is not breast-feeding

Links to third party websites are provided by Oxford in good faith and
for information only. Oxford disclaims any responsibility for the materials
contained in any third party website referenced in this work.

Preface

This book was originally motivated by a need to provide supplementary materials and additional readings to students of my lectures on vision. These lectures usually lasted a total of two to four hours. They were presented in a colloquial style in summer schools and graduate level courses on computational neuroscience. Originally, much of the material in these lectures was drawn from my review paper in 2006 (Zhaoping 2006b), which mainly focused on the topics of visual encoding and exogenous visual attentional selection.

Since then, there have been additional motivations. One is the recognition that although there are excellent books covering experimental observations in vision and special topics in theoretical or computational vision, there has been no general or systematic book on theoretical vision since *Vision* by David Marr in the early 1980s. However, there have been substantial developments in theoretical vision over the last three decades. This motivated the revised goal of providing material for the systematic teaching of computational vision. Accordingly, not only has the style of the manuscript shifted toward a textbook but also I have included material on visual decoding. Encouraged by feedback from colleagues and friends, I additionally expanded the intended readership to experimentalists in vision science and researchers in the computer and machine vision communities. I tested a draft version of the book in a vision course containing 30 lectures, each about 50 minutes long, to a student body consisting of undergraduate students, graduate students, and early career researchers.

For readers with a focused interest in just one of the topics in the book, it is feasible to read just the chapter on this topic without having read or fully comprehended the other chapters. In particular, Chapter 2 presents a brief overview of experimental observations in biological vision, Chapter 3 is on encoding of visual inputs, Chapter 5 is on visual attentional selection driven by sensory inputs, and Chapter 6 is on visual perception or decoding.

The expanded goal of the book made writing it much more challenging, although also more rewarding and educational for me. I will be most appreciative of any feedback on the book.

I have been very fortunate to have had the support and encouragement of many people and organizations over the course of writing this book, and I hereby acknowledge them with great gratitude. Firstly, David Sainsbury and the Gatsby Charitable Foundation, through the medium of Sarah Caddick, Peter Hesketh, Michael Pattison, Jess Roberts, and Gary Wilson, have most generously supported my research and teaching, especially since funding for interdisciplinary research like mine is not easily available. Secondly, many colleagues and students have read various versions or parts of my manuscript and given me useful feedback. These include Peter Dayan, Nikos Fragopanagos, Keith May, Li Zhe, Ning Qian, Richard Turner, Arnold Wilkins, Lawrence York, Yuan Jinghui, Michael Zehetleitner, Zhou Li, and Zhuge Changjing. Various colleagues have given me invaluable help with references and technical information. These include Alessandra Anglelucci, Rodney Douglas, Henry Kennedy, Dan Kersten, Simon Laughlin, Ken Nakayama, David Perrett, Jonathan Pillow, Arnd Roth, Peter Schiller, Stewart Shipp, and Keiji Tanaka. My former postdoctoral fellow Keith May particularly encouraged me to make the book accessible to researchers without a background in advanced mathematics. Many other colleagues have provided help in various forms, including discussions and kind encouragement.

During the course of writing this book, University College London (UCL), and in particular the department of computer science, has provided me with a stimulating and supportive environment intellectually and administratively. This environment enabled me to devote my time to scientific research and enjoy interactions with colleagues and students. I am also grateful to Tsinghua University where I spent my sabbatical year from 2008 to 2009 and where I continue to visit to interact with colleagues and students in the Beijing area. Enthusiasm and interests from students at UCL, Tsinghua, EU Advanced Course on Computational Neuroscience, and the Okinawa Computational Neuroscience Course provide me with additional energy.

Throughout my research career, I have benefited greatly from scientific interactions and conversations with many people including Ni Guang-Jiong, Frank Porter, Jeremy Pine, John Hopfield, James Bower, Matt Wilson, David Van Essen, Steven Frautschi, Andreas Herz, Leo van Hemmen, Gerald Westheimer, Jochen Braun, Jella Atema, Wyeth Bair, Peter Dayan, John Hertz, Jon Driver, Alex Lewis, Silvia Scarpetta, Michael Herzog, Adam Sillito, Helen Jones, Elliott Freeman, Robert Snowden, Keith May, Nathalie Guyader, Ansgar Koene, Li Jingling, Stewart Shipp, Josh Solomon, Mike Morgan, Peter Latham, Chris Frith, Uta Frith, Zhou Li, Zhao Lingyun, Gao Meng, Li Zhe, Lisa Cipolotti, Bill Geisler, Fang Fang, Zhang Xilin, Li Wu, Wang Feng, Marty Sereno, Massimiliano Oliveri, Renata Mangano, Mieke Donk, and Ingrid Scharlau. Their knowledge, perspectives, criticisms, and enthusiasm are great sources of inspiration. I am also very grateful for teaching and mentorship from people including Zhang Jingfu and John Hopfield, and for the opportunity given to me by Lee Tsung-Dao.

I would also like to thank editors Martin Baum, Charlotte Green, and Viki Mortimer from Oxford University Press for their efficient communication and support, and kind patience; Viki Mortimer and the copy editor Elizabeth Farrell have been tremendously helpful in improving the text and my English.

My family has provided tremendous material and moral support. In particular, my husband Peter read through the whole book and gave invaluable feedback, provided helpful tips for book writing, and, together with my children, has been very patient in helping with my English grammar and vocabulary. My husband and children were also very generous in giving me time, and constantly cheered me on to finish the book. I am very grateful for their love and the love of my parents and siblings.

Contents

1 Approach and scope

1.1 The approach

Vision is the most intensively studied aspect of the brain from physiological, anatomical, and behavioral viewpoints (Zigmond, Bloom, Landis, Roberts and Squire 1999). The saying that our eyes are windows onto our brain is not unreasonable since, at least in primates, brain areas devoted to visual functions occupy a large portion of the cerebral cortex—about 50% in monkeys (see Chapter 2). Understanding visual processing can, hopefully, reveal much about how the brain works. Vision researchers come from many specialist fields, including physiology, psychology, anatomy, medicine, engineering and computer science, mathematics, and physics. Each specialist field has its own distinct approach and value system. A common language is essential for effective communication and collaboration between visual scientists. One way to achieve this is to frame and define the problems and terminologies clearly in order to communicate the details. This is what I will try my best to do in this book. Clear definitions of the problems and terms are used. Meanwhile, I am without exception using a particular approach and value system. This value system is manifested particularly in this book's emphasis on linking theory with data and on linking behavior with neural substrates via theory and models. The importance of formulating research problems is especially recognized.

The book is written so that readers may skip some parts without it being too difficult to follow other parts. In particular, vision scientists unfamiliar with mathematical details should be able to conceptually follow the theoretical principles, their predictions, and their applications without going through the more mathematical pages. Meanwhile, a knowledge of linear algebra, probability theory, and sometimes nonlinear dynamics will enable readers to enjoy the technical details. For readers with a physical science background, this book serves as an analytical introduction to biological vision. Mathematical details are introduced gently in the beginning; they progress in the book to a technical level suitable for university students in physical sciences.

1.1.1 Data, models, and theory

This book aims to understand vision through the interplay between data, models, and theory, each playing its respective role, as illustrated in Fig. 1.1. All modalities of experimental data—physiological, anatomical, and behavioral—provide inspiration for, and ultimate tests of, all the theories. Theoretical studies suggest computational principles or hypotheses that elucidate physiology and anatomy in terms of visual behavior, and vice versa. They should provide non-trivial insights about the multitude of experimental observations, link seemingly unrelated data to each other, and motivate further experimental investigations. Successful theoretical hypotheses also enable us to understand the computational power of the neural substrates to solve hard problems. Often, appropriate mathematical formulations of the theories are necessary to make them sufficiently precise, so that the theories can be falsifiable and have the predictive power to impact experimental investigations.

For example, this book presents detailed expositions of two theories of early vision. One is the efficient coding hypothesis, which accounts for visual receptive fields in the early parts

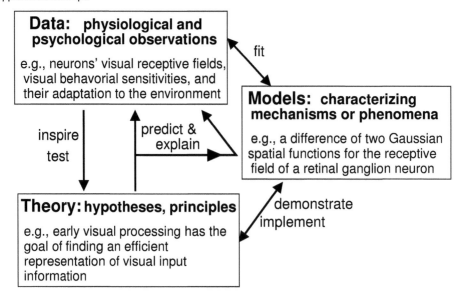

Fig. 1.1: The roles of data, models, and theory in understanding vision.

of the visual pathway (in Chapter 3). The other is the V1 saliency hypothesis, suggesting a functional role for the primary visual cortex (V1; in Chapter 5) and linking V1 mechanisms with visual behavior in attentional selection. The experimental data inspiring the theories include the receptive fields of neurons in the retina and cortex, the dependence of those receptive fields on the visual "niche" of the animal species, the adaptation of the receptive fields to the environment, the behavioral sensitivities of the animals to various visual stimuli, the particular neural circuits and properties in V1, and the behavior in visual search and segmentation tasks.

Mathematical models, including phenomenological and neural circuit models comprising neural mechanisms, are very useful tools in linking theory with data. This is particularly so when the complexity of a model is appropriately aligned with the question addressed. For example, models can be used to demonstrate theoretical hypotheses in particular instances or can be used to test the feasibility of these hypotheses through specific neural mechanisms.

In some cases, models are used not to demonstrate a theory or to link theory with data but merely to describe, organize, or categorize data. For example, if we have many pairs of data points relating input light level at a retinal location with the response of a retinal neuron, we may fit the data by using, for example, a polynomial relationship between the light levels and the responses. Such a fit could use just a few model parameters to approximate or summarize, say, 100 pairs of data. Since any fit of a model to data is never more accurate than the data themselves, a model is only worthwhile when it can describe data sufficiently accurately using far fewer parameters than the number of data points modeled. This book (in particular Chapter 2) contains many such descriptions of data. For example, one describes the receptive fields of retinal ganglion cells using a center-surround model; another describes the sensory response properties of neurons using a nonlinear function involving a few parameters. In these cases, we are merely using models to provide a simplified description or summary of data; these models are not linked to any theory or ideas.

Meanwhile, the value of a theory can often be assessed by its ability to provide the parameter values in a model. Hence, a good theory can reduce the number of adjustable, or free, parameters in a model to give a good account of the data.

Often when models are intended not merely to describe data but also to link a theory to data or to demonstrate a theory, simplifications and approximations are used. These simplifications and approximations make models quantitatively inaccurate. However, they can still enable models to serve a purpose when quantitative precision is not essential. Hence, mere quantitative imprecision of models should not be a basis to dismiss the theories that underlie the models. This is especially so when simplified toy models are used to illustrate a theoretical concept. For example, Newton's laws should not be discarded because they cannot precisely predict the trajectory of a rocket when using an imprecise model of the Earth's atmosphere. Similarly, the theoretical proposal that early visual processing aims to encode raw visual inputs efficiently (in Chapter 3) could still be correct, even if the visual receptive fields of the retinal ganglion cells are modeled simply as differences of Gaussians to illustrate the proposal. As George Box puts it, "all models are wrong, but some are useful" (Box and Draper 1987).

To illustrate a theoretical concept, the minimal model, which is the one with the smallest number of parameters or mechanisms, is preferred among models that fit equally well. For example, if the theory of efficient visual coding is equally well characterized by two models, one employing the detailed time sequences of neural action potentials and the other just the coarser firing rates, then the simpler rate model should be preferred.

1.1.2   From physiology to behavior and back via theory and models

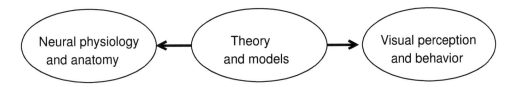

Fig. 1.2: Theory and models as the bridge between neural substrates and visual behavior.

An important criterion for understanding the brain is the ability to link physiology with animal behavior. We thus consider it essential that theory, with the aid of models, should provide this link. This means that when the "Data" component in Fig. 1.1 is decomposed into two subcomponents, physiology and behavior, theory and models should link these two subcomponents together, as indicated in Fig. 1.2. Hence, this book does not make statements such as "color vision is solved once we know the genes responsible for how photoreceptors sense light spectra," because such a statement says nothing about visual perception or behavior. For another example, we will not analyze the oscillatory temporal activities in a model neural network if the analysis does not link those activities with visual perception and cognition. For the same reason, the book is not satisfied with some of the analyses of computational and behavioral elements in Chapter 6, since they do not provide sufficient insight into which brain areas or neural mechanisms are responsible. On the other hand, this book favors experimental tests of theoretical predictions as to how visual behavior arises from neural mechanisms and which neural mechanisms are implied by visual behavior. One example is the prediction of visual gaze shifts (see Chapter 5) from neural mechanisms in V1 via the medium of the theoretical hypothesis that V1 computes a bottom-up saliency map to guide attention.

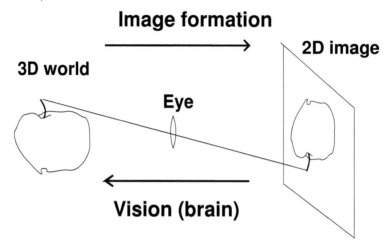

Fig. 1.3: Vision as an inverse problem of image formation.

1.2   The problem of vision

Vision could be defined as the inverse of imaging, which is the operation of transforming the three-dimensional (3D) visual world to two-dimensional (2D) images; see Fig. 1.3. A given visual world, which contains objects that reflect or emit light, gives rise to a unique image given a viewing direction, simply by projecting the light from the 3D scene onto the image plane. (For simplicity of argument, we ignore stereo vision when two or more 2D images are involved.) This imaging transform is well understood, as amply demonstrated by the success of computer graphics applied to movie making. Meanwhile, the problem that is inverse to imaging involves obtaining information about the 3D world from 2D images. Human vision is poorly understood, partly because this inverse problem is ill posed, as the 3D world is not uniquely determined given a 2D image. An image pixel value could arise from an infinite number of possible light-emitting locations along the line of sight associated with this image pixel. Nevertheless, our subjective visual experience is typically a unique and unambiguous percept (see Chapter 6) of a given visual scene. Understanding how vision chooses a unique solution to the inverse problem among the numerous possibilities is especially challenging.

However, it is insufficient to define vision merely as the inverse problem to imaging. Rather, vision should take visual input images and output a description of the visual scene that is useful to help the organism in tasks promoting its survival. Many visual tasks, such as recognizing or localizing an object, can be executed without having the full solutions to the inverse problem. For example, one can recognize some object in the foreground of a scene without recognizing anything in the background. One can also grasp an object without recognizing it or knowing all its various aspects. However, vision is also much more than the inverse problem. For instance, even if one finds the solution to inverse imaging so that one knows light reflectances and emissions as a function of locations in the 3D world, it does not mean that one can successfully recognize the person in the retinal image that these light reflectances and emissions cause. A mother needs to recognize her own infant (rather than light reflections as a function of spatial locations) to care for her child; a frog needs a description sufficient to make a decision on whether and to where to stick its tongue out to catch its prey. In both these examples, a full description of light reflectance or emission in the 3D world without recognition or decision is insufficient, in that what the animal requires is a description that could lead to a motor program resulting in orienting or acting toward the child or prey. Furthermore, in both examples, object recognition is not necessarily a pre-requisite for these motor programs. Obviously, understanding vision requires understanding what exactly are the vision problem and its subproblems.

Here we define vision as the problem of transforming visual inputs into a description which is sufficiently adequate for the specific task considered. There can be specific tasks, including immediate motor actions such as orienting and grasping, and more general tasks such as object recognition and localization for learning or memory. The latter can be used for future motor actions. For example, a description like "my child to my left" may help a mother's perceptual decision concerning childcare, and another description like "shoot tongue now five centimeters straight-ahead if hungry" may be sufficient for the frog's task. Understanding vision requires identifying and understanding the tasks and corresponding descriptions necessary.

1.2.1 Visual tasks and subtasks

Many or all visual tasks may comprise multiple subtasks which involve different visual processes. This may not be appreciated since, intuitively, visual perception appears instantaneous. Figure 1.4 demonstrates that vision involves at least two processes: one is to look, and the second is to see.[1] This example is in a visual task (Zhaoping and Guyader 2007) requiring observers to find a uniquely tilted oblique bar in an image. The image contains hundreds of "X" shapes, each made of two intersecting bars, one oblique and one cardinal (horizontal or vertical). All oblique bars except the target oblique bar are uniformly tilted 45° from vertical. The unique target bar is tilted 45° from vertical in the opposite direction; it is never at the center of the image, which is where the observer's gaze starts at the beginning of the search.

The position and shifts of gaze during a search can be observed with an eye tracker. Typically, human gaze jumps from one location to another about three times a second. These gaze jumps are called saccades, each of which takes typically 20–60 ms. The duration between two successive saccades is called a gaze fixation period. The yellow trace in Fig. 1.4 is the initial scan path of an observer's gaze during a search trial up to the moment when the gaze reached the target. In this example, the gaze became close enough to the target after only one or two saccades. This is remarkable, since there are hundreds of "X" shapes in the image, and it is very difficult to identify the tilt direction of the oblique bar in an "X" shape a few items away from one's current fixation point without moving the gaze (this is very similar to the difficulty in reading text several characters away from one's current fixation). One may imagine that the observer must have seen the target at the saccadic destination before saccading to it. However, if one removes the image as soon as the gaze reaches the target, observers are typically unsure where the target is or whether they have already looked at it.

Altogether, this suggests that before seeing something, one has first to look. This is the act of saccading to the location of the object to be seen. Hence, one of the subtasks in this visual search task is to decide where to look. Observers are apparently doing this task very well, since they can typically saccade to the target within 2–3 seconds, even though they cannot see clearly the visual inputs at their saccadic destination beforehand. Each saccade and the immediately subsequent fixation period before the next saccade may be seen as an episode of looking and seeing in vision. Apparently, for this task, observers need to see the visual input at their fixation after each saccade, to decide whether the visual input at their current fixation is the target of their search.

---

[1] In Chinese, the colloquial expression for seeing is in fact composed of two characters: 看见, one for "look", and one for "see".

The gaze scan path during the initial (yellow), subsequent (blue and then red), and final (magenta) periods of a search by an observer for a uniquely tilted bar

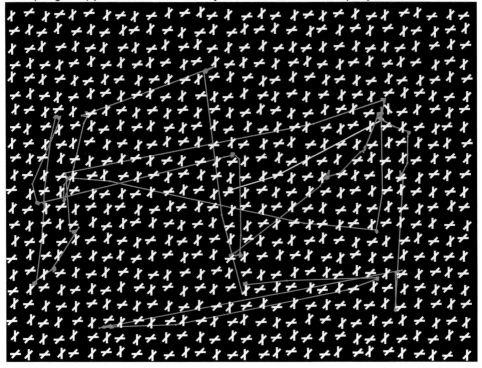

Fig. 1.4: Demonstration that vision involves separate processes of looking and seeing and that seeing can be viewpoint invariant (Zhaoping and Guyader 2007). Shown here is the gaze scan path (colored) of a human observer searching for a uniquely tilted oblique bar in an image containing many white bars which are horizontal, vertical, or oblique. Each oblique bar is intersected by a cardinal (horizontal or vertical) bar, making an "X" shape. All oblique bars have the same tilt, except for one, the search target, which is tilted in the opposite direction. Yellow, blue, red, and magenta traces represent temporally consecutive gaze scan paths from the initial (in yellow, starting from the middle of the image) to the final (in magenta) periods of this observer's search, ending when the observer reported the target (in the upper-right quadrant of the image). The gaze reached the target within 1–2 saccades after the start of the search. It abandoned the target twice, each time after arriving at and staying near the target for at least 400 milliseconds (ms). The observer only reported the target after the gaze reached it for the third time. Since the "X" shape containing the target had the same shape as the other "X"s, except for a rotation, the observer was likely confused. If the image is immediately removed when the gaze arrives at the target, observers are typically unsure where the target is.

Figure 1.4 also demonstrates that the seeing process exhibits what is known as object invariance, which means that one can recognize the same object, e.g., the same "X" shape or the same person, regardless of the viewpoint (in this case, the orientation of the object). In this search trial, after the gaze reached the target for the first or the second time, it stayed near the target for at least 400 milliseconds (ms), and then abandoned the target to search elsewhere before reapproaching the target. Apparently, although the task was to look for a unique oblique bar, the seeing process identified the "X" shape containing the target bar as being identical to the other "X"s, despite their different orientations. This confused the observer, making him abandon the target and search elsewhere. This confusion can be eliminated (Zhaoping and Guyader 2007) if the unique oblique bar is tilted only 20° from its cardinal bar partner, making the "X" shape uniquely thinner than other "X"s (this will be explained in more detail in Chapter 6).

The example above also demonstrates two of the most difficult subproblems in vision: object invariance and visual segmentation. Object invariance is exemplified by the recognition of the "X" shape regardless of its orientation. In another example, the same person can be recognized in completely different images, depending on whether the person is viewed from the front or the side, in daylight coming from the sky above or in artificial light coming from a desk lamp at the side. Object invariance is thus generally useful for visual recognition, even though it interferes with the example task in Fig. 1.4 (which we used to demonstrate its presence). The brain mechanisms responsible for it are still mysterious, despite much research.

Visual segmentation is defined as selecting the visual image area containing a particular visual object or region. In the above example, this selection is partly done by the act of looking or directing gaze at selected image locations. Visual segmentation seems puzzling, since underlying it is an apparent chicken-and-egg problem—it helps to recognize the object first in order to select the image location for it, while it helps to select the image location for the object first in order to recognize the object! We will see in this book that visual segmentation is related to visual attentional selection and that object invariance has to be addressed by the visual decoding process.

This visual search task is only an illustrative example. Other visual tasks involve other subtasks and processes. Hence, the definition of the problem of vision is task specific. Since we are still ignorant about many aspects of vision, a starting point is first to assume that some subproblems and visual processes for vision are generally applicable to many visual tasks and then to identify these general subproblems and their corresponding visual processes. We can then study vision through the medium of these visual processes.

1.2.2 Vision seen through visual encoding, selection, and decoding

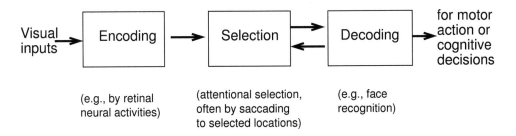

Fig. 1.5: Vision simplistically decomposed into three processes, motivating the organization of this book. Chapters 3–4 present encoding, which is the process of representing visual images in terms of neural activities such as those in the photoreceptors and retinal ganglion cells. Chapter 5 focuses on selection, often called attentional selection, which is the process of selecting a fraction of encoded information for further processing. Chapter 6 presents decoding, which is the process of inferring or making hypotheses about aspects of the visual scene based on the selected information.

Inattentional blindness — spotting the difference between the two images

Fig. 1.6: Although we have a vivid impression that we see everything clearly in these two images, it is difficult to spot the difference between them. The difference is most obvious once it is spotted or pointed out to us.[2] This demonstrates that we are blind to much of the visual input unless our attention is directed appropriately. Copyright © 2013, Alyssa Dayan.

This book presents a singular rather than an all-inclusive view. The view adopts the very simple starting point that many tasks or problems for vision involve three general visual processing stages: visual encoding, visual selection, and visual decoding (see Fig. 1.5). In simple terms, visual encoding is the process of representing or transforming visual images into neural activities, such as the sampling of visual inputs by retinal photoreceptors, and then transforming these receptor signals into the activities of the retinal ganglion cells. Because brain resources are limited, only a small fraction of the encoded information can be further processed in detail. Visual selection is the process of selecting this fraction. This selection process is often called visual attentional selection. We are therefore blind to whatever is not selected. This inattentional blindness (Simons and Chabris 1999) can be appreciated by seeing how difficult it is to spot the difference between the two images in Fig. 1.6, even though we have the impression of vividly seeing everything clearly. Selection is most obviously manifested behaviorally in directing gaze to the selected location in the scene. This is why a primary school teacher asks the pupils to look at the board. We have also seen that the target of a visual search in Fig. 1.4 is hidden from perception unless our gaze position is close to it. Visual decoding processes the selected image information to create the perception (identification and/or localization) of the visual objects in the scene so that motor actions or other cognitive decisions can be made regarding these objects. This decoding is also often called visual inference or visual recognition, the process of figuring out the nature of the visual world from the input images, as illustrated in Fig. 1.3. The inferred visual scene is often influenced by our preconceptions about, or knowledge of, the world. This is necessary to overcome the non-uniqueness associated with the many visual scenes that could possibly have given rise to a given input image. Since encoding, selection, and decoding are expected to be very approximately sequential processes, it is likely that they are roughly associated with consecutive brain regions along the visual pathway, in the feedforward direction of the flow of visual information. The vision research community is actively engaged in trying to identify which brain areas are involved in each of these stages.

It should be noted that some visual behaviors proceed without requiring all three stages. For example, involuntary saccades and blinks may not require the decoding stage, as they do not require visual objects to be localized or identified. It has also been observed that in

---

[2] The difference between the two images in Fig. 1.6 is in the background trees at the lower middle part of the images.

many animal species, much of visual selection is already implemented in the visual sampling stage. In such cases, encoding and selection stages are merged. However, as a starting point, this book adopts these three separate stages to compartmentalize the separate computational processes and keep them in sharp focus.

Viewing vision through encoding, selection, and decoding differs from many traditional approaches, which suggest decomposition in terms of low-level, mid-level, and high-level vision. Low-level vision processes visual input images to extract simple features like bars, edges, and colors. Mid-level vision typically refers to the process of obtaining representations of object surfaces from images. High-level vision often refers to visual object recognition and cognition. Low-level, and sometimes even mid-level, vision is sometimes referred to as early vision.

We eschew this traditional division of vision into levels, since it fails to highlight the problem of visual selection. Selection is highly non-trivial, since, as discussed, it suffers from a chicken-and-egg problem in relation to object recognition. Visual selection is also dramatic, since the raw data transduced by the retinal photoreceptors (Kelly 1962) come at a rate of many megabytes per second. Note that several megabytes is more than that is needed to encode the text in a typical long novel. The data rate has to be reduced by more than 99% to about 10^2 bits per second (Sziklai 1956), the capacity of the human attentional bottleneck. Thus, we can typically read no more than two sentences in a novel per second. Visual cognition presents us with the illusion that there is no such information loss. Because of this illusion, inattentional blindness as demonstrated in Fig. 1.6 was only recently recognized (Simons and Chabris 1999). The dramatic reduction in information due to selection is likely to have a profound effect on how vision works. For example, one should expect that brain areas devoted to post-selectional processing (namely, much of decoding) should be mostly blind to peripheral vision, since the selection stage should have brought whatever visual objects are to be decoded to central vision (Zhaoping 2011) using eye movements. By highlighting the "selection" stage explicitly, we hope that the encoding–selection–decoding framework will provide alternative insights as to how vision works. In particular, one may investigate whether and how visual encoding, selection, and decoding processes affect the task of building a surface representation of visual scenes, one of the unsolved mid-level vision tasks in the traditional framework and an essential component in the vision transform from 2D images to 3D scenes.

After a brief overview in Chapter 2 of the experimental data about vision, Chapters 3–4 present visual encoding. They aim to understand retinal input sampling and the receptive fields of retinal and primary cortical neurons based on the efficient coding hypothesis. The hypothesis is: neural sampling and receptive fields in early visual centers (e.g., retina and V1) aim to transform the raw visual input into a more efficient representation of information so that as much information as possible is extracted and transmitted to the next station along the visual pathway given the neural resources available (in terms of, e.g., the number of neurons and their dynamic response ranges). These chapters discuss the theory of efficient coding originally proposed half a century ago (Barlow 1961), present a mathematical formulation of this theory, derive visual receptive fields that are predicted by the theory in various developmental and ecological conditions, and highlight experimental tests of the theoretical predictions.

Chapters 4–5 present visual selection, focusing almost exclusively on selection by bottom-up or input driven factors independent of any current task goals. (Note that this bottom-up selection can however serve the evolutionary task of protecting the organism from unexpected dangers.) We adopt this focus partly because, compared to selection by top-down or goal-dependent factors (such as when children look at the teacher for learning), bottom-up selection is better understood in all three aspects of theory, neural circuit mechanisms (in V1), and visual behavior. In particular, Chapter 5 details the V1 saliency hypothesis that V1 (in primates in particular) creates a bottom-up saliency map, represented in terms of the most active V1 response to input at each visual location, which guides the selection of location. Although the efficient coding theory and the V1 theory of bottom-up selection involve different theoretical concepts and methodologies, they both concern how information bottlenecks in visual and cognitive pathways are overcome. Some experimental data have been involved in shaping the development of both theories, indicating that data exposing limitations in one theory can drive the development of another.

Chapter 6 discusses visual decoding, again emphasizing links with the neural substrates. Hence, it omits, or only very briefly describes, materials restricted to purely behavior or computer vision. This chapter is substantially abbreviated compared with Chapters 3–5, since much less is known about the neural mechanisms of decoding. The apparent huge gaps in our understanding of vision will hopefully motivate stimulating discussions and future studies.

1.2.2.1 Dependence on animal species

This book is primarily focused on vision in higher animals like primates, for which physiological and psychological experimental data are quite extensive. Hence, encoding, selection, and decoding stages of vision are examined in appropriate brain regions in these animals. The corresponding brain regions are likely very different in many lower animals such as insects. In particular, animals without a neocortex are likely to carry out more or all of the three processing stages in the retina, whereas primate retina is likely dominated by its role in encoding. It is also questionable, for example, whether a saliency map created by primate V1 to guide bottom-up selection is similarly created by rodent V1. Although variations across animal species leave many open questions, the framework of viewing vision through encoding, selection, and decoding stages can hopefully provide a starting point to understand vision and identify the corresponding brain mechanisms that will apply across many species.

1.2.3 Visual encoding in retina and V1

Often, processes occurring in retina and V1 are referred to as early visual processes. (A part of the thalamus called the lateral geniculate nucleus (LGN) is on the visual pathway from the retina to V1. Its functional role, other than that of a relay, is still controversial and is being investigated extensively (Sherman and Guillery 2004), see Section 2.2.7. We omit it for discussion here.) These two areas are better known physiologically and anatomically than most other visual regions in the brain. Hence, they afford greater opportunities for developing theories of their functional roles. This is because theoretical predictions can be more easily verified in existing data or tested in new experiments. Readers will thus find that the materials in much of this book relate to retina and V1.

Early vision creates representations at successive stages along the visual pathway from retina to LGN to V1. Its role is perhaps best understood in terms of how these representations overcome critical information bottlenecks along the visual pathway. We argue below for the process flow diagram in Fig. 1.7, in which the bottlenecks are overcome using two data reduction strategies applied sequentially: data compression and data selection.

Along the visual pathway, the first obvious bottleneck is the optic nerve from retina to LGN en route to V1. Retinal receptors can receive information at an estimated rate of 10^9 bits per second (Kelly 1962), i.e., roughly 25 frames per second of images of 2000×2000 pixels at one byte per pixel. The fact that there are approximately one million ganglion cells in humans, each transmitting information at about 10 bits/second (Nirenberg, Carcieri, Jacobs and Latham 2001), implies that the optic nerve provides a transmission capacity of only 10^7 bits/second. Hence, there is a reduction of two orders of magnitude in the transmission rate.

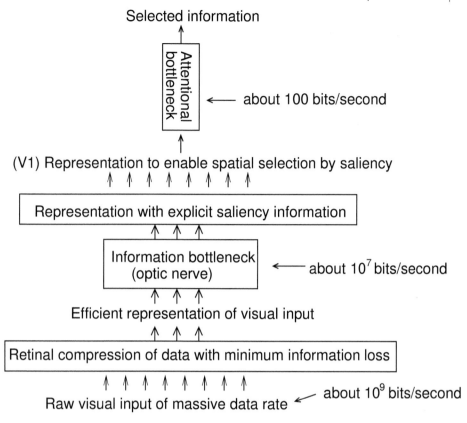

Fig. 1.7: Process flow diagram illustrating two bottom-up strategies proposed for early vision to reduce the data rate through information bottlenecks. One strategy resembles data compression, another is creating a saliency map to enable lossy selection of information. The data rates indicated are those for primates/humans. The LGN is omitted in this diagram.

Data compression, with as little loss of information as possible, should thus be the first strategy to help overcome the information bottleneck along the visual pathway.

Hence, a goal of early vision should be to compress data. Image compression methods in engineering, for instance the JPEG algorithm, can compress natural image data 20-fold without perceptually noticeable information loss, such that humans cannot easily distinguish original visual inputs from the ones reconstructed from the compressed signals. Chapter 3 presents the first data reduction strategy. Barlow (1961) and others have argued that early visual processes should take advantage of the statistical regularities or redundancy of visual inputs to represent as much input information as possible given limited neural resources. Limits may lie in the number of neurons, power consumption by neural activities, and noise; these all lead to an information bottleneck. Hence, input sampling by the cones and neural activity transforms by receptive fields (RFs), should be optimally designed to encode the raw inputs in an efficient form such that data is compressed while information loss is minimal. This is the efficient coding principle. Minimum information loss by data compression is also important since any information that is lost prior to visual selection could never subsequently be selected to be perceived.

This efficient coding principle has been shown to explain, to various extents, the color sensitivities of cones, the distributions of receptors on the retina, the properties of the receptive fields of retinal ganglion cells and V1 cells, and their behavioral manifestations in visual psychophysical performance. Since efficiency depends on the statistics of visual inputs, neural properties should adapt to prevailing visual scenes. One can thus predict from this principle the effects of visual adaptation and developmental conditions. For example, when the input signal is much stronger than the noise level, efficient coding often involves removing redundant representations of information.[3] When the input signal is weak (such as in a dim environment), efficient coding tends to preserve the input redundancy. These adaptations of the efficient code predict changes in neural receptive fields and behavioral sensitivities that can be tested experimentally.

1.2.4 Visual selection and V1's role in it

1.2.4.1 Top-down and bottom-up visual selection

The second information bottleneck along the visual pathway is more subtle but much more devastating. For humans, visual selection allows only about 40 bits/second (Sziklai 1956) to be processed further for perception. This capacity is called the attentional bottleneck, since the selected information is said to be attended. The non-selected information is thus deleted (not transmitted and/or replaced by new incoming input information), implying that our perception will be blind to it (Simons and Chabris 1999). Therefore, visual selection must decide which information to select and/or which to delete. Selection often takes the form of directing one's gaze to a spatial location, as shown in Fig. 1.4. Therefore, selection needs to decide on the location to select, so that the selected location is more likely than the ignored ones to be important or relevant to the animal.

One way is to select according to the animal's current goal; this is called goal-directed or top-down selection. For example, during reading, gaze is directed to the location of the text. In this case, the goal is to read, and selection involves directing gaze to goal locations. Top-down selection of gaze location could be based on input features such as selecting the color red when searching for one's red cup. It could also be based on knowledge of the environment, such as knowing where to look to select the text location for reading.

However, fast selection by goal-independent, autonomous mechanisms driven by external visual inputs is essential to be able to respond to unexpected events such as a sudden appearance of a stranger. Such goal-independent selection is called bottom-up selection. Bottom-up selection is known to be more potent and faster (Jonides 1981, Müller and Rabbitt 1989, Nakayama and Mackeben 1989) than top-down selection, consistent with its function of responding to unexpected events. In this book, visual locations which attract attention in the bottom-up manner are called salient, and the degree to which a location attracts bottom-up selection is called the saliency of this location.

1.2.4.2 Early or late selection?

At which stage along the visual pathway should the massively lossy information selection occur? Should it happen early along the pathway, in occipital visual cortical areas such as V1 and visual area 2 (V2), or should it occur late along the pathway, in the temporal or frontal lobe? Postponing lossy selection could postpone the deletion or non-transmission of information. Unfortunately, it also postpones the completion of cognitive processing. One may assume that data compression with minimum information loss could continue along the visual pathway until little additional efficiency could be gained. Meanwhile, it is difficult

---

[3]Barlow (1961) also argued that efficient coding could also have the cognitive role of revealing the underlying independent components of the inputs, e.g., individual objects.

to explain neural processes at the stage where lossy selection happens in terms of efficient coding (although it could still be useful after this stage; see Chapter 7).

In the psychology community, there has been an ongoing debate for at least half a century as to whether attentional selection happens early or late in visual and other (in particular auditory) sensory processing pathways. Early visual selection argues that, before selection, visual inputs are only processed according to image component features (such as luminance, color, edges) in parallel across the visual field and that visual objects are not identified (Broadbent 1958, Treisman 1985). Late visual selection argues that object identification is performed in parallel over the visual field before selection, which only subsequently limits which of the identified objects is selected to influence cognitive representations, guide actions, or be stored in memory (Deutsch and Deutsch 1963, Duncan 1980). Identifying the neural substrate and brain areas involved in lossy selection can help to resolve the debate between early versus late selection, if we can also understand the information processing that occurs up to that selection stage.

Our framework of viewing vision as being composed of encoding, selection, and decoding stages in a roughly sequential manner favors early rather than late selection, if one identifies decoding with object recognition. Meanwhile, decoding construed in this manner also influences selection, as suggested by the feedback route from decoding to selection in the diagram in Fig. 1.5. Hence, the debate between early versus late selection is perhaps difficult to resolve, as also pointed out by many other researchers. More clearly defined positions on both sides could make the debate more productive.

1.2.4.3   V1 for bottom-up selection

V1 is the largest visual cortical area in the brain. It is also the best known visual cortical area, since the pioneering investigations by Hubel and Wiesel more than half a century ago into the properties of its neural receptive fields and its underlying anatomy. However, the functional role that V1 plays in vision has been unclear for decades. Optimal visual inputs to activate a (monkey) V1 neuron are typically much smaller than the image sizes of many relevant and recognizable objects (such as faces), and V1 neurons' responses are tuned to basic features such as orientation, color, motion direction, and size of the visual inputs. Thus, it is often assumed that V1 merely plays the back-office role of preparing the image information for more important visual functions in subsequent cortical areas along the visual pathway.

Chapter 4 reviews the difficulties in understanding certain V1 properties based on the efficient coding principle. These properties include the overcomplete representation of visual inputs as well as the influence on a V1 neuron's response of contextual inputs lying outside the neuron's classical receptive field (CRF). It will be shown that these properties are consistent with the goal of information selection, the second data reduction strategy.

Chapter 5 presents the V1 saliency hypothesis (Li 1999a, Li 2002), which states that V1 creates a bottom-up saliency map of the visual field such that a location with a higher scalar value in this map is more likely to be selected in the bottom-up manner. The saliency values are proposed to be represented by the firing rates of V1 neurons such that the receptive field location of the most active V1 cell is most likely to be selected, regardless of the feature preferences of the neurons. We will show that this hypothesis links V1 physiology with the visual behavior of bottom-up selection and segmentation, provides testable predictions which are experimentally confirmed, and motivates new experimental investigations. Assuming that visual object recognition is not achieved up to V1 along the visual pathway, bottom-up selection in V1 favors early selection in the debate of early versus late selection.

Much less is known theoretically about top-down selection in the brain beyond what is already present in the experimental data. In particular, certain brain regions beyond V1 are believed to control top-down selection, and neural responses to visual inputs in various extrastriate cortex (downstream from V1 along the pathway) are modulated by top-down selection. In particular, neurons tend to be more sensitive to the attended rather than the ignored visual inputs. These observations are described as biased competition (Desimone and Duncan 1995), which means that the behavioral act of paying attention favorably biases the attended inputs in the competition for continued representation and further processing by the brain. Analysis and interpretations of these observations by the research community are still evolving. Hence, this book has only brief mentions of top-down selection in the tour of experimental facts on vision in Chapter 2 and in relationship with other presented materials.

Top-down and bottom-up selection is often studied in connection with visual decoding. For example, one can examine how behavioral performances in visual tasks depend on attention. In this book, decoding is analyzed with an implicit assumption that the decoded information is the selected information.

1.2.5 Visual decoding and its associated brain areas

Visual decoding is defined as inferring some aspects of the 3D scenes from 2D images using the selected visual information. Different aspects of the visual scene can be inferred. One is to infer the locations and movements of the surfaces and objects in the scene. Another is to infer the identities of objects; this is termed object recognition. There can be many layers of inference. For example, one can infer the location of an object without inferring its identity. This could happen for example as follows: if a saliency map contains two conspicuous locations in a scene, an observer can be aware of both locations while directing gaze to only one of them to scrutinize the object there; the object at the other location is not identified although its location is perceived. For another example of multiple layers of inference, vision can enable recognizing facial expressions as well as the identity of a face, or the face can be recognized without recognizing the expression.

Decoding presumably requires the transformation of retinal signals to other neural signals in various brain regions, and some of these neural signals represent the outcome to the decoding problem (see Fig. 1.8). For example, an outcome could be a description of an object in the scene (e.g., my child). These descriptions should be available to influence memory, decision-making, and motor actions. Alternatively, an outcome could be a direct motor command. Additionally, there can be many intervening brain areas and neural signals (representations) between the retinal input and the decoding outcome. Some of these intervening signals are related to visual encoding and selection. Therefore, the final decoding outcome is modulated by visual selection. It is desirable to understand all the brain areas involved and the signal transformation processes.

Behavioral experiments have related visual performance on object recognition and local-ization to properties of visual inputs, availabilities of visual attention, and other factors such as the preconceptions of observers. Brain lesion studies revealed that some brain areas are more involved in inferencing object location and movements whereas other brain areas are more in-volved in object identification. There is evidence suggesting that V2, a visual area immediately after V1 along the visual pathway, is involved in building a surface representation (von der Heydt, Peterhans and Baumgartner 1984, von der Heydt, Zhou and Friedman 2000, Bakin, Nakayama and Gilbert 2000, Zhou, Friedman and von der Heydt 2000, Qiu and von der Heydt 2005, Zhaoping 2005a, Qiu and von der Heydt 2007, Qiu, Sugihara and von der Heydt 2007). Cortical regions in the temporal lobe are found to contain neurons representing complex visual input features such as faces (Freiwald and Tsao 2010, Freiwald, Tsao and Livingstone 2009). Physiological data have shown that various brain regions, particularly those beyond V1, are involved in directing and controlling top-down visual attention, which is known to affect visual performance.

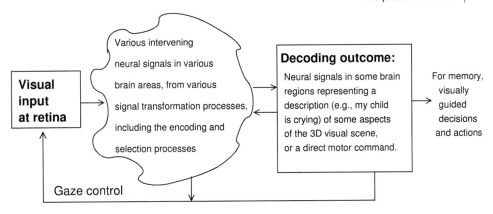

Fig. 1.8: A schematic of the presumed visual decoding.

However, it is still very unclear computationally and mechanistically how the brain solves the decoding problem. For example, although we recognize that the responses of the temporal cortical neurons tuned to specific facial features should be very useful for face recognition, and although we observe that some of these responses are desirably insensitive to the viewpoints of the faces, we do not know how these response properties are achieved by transformations of neural signals from the retina or V1. We have yet to find out how these neural responses are related to behavioral face perception such as recognizing face identities, gaze directions, and facial expressions.

Most of the brain areas devoted to vision are beyond the retina and V1, and our knowledge about each of them is far less than our knowledge about retina or V1. A starting guess is that much of visual decoding and top-down attentional control (which impacts on decoding) is carried out in these brain areas. This book has less content on visual decoding than on visual encoding and bottom-up selection associated with the retina and V1, since our understanding of visual decoding is more limited. Chapter 6 presents materials on what we do know about decoding. It includes a selection of physiological and behavioral observations that should constrain theories and models of decoding, a popular theoretical framework for decoding (Bayesian inference), and some examples directly linking neural response properties to visual perceptual behavior.

This book does not fill our theoretical ignorance about visual decoding by simply listing a selection of existing experimental data. In any case, a lack of sufficient understanding of data would make such a list immature and insufficiently meaningful. Moreover, the availability of better books on experimental facts makes this list unnecessary. However, the void in understanding made apparent in this book should hopefully motivate future investigations.

2 A very brief introduction of what is known about vision experimentally

Vision is one of the best studied functions of the brain. There is thus a vast knowledge about the neurophysiology and neuroanatomy of vision, and about particularly human visual behavior. Useful books about vision include *Foundations of Vision* by Wandell (1995), *Visual Perception, Physiology, Psychology and Ecology* by Bruce, Green, and Georgeson (2003), and *Vision Science, Photons to Phenomenology* by Palmer (1999). Chapter 2 of the book *Theoretical Neuroscience* by Dayan and Abbott (2001) offers a helpful introduction for modelers to the early stages of visual processing.

In this chapter, we present a very brief introduction to critical experimental observations, mainly in human and primate visual systems, to elucidate how theories and models in the rest of the book are inspired by the data. We also provide definitions and notations that will be useful in the rest of the book. Additional experimental results will be presented throughout the later chapters to accompany the presentation of the theories and models. For full references, please consult the books above, and also the magisterial two-volume book *The Visual Neurosciences* edited by Chalupa and Werner (2004).

First, this chapter introduces neurons, neural circuits, and brain regions, leading to an outline of brain areas comprising the visual pathway. Second, the neural properties of the retina and V1 are introduced. Third, we describe a few extrastriate areas (visual area 2 [V2], visual area 4 [V4], middle temporal visual area [MT]/visual area 5 [V5], and the inferotemporal cortex [IT]) further along the visual pathway and discuss the separation between the dorsal and ventral streams with their different processing emphases. This is followed by an overview of eye movements and visual attention, together with the brain areas and neural activities associated with these. The chapter ends with a list of typical topics in behavioral studies of vision, relating them to the theme of the book.

2.1 Neurons, neural circuits, and brain regions

2.1.1 Neurons, somas, dendrites, axons, and action potentials

Neurons are cells in the nervous system that receive, process, and transmit information. There are billions of them in the human brain. Each neuron is typically composed of dendrites, axons, and a soma or cell body; see Fig. 2.1 A. Dendrites receive inputs from other neurons or from the external sensory world through a signal transduction process. Axons send output signals to other neurons or effectors such as muscle fibers. Typically, the output signals from a neuron are in the form of electrical pulses or spikes called action potentials, each about 1 ms in duration and dozens of millivolts in amplitude. The electrical potential (called its membrane potential) of a neuron relative to the extracellular medium determines its state and its propensity to produce action potentials.

Synapses are chemical (and sometimes electrical) connections between neurons. They comprise the medium by which action potentials from a source (or presynaptic) neuron cause electric current to flow across the membrane of a target (or postsynaptic) neuron and thereby change the target neuron's membrane potential. Action potentials are nearly identical to each other and, via an active regenerative process, can propagate long distances along axons without appreciable decay before reaching their destination neurons. Hence, the information transmitted by action potentials is typically represented by the event time or rate (number of action potentials per second) at which the action potentials occur, rather than by their voltage profiles. Note that some very nearby neurons, e.g., in the retina, influence each other's states without generating action potentials.

A: A neuron drawn by Cajal

B: Two model neurons linked by a synaptic connection

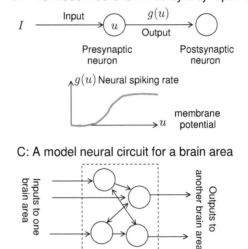

C: A model neural circuit for a brain area

Fig. 2.1: A: A neuron has a cell body (also called its soma), dendrites to receive inputs, and axons to send its outputs. This is a drawing (except for the red lines and printed words) by Cajal, a Spanish neuroscientist who died in 1934. B: A simple model of two neurons linked by a synaptic connection. C: A brain area can be modeled as a neural circuit comprising interacting neurons. It receives input from one brain area and sends output to another brain area. Neurons that do not send outputs to other brain areas are often called interneurons.

2.1.2 A simple neuron model

One can use differential equations to model both a single neuron and many interacting neurons in a neural circuit. However, readers can follow most of the book without having to understand these equations. So this paragraph, in which we introduce a neuron model, can be skipped without harm. A simple model (Hopfield 1984) of a neuron is as follows. Its internal membrane potential is modeled by a single variable u. The potential tends to stay near a resting level that is approximately 70 millivolts relative to the extracellular medium. For simplicity, and without affecting any conclusions, we consider potential differences relative to this resting potential and so write $u = 0$ at rest.

The membrane potential u can be raised by an injecting current I into the neuron. The change Δu engendered by just this current after a very small time interval Δt is $\Delta u = I \Delta t$. Hence, the potential u integrates the current I over time, implying that the neuron acts like a capacitor (with a unit capacitance), being charged up by the current. Meanwhile, consistent with its interpretation, u also returns to its resting level $u = 0$ at a rate proportional to its deviation u from the resting state. This causes another change within time Δt of $\Delta u = -(u/\tau)\Delta t$, where τ is the membrane time constant (which is defined as the time needed for u to decay to fraction $1/e$ of its initial deviation). Hence, the total change to u caused by both the injected current and the tendency to decay to the resting state is $\Delta u = [-u/\tau + I]\Delta t$. In this equation, $-u/\tau$ is like a negative current counteracting the injecting current I. Hence the neuron can be seen as integrating the input current I but with a leak current $-u/\tau$ (which is why this model is called a leaky integrator). Taking $\Delta u/\Delta t$ as du/dt as $\Delta t \to 0$, we get the differential equation

$$du/dt = -u/\tau + I$$

as a characterization of the temporal evolution of u. Given a constant input current I, the neuron's potential u eventually reaches a steady value $u = I\tau$, when the rate of change of u is $du/dt = 0$, according to the above equation.

The rate at which a neuron produces action potentials is often viewed as its output. This can be modeled as a nonlinear function $g(u) \geq 0$ of the membrane potential u such that $g(u)$ monotonically increases with u, is zero for small u, and saturates when $u \to \infty$; see Fig. 2.1 B. The output $g(u)$ contributes to the input current of postsynaptic neurons according to $w \cdot g(u)$, where w models the strength of the synaptic connection from pre- to postsynaptic neuron.

More complex treatments, such as the Hodgkin–Huxley model and cable theory (see Dayan and Abbott's textbook (2001) for an introduction) offer fuller descriptions of neurons. These include ion channels inside the membrane responsible for generating action potentials, the effects of the complex geometry of neurons' dendritic trees, and also considerations of individual output spike times. However, in this book, we do not require such complexities.

2.1.3 Random processes of action potential generation in neurons

In the simple neuron model above, the neural output is modeled as the rate $g(u)$ at which spikes or action potentials are generated. However, in general, the time intervals between spikes are irregular. This irregularity can be often modeled using a Poisson process such that the probability of a spike being generated within a very small time interval scales with the rate $g(u)$. This means that if the average number of spikes generated per second is $\bar{r} \equiv g(u)$, then within a particular duration of one second, the actual number r of spikes generated follows a Poisson probability distribution

$$P(r) = \frac{\bar{r}^r}{r!} \exp(-\bar{r}). \tag{2.1}$$

The mean of the number of spikes is duly $\sum_r rP(r) = \bar{r}$, and the variance in the number is $\sigma_r^2 \equiv \sum_r (r-\bar{r})^2 P(r) = \bar{r}$. The ratio $\sigma_r^2/\bar{r}$ is called the *Fano factor* of the neuron, and is 1 for a Poisson process (when the rate is constant). Although for most neurons σ_r^2 is proportional to $\bar{r}$, the Fano factor is rarely exactly 1, implying that neurons are not simple Poisson processes.

2.1.4 Synaptic connections, neural circuits, and brain areas

Each neuron makes synaptic connections with hundreds or thousands of other neurons, forming neural circuits that compute. There are microcircuits between nearby neurons, and there are macrocircuits between neural groups. A cortical area, such as one of the visual cortical areas in Fig. 2.2, contains locally connected groups of neurons. Nearby neurons are more likely to be connected with each other, as one might expect if the brain is to avoid devoting too much volume to axonal wiring (Mitchison 1992). Thus, generally, neurons are much more likely to be connected with each other within a cortical area than between cortical areas, and two cortical areas as a whole are more likely to be connected when they are nearer to each other (Douglas, Koch, Mahowald, Martin and Suarez 1995, Bullier 2004); although a wide bundle of fibers called the corpus callosum helps connect cortical areas in different (left and right) halves of the brain. Through these neural interactions, the brain carries out computations on sensory inputs and realizes perceptual decisions and motor actions. For instance, visual sensory inputs, after being sensed by photoreceptors in the retina, are processed by other retinal neurons and various visual areas in the brain. This processing leads to visual perception of inferred visual objects in the scene. Furthermore, by sending processed information to brain areas responsible for internal plans and motor actions, the processing guides or dictates behaviors such as orientation of attention, navigation, and object manipulation.

2.1.5 Visual processing areas along the visual pathway

The visual world is imaged on the retina, which performs initial processing on the input signals and sends the results via neural impulses along the optic nerve to the rest of the brain. Figure 2.2 shows the brain areas involved in vision. Each visual area has up to many millions of neurons; it does some information processing within itself, while receiving signals from, and sending signals to, other areas. About half of the areas in the monkey brain are involved in vision; see Fig. 2.2. Most brain regions are denoted by their abbreviated names in Fig. 2.2. For instance, LGN is the lateral geniculate nucleus, which is part of the thalamus, and is often viewed as the relay station between the retina and cortex, although this view perhaps reflects our ignorance about its role. V1 denotes visual area one, the primary visual cortex. It is the first cortical area to receive visual inputs and is also the largest visual cortical area in the brain. V2 is visual area 2, which receives most of its inputs from V1. V4 is visual area 4 further along the visual pathway. FEF is the frontal eye field in the frontal areas. SC is the superior colliculus, which is in the midbrain located below the cerebral cortex. Both FEF and the SC are involved in the control of eye movements. IT is the inferotemporal cortex, whose neurons respond to complex shapes in visual inputs. MT (also called V5 (Dubner and Zeki 1971)) stands for the middle temporal area (using the terminology from the macaque); neurons in this region are particularly sensitive to visual motion. LIP is the lateral intraparietal area and is implicated in decision making for eye movements. In Fig. 2.2, the lower case letters at the end of some of the abbreviations often denote the spatial locations of the cortical areas, e.g., v for ventral, d for dorsal.

The term visual pathway implies that there is a hierarchy of levels for information processing, starting from the retina, as shown schematically in Fig. 2.3. Information processing progresses from lower stages, starting at the retina (and excluding the SC and gaze control stages in the yellow shaded area), to higher stages, ending at FEF within this figure. Each neuron typically responds to visual inputs in a limited extent of the visual space called its *receptive field* (RF). The receptive fields of retinal neurons are small, having a diameter of only 0.06 degree in visual angle near the center of vision (Shapley and Perry 1986). (The extent of visual space is measured in visual angles, which is defined by the spatial extent in the visual scene perpendicular to the viewing direction divided by the distance to the viewer; the thickness of one's thumb held at an arm's length gives roughly one degree of visual angle on one's retina.) It is too small to cover most recognizable or relevant visual objects, e.g., an apple, in a typical scene. As one ascends the visual hierarchy, the receptive fields of neurons get progressively larger. Their diameters are about (in order of magnitude) 10 degrees of visual angle in V4 and 20–50 degrees in IT (Rolls 2003). Hence, if each neuron's response is selective to the spatial form of the visual input in its receptive field, one could possibly

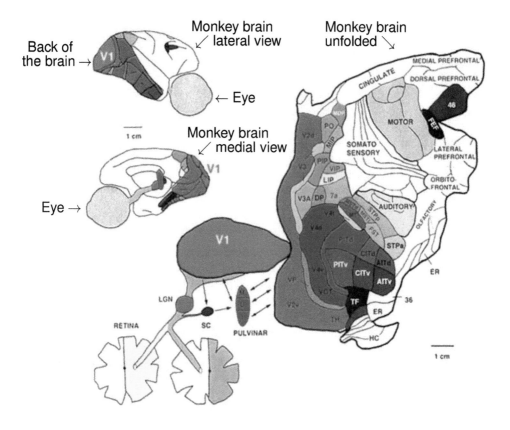

Fig. 2.2: Areas of the primate brain involved in vision. Shown is a monkey brain from the lateral view and medial view of its right hemisphere and from its unfolded representation. Cortical areas are about 1–3 mm thick and are folded like sheets to fit into the skull. In the digitally unfolded representation, for a clearer view, area V1 has been cut away from the other areas (notably area V2) in the unfolding process. Visual input enters the eye, is processed initially by the retina, and then is sent to the rest of the brain. The colored cortical areas are the ones exclusively or primarily focused on visual processing. Adapted with permission from Van Essen, D. C., Anderson, C. H., and Felleman, D. J., Information processing in the primate visual system: an integrated systems perspective, *Science*, 255 (5043): 419–23, copyright © 1992, AAAS.

hope that a single neuron in IT could signal the recognition of a visual object, e.g., one's grandmother. In the early areas such as the retina and V1, receptive fields are relatively fixed and independent of the animal's attention. They become increasingly variable in the later areas. For instance, the sizes of the receptive fields depend on the animal's focus of attention and on the complexity of the visual scene (Moran and Desimone 1985).

The connections between brain regions in Fig. 2.3 indicate the existence of neural connections between the regions. Most of these connections are bidirectional, such that each of the two areas connected receives signals from the other. This figure shows not only the flow of sensory information through various areas along the visual hierarchy but also the flow of information toward the control of eye movements. It reflects the view shared by many others

A schematic of the visual processing hierarchy

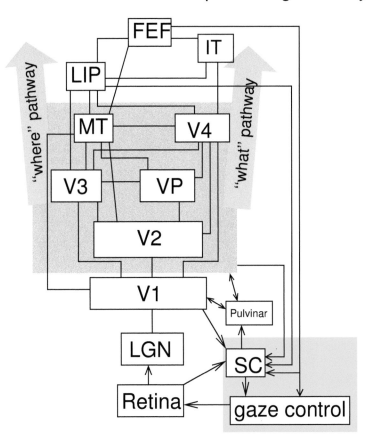

Fig. 2.3: The hierarchy of levels of visual processing in the brain, simplified from information in Felleman and Van Essen (1991), Bruce et al. (2004), Shipp (2004), and Schiller and Tehovnik (2005). Various labeled areas can be located in the brain map in Fig. 2.2. V1 is also called the striate cortex. The gray shaded area encloses many of the visual areas collectively called the extrastriate cortex. The yellow shaded area outlines the areas carrying out gaze movements caused by visual inputs or other factors. Cortical areas downstream from V1 are differentially involved in processing "what" and "where" information about visual objects.

(Findlay and Gilchrist 2003) that understanding the motor actions associated with vision is very important for understanding sensory processing. After all, the main purpose of recognizing and localizing objects in the scene is to act on them. Although eye movements are not the actions that are ultimately important, such as grasping, manipulating, and navigating, they can be seen as important intermediate actions. It is therefore noteworthy that signals from as early as the retina and V1 in the hierarchy already influence the motor outputs of vision.

Physiologically and anatomically, much more is known about the earlier visual areas, in particular the retina, LGN, and V1, than the higher visual areas. This is partly because it is often easier to access these early areas and to determine how their neural responses depend on the visual inputs (see Fig. 2.4). Behaviorally, one can measure the sensitivity of an animal to various simple or complex visual inputs, ranging from images of simple small bars to those of emotionally charged faces. One can also measure the speed of object recognition and

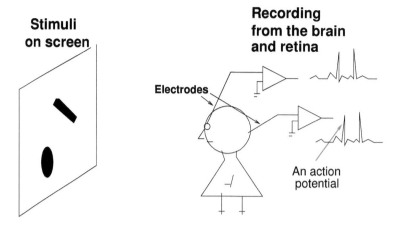

Fig. 2.4: A schematic of visual physiological experiments. Electrical activities in the neurons, including the action potentials (spikes), can be recorded while visual stimuli are presented to the animal.

localization, e.g., to find a tomato among many apples. Data associated with simple visual behaviors are more easily linked with physiological and anatomical findings from early visual areas. In contrast, data associated with complex visual behaviors are not as easily or clearly linked with findings from higher visual areas. In particular, the hierarchy of visual cortical areas shown in Fig. 2.3 is inferred mostly from anatomical observations, which identify some neural connections as feedforward and others as feedback along the visual pathway. This suggests but does not precisely determine the hierarchy of information processing. Different anatomical or physiological evidence (Bullier 2004) can be interpreted differently as to the level in the hierarchy to which a particular visual cortical area should be assigned.

2.2   Retina

The 3D visual scene is imaged on the retina, where photons coming from the scenes are absorbed by the photoreceptors at the image plane in the retina; see Fig. 2.5. In the primate retina, there are about 5×10^6 cones responsible for color vision in the daytime, and 10^8 rods, which are mainly functional in dim light (Wandell 1995). Each photoreceptor absorbs the light in a tiny image area to produce electrical response signals. These signals are transformed through several intermediate cell types called bipolar cells, horizontal cells, and amacrine cells before they are finally received by about 10^6 retinal ganglion cells, which are the output neurons of the retina. Each ganglion cell fires action potentials at a rate of up to approximately 100 spikes/second. They send signals to the brain via a bundle of axons known as the optical nerve. The blood vessels in the eyeball are also imaged onto the back of the retina, together with the visual scene. Nevertheless, we seldom see them since human vision is insensitive to static non-changing inputs. Voluntary and involuntary eye movements, including the constant jitters of the eye, keep the motionless parts of our visual scene visible.

2.2.1   Receptive fields of retinal ganglion cells

The response of a retinal ganglion cell will be quantified as the rate at which it fires action potentials, i.e., in spikes per second. Visual inputs produces responses from the photoreceptors. From most ganglion cells (called P cells in monkeys and X cells in cats), the response of

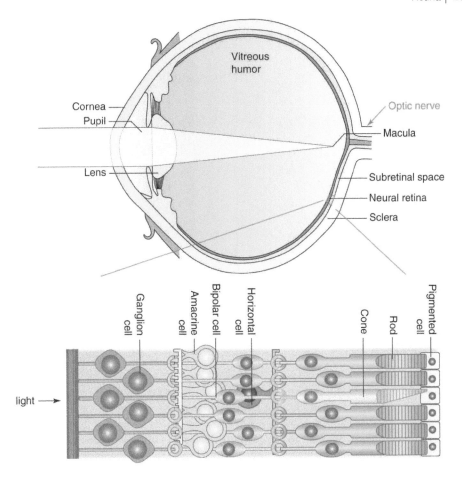

Fig. 2.5: A schematic illustration of the retina and its neurons. Light enters the eye, and retinal neural responses are transmitted by the optic nerve to the rest of the brain. The bottom half of this figure is a zoomed-up view of a patch of the retina: visual input light passes through the retinal ganglion cells and other cell layers before hitting the rods and cones. Adapted with permission from Streilein, J. W., Ocular immune privilege: therapeutic opportunities from an experiment of nature, *Nature Reviews Immunology*, 3: 879–889, Fig. 1, copyright © 2003, Macmillian Publishers Ltd, http://www.nature.com/, and from Dyer, M. A. and Cepko, C. L., Regulating proliferation during retinal development, *Nature Reviews Immunology*, 2: 333–342, Fig. 4, copyright © 2001, Macmillian Publishers Ltd, http://www.nature.com/.

each is approximately a linear weighted sum of the receptor responses (Enroth-Cugell and Robson 1966, Shapley and Perry 1986).

To formalize this input–output relationship, let $S(x)$ denote a stationary signal, the response of the photoreceptor, as a function of the photoreceptor locations x. Let O denote the steady state response from a ganglion cell after an initial phase of transient responses. This steady response can be described by a model

$$O = \sum_x K(x)S(x) + \text{spontaneous firing rate.} \tag{2.2}$$

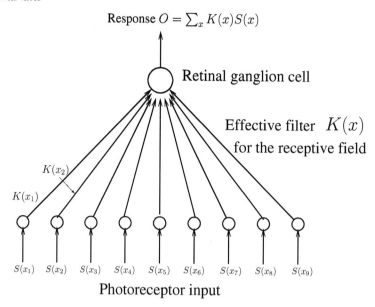

Fig. 2.6: Schematic of how the response $O = \sum_x K(x)S(x)$ of a retinal ganglion depends linearly on the photoreceptor response signal $S(x)$.

The neural responses are sometimes also referred to as neural outputs, hence the notation O. In the above equation, $K(x)$ is called the effective linear weight from the input receptor at x to the ganglion cell (see Fig. 2.6). The effect of $S(x)$ on O is the same as if the photoreceptor at x has a direct connection $K(x)$ to the ganglion cell, even though the photoreceptor signal actually passes through intermediate cell layers before reaching the ganglion cell. Often, the linear sum $\sum_x$ above is written as an integral $\int dx$, as if the input $S(x)$ and weights $K(x)$ are functions of continuous space x. In this book, we often use these two different symbols for summation interchangeably, and, for most of the book, readers may view $\sum_x$ and $\int dx$ as being equivalent (except for a constant scale factor). Furthermore, the bounds of summations and integrations are typically not explicitly indicated when they are clear from the context (we typically integrate over the whole range of admitted values for the variables concerned). Hence, we can also write

$$O = \int dx K(x)S(x) + \text{spontaneous firing rate.} \tag{2.3}$$

The function $K(x)$ is often called a filter, or a kernel (hence the letter K). In this case, it is a spatial filter. In physiology it is called the receptive field of the neuron.

Image space is actually two-dimensional (2D). Hence, the image or photoreceptor location x is a 2D vector with two components, often x and y along the horizontal and vertical axes, respectively. For simplicity, unless emphasis is necessary, the vector nature of image location x is not explicitly denoted. The notation simplicity is sometimes used to suggest that the mathematical expression applies whether space x is one-dimensional (1D) or two-dimensional. Thus, when x is 2D, integration $\int dx$ (e.g., in equation (2.3)) should be understood as integrating over 2D space.

The filter value $K(x)$ is non-zero for a limited spatial range of x, which defines the size of the receptive field of the neuron. This size is typically only a fraction of a degree for retinal and V1 neurons (in primates). The center of the receptive field varies from neuron to neuron, such that the whole population of the retinal ganglion cells can adequately sample the whole visual field. The receptive fields of retinal ganglion cells can often be modeled as a difference between two Gaussian functions (Enroth-Cugell and Robson 1966) which have the same mean location, but different standard deviations σ_c and σ_s and strengths w_c and w_s for center (c) and surround (s) respectively. If the mean location of the receptive field is at $x = 0$, then

$$K(x) = \frac{w_c}{\sigma_c} \exp\left(-\frac{x^2}{2\sigma_c^2}\right) - \frac{w_s}{\sigma_s} \exp\left(-\frac{x^2}{2\sigma_s^2}\right),$$

in 1D space x,

$$K(x,y) = \frac{w_c}{\sigma_c^2} \exp\left(-\frac{x^2+y^2}{2\sigma_c^2}\right) - \frac{w_s}{\sigma_s^2} \exp\left(-\frac{x^2+y^2}{2\sigma_s^2}\right),$$

in 2D space (x,y).

$$(2.4)$$

In each equation above, the first and the second terms denote the two Gaussian shapes respectively, as illustrated in Fig. 2.7. Typically, $\sigma_c < \sigma_s$, and in many situations $w_c \approx w_s$ such that $K(x)$ has a spatially opponent shape. The signs for w_c and w_s are meant to be the same. In the example in Fig. 2.7, they are both positive. As a result, this ganglion neuron increases its output O when presented with a bright spot near the center of the receptive field but decreases its output when this bright spot is farther from the center. Such a receptive field is called a *center-surround receptive field*. The optimal visual input, defined as the stimulus that excites the cell the most, is a bright central disk surrounded by a dark ring.

If both w_c and w_s are negative, then the optimal stimulus is a dark central spot surrounded by a bright ring; a bright central spot in a dark ring would decrease the neural response instead. The two kinds of receptive fields with positive or negative values for w_c and w_s are called *on-center* and *off-center* cells respectively. The receptive field region in which $K(x)$ is positive or negative is accordingly called the "on" or "off" region of the receptive field.

The spontaneous firing rate in equation (2.3) denotes the value of O with zero input $S(x) = 0$. This spontaneous rate is around 50 and 20 spikes/second, respectively, for the majority (i.e., the X or P cells, see later) of ganglion cells in the cat and monkey (Troy and Robson 1992, Troy and Lee 1994). As firing rates are never negative, the substantial spontaneous firing rate helps maintain the validity of the linear receptive field model. Non-optimal visual inputs simply decrease the neural outputs from the spontaneous level without making the outputs negative.

2.2.2 Sensitivity to sinusoidal gratings, and contrast sensitivity curves

One can also investigate how a ganglion cell responds to a sinusoidal input pattern

$$S(x) = \mathcal{S}_k \cos(kx + \phi) + \text{constant}, \qquad (2.5)$$

which is a grating of spatial frequency $k/(2\pi)$ cycles/degree, or frequency k radians/degree, with an amplitude $\mathcal{S}_k$ and phase ϕ. For simplicity, we consider space as 1D. In 2D space, the grating above can be seen, without loss of generality, as being vertical and propagating along the horizontal dimension x with a wave vector $(k, 0)$.

Let us first decompose the spatial receptive field $K(x)$ as a sum of cosine and sine waves:

$$K(x) = \int dk [g_c(k) \cos(kx) + g_s(k) \sin(kx)]. \qquad (2.6)$$

The coefficients $g_c(k)$ and $g_s(k)$ of the waves are obtained from the Fourier transform of $K(x)$ (if needed, see Box 2.1 for an introduction to Fourier transforms and Box 2.2 for an

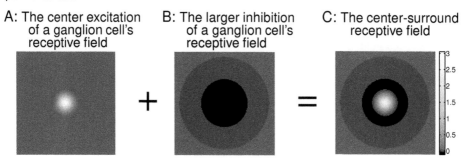

A: The center excitation of a ganglion cell's receptive field

B: The larger inhibition of a ganglion cell's receptive field

C: The center-surround receptive field

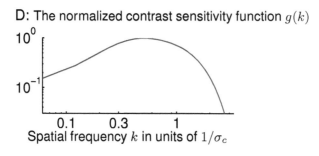

D: The normalized contrast sensitivity function $g(k)$

Spatial frequency k in units of $1/\sigma_c$

Fig. 2.7: A center-surround receptive field of a retinal ganglion cell modeled as a difference between two Gaussians. A: Gaussian excitatory field; B: larger Gaussian inhibitory field; and C: their combination as a center-surround shape of the receptive field. In each plot, the value of the receptive field $K(x)$, or one of its Gaussian components, is visualized by the grayscale value of the pixel at image location x. Parameters used are: $\sigma_s/\sigma_c = 5$, $w_c/w_s = 1.1$. D: Normalized contrast sensitivity $g(k)$ versus spatial frequency k in the units of $1/\sigma_c$ for the receptive field in C. Also see Fig. 3.19.

introduction to complex variables. This book does not distinguish between variants of Fourier analysis, such as discrete Fourier transforms, continuous Fourier transforms, and Fourier series.)

$$g_c(k) = \int dx K(x) \cos(kx) \quad \text{and} \quad g_s(k) = \int dx K(x) \sin(kx). \quad (2.7)$$

For simplicity, we typically omit scale factors in the expressions for Fourier transforms and inverse transforms. These are powers of (2π). Similarly, we use $\sum_x$ and $\int dx$ (and analogously $\sum_k$ and $\int dk$) interchangeably in the expressions for these transforms, since these different expressions differ also by a scale factor. These simplifications make mathematical expressions less cumbersome, especially in Chapter 3. They do not make any difference to our key questions of interest: how sensitivity $g_{c,s}(k)$ varies with frequency k and how receptive field $K(x)$ varies with space x.

If $K(x)$ is an even function of space, $K(x) = K(-x)$, it is symmetric with respect to the origin of the coordinate system. Then $g_s(k) = 0$ for all k, and the asymptotic response to the sinusoidal input is given by (omitting the constant)

$$O = \int dx K(x) \mathcal{S}_k \cos(kx + \phi) \propto g_c(k) \mathcal{S}_k \cos(\phi). \quad (2.8)$$

Let us assume that $g_c(k) > 0$. Then the response of the neuron is largest when $\phi = 0$, which occurs when the peak of the grating is centered at the on-region of the receptive field; see

Exposing a receptive field to
an input of a sinusoidal wave

Contrast sensitivity of a ganglion cell in cat

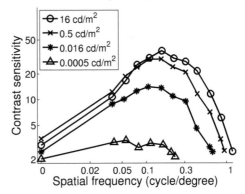

Fig. 2.8: Measuring a neuron's response to spatial gratings. A: Illustration of a spatial center-surround receptive field $K(x)$ exposed to a sinusoidal wave $S(x)$. The neural response $O = \int dx S(x) K(x)$ is largest when the center-surround receptive field is exactly centered on the peak of the sinusoidal wave $S(x)$. In general the response is proportional to $\cos(\Delta\phi)$, the cosine of the phase value $\Delta\phi = \phi - \theta$ of the sinusoidal wave at the center location of the receptive field; see equation (2.9). B: Contrast sensitivity of a retinal ganglion cell in cat. Note the change with the mean light level to which the animal is adapted. The cat (whose visual acuity is typically 10 times worse than humans) was anesthetized, with a pupil diameter 3.5 mm, and the grating was drifting such that input temporal frequency was 1 Hz (cycle/second). Data from Enroth-Cugell, C. and Robson, J. G., The contrast sensitivity of retinal ganglion cells of the cat, *The Journal of Physiology*, 187 (3): 517–552, Fig. 15, 1966.

Fig. 2.8. Given this spatial coincidence, the response level is proportional to $g_c(k)$, which is the sensitivity of the neuron to a cosine grating of frequency k. One can see that this sensitivity is higher when the size of the center of the receptive field is closer to half of the wavelength of the grating. Hence, there exists an optimal spatial frequency k to which the neuron responds most vigorously.

For general $g_c(k)$ and $g_s(k)$ (e.g., when the center-surround receptive field is not centered at the origin of the coordinate system), the ganglion cell's response to the sinusoidal input wave is

$$O = \int dx K(x) \mathcal{S}_k \cos(kx + \phi)$$

$$\propto \int dx \mathcal{S}_k \left[g_c(k) \cos(kx) \cos(kx + \phi) + g_s(k) \sin(kx) \cos(kx + \phi) \right]$$

$$\propto \mathcal{S}_k \left[g_c(k) \cos(\phi) - g_s(k) \sin(\phi) \right] = |g(k)| \mathcal{S}_k \cos(\phi - \theta), \tag{2.9}$$

in which $\quad |g(k)| \equiv \sqrt{g_c^2(k) + g_s^2(k)} \quad$ and $\quad \theta = \tan^{-1}(-g_s(k)/g_c(k))$.

Here, we defined a 2D vector $[g_c(k), -g_s(k)]^T$ (superscript T denotes vector or matrix transpose), which has length $|g(k)| \equiv \sqrt{g_c^2(k) + g_s^2(k)}$ and angle θ relative to the horizontal axis; see Fig. 2.11. In the case above, when the center of the receptive field is at the origin of the coordinate system, we have $g_s(k) = 0$ and $\theta = 0$; so $|g(k)| = |g_c(k)|$ alone (see equation (2.8)). In general, $|g(k)|$ is the sensitivity of the neuron to the sinusoidal wave of frequency k. Define a complex variable $g(k) = g_c(k) - i g_s(k)$, which has real part $g_c(k)$ and imaginary part $-g_s(k)$, with $i = \sqrt{-1}$. Then $g(k) = |g(k)| e^{i\theta}$ is said to have a magnitude $|g(k)|$ and phase θ. It can be obtained by taking the Fourier transform

Box 2.1: **Fourier analysis**

If $f(x)$ is a function for integer values $x = 1, 2, ..., N$, it can be seen as a vector with N components $f(1)$, $f(2)$, ... , $f(N)$, or a weighted sum $f(x) = \sum_{i=1}^{N} f(i) b_i(x)$ of N basis functions $b_i(x)$, with the ith basis function $b_i(x) = 1$ when $x = i$ and $b_i(x) = 0$ otherwise. Different bases can be used instead; thus, all functions $f(x)$ can also be written as a weighted sum $f(x) = \sum_k g_c(k) \cos(kx) + g_s(k) \sin(kx)$ of cosine and sine waves $\cos(kx)$ and $\sin(kx)$ of different frequencies k by weights $g_c(k)$ and $g_s(k)$. For instance, with $N = 100$, the $f(x)$ in the plot to the right is made by summing two waves: $\sin(2\pi x/N)$, which has a low frequency $k = 2\pi/N$, or equivalently a long wavelength, and contributes to the sum with weight 1; and $\cos(40\pi x/N)$, which has a high frequency $k = 40\pi/N$, a short wavelength, and weight 0.2. Hence, $f(x)$ appears like the sine wave on a coarse scale, but has fine scale ripples due to the cosine wave. All functions can be made this way, using N weights on the N sinusoidal wave basis functions: $N/2 + 1$ cosine waves with $k = 2\pi n/N$ for integer $n = 0, 1, 2, ..., N/2$ and $N/2 - 1$ sine waves with $k = 2\pi n/N$ for integer $n = 1, 2, ..., N/2 - 1$. We say that these sine and cosine waves constitute a complete set of basis functions.

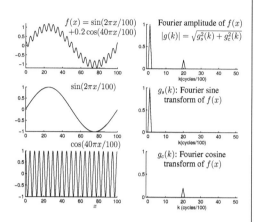

Fig. 2.9

This $f(x)$ can be seen as a vector in a N-dimensional space spanned by N orthogonal axes, each defined by one of the sinusoidal functions above. The projection of $f(x)$ onto each axis, i.e., the dot product of $f(x)$ and the basis function, $g_c(k) = (2/N) \sum_x f(x) \cos(kx)$ or $g_s(k) = (2/N) \sum_x f(x) \sin(kx)$, is the corresponding weight. Hence, our $f(x)$ in the figure has non-zero projections only onto two axes, one for $\sin(2\pi x/N)$ and the other for $\cos(2\pi 20 x/N)$. Since phase shifting a sine wave gives a cosine wave of the same frequency k, the quantity $\sqrt{g_c^2(k) + g_s^2(k)}$ is called the amplitude, and the ratio $g_s(k) : g_c(k)$ characterizes the phase, of the Fourier transform $g(k) = g_c(k) - i g_s(k)$ of $f(x)$ (see Box 2.2 for Complex variables). Obtaining $f(x) = \sum_k g_c(k) \cos(kx) + g_s(k) \sin(kx)$ from $g_c(k)$ and $g_s(k)$, or obtaining $f(x) = \sum_k g(k) e^{ikx}$ from $g(k)$, is called the inverse Fourier transform. This can be generalized to continuous functions $f(x)$ or to $N \to \infty$ basis functions. A smooth function contains more contributions from low frequency waves than a more rapidly changing function. Smoothing a function filters out higher frequency waves and so is called a low-pass operation. Conversely, a high-pass operation filters out the slowly changing, or low frequency, components.

$$g(k) = \int dx K(x)[\cos(kx) - i \sin(kx)] \equiv \int dx K(x) e^{-ikx}. \tag{2.10}$$

For the center-surround receptive field in equation (2.4), this gives

$$g(k) \propto w_c \exp\left(-\frac{|k|^2 \sigma_c^2}{2}\right) - w_s \exp\left(-\frac{|k|^2 \sigma_s^2}{2}\right), \tag{2.11}$$

which is another difference of two Gaussians. The above equation holds for both 1D and 2D space. In the latter case, k should be understood as a 2D vector (k_x, k_y) with compo-

Box 2.2: **Complex variables**

A complex number may be seen as a vector of two components, one along a horizontal or "real" axis having (a real) unit 1, and the other along a vertical or "imaginary" axis and has another (imaginary) unit defined as $i \equiv \sqrt{-1}$ (hence $i^2 = -1$). A complex number having x units along horizontal axes and y units along vertical axes is written as $z = x + iy$. It is said to have a real part x and an imaginary part y. As it has an angle θ from the x axis, it may be written as

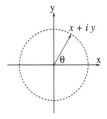

Fig. 2.10

$$z = |z|[\cos(\theta) + i\sin(\theta)] = |z|e^{i\theta},$$

making explicit the magnitude $|z| \equiv \sqrt{x^2 + y^2}$ and angle θ (also called phase; see Fig. 2.10) of this vector relative to the horizontal axis. Note that $|z| = \sqrt{(x + iy)(x - iy)}$. Also, $z^{\dagger} \equiv x - iy$, a vector with the same magnitude but phase $-\theta$ is called the complex conjugate of z. The equality $\cos(\theta) + i\sin(\theta) = e^{i\theta}$ can be verified by noting the Taylor expansions in θ of

$$\cos(\theta) = 1 - \theta^2/2 + \theta^4/4! - \theta^6/6! \dots ,$$
$$\sin(\theta) = \theta - \theta^3/3! + \theta^5/5! \dots$$
$$\text{and, since } i^2 = -1$$
$$e^{i\theta} = 1 + (i\theta) + (i\theta)^2/2! + (i\theta)^3/3! + \dots$$
$$= 1 + i\theta - \theta^2/2! - i\theta^3/3! + \theta^4/4! + i\theta^5/5! - \theta^6/6! \dots$$

Since $e^{ikx} = \cos(kx) + i\sin(kx)$, the Fourier transform of $K(x)$ to obtain $g_c(k)$ and $g_s(k)$ can be conveniently combined as in equation (2.10) to obtain a complex Fourier component $g(k) \equiv g_c(k) - ig_s(k)$. The inverse Fourier transform to obtain $K(x)$ from $g(k)$ is then $K(x) = \int dk g(k)e^{ikx}$, an integral of complex waves e^{ikx} with complex weights $g(k)$. If $K(x)$ is a real-valued function, $g(k) = [g(-k)]^{\dagger}$ must be satisfied for all k.

nents k_x and k_y along the horizontal and vertical dimensions, respectively, and a magnitude $|k| = \sqrt{k_x^2 + k_y^2}$. Just as for location x, when there is no confusion, we often denote spatial frequency as k without explicitly pointing out whether it is for 1D or 2D space. For $|w_c| \geq |w_s|$ (remember that w_c and w_s have the same sign), $|g(k)|$ slowly increases and then decreases with $|k|$, reaching a peak value at some frequency $|k| = k_p$. Thus $K(x)$ is a band pass filter, i.e., it is most sensitive to a particular, intermediate, frequency band. The neuron is relatively insensitive to low spatial frequency signals that vary smoothly in space. It is also insensitive to high frequency signals that vary over a spatial scale much finer than the scales σ_c and σ_s of the receptive field. In contrast, it is most sensitive to spatial frequencies on the order of $k_p \sim 1/\sigma_c$, when the wavelength of the grating is comparable to the size of the center of the receptive field; see Fig. 2.7 ACD and Fig. 2.8. The exact peak k_p occurs when $d|g(k)|/dk = 0$. Hence, from equation (2.11),

$$k_p = \left[\frac{2}{\sigma_s^2 - \sigma_c^2} \ln\left(\frac{w_s \sigma_s^2}{w_c \sigma_c^2}\right)\right]^{1/2}. \tag{2.12}$$

The sensitivity $|g(k)|$ of a neuron's response to gratings as a function of the grating frequency k is related to the contrast sensitivity curves measured experimentally. In experiments, this contrast sensitivity is measured by a stimulus

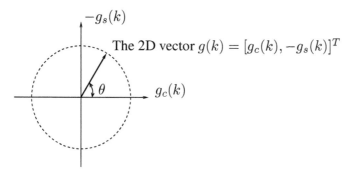

Fig. 2.11: A 2D vector $g(k)$, with the horizontal component $g_c(k)$ and vertical component $-g_s(k)$, forming an angle θ from the horizontal axis.

$$S(x) = S(1 + c\cos(kx + \phi)), \text{ hence, } c = S_k/S, \qquad (2.13)$$

where S is the mean input level and c is the contrast value. The contrast sensitivity is the sensitivity to the contrast c. Since this contrast c scales with the amplitude S_k for the grating by $c = S_k/S$, $|g(k)|$ scales with the contrast sensitivity. In this book, $|g(k)|$ as a function of k is often referred to as the contrast sensitivity function for simplicity; see Fig. 2.7 D.

Figure 2.8 B shows that the contrast sensitivity changes with the mean light level to which the animal is adapted. As the light level decreases, the band-pass function becomes progressively low-pass, and the k_p value decreases. This implies that the size of the receptive field, in particular σ_c, increases. In Chapter 3, we will see how this adaptation can be understood from the principle of efficient coding.

2.2.3 Responses to spatiotemporal inputs

Visual inputs are actually spatiotemporal, being described as a function $S(x, t)$ that depends on both space x and time t. A ganglion cell's response at time t can be affected by inputs at an earlier time $t' < t$ in a way that depends on the time difference $t - t'$. The spatial filter $K(x)$ should thus be generalized to a spatiotemporal filter $K(x, t - t')$ that sums inputs over both space and time for a general input pattern $S(x, t')$

$$O(t) = \int dt' dx K(x, t - t') S(x, t') + \text{spontaneous firing rate.} \qquad (2.14)$$

Due to causality, $K(x, t - t') = 0$ when $t - t' < 0$.

A particular class of spatiotemporal inputs has $S(x, t) = S_x(x)S_t(t)$, which can be written as a product between a spatial function $S_x(x)$ and a temporal function $S_t(t)$. These are called spatiotemporally separable inputs. Although natural visual inputs are seldom spatiotemporally separable, separable inputs are useful to examine neural properties experimentally. For example, if a spatial input $S_x(x)$ turns on at time $t = 0$ and stays unchanged thereafter, it can be written as

$$S(x, t) = S_x(x)H(t), \qquad (2.15)$$

$$\text{where } H(t) \text{ is a step function with } \quad H(t) = \begin{cases} 1, t \geq 0, \\ 0, \text{ otherwise.} \end{cases} \qquad (2.16)$$

The neural response at any time $t > 0$ is, omitting the spontaneous firing rate,

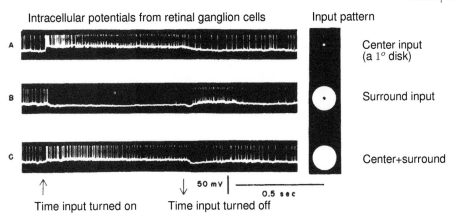

Intracellular potentials from retinal ganglion cells   Input pattern

A — Center input (a 1° disk)

B — Surround input

C — Center+surround

↑ Time input turned on   ↓ 50 mV   Time input turned off   0.5 sec

Fig. 2.12: Responses from a retinal ganglion cell in cat to inputs as recorded by T. Wiesel (1959). Shown are three temporal traces of intracellular potentials in response to three different input patterns. The center input pattern in A is a disk of one degree diameter. Adapted with permission from Wiesel, T. N., Recording Inhibition and Excitation in the Cat's Retinal Ganglion Cells with Intracellular Electrodes, *Nature*, 183 (4656): 264–265, Fig. 1, copyright © 1959, Macmillan Publishers Ltd, http://www.nature.com/.

$$O(t) = \int dx S_x(x) \int_{-\infty}^{t} dt' K(x, t - t') H(t')$$

$$= \int dx S_x(x) \int_{0}^{t} dt' K(x, t - t'). \qquad (2.17)$$

At $t > 0$, soon after the onset time $t' = 0$, the response $O(t)$ depends sensitively on time t. It is called the transient component of the responses. When $t \to \infty$, the response reaches an asymptote, which is its sustained component:

$$O(t \to \infty) = \int dx S_x(x) \int_{0}^{t \to \infty} dt' K(x, t - t') dt'. \qquad (2.18)$$

The spatiotemporal filter $K(x, t - t')$ has only a limited temporal span. Hence, the temporal integral $\int_{0}^{t \to \infty} dt' K(x, t - t') dt'$ is finite, and can be denoted as a spatial function

$$\bar{K}(x) \equiv \int_{0}^{\infty} K(x, t) dt. \qquad (2.19)$$

Hence the asymptotic response to a static spatial input $S_x(x)$ (after its onset) is

$$O(t \to \infty) = \int dx \bar{K}(x) S_x(x). \qquad (2.20)$$

This sustained response level should correspond to the steady state response of the ganglion cells to static input in equation (2.3). Hence, the spatial filter in equation (2.3) can be seen as $\bar{K}(x)$, the temporal integration of the whole spatiotemporal filter, shown in equation (2.19).

If the spatial pattern $S_x(x)$ is turned on at time $t = 0$ and turned off at time $t = T > 0$, $S(x, t) = S(x) H'(t)$ where $H'(t) = 1$ for $0 \leq t \leq T$ and $H'(t) = 0$ otherwise. Modifying from equation (2.17), the neural response is then,

$$O(t) = \int dx S_x(x) \int_{0}^{\min(t, T)} dt' K(x, t - t'), \qquad (2.21)$$

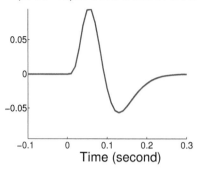

Fig. 2.13: The impulse response function of a neuron modeled by equation (2.24).

in which the temporal integration is only up to $\min(t, T)$, which is the smaller of t and T. The response very soon after the onset time $t = 0$ or offset time $t = T$ is usually called the *on-response* or *off-response*, respectively. Figure 2.12 shows example responses from a retinal ganglion cell in cats to such inputs.

As another example, consider an input temporal profile $S_t(t)$ as a Gaussian function centered at time $t = 0$ with a temporal spread σ. When σ is infinitesimally small, $S_t(t)$ is infinitesimally large within an infinitesimally narrow time window, and $S_t(t) = 0$ outside this time window, but so that $\int dt S_t(t) = 1$ is maintained—such a function is called a *delta function* $\delta(t)$ which can be written as

$$\delta(t) = \lim_{\sigma \to 0} \frac{1}{\sqrt{2\pi}\sigma} \exp\left(\frac{-t^2}{2\sigma^2}\right). \tag{2.22}$$

The input $S(x, t) = S_x(x)\delta(t)$ describes an input impulse, i.e., a very brief stimulus presentation of a spatial pattern $S_x(x)$. When the spatial pattern $S_x(x)$ matches the shape of the spatial receptive field $\bar{K}(x)$, the ganglion cell's response O is typically an increase followed by a decrease of response (relative to the spontaneous response level) lasting for tens of milliseconds. The deviation (as a function of time) of neural response from the spontaneous rate, due to an input impulse at time $t = 0$, is called the *impulse response function* (of time)

$$\text{impulse response function} = \int K(x, t - t')S_x(x)\delta(t')dx dt'$$

$$= \int dx K(x, t)S_x(x), \tag{2.23}$$

which depends on the spatial input pattern $S_x(x)$. It has a temporal shape like that shown in Fig. 2.13 and can be approximated by (for $t \geq 0$) (Adelson and Bergen 1985, Watson 1986)

$$O(t) = e^{-\alpha t} \left[\frac{(\alpha t)^5}{5!} - \frac{(\alpha t)^7}{7!}\right], \tag{2.24}$$

with, e.g., $\alpha \approx 70$ second^{-1}. The impulse response of a neuron should resemble the onset response of this neuron.

Extending equation (2.4), a good model of a retinal neuron's spatiotemporal receptive field is:

$$K(x, t) = \frac{K_c(t)w_c}{\sigma_c^d} \exp\left(-\frac{|x|^2}{2\sigma_c^2}\right) - \frac{K_s(t)w_s}{\sigma_s^d} \exp\left(-\frac{|x|^2}{2\sigma_s^2}\right), \tag{2.25}$$

where $d = 1$ or $d = 2$ for 1D or 2D image space x respectively, and $K_c(t)$ and $K_s(t)$ are the temporal impulse response functions of the center and surround components of the receptive fields, respectively. The temporal profiles of $K_c(t)$ and $K_s(t)$ are qualitatively similar to the expression in equation (2.24).

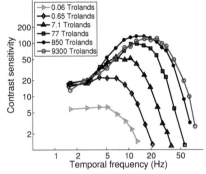
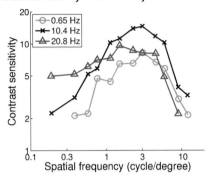

A: Human temporal contrast sensitivity

B: Contrast sensitivity of a monkey LGN neuron

Fig. 2.14: Spatiotemporal contrast sensitivity. A: Human temporal contrast sensitivity function (to a uniform light field) changes with the mean light level to which humans are adapted. Data from Kelly, D. H., Information capacity of a single retinal channel, *IEEE Transactions Information Theory*, 8: 221–226, 1962. B: The spatial contrast sensitivity curves change with the temporal frequency ω. Data from Derrington, A. M. and Lennie, P., Spatial and temporal contrast sensitivities of neurones in lateral geniculate nucleus of macaque, *Journal of Physiology*, 357 (1): 219–240, Figs. 8a, 8c, and 8d, 1984.

Since $K_c(t)$ and $K_s(t)$ have positive and negative parts, equation (2.25) suggests that an on-center cell at an earlier phase of the response can turn into an off-center cell in the later phase. The time constant for the surround component $K_s(t)$ can be longer than that for the center component $K_c(t)$ (DeAngelis, Ohzawa and Freeman 1995, Dayan and Abbott 2001). This also means that the spatiotemporal $K(x, t)$ cannot in general be space-time separable, i.e., it cannot be written as $K(x, t) = K_x(x)K_t(t)$, a product between a spatial function and a temporal function.

Equation (2.20) and equation (2.9) together imply that the sustained response level of the cell to a spatial grating $\cos(kx + \phi)$ should be $O(t \to \infty) \sim |g(k)| \cos(\phi - \theta)$, in which $g(k) = |g(k)|e^{i\theta}$. Hence, by using spatial grating with various k and ϕ, one can obtain $g(k)$, from which one can quite easily construct the shape of the spatial filter (averaged over time)

$$\bar{K}(x) = \int dt K(x, t) = \int dk g(k) e^{ikx}, \qquad (2.26)$$

which is the inverse Fourier transform of $g(k)$.

Experiments often use a drifting grating (which is not space-time separable) as visual input

$$S(x, t) \propto \cos(kx + \omega t) + \text{constant}. \qquad (2.27)$$

As the input changes in time, the response $O(t)$, as $t \to \infty$, does not approach a steady sustained level. Instead, after the initial transient phase, $O(t)$ follows the input, oscillating over time with the same frequency ω, but a different temporal phase ϕ_t,

$$O(t \to \infty) \propto \cos(\omega t + \phi_t). \qquad (2.28)$$

The amplitude of the oscillation scales with

$$g(k,\omega) = \int dx dt K(x,t) e^{-ikx-i\omega t}, \tag{2.29}$$

which is the Fourier transform of the spatiotemporal filter $K(x,t)$:

$$K(x,t) \propto \int dk d\omega g(k,\omega) e^{ikx+i\omega t}. \tag{2.30}$$

The above equation is the generalization of equation (2.26) to include the dimension of time. The response to the static grating is simply the special case when $\omega = 0$ (hence $g(k,0)$ should be used in equation (2.26)). Typically, monkey retinal ganglion cells are most sensitive to temporal frequencies on the order of $\omega/(2\pi) = 10$ Hz. This means that the impulse response to a momentary sinusoidal spatial wave is typically a transient wave form lasting about ~ 100 ms. The contrast sensitivity functions of monkey ganglion cells correspond quite well to human observers' sensitivity to the same gratings (Lee, Pokorny, Smith, Martin and Valberg 1990).

Figure 2.14 A shows the temporal contrast sensitivity in humans. Like the spatial contrast sensitivity function in Fig. 2.8 B, it changes from band-pass to low-pass as adaptation light levels decrease. Figure 2.14 B shows that the shape of the spatial contrast sensitivity function $g(k)$ depends on the temporal frequency ω. This means that $g(k,\omega)$ is not spatiotemporally separable, or, equivalently, that the dependence of $g(k,\omega)$ on k is influenced by ω. This can be understood from the observation that the center and surround components of the spatiotemporal receptive field have different temporal impulse response functions $K_c(t)$ and $K_s(t)$. According to equation (2.25), the contrast sensitivity function should be

$$g(k,\omega) \propto g_c(\omega) w_c \exp\left(-\frac{|k|^2\sigma_c^2}{2}\right) - g_s(\omega) w_s \exp\left(-\frac{|k|^2\sigma_s^2}{2}\right), \tag{2.31}$$

where $g_c(\omega)$ and $g_s(\omega)$ are the *temporal Fourier transforms* of $K_c(t)$ and $K_s(t)$ respectively, i.e.,

$$g_c(\omega) = \int dt K_c(t) e^{-i\omega t}, \quad g_s(\omega) = \int dt K_s(t) e^{-i\omega t}. \tag{2.32}$$

Hence, as long as $g_c(\omega)$ and $g_s(\omega)$ differ by more than a scale factor, a neuron's spatial contrast sensitivity function should depend on the temporal frequency of the grating used.

2.2.4   P and M cells

The two best known classes of retinal ganglion cells in primates are called parvocellular and magnocellular, or P and M for short. Cats have X and Y cells, which are similar to the P and M cells respectively. P cells are about 10 times more numerous than M cells, and have smaller receptive fields and longer impulse responses. Hence, P cells have a better spatial resolution, i.e., sample space on a finer scale, whereas M cells have a better temporal resolution. P cells are more sensitive to higher spatial frequencies whereas M cells are more sensitive to higher temporal frequencies. The responses of the M cells (and the Y cells in cats) to drifting gratings, contain two substantial temporal frequency components: one at frequency ω of the drifting grating (equation (2.28)), which is called the *fundamental frequency* response, and the other at frequency 2ω, which is called the *second harmonic* response. The second harmonic response arises from the nonlinear response properties of these cells. In addition, M cells are more input sensitive than P cells (to luminance inputs); see Fig. 2.15.

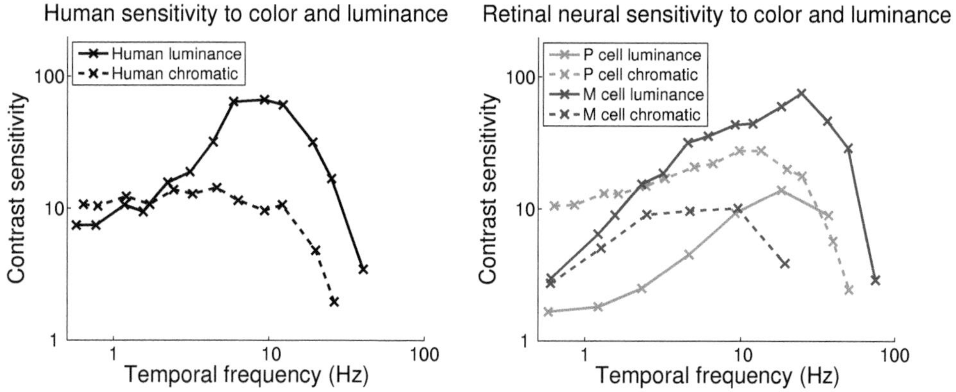

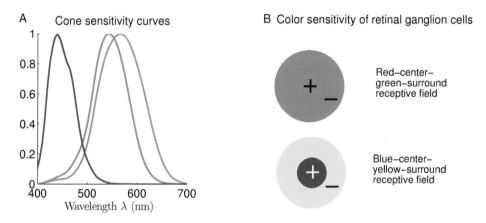

Fig. 2.15: Sensitivities to luminance and chromatic temporal modulation for human observers (left) and monkey retinal ganglion cells (right). Note the following characteristics: the filter for the luminance signal prefers higher frequencies than the chromatic signal; M cells are more sensitive to the luminance signals than P cells; and P cells are more sensitive to the chromatic signal than M cells. Data from Lee, B., Pokorny, J., Smith, V., Martin, P., and Valberg, A., Luminance and chromatic modulation sensitivity of macaque ganglion cells and human observers. *Journal of the Optical Society of America, A,* 7 (12): 2223–2236, 1990.

Fig. 2.16: A: Spectral sensitivities of the cones as a function of the wavelength of light. B: Schematics of two retinal ganglion cells with center-surround color opponency in their receptive fields. These are called single-opponent cells since each subregion of a receptive field involves only one color type.

2.2.5 Color processing in the retina

Cones are the photoreceptors activated by daylight. In human vision, there are red, green, and blue cone types, defined by their respective preferences for the predominantly red, green, and blue parts of the visible light spectrum; see Fig. 2.16 A. They are also called L, M, or S cones respectively, since they are tuned to long, medium, or short wavelengths of the input light. Let $I(\lambda)$ denote input light intensity as a function of the light wavelength λ, ignoring space and time. Each class of cones, denoted by $c = r, g, b$ for the red, green, and blue cones, has a particular sensitivity $\hat{f}_c(\lambda)$ to the light $I(\lambda)$ at every λ, such that the cone absorption (describing the amount of photons absorbed by the cone) by the cone class c is on average

$$r_c = \int d\lambda I(\lambda)\hat{f}_c(\lambda). \tag{2.33}$$

These sensitivity curves are also called cone fundamentals. The curves $\hat{f}_c(\lambda)$ of the red and green cones overlap substantially. The absorption values in these two cones are therefore highly correlated.

The cone absorption r_c evokes a cone response S_c, which is a nonlinear function of r_c (Lennie 2003b). Hence, for each cone class c, there is a retinal cone response image described by $S_c(x)$ as a function of space x. This image $S_c(x)$ is then sent to the retinal ganglion cells via interneurons. Hence, a ganglion cell's response $O = \int dx K(x)S(x)+$ spontaneous firing rate, described in equation (2.3), should be extended to (omitting the temporal dimension for simplicity)

$$O = \sum_{c=r,g,b} \int dx K_c(x)S_c(x) + \text{spontaneous firing rate}. \tag{2.34}$$

In general, the receptive field $K_c(x)$ for the image $S_c(x)$ from cone type c is different from $K_{c'}(x)$ for the image from another cone type c'. Different ganglion cells differ as to how $K_c(x)$ depends on c. For example, in one type of ganglion cells, the red cones excite the center of the receptive field, and the green cones inhibit the surround. This gives a red-on-center and green-off-surround receptive field; see Fig. 2.16 B. This cell is thus most sensitive to a small red disk of light. Another type of ganglion cell is blue-center-yellow-surround, giving a blue-yellow opponency. We will explain later (in Section 3.6.3) that such receptive fields decorrelate the responses from different cones to make for more efficient color coding.

Instead of viewing visual inputs as arising from three parallel cone channels with respective inputs $S_c(x)$, one can equivalently view them as arising from another set of three parallel channels $i = 1, 2, 3$, each made of a weighted combination of the cone images $S_c(x)$ as (more details in Section 3.6.3)

$$S_i(x) = \sum_{c=r,g,b} w_{ic}S_c(x). \tag{2.35}$$

One i of these channels is the luminance channel—it is a particular weighted sum of images $S_c(x)$ from all the cones c, comprising the colorless, grayscale, inputs. The other two i channels are chromatic channels, one being approximately a weighted difference between the images $S_r(x)$ and $S_g(x)$ from the red and green cones respectively; the other approximately a weighted difference between blue $S_b(x)$ and yellow $S_r(x) + S_g(x)$ images. By presenting only luminance signals (which do not excite the chromatic channels) to the visual system, one can measure the spatiotemporal contrast sensitivity curves just for the luminance channel. This was in fact what we did in Section 2.2.3. Figure 2.15 shows the sensitivity of the ganglion cells and human observers to temporal modulations of the luminance or chromatic signals presented in a light patch (Lee et al. 1990). (The relationship between the sensitivity of an observer and those of the retina ganglion cells will be discussed in Chapter 6.) A few notable features can be observed. First, the sensitivity to high temporal frequency signals is higher when the signals come from the luminance than the chromatic channel. Luminance signals are thus described as being band-passed, while the chromatic signals are more low-passed. Second, M cells are more sensitive than P cells to the luminance channel, whereas P cells are more devoted to the chromatic channels. Third, retinal ganglion cells can respond to input modulations at temporal frequencies which are too high to be perceived by humans.

A: Density of photoreceptors ($\times 10^3$ /mm^2) versus eccentricity

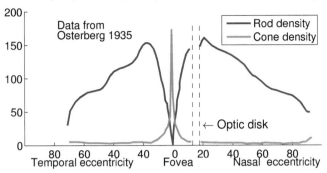

B: Visual acuity illustrated in an eye chart

Fig. 2.17: A: The density of cones and rods in the human retina versus visual angle from the center of vision. Note that the sampling density of the cones drops dramatically with eccentricity. It is densest at the fovea, where there is no room for the rods. The density of rods peaks slightly off fovea. The optic disc is where axons from the ganglion cells exit the eye to form the optic nerve, causing a blind spot, which is not noticeable in binocular viewing when the blind location in one eye's visual field is not blind in the other eye's visual field. Data from Osterberg, G., Topography of the layer of rods and cones in the human retina, *Acta Ophthalmology*, 6, supplement 13: 1–102, 1935. B: Visual acuity drops dramatically with increasing eccentricity: all the letters are equally visible when one fixates at the center of the eye chart (Anstis 1974). Copyright ©Stuart Anstis, 2013.

2.2.6 Spatial sampling in the retina

Figure 2.17 A shows that the density D of cones per unit area decreases rapidly with eccentricity e, the distance in visual angle from the center of vision. Roughly,

$$D \propto \frac{1}{e + e_o},$$

(2.36)

with e_o ~1–2 degrees (van Essen and Anderson 1995). Thus visual acuity drops drastically with eccentricity e, as demonstrated in Fig. 2.17 B; the size of the smallest recognizable letter increases roughly linearly with e. The sizes of the receptive fields of the ganglion cells also scale up with e accordingly (van Essen and Anderson 1995). As a result, humans have to use eye movements to bring objects of interest to the fovea in order to scrutinize them. This is directly linked with the problem of visual attention, which is typically directed at the center of gaze (see Section 2.5).

Rods belong to another class of photoreceptors that function mainly in dim light, due to their higher sensitivity to light. In humans, because the cones are packed so densely in the fovea, there is no room for rods in the center of fovea, and rod density peaks at an eccentricity of around 20°, as shown in Fig. 2.17 A. In dim light, cones are not functional, so one often has to avoid staring at an object directly in order to see it. Looking slightly away brings its image onto the rod-rich part of the retina. This is the trick often used to see a dim star in the night sky.

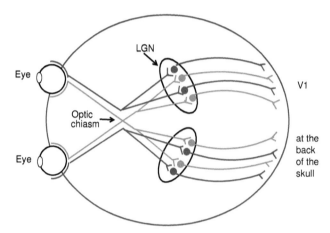

Fig. 2.18: A schematic of the visual pathway from the retina to V1 via the LGN. Information from the two eyes is separated in separate layers within LGN, but combined in V1. Information from the left and right hemifields of the visual space are sent to V1 in the right and left hemispheres of the brains, respectively.

2.2.7 LGN on the pathway from the retina to V1

The optic nerve carries the responses of the retinal ganglion cells to a region of the thalamus called the LGN; see Fig. 2.18 and Fig. 2.2. The signals from the P and M cells project to separate layers in the LGN. Signals from different eyes also project to separate layers. In each eye, signals from the left and right visual fields are projected to the right and left cortical hemispheres, respectively, via the LGN.

LGN has been seen as a relay station for retinal signals en route to V1, because the receptive fields of the LGN cells resemble those of the retinal ganglion cells in anaesthetized animals. However, it is a common idea that the LGN is very likely to be much more than a relay station. This idea is not only motivated by the expectation that the brain is unlikely to waste resources on a relay station without any reason but also by the observation that feedback neural connections from V1 to LGN, also including an additional structure called thalamic reticular complex, are much more numerous than feedforward connections from LGN to V1.

Further, the LGN also receives inputs from the brain stem, which also processes other sensory inputs. It has been suggested that the cortical and brain stem inputs jointly control the visual information relayed to the cortex from the retina and that this control depends on the state of the animal, such as the degree of arousal and visual attention (Sherman and Koch 1986). However, there is a lack of consensus as to the exact functional role of the LGN.

2.3 V1

2.3.1 The retinotopic map

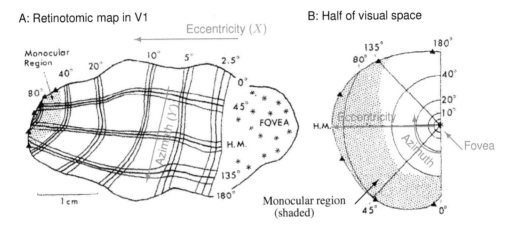

Fig. 2.19: The retinotopic map in V1, binocular and monocular visual space, and the definitions of eccentricity e and azimuth θ. A: Half of the visual field depicted in B is mapped to half of the V1 cortical area (in one hemisphere of the brain). More V1 surface area is devoted to central vision. B: The visual space mapped to cortical space in A. H.M. means horizontal meridian. Adapted from *Vision Research*, 24 (5), David C. Van Essen, William T. Newsome, and John H.R. Maunsell, The visual field representation in striate cortex of the macaque monkey: Asymmetries, anisotropies, and individual variability, pp. 429-48, figure 4, Copyright (1984), with permission from Elsevier.

V1 neurons also have small receptive fields. Neighboring V1 neurons have their receptive fields at or near the same location in visual space, and neighboring points in a visual image evoke activities at neighboring locations in V1. The retinotopic map refers to the transformation from the coordinates of the input image to those on the cortical surface. Figure 2.19 shows that the cortex devotes a larger surface area (per square degree of visual field) to the central part of the visual field, just as the retina has a higher density of receptors and ganglion cells in the foveal region. Let X and Y denote Cartesian coordinates along the horizontal and vertical directions, respectively, on the unfolded cortical surface. Let e and θ (in unit of degrees) denote eccentricity and azimuth, respectively, in the input image. The correspondence between the visual image in degrees e and θ and the cortical surface X and Y in millimeters (mm) is approximately:

$$X = \lambda \ln\left(1 + \frac{e}{e_0}\right), \quad Y = -\frac{\lambda e \theta \pi}{(e_0 + e)180^\circ}, \tag{2.37}$$

where $\lambda \approx 12$ mm, $e_0 \approx 1^\circ$, and the negative sign in the expression for Y comes from the inversion of the visual image in the image formation process. For visual locations much beyond the foveal region, i.e., $e \gg e_0$, we have $X \approx \lambda(\ln e - \ln e_0)$ growing linearly with log eccentricity $\ln e$ and $Y \approx -\lambda\pi\theta/180°$ growing linearly with azimuth θ. Denoting

$$z \equiv \frac{e}{e_0} \exp\left(-i\frac{\pi\theta}{180°}\right) \quad \text{and} \quad Z \equiv X + iY, \quad \text{(with } i = \sqrt{-1}\text{)}, \tag{2.38}$$

we have

$$Z \approx \lambda \ln(z) \quad \text{for large eccentricity locations.} \tag{2.39}$$

Hence, the cortical map is sometimes called a complex logarithmic map (Schwartz 1977). For large e, if the image on the retina scales as $e \to \gamma e$, its projection onto the cortex approximately shifts $X \to X + \lambda \ln(\gamma)$. The cortical magnification factor

$$M(e) \equiv \frac{dX}{de} = \frac{\lambda}{(e + e_0)} \tag{2.40}$$

characterizes how much cortical space is devoted to how much visual space at different eccentricities e. We note that this magnification factor $M(e)$ and the retinal receptor density $D \propto \frac{1}{e+e_0}$ depend similarly on the eccentricity e.

2.3.2 The receptive fields in V1—the feature detectors

There are about 100 times as many neurons in V1 as there are ganglion cells in the retina (Barlow 1981), making V1 the largest visual area in the brain. The receptive fields of the neurons have largely been known since the pioneering works of Hubel and Wiesel about half a century ago. The center-surround type of stimulus that retinal ganglion cells prefer is not the preferred type of stimulus for most V1 neurons. Instead, V1 neurons typically prefer a stimulus which resembles a bright or dark bar, or a luminance edge. Hubel and Wiesel proposed that such a preference could arise from a V1 neuron receiving inputs from several LGN neurons in a structured array. For instance, if three on-center LGN neurons have the centers of their receptive fields placed next to each other horizontally and their outputs are sent to a V1 neuron, this V1 neuron would then prefer a horizontal bright bar flanked by two horizontal dark bars; see Fig. 2.20. Different V1 neurons prefer different orientations of the bar or edge.

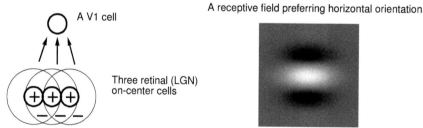

Fig. 2.20: A schematic of how three on-center retinal ganglion neurons, or LGN neurons, feeding into a V1 cell could make a V1 cell tuned to a bright oriented bar, according to Hubel and Weisel.

Apart from orientation tuning, V1 neurons can also be tuned to spatial scale (i.e., size), color, direction of motion, disparity, eye of origin of inputs, and combinations of these features. Hence, V1 neurons are often called feature detectors. V1 can be divided into patches called *hypercolumns*, which are cylindrical or cuboid volumes with a diameter of about 1 mm and spanning all six layers of the cortex. Each hypercolumn contains many neurons whose receptive fields are centered at similar locations in visual space but whose preferred features span a full range of values across the various feature dimensions.

To study the feature tunings of V1 neurons, let us describe visual inputs by $S(x, y, t, e, c)$, a signal that depends on space (x, y), time t, eye of origin $e = L, R$ for the left or right eye, respectively, and color $c = r, g, b$ for red, green, and blue cones, respectively. As above, a linear receptive field can be described by a filter K acting on the various feature dimensions, giving

$$\text{linear response } L \text{ at time } t = \sum_{x,y,t',e,c} K(x, y, t - t', e, c) S(x, y, t', e, c). \qquad (2.41)$$

2.3.2.1 Simple cells

In reality, V1 neurons are nonlinear. One class of neurons comprises what is known as simple cells. Their responses to inputs are essentially linear, except for a static nonlinearity, such that a response O rises with input strength and then saturates. Such a response O can be modeled by

$$O = F(L) + \text{spontaneous firing rate}, \qquad (2.42)$$

where L is the linear response from equation (2.41), and F is a nonlinear function such as

$$F(L) \propto \frac{[L]_+^n}{A_{1/2}^n + [L]_+^n}, \qquad (2.43)$$

where $A_{1/2}$ and $n \approx 2$ are parameters and $[.]_+$ is a rectification function such that $[x]_+ = x$ for $x > 0$ and 0 otherwise. The tuning properties of a simple cell can be well characterized by its linear filter K. Hence, in many models, simple cells are approximately seen as linear filters with response L.

Often, the property of the linear filter K in one feature dimension, e.g., space, time, color, or eye of origin, is studied separately from other feature dimensions. This is done by fixing the feature values of the input stimulus in the other dimensions or integrating out the other dimensions as we did when we obtained the retinal spatial filter $\bar{K}(x) = \int K((x, t)dt$ by integrating out the temporal dimension in a spatiotemporal filter $K(x, t)$ in equation (2.19). For simplicity, properties of a receptive field in the feature dimension(s) of interest is often described by omitting the other feature dimensions.

2.3.3 Orientation selectivity, bar and edge detectors

If the neurons are (approximated as) linear, the spatial receptive field a V1 cell can be modeled by a Gabor function of space (x, y), defined as

$$K(x, y) \propto \exp\left(-\frac{x^2}{2\sigma_x^2} - \frac{y^2}{2\sigma_y^2}\right) \cos(\hat{k}x + \phi), \qquad (2.44)$$

with parameters σ_x, σ_y, $\hat{k}$, and ϕ. This spatial filter can be seen as a vertically oriented receptive field. Typically, in such a case, $\sigma_x < \sigma_y$. If $\phi = 0$, this neuron prefers a vertical bar of width $\sim 1/\hat{k}$ centered at $x = 0$; if $\phi = \pi/2$, it prefers instead a vertical luminance edge; see Fig. 2.21, for an illustration with the preferred phase $\phi = \pi/4$. Hence, V1 neurons are often called bar and edge detectors.

Presented with a spatial grating stimulus $S(x, y) \propto \cos(kx + \phi')$, the response is

$$O = \int dx dy K(x, y) S(x, y)$$

$$\propto \int dy \exp\left(-\frac{y^2}{2\sigma_y^2}\right) \int dx \exp\left(-\frac{x^2}{2\sigma_x^2}\right) \left[\cos\left((k - \hat{k})x + (\phi - \phi')\right)\right.$$
$$\left. + \cos\left((k + \hat{k})x + (\phi + \phi')\right)\right]$$

$$\propto \int dx \exp\left(-\frac{x^2}{2\sigma_x^2}\right) \left[\cos\left((k - \hat{k})x\right) \cos(\phi - \phi')\right.$$
$$\left. + \cos\left((k + \hat{k})x\right) \cos(\phi + \phi')\right]$$

$$\propto \exp\left(-\frac{(k - \hat{k})^2 \sigma_x^2}{2}\right) \cos(\phi - \phi') + \exp\left(-\frac{(k + \hat{k})^2 \sigma_x^2}{2}\right) \cos(\phi + \phi')$$

$$\approx \exp\left(-\frac{(k - \hat{k})^2 \sigma_x^2}{2}\right) \cos(\phi - \phi'). \tag{2.45}$$

The last approximation holds when $\exp\left(-(k + \hat{k})^2 \sigma_x^2/2\right) \ll \exp\left(-(k - \hat{k})^2 \sigma_x^2/2\right)$. This response is sensitive to the position, or phase ϕ', of the grating, giving a maximum response when the phase ϕ' of the grating is the same as the phase ϕ of the receptive field (which happens when the center of the on-region of the receptive field coincides with a peak in the grating).

Receptive fields with a preferred orientation of θ away from vertical can be obtained by rotating $K(x, y)$ in equation (2.44) such that in the expression for $K(x, y)$,

$$x \text{ is replaced by } x \cos(\theta) + y \sin(\theta),$$
$$y \text{ is replaced by } -x \sin(\theta) + y \cos(\theta). \tag{2.46}$$

The preferred orientations of different V1 neurons span the whole range of orientations. Within a hypercolumn, neurons preferring similar orientations tend to cluster together. An *orientation column* describes this cluster across different hypercolumns. Orientation columns whose preferred orientations span the full orientation range should pass each hypercolumn.

2.3.4 Spatial frequency tuning and multiscale coding

According to equation (2.45), a grating $S(x, y) \propto \cos(kx + \phi')$ evokes the highest response from the Gabor filter $K(x, y) \propto \exp\left(-\frac{x^2}{2\sigma_x^2} - \frac{y^2}{2\sigma_y^2}\right) \cos(\hat{k}x + \phi)$ when the grating's frequency is $k = \hat{k}$. More specifically, neural sensitivity to frequency k can be described by

$$\text{frequency tuning curve} \propto \exp\left(-\frac{(k - \hat{k})^2 \sigma_x^2}{2}\right). \tag{2.47}$$

Therefore, the cell is tuned to spatial frequency k. Let k_h and k_l be the highest and lowest frequencies satisfying $(k - \hat{k})^2 \sigma_x^2 \leq 1$ or $\exp\left(-(k - \hat{k})^2 \sigma_x^2/2\right) \geq \exp(-0.5)$. The width of frequency tuning can be defined as $\Delta k \equiv k_h - k_l$, which is proportional to $1/\sigma_x$. Hence, a spatial filter with a larger width σ_x is more narrowly tuned in frequency. Another measure of the frequency bandwidth uses octaves:

A: A Gabor receptive field

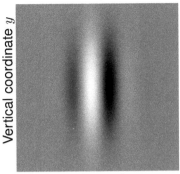

Vertical coordinate y

Horizontal coordinate x

B: Frequency tuning of the filter in A

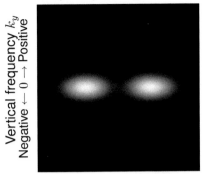

Vertical frequency k_y
Negative $\leftarrow 0 \rightarrow$ Positive

Negative $\leftarrow 0 \rightarrow$ Positive
Horizontal frequency k_x

Fig. 2.21: The Gabor filter $K(x,y) \propto \exp\left(-\frac{x^2}{2\sigma_x^2} - \frac{y^2}{2\sigma_y^2}\right) \cos(\hat{k}x + \pi/4)$ in A is tuned to the vertical orientation. B shows the frequency tuning of this filter; note that the peaks are on the horizontal axis. The brightness levels in both pictures scale with the numerical values of the functions in space (x,y) or in frequency (k_x, k_y).

$$\text{bandwidth in octaves} \equiv \log_2(k_h/k_l) \quad \text{octaves.} \tag{2.48}$$

The spatial frequency bandwidth of V1 neurons is typically around 1.5 octaves (De Valois, Albrecht and Thorell 1982), which means $k_h \approx 3k_l$. Different neurons prefer different spatial frequency bands, such that they collectively span the whole frequency range covered by the contrast sensitivity curve of the retinal ganglion cells (see Fig. 2.8).

In octaves, the bandwidths of V1 neurons are almost independent of their preferred spatial frequency $\hat{k}$. This means that the widths σ_x of their receptive fields are inversely proportional to $\hat{k}$; or, equivalently, σ_x is proportional to the wavelength of the Gabor filter. Consequently, receptive fields tuned to different frequency bands can have identical shapes, even though their sizes differ. The overall sensitivity of a V1 neuron scales with the contrast sensitivity of the visual system at that neuron's preferred frequency $\hat{k}$.

In 2D space, the spatial frequency k is a 2D vector (k_x, k_y) with components k_x and k_y along the horizontal and vertical dimensions, respectively. Hence, the Gabor filter above should in fact be written as $K(x,y) \propto \exp\left(-\frac{x^2}{2\sigma_x^2} - \frac{y^2}{2\sigma_y^2}\right) \cos(\hat{k}_x x + \phi)$. Its preferred frequency is $(\hat{k}_x, 0)$, which lies on the horizontal axis in 2D frequency space (remembering that the filter is tuned to vertical orientation). The Fourier transform of this spatial filter indicates its frequency tuning as a function of the 2D spatial frequency (k_x, k_y). This tuning function is approximately two Gaussians centered at $(\hat{k}_x, 0)$ and $(-\hat{k}_x, 0)$, respectively. Its magnitude is

$$|g(k)| \propto \exp\left(-\frac{(k_x - \hat{k}_x)^2 \sigma_x^2}{2} - \frac{k_y^2 \sigma_y^2}{2}\right) + \exp\left(-\frac{(k_x + \hat{k}_x)^2 \sigma_x^2}{2} - \frac{k_y^2 \sigma_y^2}{2}\right); \tag{2.49}$$

see Fig. 2.21. The frequency-tuning profile is mirror symmetric across the origin $k = (0,0)$. This is a property of the Fourier transform of a real function of spatial location.

2.3.5 Temporal and motion direction selectivity

A space-time separable receptive field of a neuron can be written as

$$K(x, \tau = t - t') = K_x(x)K_t(\tau = t - t'). \tag{2.50}$$

The $K_x(x)$ is like that in equation (2.44) except that the vertical dimension y is omitted for simplicity, and $K_t(\tau)$ is like that in equation (2.24). Because it is space-time separable, the receptive field is not selective to the direction of spatial motion such as that of a drifting grating. To see this, let us write a vertical grating, $S_\pm(x, t) = \cos(kx \pm \omega t)$, drifting left $(S_+(x, t))$ or right $(S_-(x, t))$ as the sum of two temporally oscillating (flashing) gratings:

$$S_\pm(x, t) = \cos(kx \pm \omega t) = \cos(kx) \cos(\omega t) \mp \sin(kx) \sin(\omega t). \tag{2.51}$$

The speed of drift is $v = \omega/k$. Let $K_t(t) \equiv \int_0^\infty d\omega g(\omega) \cos[\omega t + \phi(\omega)]$. Then, the neural response to the grating is

$$
\begin{aligned}
O(t) &= \int dx K_x(x) \int dt' K_t(t - t') S_\pm(x, t') \\
&\propto \left[\int dx K_x(x) \cos(kx) \right] g(\omega) \cos[\omega t + \phi(\omega)] \\
&\mp \left[\int dx dy K_x(x) \sin(kx) \right] g(\omega) \sin[\omega t + \phi(\omega)] \tag{2.52} \\
&= g(\omega) \left\{ \left[\int dx K_x(x) \cos(kx) \right]^2 \right. \\
&\quad \left. + \left[\int dx K_x(x) \sin(kx) \right]^2 \right\}^{1/2} \cos(\omega t + \theta_\pm), \tag{2.53}
\end{aligned}
$$

where

$$\theta_\pm = \phi(\omega) \pm \psi, \quad \psi = \tan^{-1} \frac{\left[\int dx K_x(x) \sin(kx) \right]}{\left[\int dx K_x(x) \cos(kx) \right]}. \tag{2.54}$$

Hence, the response oscillates with frequency ω, and the direction in which the grating drifts only affects the phase, but not the amplitude, of this oscillation.

The spatiotemporal receptive field of a directionally selective V1 neuron is not space-time separable. It can be made so by adding two space-time separable ones as follows; see Fig. 2.22. One builds the following four filters, two in space and two in time:

$$K_x(x) \equiv \int_0^\infty dk g_x(k) \cos[kx + \phi(k)], \quad K_t(t) \equiv \int_0^\infty d\omega g_t(\omega) \cos[\omega t + \psi(\omega)], \tag{2.55}$$

$$\hat{K}_x(x) \equiv \int_0^\infty dk g_x(k) \sin[kx + \phi(k)], \quad \hat{K}_t(t) \equiv \int_0^\infty d\omega g_t(\omega) \sin[\omega t + \psi(\omega)]. \tag{2.56}$$

In the above filters, $g_x(k)$ and $g_t(\omega)$ are the frequency sensitivity functions in spatial and temporal dimensions, respectively, and $\phi(k)$ and $\psi(\omega)$ are the corresponding phase functions. The functions $K_x(x)$ and $\hat{K}_x(x)$ are said to be in quadrature of each other, as are $K_t(t)$ and $\hat{K}_t(t)$, because the two functions in each pair are related by a phase shift of $90°$ in each frequency of their spectra. One sees from Fig. 2.22 that the spatial or temporal on-regions of the two filters in each pair are shifted from each other in space or time respectively. Note that

$$\int dx K_x(x) \cos(kx) = \int dx \hat{K}_x(x) \cos(kx - \pi/2),$$

$$\int dt' K_t(t - t') \cos(\omega t') = - \int dt' \hat{K}_t(t - t') \cos(\omega t' - \pi/2). \tag{2.57}$$

Quadrature model of motion direction selectivity

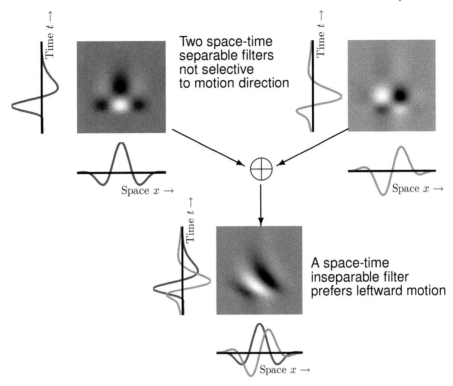

Two space-time
separable filters
not selective
to motion direction

A space-time
inseparable filter
prefers leftward motion

Fig. 2.22: Using a quadrature model to build a spatiotemporal receptive field tuned to motion direction by summing two filters not selective to direction. Each of the three grayscale images plots a spatiotemporal receptive field, with space and time along horizontal and vertical axes, respectively. The two images at the top are space-time separable and not direction selective; each is the product of a spatial filter (the curve to its bottom) and a temporal filter (the curve to its left). The receptive field in the grayscale image at the bottom is made by superposing the top two grayscale images. The blue and red curves superposed in each graph, to the bottom and left of the bottom grayscale image, depict two filters which are quadratures of each other; see equations (2.55–2.61).

Hence, when $S_\pm(x,t) = \cos(kx \pm \omega t) = \cos(kx)\cos(\omega t) \mp \sin(kx)\sin(\omega t)$ is a drifting grating, one can derive from the above that

$$\int dx dt' \, K_x(x) K_t(t-t') S_\pm(x,t') = \pm \int dx dt' \, \hat{K}_x(x) \hat{K}_t(t-t') S_\pm(x,t'). \qquad (2.58)$$

This means that the following two spatiotemporal filters

$$K_x(x)K_t(t) - \hat{K}_x(x)\hat{K}_t(t)$$
$$= \int_0^\infty dk d\omega \, g_x(k) g_t(\omega) \cos[kx + \omega t + \phi(k) + \psi(\omega)], \quad \text{and} \qquad (2.59)$$
$$K_x(x)K_t(t) + \hat{K}_x(x)\hat{K}_t(t)$$
$$= \int_0^\infty dk d\omega \, g_x(k) g_t(\omega) \cos[kx - \omega t + \phi(k) - \psi(\omega)] \qquad (2.60)$$

are exclusively direction tuned. In particular, a grating $\cos(kx - \omega t)$ or $\cos(kx + \omega t)$ drifting right or left, respectively, whose spatiotemporal frequency (k, ω) is within the span of the sensitivity profile $g_x(k)g_t(\omega)$, evokes response only in the first or second filter respectively. This direction selectivity is maintained when the stimuli are generalized to moving objects $S(x, t) = \int dkd\omega S(k, \omega) \cos[kx \mp \omega t + \alpha(k, \omega)]$ with general amplitude and phase profiles $S(k, \omega)$ and $\alpha(k, \omega)$, moving to the right or left, respectively, when the sign in front of ωt is negative or positive.

In general, let $A^+ \neq A^-$ be two real numbers; then, a space-time inseparable filter

$$K(x, t) = (A^+ + A^-)K_x(x)K_t(t) + (A^- - A^+)\hat{K}_x(x)\hat{K}_t(t) \tag{2.61}$$

$$= A^+ \int_0^\infty dkd\omega g_x(k)g_t(\omega) \cos[kx + \omega t + \phi(k) + \psi(\omega)]$$

$$+ A^- \int_0^\infty dkd\omega g_x(k)g_t(\omega) \cos[kx - \omega t + \phi(k) - \psi(\omega)] \tag{2.62}$$

is a superposition of two filters preferring motion in opposite directions, with unequal sensitivities A^+ and A^-. It is made by adding two space-time separable filters, which are space-time quadratures of each other, with weights $(A^+ + A^-)$ and $(A^- - A^+)$. The model spatiotemporal filter in equations (2.55–2.62) is called a linear quadrature model (Watson and Ahumada 1985). The receptive fields of the primary simple cells can be well approximated by such models (Hamilton, Albrecht and Geisler 1989).

The space-time inseparable filter in Fig. 2.22 can also be seen as having a particular tilt in the space-time plane. The orientation of this tilt determines the preferred direction and speed of the motion. The preferred drifting speed $v = \omega/k$ of a filter depends on its preferred spatial frequency k and preferred temporal frequency ω. However, since V1 neurons are only rather coarsely tuned to temporal frequency ω, their speed tuning is weak.

2.3.6 Ocular dominance and disparity selectivity

In general, a V1 neuron's receptive field and sensitivity depend on the eye of origin of the input. Let $e = L$ and $e = R$ denote the left and right eye, respectively. The neural receptive fields for inputs from the two eyes can be (omitting other feature dimensions)

$$K_L(x, y) = g_L \exp\left(-\frac{x^2}{2\sigma_x^2} - \frac{y^2}{2\sigma_y^2}\right) \cos[k(x - x_L)], \quad \text{and}$$
$$K_R(x, y) = g_R \exp\left(-\frac{x^2}{2\sigma_x^2} - \frac{y^2}{2\sigma_y^2}\right) \cos[k(x - x_R)]. \tag{2.63}$$

This pair of receptive fields has sensitivities g_L and g_R to the inputs from the two eyes. Although the centers of the receptive fields are the same, the on-region of the receptive field is at $x = x_L$ for the left eye input and $x = x_R$ for the right eye input. Figure 2.23 A shows that the location of the image of an object differs between the two eyes. This difference is called disparity and is zero when gaze fixates on the object. According to this terminology, the neuron with receptive fields given in the equations above is said to be tuned to an input disparity $d = x_L - x_R$ (provided that the input position and contrast polarity are also appropriate). An object at a depth nearer than the fixation point is said to have a *crossed disparity*, since the two eyes have to cross more to fixate on it. An object further than the fixation point has an *uncrossed disparity*. The disparity of an object along the horizontal or vertical axis in the visual image is called its horizontal or vertical disparity, respectively. The equations above model a neuron preferring a zero vertical disparity. V1 neurons can also prefer non-zero vertical disparities.

A: Visual objects and disparities   B: Two receptive fields for a single neuron

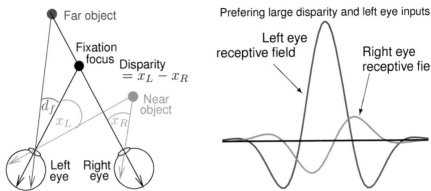
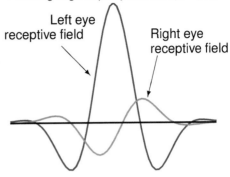

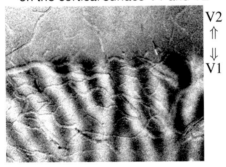

Fig. 2.23: Disparity tuning and ocular dominance of neural receptive fields. A: Each object in the visual scene has a disparity. The visual angles x_L and x_R give image locations on the retina. B: A schematic of the receptive fields for the left and right eyes, respectively, for a single V1 cell. The two receptive fields (RFs) can differ in both amplitude and phase. This single neuron is said to be left eye dominant since the RF for left eye input has a higher amplitude. The peak location x_R of its on-region in the right eye is right shifted relative to its counterpart x_L in the left eye. C: Ocular dominance columns revealed by optical imaging (copyright © 2013, Anna Roe). Black and white stripes mark V1 cortical surface locations for neurons preferring inputs from one and the other eye, respectively. Each column is about 400 μm thick in monkeys; they abruptly stop at the border between V1 and V2, in the upper part of the image, since V2 neurons are mostly binocular.

Differential sensitivities $g_L \neq g_R$ make a neuron favor inputs from one eye over the other. The population of V1 neurons covers a spectrum of relative ocular preference and preferred disparities. (Stereo coding, manifested in these data, will be a case study for the efficient coding principle in Sections 3.5 and 3.6.7.) Some are tuned to near zero disparities, others to crossed or uncrossed disparities, and still others are suppressed by near zero disparities. Neurons similarly sensitive to inputs from either eye are called *binocular cells*; otherwise they are called *monocular cells*. Hubel and Wiesel defined ocular dominance index values ranging from 1 to 4 to 7 to denote, respectively, neurons preferring the contralateral eye, binocular neurons, and neurons preferring the ipsilateral eye. Ocular preference is a special property of V1 neurons, since neurons in cortical areas farther along the visual pathway are mostly binocular. This property is often used to identify V1 as the neural substrate for various visual processes. Meanwhile, disparity tuning is a property shared by neurons in other visual cortical areas.

In V1, neurons tuned to the same eye of origin tend to cluster together. If one colors the cortical surface black or white, according to whether the neurons around a cortical position prefer inputs from the left or right eye, black and white stripes appear to visualize the underlying *ocular dominance columns* across cortical layers. These columns, which are about 400 μm thick in monkeys, are indeed seen when imaging cortical activities while stimulating one eye only with strong visual inputs; see Fig. 2.23 C. Binocular neurons cluster at the boundaries between neighboring columns tuned to different eyes, making these boundaries appear fuzzy in such images.

2.3.7 Color selectivity of V1 neurons

A V1 neuron can have both spatial and chromatic selectivities by having different receptive fields for inputs of different color. Some V1 neurons are called *double-opponent* (Livingstone and Hubel 1984), since their receptive fields exhibit both chromatic and spatial opponencies such that color preference at one input location is opposite to that at another location.

V1 neurons tuned to color are often not tuned to orientation, in which case their spatial receptive fields may have center-surround shapes like (but much larger than) those of the retinal ganglion cells; see Fig. 3.33. Let $K_c(x, y)$ denote the spatial receptive field for color c of the input. The filter for red input $K_{c=r}(x, y)$ can, for example, be on-center, while the filter for green input $K_{c=g}(x, y)$ is off-center, making the neuron favor a red dot in a green background. This neuron is a double-opponent cell. Some V1 neurons can be tuned to both orientation and color. Other V1 neurons, especially those tuned to high spatial frequencies, are tuned to orientation only without significant color preference. This coupling between spatial and color coding will be interpreted in the efficient coding framework in Section 3.6.6.

2.3.8 Complex cells

Complex cells are neurons whose responses O cannot be approximated as a static nonlinear transform of the outcome L of a linear filter of the visual input (i.e., by equation (2.42)). Unlike simple cells, the response of a complex cell is somewhat insensitive to the position of visual inputs, such that its response to a grating is insensitive to the grating's phase ϕ. A complex cell is often modeled as receiving inputs from two simple cells whose spatial receptive fields are components in a quadrature pair. For example, let the spatial receptive fields of these two simple cells be

$$K_1(x) = \exp\left(-\frac{x^2}{2\sigma^2}\right)\cos(\hat{k}x + \psi) \quad \text{and} \quad K_2(x) = \exp\left(-\frac{x^2}{2\sigma^2}\right)\sin(\hat{k}x + \psi). \quad (2.64)$$

According to equation (2.45), their respective linear responses to an input grating $\cos(kx + \phi)$ are approximately,

$$L_1 \propto \int dx K_1(x)\cos(kx + \phi) = \exp\left(-\frac{(k - \hat{k})^2\sigma^2}{2}\right)\cos(\phi - \psi), \quad \text{and} \quad (2.65)$$

$$L_2 \propto \int dx K_2(x)\cos(kx + \phi) = -\exp\left(-\frac{(k - \hat{k})^2\sigma^2}{2}\right)\sin(\phi - \psi). \quad (2.66)$$

A complex cell's response can be modeled as combining the linear responses of the simple cells (i.e., before the application of the static nonlinearity) in what is called the *energy model*:

$$E \equiv L_1^2 + L_2^2 \tag{2.67}$$
$$= [L_1]_+^2 + [-L_1]_+^2 + [L_2]_+^2 + [-L_2]_+^2. \tag{2.68}$$

We remind ourselves that $[.]_+$ enforces rectification. Hence, L_i^2 can be seen as arising from two (rectified) simple cells whose receptive fields are negatives of each other, whereas the complex cell can thus be seen as made from four rectified simple cells. An additional static nonlinearity, like that of the simple cells in equation (2.43), models the actual response of a complex cell as

$$O = F(E) + \text{spontaneous firing rate}, \quad \text{where} \quad F(E) \approx \frac{E}{\text{constant} + E}. \tag{2.69}$$

Since O is monotonically related to E, the approximation $O = E$ is often adopted in some applications, just like a simple cell is often approximated by a linear filter of input images.

The response of the complex cell to a static input grating is (approximately)

$$O = E = L_1^2 + L_2^2 \propto \exp\left(-\frac{(k - \hat{k})^2\sigma^2}{2}\right), \tag{2.70}$$

which is independent of the phase ϕ or position of the grating. Let us examine the degree of position invariance in its response to a bar. The input from a bar centered at location x_o is modeled as

$$S(x) = \exp\left[-\hat{k}^2(x - x_o)^2\right]. \tag{2.71}$$

The width of this bar is comparable to that of the subfields of the Gabor filter in each linear subunit of the complex cell. The linear response from the first simple cell component is

$$L_1 = \int dx \exp\left(-\frac{x^2}{2\sigma^2}\right) \cos(\hat{k}x + \psi) \exp\left[-\hat{k}^2(x - x_o)^2\right]$$
$$\propto \exp\left(\frac{-\hat{k}^2x_o^2}{1 + 2\hat{k}^2\sigma^2}\right) \cos\left[\frac{2(\hat{k}^2\sigma^2)\hat{k}x_o}{1 + 2\hat{k}^2\sigma^2} + \psi\right]. \tag{2.72}$$

Similarly, $L_2 \propto \exp\left(\frac{-\hat{k}^2x_o^2}{1+2\hat{k}^2\sigma^2}\right) \sin\left[\frac{2(\hat{k}^2\sigma^2)\hat{k}x_o}{1+2\hat{k}^2\sigma^2} + \psi\right]$. In V1 cells, $\hat{k}$ and σ typically satisfy $\hat{k}\sigma \approx 2.5$. Hence, the sine and cosine factors are approximately $\sin(\hat{k}x_o + \psi)$ and $\cos(\hat{k}x_o + \psi)$, respectively. This implies that responses L_1 and L_2 both oscillate with the position x_o of the bar through the cosine and sine factors, respectively. However, the complex cell's response is

$$O = L_1^2 + L_2^2 \propto \exp\left(\frac{-2\hat{k}^2x_o^2}{1 + 2\hat{k}^2\sigma^2}\right) \approx \exp\left(\frac{-x_o^2}{\sigma^2}\right). \tag{2.73}$$

This means, as long as the position x_o of the bar remains securely within the envelope of the complex cell's receptive field, i.e., $|x_o| < \sigma$, the cell's response will not be too sensitive to x_o relative to the on and off subfields of the simple cell components.

2.3.8.1 Disparity tuning of a complex cell

The insensitivity to input position in the responses of the complex cells can be useful for signaling input disparity without being too sensitive to the position and polarity of the input within the envelope of the complex cell's receptive field (Ohzawa, DeAngelis and Freeman 1990, Qian 1994). Let a simple cell have the following receptive fields for the two eyes (omitting the y dimension in space for simplicity):

$$K_L(x) = \exp\left(-\frac{x^2}{2\sigma^2}\right)\cos\left[\hat{k}(x - x_L)\right], \quad K_R(x) = \exp\left(-\frac{x^2}{2\sigma^2}\right)\cos\left[\hat{k}(x - x_R)\right].$$

(2.74)

Let the inputs to the left and right eyes be $S_L(x) = I(x)$ and $S_R(x) = I(x + D)$, which are shifted from each other by a disparity D. If one assumes that the width σ of the receptive field is much larger than the D, the linear response from the simple cell is (approximately, noting that $\hat{k}\sigma \approx 2.5$ for V1 cells)

$$L = \int_{-\infty}^{\infty} [K_L(x)S_L(x) + K_R(x)S_R(x)]\, dx$$

$$\propto \rho\cos\left(\theta + \frac{\hat{k}(x_L + x_R)}{2} + \frac{\hat{k}D}{2}\right)\cos\left(\frac{\hat{k}(x_L - x_R)}{2} - \frac{\hat{k}D}{2}\right),$$

(2.75)

in which ρ and θ are the amplitude and phase, respectively, of the Fourier transform of the image $I(x)$ at frequency $\hat{k}$ (Qian 1994). If the input contains only a bar at disparity D, the response L is tuned to this disparity D and also to θ, which is determined by the position and contrast polarity of the input bar. The disparity tuning of a complex cell combining two simple cells which are identical to each other in σ, $\hat{k}$, and $x_L - x_R$ but differ by $90°$ in their average receptive field phase $\hat{k}(x_L + x_R)/2$ will be insensitive to the position and contrast polarity of the input bar. Given the linear responses L_1 and L_2 of these two simple cells to the above input, the response from the complex cell is then

$$O = L_1^2 + L_2^2 \approx O_{\max}\cos^2\left(\frac{\hat{k}(x_L - x_R)}{2} - \frac{\hat{k}D}{2}\right),$$

(2.76)

where $O_{\max}$ is the response to the optimal disparity.

Such a complex cell prefers a disparity $D = x_L - x_R$ and is duly insensitive to the exact position and contrast polarity of the visual input.

2.3.8.2 Motion energy model of the complex cells

From equations (2.52), (2.53), and (2.58–2.60), we see that the response of the direction-selective linear filters in equations (2.59–2.60) to a grating drifting in their preferred directions oscillates over time $L_1 \propto \cos(\omega t + \theta_{\pm})$. Consider, for each of these filters, constructing another linear filter that is tuned to the same direction but responds over time according to $L_2 \propto \sin(\omega t + \theta_{\pm})$, i.e., in a quadrature phase relationship with L_1. Then, one can combine the two linear filters to make a complex cell whose response $O = L_1^2 + L_2^2$ does not oscillate over time but still signals motion in the preferred direction. This can be done as follows (Adelson and Bergen 1985, Watson and Ahumada 1985). Using the spatial and temporal filters in equations (2.55) and (2.56), one can construct another pair of filters

$$\hat{K}_x(x)K_t(t) + K_x(x)\hat{K}_t(t)$$
$$= \int_0^\infty dk d\omega g_x(k)g_t(\omega)\sin[kx + \omega t + \phi(k) + \psi(\omega)], \quad \text{and} \tag{2.77}$$

$$\hat{K}_x(x)K_t(t) - K_x(x)\hat{K}_t(t)$$
$$= \int_0^\infty dk d\omega g_x(k)g_t(\omega)\sin[kx - \omega t + \phi(k) - \psi(\omega)], \tag{2.78}$$

which are the same as the pair in equations (2.59–2.60), except each spatiotemporal component of the filters has changed phase by $\pi/2$. This phase difference translates to the same phase difference in the oscillating responses. For example, each of the following pair of spatiotemporal filters

$$K_x(x)K_t(t) - \hat{K}_x(x)\hat{K}_t(t) \quad \text{and} \quad \hat{K}_x(x)K_t(t) + K_x(x)\hat{K}_t(t) \tag{2.79}$$

is tuned to a rightward drifting grating $S(x,t) = \cos(kx - \omega t)$. Their respective responses L_1 and L_2 can be combined to give $O = L_1^2 + L_2^2$, which retains the direction selectivity of the filters but whose response does not oscillate in time. This construction is called the *motion energy model* of complex cells (Adelson and Bergen 1985). Many V1 complex cells have this property (Emerson, Bergen and Adelson 1992).

Let a second complex cell prefer the opposite direction, having as subunits the following pair of linear receptive fields (analogous to equation (2.79))

$$K_x(x)K_t(t) + \hat{K}_x(x)\hat{K}_t(t) \quad \text{and} \quad \hat{K}_x(x)K_t(t) - K_x(x)\hat{K}_t(t). \tag{2.80}$$

To any visual input $S(x,t)$, let L_1 and L_2 denote the responses of the two filters in equation (2.79), and L_3 and L_4 of those from the two filters in equation (2.80). The difference between the responses from these two complex cells (which prefer opposite directions) is (Adelson and Bergen 1985):

Response from a *Reichardt* motion direction detector
$$= (L_1^2 + L_2^2) - (L_3^2 + L_4^2)$$
$$\propto (K_x\hat{K}_t * S) \cdot (\hat{K}_x K_t * S) - (K_x K_t * S) \cdot (\hat{K}_x\hat{K}_t * S). \tag{2.81}$$

The $*$ in, e.g., $K_x\hat{K}_t * S$, indicates a convolution between a filter (e.g., $K_x\hat{K}_t$) and a visual input S (whereas a simple $\cdot$ in $a \cdot b$ denotes the normal multiplication). The above mechanism is (a variation of what is) commonly known as the *Reichardt detector*, after the vision scientist Reichardt (Adelson and Bergen 1985). This can be understood by noting the following: (1) K_x and $\hat{K}_x$ are two spatial filters with a spatial offset between their on-regions and (2) K_t and $\hat{K}_t$ are two temporal filters which differ in impulse response latencies; see Fig. 2.22 for an example of a relationship between filters in a quadrature pair. Hence, in equation (2.81), the first term, $(K_x\hat{K}_t * S) \cdot (\hat{K}_x K_t * S)$, is a correlation of two signals; one, $K_x\hat{K}_t * S$, signals an event at one location by one particular latency, and the other, $\hat{K}_x K_t * S$, signals another event at a neighboring location by another latency. This correlation of two space-time events thus signals a particular motion direction. The term $(K_x K_t * S) \cdot (\hat{K}_x\hat{K}_t * S)$ signals the opposite motion direction. The opponency between these two detectors of opposite motion gives, still, a detector of a particular motion direction. One can imagine that higher visual areas can combine responses from two complex cells in V1 to make such a Reichardt detector.

2.3.9 The influences on a V1 neuron's response from contextual stimuli outside the receptive field

Contextual stimuli, which are presented outside the *classical receptive field* (CRF) of a V1 neuron, can substantially, and selectively, influence the neuron's response to a stimulus within its CRF. This happens even though they cannot by themselves excite the neuron (Allman, Miezin and McGuinness 1985). Typical types of contextual influences are sketched in Fig. 2.24. The CRF of a V1 neuron is roughly the size of the linear spatial filter in equation (2.44), such that an isolated small stimulus (e.g., a bar) can excite or influence the neuron's response only when it is within its CRF; see Fig. 2.24. Typically, a V1 neuron's response to an optimally oriented bar within its CRF is suppressed by contextual bars surrounding the CRF. This suppression is strongest when the contextual bars are parallel to the central bar, reducing the response by up to about 80% (Knierim and Van Essen 1992). This is called *iso-orientation suppression* (Fig. 2.24 C). When the surrounding bars are orthogonal to the central bar, this suppression is often reduced (Fig. 2.24 E), and, in some circumstances (discussed later in Fig. 5.35 C), may even appear to be replaced with facilitation (Sillito, Grieve, Jones, Cudeiro and Davis 1995). If the surrounding bars are randomly oriented, they also appreciably suppress the response to the central bar. However, when the neural response is weak, such as when the central bar has a low contrast, or when suppression from other contextual stimuli is sufficiently strong, high contrast contextual bars aligned with the central bar can increase the response by several fold (Kapadia, Ito, Gilbert and Westheimer 1995); see Fig. 2.24 F. This contextual facilitation can become suppressive when the neuron's response is increased by a stronger central bar.

2.3.9.1 Iso-feature suppression

Iso-color, iso-motion-direction, iso-scale, and iso-eye-of-origin suppression generalize iso-orientation suppression (Li and Li 1994). We refer to them in general as iso-feature suppression (Li 1999a), encompassing all cases in which a neuron's response to an optimal stimulus within its CRF is suppressed by contextual stimuli which have the same or similar feature value as the stimulus within the CRF.

2.3.9.2 Feature-unspecific suppression

Although iso-feature suppression is a dominant form of contextual influence, a component of contextual suppression is feature unspecific such that it occurs regardless of the degree of feature similarity between central and surround features. Feature-unspecific suppression should be at least partly responsible for the suppression shown in Fig. 2.24 C–E. It has been modeled as a form of normalization (Heeger 1992) as follows. Let us denote a complex cell as i, for $i = 1, 2, ...$ From equation (2.68), the energy model of this complex cell has

$$E_i = [L_{i,1}]^2_+ + [-L_{i,1}]^2_+ + [L_{i,2}]^2_+ + [-L_{i,2}]^2_+ \equiv \sum_{\phi=1}^{4} [A_{i,\phi}]^2, \qquad (2.82)$$

where again $[.]_+$ means the rectified outcome, and $A_{i,\phi} = L_{i,1}, -L_{i,1}, L_{i,2}$, and $-L_{i,2}$, for $\phi = 1, 2, 3$, and 4, are the linear components of the four simple cell subunits which feed into this complex cell i. The normalized response of the complex cell i and that of the simple cell (i, ϕ) are then modeled as

$$O_i = \frac{E_i}{\text{constant} + \sum_j E_j}, \qquad \text{for a complex cell } i, \qquad (2.83)$$

$$O_{i,\phi} = \frac{[A_{i,\phi}]^2_+}{\text{constant} + \sum_j E_j}, \qquad \text{for a simple cell } (i, \phi). \qquad (2.84)$$

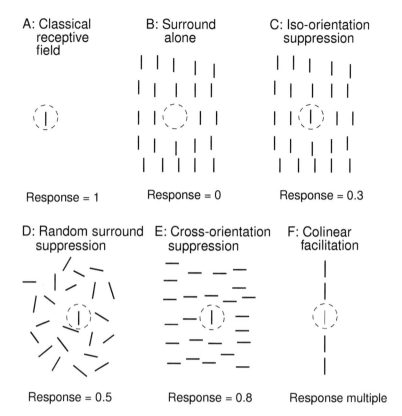

A: Classical receptive field

Response = 1

B: Surround alone

Response = 0

C: Iso-orientation suppression

Response = 0.3

D: Random surround suppression

Response = 0.5

E: Cross-orientation suppression

Response = 0.8

F: Colinear facilitation

Response multiple

Fig. 2.24: A schematic of typical types of contextual influence on the response of a V1 neuron. The CRF is marked by the dashed oval, which is not part of the visual stimuli. In A, the neuron responds to its optimal stimulus, a vertical bar, within the CRF. The surrounding vertical bars outside the CRF do not excite the cell (in B) but can suppress the response to the vertical bar within the CRF by up to about 80% (in C)—*iso-orientation suppression*. This suppression is weaker when the contextual bars are randomly oriented (D); it is weakest when the contextual bars are orthogonal to the central bar (E)—cross-orientation suppression. F: *Colinear facilitation*—when the contextual bars are aligned with a central, optimally oriented, low contrast bar, the response level could be a multiple of that when the contextual bars are absent.

Through the terms E_j in the denominators of the equations above, the contextual input which activates other neurons j nearby can affect neural response O_i and $O_{i,\phi}$. The neurons contributing to the $\sum_j E_j$ summation are understood as having receptive fields that are nearby, but not necessarily overlapping with, the receptive fields of the central neurons i and (i, ϕ). The similarity between the equations above and equations (2.43) and (2.69) suggest that the same root cause is behind the nonlinearity of the neural responses and this normalization model.

2.3.9.3 Origins and functions of contextual influences

Contextual influences are seen in V1 of cats and monkeys whether the animals are awake or anesthetized. When contextual inputs are presented simultaneously with the stimulus within the CRF, their influence on the neural response of a cell occurs within 10–20 ms of the onset of its initial response to stimulation. Extensive axon collaterals have been observed (Rockland

and Lund 1983, Gilbert and Wiesel 1983) from presynaptic V1 neurons to postsynaptic V1 neurons whose receptive fields are nearby but not necessarily overlapping with those of the presynaptic neurons. These axons mediate interactions between neurons which are typically in layer 2 and layer 3 of the six layers of the cortex. These intracortical connections have been postulated to be the neural substrates for contextual influences, mediating monosynaptic excitation between the excitatory neurons or disynaptic inhibition between them via the inhibitory interneuron (Hirsch and Gilbert 1991). Meanwhile, since V1 also receives feedback signals from higher cortical areas, it has been difficult to dissect which mechanism—intracortical interactions in V1 or feedback signals from higher areas—is more responsible for the contextual influences. However, contextual influences persist after V2 inactivation (Hupé, James, Girard and Bullier 2001) or anesthesia (Nothdurft, Gallant and Van Essen 1999), suggesting that V1 mechanisms alone can be sufficient.

In the classical framework of (classical) receptive fields, contextual influences appear to be a nuisance. The classical framework was supposed to capture all or most of the response properties of the cortical neurons and lies behind the popular notion that V1, with its population of feature detectors, mainly serves to supply feature values at local points in visual space to higher visual areas. However, the magnitudes of the contextual influences make it difficult to think that they are mere perturbations of the classical framework. Chapter 5 will show that the contextual influences are no longer puzzling, due to a recently proposed theory that V1 computes a map of visual saliency from local and global input features by intracortical interactions that mediate contextual influences.

2.4 Higher visual areas

Cortical areas downstream from V1, including V2, V3, V4, MT, the medial superior temporal area (MST), and IT, are numerous, complex, and less well understood than V1. Neighboring cortical areas along the visual pathway are almost always reciprocally connected, such that there are feedforward as well as feedback connections. However, a hierarchy is still maintained. Along the visual pathway in the downstream direction, the receptive field sizes become progressively larger. Meanwhile, compared to V1, it is often difficult to find stimuli that drive neurons in these regions effectively, let alone the stimulus that drives a neuron most strongly. Some complex visual inputs that excite cortical cells are shown in Fig. 2.25. Retinotopic organization becomes obscured or lost going from V4 to the anterior part of IT.

2.4.1 Two processing streams

Ungerleider and her colleagues proposed (Ungerleider and Mishkin 1982, Ungerleider and Pasternak 2004) that there are two functional processing streams along the visual pathway: one for visual object identification ("what" an object is) and the other for object localization ("where" an object is or "how" to act on them (Goodale and Milner 1992)); see Fig. 2.3.

The "what" stream is the ventral (or occipitotemporal) stream, which includes V1, V2, V4, and the inferior temporal areas TEO and TE. The "where" stream is the dorsal (or occipitoparietal) stream, which includes V1, V2, V3, V3A, MT, MST, and additional areas in the inferior parietal cortex. Lesioning cortical areas in the ventral stream leads to substantial difficulties in recognizing objects but leaves visuospatial tasks such as reaching for an object relatively spared. Conversely, lesions in areas in the dorsal stream do not affect visual discrimination as much as visuospatial tasks. Neurons in areas along the ventral stream are more sensitive to color, shape, and surface texture, all of which are useful for object identification,

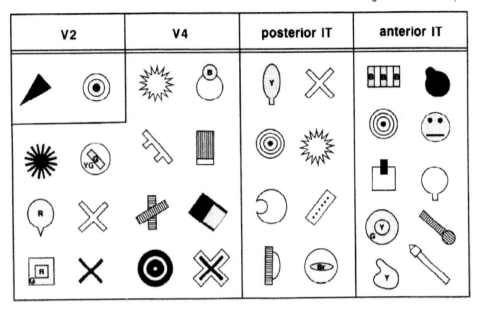

Fig. 2.25: Examples of visual inputs to which some visual cortical neurons in monkey are selective. Posterior IT and anterior IT are also referred to in the literature as areas TEO and TE respectively (Tanaka 1996). Reproduced from *Journal of Neurophysiology*, Kobatake, E. and Tanaka, K., Neuronal selectivities to complex object features in the ventral visual pathway of the macaque cerebral cortex, 71 (3): 856–867, Fig. 11, copyright © 1994, The American Physiological Society, with permission.

whereas neurons in the dorsal stream are sensitive to visual motion, speed, and other spatial aspects of visual inputs.

Functional magnetic resonance imaging studies in humans show that some brain areas respond more to images of objects than to scrambled versions of the same images (made by cutting the original images into small regions and randomly rearranging these regions across space to form new images that share at least some primitive properties of image features with the originals). These brain areas concerned are collectively called lateral occipital complex (LOC) because of their anatomical location; this includes the classical macaque IT cortex.

In general, the span of the visual field from which visual inputs can evoke significant responses in a given cortical area becomes progressively smaller downstream along the visual pathway, especially in the ventral ("what") stream. V1 neurons collectively respond to the whole visual field, up to eccentricities beyond 80° whereas V4 neurons seldom respond to visual inputs beyond 40°. There are exceptions along the dorsal pathway, in which some cortical areas, particularly those which are also involved in sensory motor functions, respond to the peripheral visual field. This is consistent with their playing a role in "where" or "how" vision, i.e., in guiding the performance of sensory motor tasks (Goodale and Milner 1992).

2.4.2 V2

V2 is the cortical area immediately downstream from V1 and is similar in size. The sensory response properties of V2 neurons are also superficially similar to those of V1 neurons. However, at a given eccentricity, receptive fields of V2 neurons are slightly bigger than those of V1 neurons. In addition, V2 neural responses suggest that the neurons can sense object

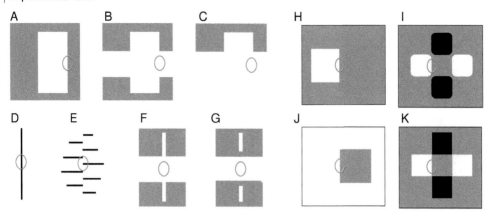

Fig. 2.26: V2 neurons respond to illusory contours and can signal border ownership, as observed by von der Heydt and colleagues. A–K sketch the visual inputs used to probe V2 neural properties. The red ellipse in each plot marks the classical receptive field (CRF) of a V2 neuron of interest and is not part of the visual input. In A and D, the CRF contains a vertical contour or surface border defined by actual contrast in the image; in B, E, and F, the CRF contains only an illusory vertical border or contour. A neuron tuned to vertical orientation can be activated by these real or illusory contours or borders. However, the neuron will not be activated at its CRF position in C and G, since the CRF contains no real or illusory contour. In H–K, the images contained within the receptive field are identical, depicting a transition from a white region from the left to a darker region on the right. This transition may signal a border to a surface to the left of the border (in H and I) or a border to a surface to the right of the border (in J and K). A V2 neuron preferring a border to a surface on the left of the border will be more activated by H and I than by J and K.

surface properties which are not apparent in the image features within the receptive fields but have to be inferred from context.

2.4.2.1 Response to illusory contours

In particular, V2 neurons can respond to illusory contours which are not manifest in the image pixel or contrast values. For example, a vertical edge or contour is defined directly by the image pixel values within a receptive field (marked by a red ellipse) in Fig. 2.26 AD, but can only be inferred from the context of the receptive field in Fig. 2.26 BEF. Such inferred contours are called illusory contours. Many V2 neurons can be activated by both real and illusory contours, which share the same preferred orientation (von der Heydt et al. 1984). However, these neurons will not be activated by inputs such as those in Fig. 2.26 C and Fig. 2.26 G, even though the image content within each receptive field is identical to that in Fig. 2.26 B and Fig. 2.26 F respectively. If an stimulus such as Fig. 2.26 F is augmented with disparity features, such that the two white vertical bar segments lie in a different depth plane from the background gray surface, then the illusory contour is only present when the two bar segments are in front of the background surface. V2 neurons can discriminate the depth order, such that they correctly signal illusory contours only when this order is appropriate (Bakin et al. 2000). Unlike V1 neurons, V2 neurons can also signal the relative depths of two surfaces in a way that is insensitive to the absolute depth of both surfaces (Thomas, Cumming and Parker 2002).

2.4.2.2 Discriminating border ownership

V2 neurons can also signal whether a contour is more likely to have been caused by a surface border on one or the other side of its receptive field (Zhou et al. 2000, Qiu and von der Heydt 2005). For example, Fig. 2.26 H suggests a scene containing a white square on a dark background whereas Fig. 2.26 J suggests a dark square on a white background. However, the images contained within the receptive fields in these plots are identical. A V2 neuron with that RF and preferring vertical orientation can respond differently to these two inputs, preferring the surface to be either to the left or right of its RF. A V2 neuron preferring Fig. 2.26 H over Fig. 2.26 J will also prefer Fig. 2.26 I over Fig. 2.26 K, when its RF is placed at the red ellipse in Fig. 2.26 HIJK. Thus, this V2 neuron has a consistent preference for surface border ownership across various contextual inputs.

2.4.3 MT (V5)

MT is also called V5 (Dubner and Zeki 1971). It is a visual area in the dorsal stream of the visual processing pathway, in the middle temporal area; see Fig. 2.2 and Fig. 2.3. It has the following properties (Born and Bradley 2005). It is much smaller than V1 and V2, receives direct input from V1, V2, and V3, and some limited direct input from LGN and the superior colliculus. Most MT neurons are tuned to motion direction and disparity. Their receptive fields have diameters about 10 times larger than those of V1 neurons. The receptive fields are biased such that they cover the lower visual field more extensively. MT neurons also exhibit the following properties not present in V1 neurons.

2.4.3.1 Responses to plaid and component motions

For example, consider an input stimulus comprising two superposed large gratings that move in different directions and are seen through an aperture; see Fig. 2.27 A. This can lead to two percepts, either as a single plaid pattern moving rigidly in a direction that is in-between the component directions (vertically; lower right in the figure), or as two transparent gratings moving in their respective (original) directions (upwards to the left and upwards to the right; lower left). A V1 neuron is generally activated when either grating moves in its preferred direction, even when a single rigid motion is perceived to move in another direction. Some MT neurons behave like the V1 neurons, and these are called *component cells*. Other MT neurons, called *pattern cells*, are most activated when the whole pattern moves in their preferred directions (Movshon et al. 1985).

One may also manipulate the intersection points of the two gratings to favor either the plaid or component percept. The lower part of Fig. 2.27 A displays two example manipulations: the left example appears more like two transparent gratings than the right example. Some MT cells can behave more like pattern cells or component cells according to whether the input appears like a whole pattern moving singly or otherwise (Stoner and Albright 1992).

2.4.3.2 Responses to groups of moving dots

MT neurons are also tuned to the moving direction of a group of dots moving in the same speed and direction within their receptive fields. However, their responses to their preferred directions are greatly suppressed when another group of dots moves in the opposite direction in the same visual location and same depth plane; see Fig. 2.27 B. The suppression is stronger when the two motion directions are balanced at a small spatial scale, e.g., when each moving dot is paired with another nearby dot moving in the opposite direction (Qian and Andersen 1994); this input pattern does not lead to the perception of transparency between the two groups (Qian, Andersen and Adelson 1994). When the two groups of dots are separated in depth, the suppression is much reduced or eliminated (Bradley et al. 1995).

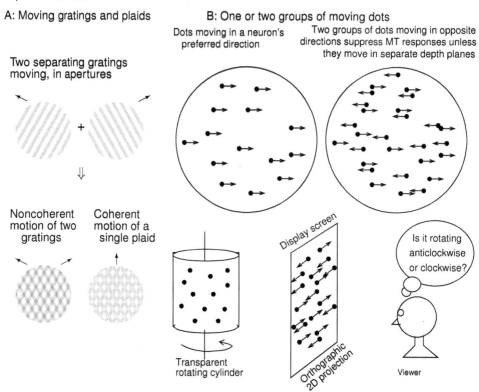

A: Moving gratings and plaids

Two separating gratings
moving, in apertures

Noncoherent motion of two gratings   Coherent motion of a single plaid

Transparent rotating cylinder

B: One or two groups of moving dots

Dots moving in a neuron's preferred direction

Two groups of dots moving in opposite directions suppress MT responses unless they move in separate depth planes

Display screen

Orthographic 2D projection

Is it rotating anticlockwise or clockwise?

Viewer

Fig. 2.27: MT properties and perception. A: Moving gratings and patterns to probe MT neurons. Superposing two drifting gratings (top, seen through apertures), each moving in the direction indicated by the arrow, gives rise to the percept either of a single pattern moving upwards or two transparent gratings moving in their respective directions. Manipulating the intersection points of the two gratings in the superposition can bias this percept, as demonstrated in the two example superpositions. Some MT cells (pattern cells) are tuned to the moving direction of the (single) whole pattern, whereas others (component cells) are tuned to the directions of the component gratings (Movshon et al. 1985). Furthermore, some MT neurons behave more like a pattern or component cell, respectively, according to whether the perception is more like a coherent pattern or otherwise (Stoner and Albright 1992). B: Dots moving in the preferred direction of an MT neuron are also effective stimuli, but the responses are typically suppressed when adding another group of dots moving in the opposite direction in the same depth plane (Qian and Andersen 1994). Separating the two groups of dots in depth can release the suppression (Bradley et al. 1995). When an input is consistent with more than one depth order (bottom), the responses of some neurons in MT are correlated with what the animal perceives, according to the preferred directions and disparities of the neurons concerned (Bradley et al. 1998).

If a scene of a transparent, dotted, and rotating cylinder is projected onto a display screen, as in the lower part of Fig. 2.27 B, it gives rise to a 2D spatial pattern of moving dots moving in opposite directions in the same depth (of the display screen). Since the lateral dots move more slowly than the central ones, one can perceive the pattern as a rotating cylinder—this is called *structure-from-motion*. However, because there is no depth difference between dots moving in opposite directions, the cylinder can be seen as rotating in either of the two possible directions, with the rightward moving dots appearing either in front of, or behind, the leftward moving dots. As for many ambiguous percepts, these two interpretations switch occasionally, apparently at random. MT neural responses can be correlated with the percept, such that, e.g., a neuron tuned to rightward motion and near (crossed) disparity responds more strongly when the perception has the rightward motion in front (Bradley et al. 1998).

2.4.4 V4

V4 is about half the size of V2 (Lennie 1998). Neurons have receptive fields that are about 4–7 times larger in diameter than those in V1 but are tuned to many of the same visual features as V1 neurons: color, orientation, motion direction, spatial frequency of gratings, size, and disparity (Desimone and Schein 1987, Watanabe, Tanaka, Uka and Fujita 2002). Lesions to V4 only mildly or moderately impair simple aspects of visual perception such as contrast sensitivity, motion perception, and stereopsis (Schiller 1993, Merigan 1996). However, they severely impair shape discrimination and the ability to select less salient objects in the scene. For example, lesions make a monkey unable to discriminate a foreground shape in Fig. 2.28 A (Merigan, Nealey and Maunsell 1993). V4-lesioned monkeys can pick out the odd shape in Fig. 2.28 B when the odd shape is the most salient in the scene but not when it is not (in the left ring of Fig. 2.28 B). V4 receives most of its input from V2; hence, it is not surprising that many V4 neurons can respond according to visual surface properties such as border ownership (see Fig. 2.26).

A: is the foreground in each texture horizontal or vertical?

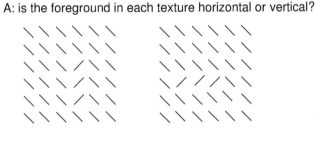

B: which one in each ring of 8 items is odd?

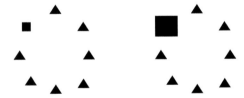

Fig. 2.28: A V4 lesion makes a monkey unable to perform the task in A, even though it can still do this task when the foreground bars are thicker or distinct in color (Merigan 1996). It also makes a monkey unable to detect the odd object in the left ring in B, when the square is insufficiently salient; although performance in the right ring, in which the square is most salient, remains sound (Schiller 1993).

2.4.5 IT and temporal cortical areas for object recognition

Unlike most cells upstream along the ventral pathway, IT neurons typically have large receptive fields which include the central fovea. Cells are most strongly driven by moderately complex spatial shapes such as a face, a brush-like shape, or a hand; see Fig. 2.25. However, many cells respond well to many different shapes, rather unselectively (Tanaka 1996, Desimone, Albright, Gross and Bruce 1984). The response to the optimal shape is often insensitive to its position within the receptive field, and most cells also tolerate a moderate change in the size of the shape (e.g., a factor of two). Nevertheless, most cells reduce their response by more than half in the face of a $90°$ rotation of this optimal shape (Tanaka 1996). The posterior part of IT is also called TEO in the literature (Felleman and van Essen 1991, Tanaka 1996). Compared to the neurons in the anterior part of IT, often referred to as area TE, neurons in TEO respond to simpler features (more like those preferred by V4 cells) and have smaller receptive fields (Kobatake and Tanaka 1994); see Fig. 2.25. Accordingly, TEO is sometimes distinguished from IT in the literature.

IT cells responding to common visual features are grouped together into cortical columns. An IT neuron's relative preferences (rather than absolute response levels) to object identities are not very sensitive to image changes caused by, e.g., changes in object size and background clutter (Li, Cox, Zoccolan and DiCarlo 2009).

Localized cortical patches within IT, even within area TEO, have been found to contain almost exclusively neurons which prefer visual images of faces over non-faces (Freiwald et al. 2009). Most such neurons are tuned to one or more quantifiable facial features such as face aspect ratio, face direction, iris size, eyebrow slant, inter-eye distance, and hair thickness. Neural activities tend to change monotonically with an input feature value, such that the highest and lowest responses of a neuron are typically for the extreme values of the face features (e.g., face aspect ratio). This representation provides a rich population code for facial features.

Functional magnetic imaging in humans has found that some moderately localized cortical patches in the temporal lobe respond most vigorously to inputs coming from specific categories: faces (Kanwisher, McDermott and Chun 1997), scenes (Epstein and Kanwisher 1998), and body parts (Downing, Chan, Peelen, Dodds and Kanwisher 2006).

2.5 Eye movements, their associated brain regions, and links with attention

Eye gaze is typically directed to the spatial focus of visual attention (Hoffman 1998). Hence, knowing the brain circuits which control eye movements can shed light on the control of visual attention. Gaze moves either in smooth pursuit, when following an attended object that moves relative to the observer, or via saccades, when shifting from one object to another. We are typically unaware that we saccade as often as about three times each second, suggesting that many of the saccades are carried out more or less involuntarily. Eye movements are also very important for coordinating motor actions, body motion, and balance. Figure 2.29 depicts a summary of brain centers involved in controlling eye movements.

In smooth pursuit, the *vestibulo-ocular reflex* helps to keep our gaze fixed on an object when we move our body. Extrastriate areas MT and MST and the frontal area FEF are particularly involved in smooth pursuit (Schiller 1998).

Saccades involve a host of brain centers. The superior colliculus (SC) receives inputs from the retina and visual cortical areas including V1, and other brain centers like LIP and FEF; it sends saccadic commands to motor regions.

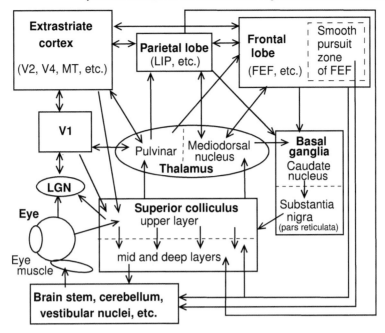

Fig. 2.29: Brain circuit for the control of eye movements. This schematic summarizes and simplifies information in the literature (Felleman and van Essen 1991, Bruce et al. 2004, Shipp 2004, Schiller and Tehovnik 2005).

The upper layer of the SC has a retinotopic map of the visual field; its neurons prefer small visual stimuli, although their receptive fields are much larger than those of the V1 cells. These neurons are insensitive to input differences in shape, orientation, color, and motion direction (Schall 2004). The deeper layers of the SC contain neurons which are active before and during saccades toward visual field regions that are called the *movement fields* of the neurons. These movement fields coincide with the receptive field locations of the neurons in the corresponding upper layers (Schiller 1998, Schall 2004). Some deeper layer neurons also have visual receptive fields; these then coincide with the movement fields.

Most neurons in areas LIP and FEF have visual receptive fields. Many neurons respond before or during a saccades toward their movement fields (Colby and Goldberg 1999, Schall 2004), which coincide with the receptive fields for the visually responsive neurons.

Electrical stimulation of V1, V2, the SC, LIP, and FEF can produce saccadic eye movements toward the receptive field or the movement field of the stimulated neurons (Schiller and Tehovnik 2005). If the SC is lesioned, then stimulation of the visual cortex or LIP no longer evokes saccades. Neurons in the SC, LIP, and FEF increase their responses to a stimulus within their receptive field if the stimulus is the target of an upcoming saccade. If the animal covertly attends to a stimulus without actually making a saccade to it (for instance, if it has to respond to the stimulus manually but is prevented from directing gaze toward it), then LIP neurons have enhanced responses, but the SC and FEF neurons do not (Wurtz, Goldberg and Robinson 1982, Colby and Goldberg 1999, Bislay and Goldberg 2011).

The FEF and the SC both have access to motor regions involved in executing eye movements. Thus, simultaneously lesioning both regions makes a monkey unable to direct eye movements to visual targets, whereas lesioning either alone does not give rise to long-lasting serious deficits (Schiller 1998). Lesioning the SC reduces the number of spontaneous saccades and eliminates a class of so-called express saccades, which have much shorter latencies than the bulk of saccades following the onset of a saccadic target (Schiller 1998). Lesioning FEF makes memory-guided saccades difficult and makes an animal unable to move its eyes in response to verbal commands, even when the animal can follow visual objects with saccades and can understand the verbal command (Bruce et al. 2004). Lesioning V1 in monkeys abolishes all visually evoked saccades until after two months of training and recovery (Isa and Yoshida 2009). This suggests that the direct pathway from the retina to the SC is normally not responsible for saccades in primates. Adaptive changes in the organization of neural circuits may be responsible for visually guided saccades after V1 lesion. Slightly before saccades, neurons in LIP, FEF, and in the deeper layers of the SC have been found to shift their receptive fields such that they respond to stimuli that are about to be brought into their classical receptive fields by the impending saccade (Colby and Goldberg 1999, Walker, Fitzgibbon and Goldberg 1995, Umeno and Goldberg 1997). This is called *receptive field remapping*; it is presumed to update the representation of the visual space that is shifted by the saccades. Some neurons in the extrastriate cortex exhibit similar behavior (Nakamura and Colby 2002). Neurons in the mediodorsal nucleus in thalamus relay a corollary discharge associated with saccades from the SC to FEF. These signals have been shown to be responsible for the receptive field remapping in FEF (Sommer and Wurtz 2006). If these relay neurons are inactivated and if the monkey is required to make two saccades, one after another, to two briefly flashed saccadic targets which disappear before the first saccade starts, then the second saccade is not aimed correctly, since the spatial shift of the visual stimulus in the retinal coordinates that results from the first saccade is not compensated for properly (Sommer and Wurtz 2004).

2.5.1 Close link between eye movements and attention

Experimental data indicate that eye movements and attention are closely linked. Although it is possible to keep one's gaze fixed while directing attention elsewhere, it is very difficult, if not actually impossible, to look somewhere without moving attention to the same location just before or at around the same time (Hoffman 1998). Electrically stimulating two locations within the SC evokes a saccade which is the vector average of the saccades evoked by stimulating each location alone (Schiller 1998). If one location in the SC is stimulated while the animal is covertly paying attention to a peripheral visual location, the direction of the evoked saccade deviates (from that evoked by the SC stimulation alone) in the direction of attended location (Kustov and Robinson 1996). This observation suggests that paying attention to a location resembles stimulating the SC cells whose movement fields cover this attended location; or, in other words, paying attention somewhere is closely associated with preparing for an eye movement toward the attended location.

When human observers are about to make a saccade, discrimination of visual inputs appearing and disappearing quickly before the saccadic onset is best at the location of the saccadic target (Deubel and Schneider 1996). Equally, if observers are specifically asked to discriminate an object at one location at around the same time as making a quick saccade to somewhere else, they are unable to discriminate the object as well as when they do not have to make the saccade, unless they slightly delay the onset of their saccade (Kowler, Anderson, Dosher and Blaser 1995). If discrimination performance at a location is a measure of the amount of attention directed to that location, then we can conclude that attention is focused on the saccadic target before the saccade. In monkeys, visual discrimination is better at the receptive field of the more activated neuron in LIP, regardless of whether the neural activation is caused by bottom-up visual inputs at that location or by the top-down demands of a behavioral task (Bislay and Goldberg 2011).

2.6 Top-down attention and neural responses

Most physiological experiments on visual attention involve top-down attention, which is voluntary, non-reflexive, or task dependent. Neural responses to the same visual inputs are compared under different attentional states defined by different locations of attentional focus. Often, neural responses to stimuli within receptive fields (RFs) are enhanced when the animal is paying attention to those inputs or performing tasks related to them. More enhancement tends to occur further downstream along the visual pathway. For example, enhancements are observed in V2, V4, and MT but are often weak or absent in V1 (Motter 1993, Luck, Chelazzi,

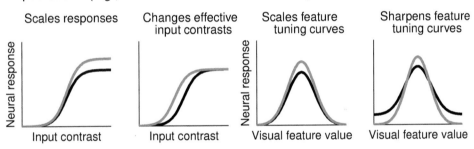

A: Two designs of stimulus and task to study the effects of top-down attention

Attention inside or outside the RF, to preferred or non-preferred stimulus

RF

A stimulus outside the RF   Two visual stimului inside the RF one preferred, one non-preferred

Attention inside or outside the RF

Fixation point   RF

A stimulus outside the RF   A visual stimulus inside the RF

B: Various effects of top-down attention on neural responses to input contrast or input feature (e.g., orientation and motion direction) in extrastriate cortices

Scales responses   Changes effective input contrasts   Scales feature tuning curves   Sharpens feature tuning curves

Neural response   Input contrast   Input contrast   Neural response   Visual feature value   Visual feature value

Fig. 2.30: Effects of top-down attention on extrastriate (e.g., in V2, V4, MT, MST) neural responses in behaving monkeys. A: Sketches of two designs to compare neural responses to the same inputs in different attentional states. Monkeys fixate on the central fixation point (cross) while doing a task based on a stimulus (e.g., bar, grating, moving dots) inside or outside the receptive field (RF, marked by the dashed rectangle) of the neuron whose response is being measured. Their attention is assumed to be covertly directed to the task-relevant stimulus. In the left design, two stimuli are inside the RF, one (preferred) evokes a higher response than the other if each is presented by itself within the RF. In the right design, only one stimulus is within the RF. Attentional modulation is manifest in the difference between the responses when attention is inside versus outside the RF; it is also manifest in the difference between the responses when attention is to the preferred (e.g., preferred orientation or motion direction) versus the non-preferred stimulus (inside or outside the RF). B: Four examples of possible effects of top-down attention on neural response curves to input contrast and input features. In each plot, the red and black curves are two response curves to the same visual input when attention is directed respectively to two different targets, as schematized in A. Attention can scale the response curves (in the first and third plot), act as if the external input contrast is changed (the second plot), or change the shape of the feature tuning curves (the fourth plot).

Hillyard and Desimone 1997). Further, when there are two input items within a single receptive field of a neuron and when the animal (typically a monkey) is not paying attention to either item, the neuron's response tends to be close to the average of the responses evoked when either item alone is in the receptive field. However, when the animal's attention is directed toward one of two items presented simultaneously within the receptive field, the response is closer to that when the attended item alone is within the receptive field, as if the unattended item were absent (Moran and Desimone 1985, Chelazzi, Miller, Duncan and Desimone 1993, Luck et al. 1997, Reynolds, Chelazzi and Desimone 1999). When there is only one item within the receptive field, evoked neural response is higher when attention is directed to the item inside the RF rather than to somewhere outside the RF (Treue and Maunsell 1999, Treue and Martínez-Trujillo 1999, Reynolds, Pasternak and Desimone 2000, Williford and Maunsell 2006).

Visual attention can change input contrast response curves and feature tuning curves of the neurons. For example, attention can scale up a contrast response curve (the leftmost plot in Fig. 2.30 B), or scale up a feature tuning curve (the third plot in Fig. 2.30 B). Attention can also change the input contrast response curve as if the external input contrast had been increased (the second plot in Fig. 2.30 B).

When attention is directed to a stimulus outside the receptive field, the neural response is higher when the attended stimulus has the preferred feature (e.g., motion direction) of the neuron rather than a non-preferred feature (Martínez-Trujillo and Treue 2004, Treue and Martínez-Trujillo 1999). When the stimulus feature within the receptive field is kept identical to the stimulus feature outside the receptive field, the animal's attention to the stimulus outside the receptive field can cause the feature tuning curve to sharpen (see the last plot in Fig. 2.30 B), because the response to the preferred feature is relatively enhanced by attention and the response to the non-preferred feature is relatively suppressed by attention (Martínez-Trujillo and Treue 2004). These observations are often interpreted as biased competition (Desimone and Duncan 1995), which means that the brain devotes more resources to process the attended rather than non-attended visual inputs when there are multiple input items in the scene.

While attending to task-relevant stimulus can improve visual performance, electrical stimulation of brain areas associated with eye movements and attention can also improve performance. For example, when monkeys are performing the task of detecting a brief dimming of a target, subthreshold stimulation (which is insufficiently strong to evoke saccade) of FEF neurons whose movement fields cover the target location can improve task performance (Awh, Armstrong and Moore 2006). A separate experiment found that V4 neurons can increase their responses to the visual inputs in their receptive fields by subthreshold stimulation of the FEF neurons with the corresponding movement fields (Awh et al. 2006). Monkeys can also better detect a change in the direction of motion and better discriminate motion directions given subthreshold stimulation of the SC neurons whose receptive fields are at the task relevant location (Awh et al. 2006).

Although the spatial location of the top-down attentional focus has no, or only a weak, effect on the sensitivities of V1 neurons, the influences of the contextual inputs on a V1 neuron's responses can be altered by behavioral tasks after monkeys are trained on the tasks. For example, if two gratings, grating A and grating B, are presented in a regular temporal sequence, ABABAB ..., and the monkey has to detect an irregular repetition of either grating, such as the repetition of A in ABABAABAB..., then 31% of the V1 cells increase their response to this repetition by 20% or more (Haenny and Schiller 1988).

Another example is shown in Fig. 2.31 A. Take a V1 neuron preferring vertical bars. The response of this neuron to a vertical bar within its RF can be modulated by the distances from the central bar to left and right neighboring vertical bars outside its RF. This modulation is stronger when the monkey is performing a bisection task, i.e., deciding whether the left or

A: Task determines whether V1 response to the center bar depends more on its distance to the left/right or top/down neighbors

B: Shading direction of the contextual balls affects V1 responses to the central ball after task experience

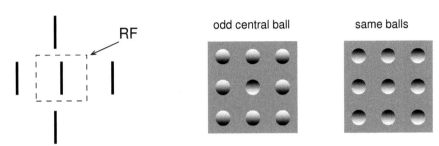

odd central ball   same balls

Fig. 2.31: Influences on the responses of V1 neurons of top-down attention, task, and experience. A: the contextual influence of the bars surrounding the RF on the response of the V1 neuron (with that RF) depends on the task the monkey is performing (Li et al. 2004). The monkey is trained to do two tasks, a bisection task involving deciding whether the left or right neighbor is closer to the central bar, and a vernier task requiring a decision as to whether the central bar is to the left or right of the top and bottom bars (which are vertically aligned). During the bisection task, the neuron's response to the central bar is typically more strongly modulated by the task-relevant distance between this bar and the center of mass of the left and right neighbors, and less modulated by the task-irrelevant distance between the central bar and the center of mass of the top and bottom neighbors. During the vernier task, the reverse holds. B: A V1 neuron's response to the central ball in each image does not depend on whether the surrounding balls have the same shading direction as the central ball. This ceases being true if the monkey has had sufficient experience in the task of detecting an odd-ball by its shading (Lee et al. 2002). The shading direction of the context only affects the longer latency responses after the initial response to the image.

the right neighbor is closer to the central vertical bar. The modulation is weaker when the monkey is performing another task (a vertical vernier task) in which these neighboring bars are task-irrelevant. The effect of the behavioral context is present in the very initial component of the neural responses. The task-dependence of these contextual influences is present only after the monkeys are trained on the task; it is also absent at the visual locations where the monkey has not experienced the trained task.

Dependence on task experience in V1 responses has also been seen in the task of detecting an odd-ball among a group of balls based on image shading (Lee et al. 2002); see Fig. 2.31 B. The V1 response is independent of whether the shading of balls outside the RF are identical to that of the ball within until the monkey has learned the task. This dependence is only present in longer latency responses after the immediate activities evoked by the image. In contrast, V2 neural responses are sensitive to the shading direction without any task experience.

2.7 Behavioral studies on vision

Behavioral studies, typically involving human observers, aim to link visual inputs to sensation and perception. According to the theme of this book, to view vision through encoding, selection, and decoding, these behavioral studies may be divided into three corresponding categories. Studies most related to the encoding stage measure behavioral sensitivities to basic visual image features such as the contrast of a grating or the color differences between two inputs. These studies often measure discrimination thresholds, defined as the minimum difference between two visual inputs that observers can reliably detect. In studies of visual selection or attention, one probes how speed and accuracy in visual task performance are affected by attention, and one examines how attention is directed (e.g., by visual inputs or top-down requirements, with or without eye movements). In studies most related to decoding, one probes whether certain visual inputs are seen in one category or another, such as whether one sees a face or a cow in an image or whether one sees an animal in a group of images.

Typical topics in conferences of visual psychology can give some idea of the active questions of current interest. The topics include: adaptation and aftereffects, motion processing (e.g., optic flow), temporal processing, brightness and lightness, perceptual learning, 3D perception, binocular vision (binocular rivalry, stereopsis), texture perception, face processing, object recognition, shape from shading (or textures, motion, or contour), illusions, crowding (impairment of visual recognition in the periphery in the presence of neighboring objects), color perception, scene perception, perceptual organization, visual masking, visual grouping, attention (and many of its subcategories), eye movements, active vision, perception and action (e.g., navigation, grasping, saccades), multisensory processing (interactions between vision and other senses such as hearing, smelling, touching), visual memory, and visual awareness (or consciousness). Readers should not be worried if they are unfamiliar with some terms in the above list (they have not yet been defined in this book).

There are many excellent textbooks and websites on visual behavior. Many visual phenomena inspire wonder even to people outside the field. I encourage readers to look them up, as this book is unable to offer a comprehensive introduction to behavioral studies in vision. In the rest of the book, some facts of visual behavior will be introduced in connection with the theories, models, and physiological data. An aim of theoretical or computational vision is to understand why vision is.

It is increasingly the case that behavioral studies are performed in conjunction with physiological measurements. These measurements include electroencephalography, imaging (e.g., functional magnetic imaging (fMRI)), and even physiological recordings from neurons directly (in monkeys or human volunteers who have electrodes in the brain typically due to impending or ongoing brain surgery). Such studies provide direct links between behavior and neural substrates; these links are invaluable for developing theories whose role is to link physiology with behavior (see Fig. 1.2).

3 The efficient coding principle

In this chapter, we focus on encoding, the first of the three stages of vision: encoding, selection, and decoding (see Fig. 1.5). In particular, we present the efficient coding principle (Barlow 1961), which states that one of the goals of early visual processing is to transform raw visual inputs into an efficient representation such that as much information as possible can be extracted from them given limited neural resources (see Fig. 3.1). We present a detailed formulation of this principle, using information theoretic measures to quantify the process of encoding. We apply the principle to understand many properties of neural sampling and receptive fields in the retina, LGN, and part of the primary visual cortex, and to understand how these properties adapt to the visual environment and depend on the species of animals involved. We describe some non-trivial and testable predictions that follow from this principle, along with some experimental confirmations.

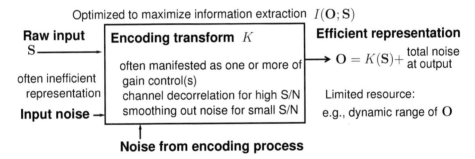

Fig. 3.1: Efficient coding K transforms the signal $\mathbf{S}$ to neural responses $\mathbf{O}$ to extract the maximum amount of information $I(\mathbf{O}; \mathbf{S})$ about signal $\mathbf{S}$, given limited resources, e.g., capacity (dynamic range) or energy consumption of the output channels. Often, the signal is kept within an appropriate dynamic range via gain control. With a high input signal-to-noise ratio (S/N), efficient coding requires the removal of correlations (of all orders, including pairwise correlation) between input channels to eliminate redundancy in the information transmitted by different output channels. With low S/N, efficient coding demands that input channels be averaged to help smooth out noise and recover inputs from correlated responses. Adapted with permission from Li Zhaoping, Theoretical understanding of the early visual processes by data compression and data selection, *Network: Computation in Neural Systems*, 2006; 17 (4), pp. 301–34, figure 2, copyright © 2006, Informa Healthcare.

More specifically, the questions that can be addressed by the efficient coding principle include: why should the input contrast response function of a neuron take its particular form? Why do retinal ganglion cells have center-surround receptive fields? Why do these receptive fields increase their sizes in dim light conditions? What is the optimal distribution of cones on the retina and how should this depend on the characteristics of eye movements? Why does the contrast sensitivity curve associated with chromatic inputs suppress higher spatial frequencies more than that for luminance inputs? Why are color selective V1 neurons less sensitive to motion signals (which have higher temporal frequencies)? How can one predict the ocular dominance properties of V1 neurons from developmental conditions? Why are some V1 neurons double-opponent in space and color whereas retinal neurons are not double-opponent?

Our mathematical formulation of the efficient coding hypothesis enables us to derive how signals should be transformed as they are encoded, thereby allowing us to answer a host of questions such as those above. Since the most efficient encoding transform depends on the statistics of visual inputs, these derivations can predict how neural properties should adapt to changes in visual environment or developmental conditions. In particular, the principle predicts that input signal redundancy should be more or less reduced by the encoding transform in input environments with low or high noise, respectively. These encoding transforms can be compared with known experimental data or tested by new experiments, thereby demonstrating the applicability and limitation of the principle. It will be shown that this principle can account for many encoding processes in the retina and aspects of the diversity of neural receptive fields in V1. However, it fails to account for other prominent properties of V1 neurons.

3.1 A brief introduction to information theory

This brief introduction to information theory (Shannon and Weaver 1949) is intended to provide sufficient intuition to understand sensory information coding and transmission. To keep it brief, many concepts are introduced using intuition and analogies rather than rigorous definitions and derivations. Readers with a basic knowledge of information theory can skip this section.

3.1.1 Measuring information

Many readers will be familiar with the computer terminology "bits". For instance, one can represent any integer a between 0–255 using 8 binary digits, so the integer 15 is 00001111 in binary, or base 2, because

$$15 = 0 \cdot 128 + 0 \cdot 64 + 0 \cdot 32 + 0 \cdot 16 + 1 \cdot 8 + 1 \cdot 4 + 1 \cdot 2 + 1 \cdot 1. \quad (3.1)$$

Accordingly, we say that we need 8 bits to convey or represent an integer a between 0–255. If you know nothing about integer a except its range $a \in [0-255]$ (this notation means that a takes a value within the range from 0 to 255 inclusive), you may assume that it is equally likely to be any integer in the range, with a probability of $P(a) = 1/256$ for any a. However, once you know the exact number, say $a = 10$, this integer has a probability $P(a) = 1$ for $a = 10$ and $P(a) = 0$ otherwise, and there is nothing more (i.e., no more bit) to know.

Note that $- \log_2(1/256) = 8$, and $- \log_2 1 = 0$. Hence, before you know which one a is among the 256 possibilities, it has

$$- \log_2 P(a) = \log_2 256 = 8 \text{ bits} \quad (3.2)$$

of information missing from you. Once you know $a = 10$, a is not allowed to be any other number, you miss no bit of information, and $- \log_2 P(a = 10) = 0$.

In general, given a probability distribution $P(a)$ over the allowed values of a random variable a, the amount of information one needs in order to know its exact value, quantified by the number of bits, is

$$I(a) = - \sum_a P(a) \log_2 P(a) \text{ bits.} \quad (3.3)$$

This formula applies only when the random variable a takes discrete values. In such a case, the formula for information is the same as that for entropy, which we denote by $H(a)$. When signals are represented as discrete quantities, we often use entropy H and information I interchangeably to mean the same thing. Entropy is a measure of the uncertainty about a variable; it is the amount of information missing before one knows the exact value of this variable.

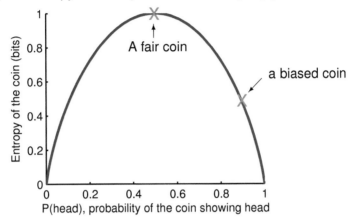

Fig. 3.2: Entropy of outcomes a from the toss of a biased coin as a function of the probability $P(a = \text{head})$ for the outcome "head." A fair coin, with equal probability for the two outcomes head and tail, has the largest entropy.

Note that $I(a) = -\sum_a P(a) \log_2 P(a) = \sum_a P(a)[-\log_2 P(a)]$ bits is an *average* over the quantity $-\log_2 P(a)$ for the variable a in the probability distribution $P(a)$. This quantity $-\log_2 P(a)$ can also be called the *surprise* of a particular value a. Hence, information may be viewed as the average surprise. When one value $a = a_1$ is more probable than another $a = a_2$, the surprise $-\log_2 P(a_1)$ of the more probable a_1 is smaller than that, $-\log_2 P(a_2)$, of the less probable a_2. For instance, consider a biased coin for which the probabilities for seeing a head and tail are $P(a = \text{head}) = 9/10$ and $P(a = \text{tail}) = 1/10$. Before flipping the coin, one can already guess that the outcome is most likely to be "head." So observing what actually happened provides you with less new information than if the outcomes were equally likely. For instance, if the flipping outcome is "head," then one could say "well, that is what I guessed," and the surprise provided by the coin flip is of only a small amount, merely to confirm the guess. Observing a "tail" is surprising, and so this outcome is more informative. Explicitly,

$$-\log_2 P(\text{head}) = -\log_2 9/10 \approx 0.152 \text{ bit,}$$
$$-\log_2 P(\text{tail}) = -\log_2 1/10 \approx 3.3219 \text{ bit.}$$

Thus, an outcome of "head" provides only 0.152 bit of surprise, but a "tail" gives 3.3219 bits. Averaged over many coin flips, each flip gives

$$P(\text{head})[-\log_2 P(\text{head})] + P(\text{tail})[-\log_2 P(\text{tail})]$$
$$= 0.9 \cdot 0.152 + 0.1 \cdot 3.3219 = 0.469 \text{ bit} \qquad (3.4)$$

of information. By comparison, if head and tail are equally likely, $P(\text{head}) = P(\text{tail}) = 1/2$, the average is

$$P(\text{head})[-\log_2 P(\text{head})] + P(\text{tail})[-\log_2 P(\text{tail})]$$
$$= 0.5 \cdot 1 + 0.5 \cdot 1 = 1 \text{ bit.} \qquad (3.5)$$

This is more than the average when the two outcomes are not equally likely. In general, the amount of entropy or information associated with a random variable a is higher when a is more evenly distributed across its allowed values. It is highest when $P(a) = $ constant, i.e., when all a are equally likely. If variable a can take n possibilities, each equally likely, the amount of information is $I = \log_2 n$ bits. This gives 8 bits for an integer $a \in [0, 255]$. A more evenly distributed $P(a)$ means more variability, randomness, or ignorance about a typical a before one knows its exact value.

More intuition about "bits" of information can be gained through the following game. Suppose that you can randomly pick an integer $a \in [0, 255]$ by flipping a fair coin for which head and tail have equal probabilities. Let us say that the first coin flip determines whether $a \in [0, 127]$ (for a head) or $a \in [128, 255]$ (for a tail). Consider it being a head, so $a \in [0, 127]$. Then you flip the coin the second time to determine whether $a \in [0, 63]$ (head) or $a \in [64, 127]$ (tail). You flip the third time to determine whether a is in the bottom or top half of the numbers in the range that remains from the previous flips, and so on. In total, eight coin flips are required to determine the number a exactly. Thus, an integer between [0,255] needs 8 bits of information. Here, one bit of information means an answer to one "yes–no" question, and m bits of information means answers to m "yes–no" questions. This is true only for optimal "yes–no" questions. On average, more than m suboptimal "yes–no" questions, such as in asking whether the number is each integer one by one in order from 0 to 255, are needed to get m bits of information.

3.1.2 Information transmission, information channels, and mutual information

Let a signal S be transmitted via a noisy channel to a destination. Given noise N, the output O from the channel can be written

$$O = S + N. \qquad (3.6)$$

For instance, S could be the input at a sensory receptor, and O could be the output (e.g., an average firing rate) received at a destination neuron. Before O is received, knowledge about S is just its probability distribution $P_S(S)$. Hence, there are

$$H(S) = -\sum_S P_S(S) \log_2 P_S(S) \text{ bits} \qquad (3.7)$$

of ignorance, or missing information, about S. As long as noise N is not too large, receiving O allows us to guess S more proficiently, based on a probability distribution $P(S|O)$, which is the conditional probability of S given O.

To minimize notational clutter, we often write e.g., $P(S)$ and $P(N)$ rather than $P_S(S)$ and $P_N(N)$, to indicate probabilities of variables S and N respectively. Similarly, the conditional probability $P(S|O)$ is not written as $P_S(S|O)$ when it is clear from context the meaning and role of each variable involved. Throughout this book, many such similarly or analogously simplified notations will be used when the meaning is clear from the context.

One expects $P(S|O)$ to have a narrower distribution than $P_S(S)$. For instance, let $S \in [0, 255]$ according to $P_S(S)$, and let noise be $N \in [-1, 1]$ according to $P(N)$. Then knowing $O = 5$ enables one to narrow down the guess about S from $[0, 255]$ (before receiving O) to $[4, 6]$. If S takes 256 integer values with equal probability $P_S(S) = 1/256$ (for instance), then

$$H(S) = -\sum_S P_S(S) \log_2 P_S(S) = \log_2 256 = 8 \text{ bits} \tag{3.8}$$

gives the amount of information about S missing before O is known. After $O = 5$ is known, if S is narrowed down to 4, 5, or 6 with equal probability $P(S|O) = 1/3$, the amount of information still missing is

$$H(S|O)|_{O=5} \equiv -\sum_S P(S|O) \log_2 P(S|O) = \log_2(3) \tag{3.9}$$

$$= 1.59 \text{ bits in the example above.} \tag{3.10}$$

Here $H(S|O)|_{O=5}$ means the residual entropy of S on the condition that $O = 5$. Much less information is still missing than the 8 bits that preceded knowledge of O. So, knowing $O = 5$ provides the following amount of information about S

$$H(S) - H(S|O)|_{O=5} = [-\sum_S P_S(S) \log_2 P_S(S)]$$

$$-[-\sum_S P(S|O) \log_2 P(S|O)] \tag{3.11}$$

$$= 8 - 1.59 = 6.4 \text{ bits, in the example above.}$$

The amount of information provided depends on the value of O. If the distribution of possible values for O is $P_O(O)$, then the amount of information O can tell us about S when averaged over all of these values is called the *mutual information* $I(O; S)$ between O and S. It is

$$I(O; S) \equiv H(S) - \sum_O P_O(O) H(S|O)|_O$$

$$= \left[-\sum_S P_S(S) \log_2 P_S(S)\right] - \sum_O P_O(O) \left[-\sum_S P(S|O) \log_2 P(S|O)\right]$$

$$= \left[-\sum_S P_S(S) \log_2 P_S(S)\right] - \left[-\sum_{O,S} P(O, S) \log_2 P(S|O)\right]. \tag{3.12}$$

Here $P(O, S) = P_O(O)P(S|O)$ is the joint probability of O and S. It can also be written as $P(O, S) = P_S(S)P(O|S)$, where $P(O|S)$ is the conditional probability of O given S. The second term in equation (3.12) is the conditional entropy

$$H(S|O) \equiv \sum_O P_O(O) H(S|O)|_O = -\sum_{O,S} P(O, S) \log_2 P(S|O).$$

Hence

$$I(O; S) = H(S) - H(S|O). \tag{3.13}$$

Since

$$-\sum_S P_S(S) \log_2 P_S(S) = -\sum_{O,S} P(O, S) \log_2 P_S(S),$$

the mutual information becomes

$$I(O; S) = \left[-\sum_{S,O} P(O, S) \log_2 P_S(S)\right] - \left[-\sum_{O,S} P(O, S) \log_2 P(S|O)\right]$$

$$= \sum_{O,S} P(O, S) \log_2 \frac{P(S|O)}{P(S)}. \tag{3.14}$$

The mutual information $I(O; S)$ above is non-zero because O and S are not mutually independent. The difference between O and S is caused by noise, which affects the former and not the latter. The mutual information is symmetric between O and S, i.e., the amount of information O provides about S is the same as the amount of information S provides about O. This can be seen by noting that $P(S|O) = \frac{P(O,S)}{P(O)}$, and $P(O|S) = \frac{P(O,S)}{P(S)}$. From equation (3.14), we have

$$I(O; S) = \sum_{O,S} P(O, S) \log_2 \frac{P(S|O)}{P(S)}$$

$$= \sum_{O,S} P(O, S) \log_2 \frac{P(O, S)}{P(S)P(O)} \tag{3.15}$$

$$= \sum_{O,S} P(O, S) \log_2 \frac{P(O|S)}{P(O)} = I(S; O). \tag{3.16}$$

In other words, $I(O; S) = H(S) - H(S|O) = H(O) - H(O|S)$. The symmetry between O and S is also clear in equation (3.15). We can also define the joint entropy of a joint probability distribution $P(O, S)$ as

$$H(O, S) \equiv -\sum_{O,S} P(O, S) \log_2 P(O, S). \tag{3.17}$$

Then $I(O; S)$ can also be written as

$$I(O; S) = \sum_{O,S} P(O, S) \log_2 \frac{P(O, S)}{P(S)P(O)}$$

$$= -\sum_{O,S} P(O, S) \log_2 P(S) - \sum_{O,S} P(O, S) \log_2 P(O) - H(O, S)$$

$$= -\sum_{S} P(S) \log_2 P(S) - \sum_{O} P(O) \log_2 P(O) - H(O, S)$$

$$= H(S) + H(O) - H(O, S). \tag{3.18}$$

This equation is another apparent demonstration of the symmetry $I(O; S) = I(S; O)$.

If an information channel transmits $I(O; S)$ bits of information from source S to output O per unit time, then this channel is said to have a *capacity* of at least $I(O; S)$ bits per unit time.

A particularly useful example is when S and N are both Gaussian random variables, and are mutually independent:

$$P(S) = \frac{1}{\sqrt{2\pi}\sigma_s} \exp\left(-\frac{S^2}{2\sigma_s^2}\right), \quad P(N) = \frac{1}{\sqrt{2\pi}\sigma_n} \exp\left(-\frac{N^2}{2\sigma_n^2}\right) \tag{3.19}$$

with zero means and variances σ_s^2 and σ_n^2 respectively. Since $O = S + N$, the probability $P_O(O)$ of O can be obtained from the probability $P_S(S)$ of S and the probability $P_N(N)$ of noise N. Each input S gives rise to a conditional probability distribution $P(O|S)$ (which is the probability of O given S) of the output O. We know that $O = S + N$ should differ from S by an amount dictated by the noise which follows a probability $P_N(N)$. Hence $P(O|S) = P_N(O - S)$. This means, the probability $P(O|S)$ that O occurs given S is equal to the probability $P_N(N = O - S)$ that the noise value $N = O - S$ occurs. Over many trials, many different output values O can occur, arising from randomly drawn inputs S from its probability distribution $P_S(S)$ and from randomly drawn noise N. Hence, the overall probability distribution of $P_O(O)$, which is called the *marginal distribution*, can be obtained as a weighted sum of the conditional probability $P(O|S)$ according to the probability $P_S(S)$ that S occurs, i.e.,

$$P_O(O) = \sum_S P_S(S)P(O|S) = \sum_S P_S(S)P_N(O - S). \tag{3.20}$$

Using the Gaussian distributions for P_S and P_N from equation (3.19), we have

$$P_O(O) = \int dS P(S)P(O|S) = \int dS P(S)P_N(O - S)$$

$$= \frac{1}{2\pi\sigma_s\sigma_n} \int dS \exp\left(-\frac{S^2}{2\sigma_s^2}\right) \exp\left(-\frac{(O - S)^2}{2\sigma_n^2}\right)$$

$$= \frac{1}{\sqrt{2\pi(\sigma_s^2 + \sigma_n^2)}} \exp\left(-\frac{O^2}{2(\sigma_s^2 + \sigma_n^2)}\right)$$

$$\equiv \frac{1}{\sqrt{2\pi}\sigma_o} \exp\left(-\frac{O^2}{2\sigma_o^2}\right), \tag{3.21}$$

$$\text{where } \sigma_o^2 \equiv \sigma_s^2 + \sigma_n^2. \tag{3.22}$$

Hence, O is also a Gaussian random variable, with a mean zero and a variance $\sigma_o^2 \equiv \sigma_s^2 + \sigma_n^2$. Here we are dealing with continuous, rather than discrete, variables, S, N, and O. Entropy can also be defined for continuous variables, analogously to the definition for discrete variables. The entropy of a Gaussian signal is always the log of the standard deviation plus a constant, as shown for instance for $P(S)$ as

$$H(S) = -\int dS P(S) \left\{ \log_2\left[\frac{1}{\sqrt{2\pi}\sigma_s}\right] + \log_2\left[\exp\left(-\frac{S^2}{2\sigma_s^2}\right)\right] \right\}$$

$$= \log_2 \sigma_s + \frac{1}{2}\log_2(2\pi) + (\log_2 e) \int dS P(S)\frac{S^2}{2\sigma_s^2}$$

$$= \log_2 \sigma_s + \frac{1}{2}\log_2(2\pi) + (\log_2 e)\frac{1}{2} = \log_2 \sigma_s + \text{constant}. \tag{3.23}$$

For continuous variables, the entropy value depends on the units in which each variable is measured. For example, if the numerical value of σ_s is increased because S is measured in different units, e.g., centimeters versus meters, the entropy $H(S)$ increases accordingly. In general, let $\mathbf{S} = (S_1, S_2, ...)$ be a multi-dimensional variable (a scalar is a special case), if $\mathbf{S}$ is transformed to $\mathbf{S}' = \mathbf{US}$ by a transform matrix $\mathbf{U}$ (let us say because $\mathbf{S}$ and $\mathbf{S}'$ are measured in different units or along a different set of axes in a multi-dimensional space), we have

$$P(\mathbf{S})d\mathbf{S} = P(\mathbf{S}')d\mathbf{S}', \quad \text{hence} \quad P(\mathbf{S}) = P(\mathbf{S}')|\det|d\mathbf{S}'/d\mathbf{S}|| = P(\mathbf{S}')|\det \mathbf{U}|,$$

where det means the determinant of a matrix. Continuing,

$$H(\mathbf{S}) = -\int d\mathbf{S} P(\mathbf{S}) \log_2 P(\mathbf{S}) = -\int d\mathbf{S}' P(\mathbf{S}') \left[\log_2 P(\mathbf{S}') + \log_2(|\det \mathbf{U}|)\right]$$

$$\underrightarrow{\text{when } |\det \mathbf{U}| = 1}$$
$$= H(\mathbf{S}') + \log_2|\det \mathbf{U}| \longrightarrow= H(\mathbf{S}'). \tag{3.24}$$

Hence, linear transformations $\mathbf{S}' = \mathbf{US}$ of variables change the entropy by $\log_2|\det \mathbf{U}|$, a constant. If the transformation leaves the unit volume unchanged, i.e., $|\det \mathbf{U}| = 1$, as for instance for coordinate rotations, then the entropy is unchanged.

However, the mutual information $I(O; S)$ between O and S is not affected by the units in which the variables are measured. For the Gaussian signals above,

$$\begin{aligned} I(O; S) &= H(S) - H(S|O) \\ &= H(O) - H(O|S) = H(O) - H(N) \\ &= \log_2 \frac{\sigma_o}{\sigma_n} \end{aligned}$$ (3.25)

$$= \frac{1}{2} \log_2 \left(1 + \frac{\sigma_s^2}{\sigma_n^2}\right),$$ (3.26)

which depends on the signal-to-noise ratio (S/N) σ_s^2/σ_n^2, a unitless quantity! The equality $H(O|S) = H(N)$ used above derives from the observation that the conditional probability $P(O|S)$ is the same as the probability that the noise N takes the value $N = O - S$, i.e., $P(O|S) = P(N)$. Thus any uncertainty about O given S, i.e., conditional entropy $H(O|S)$, is the same as the uncertainty about, or entropy $H(N)$ of, the noise N.

Equation (3.26) provides an intuitive understanding of the mutual information $I(O; S)$ for Gaussian signals. We may very coarsely approximate the Gaussian distribution of O by a constant distribution $P(O) = $ constant within a range σ_o of possible O values (and $P(O) = 0$ for O values outside this range). Furthermore, we discretize O into σ_o/σ_n possible values, with a quantization step size σ_n determined by the size of the noise, which relates O to S. Then, since each discrete value is equally likely, the information provided by each discrete value is $\log_2(\sigma_o/\sigma_n) = I(O; S)$.

3.1.3 Information redundancy, representation efficiency, and error correction

Let us apply the concept of mutual information to the case when information is shared between neighboring pixels in images. Let S_1 and S_2 be the image intensities in these two pixels. They are likely to be similar to each other in most natural images, since coherent image structures are typically much larger than the distance between the two pixels. For example, in an image whose pixel intensities range from 0 to 255, a value $S_1 = 10$ makes one expect a S_2 value close to $S_2 = 10$ rather than taking a value anywhere between 0 and 255. In other words, knowing S_1 enables one to guess S_2 more proficiently, in just the same way in the previous example that knowing O provides some information about S, when $O = S + N$ through a noisy channel. This can be expressed as $P(S_2|S_1) \neq P(S_2)$, meaning that S_1 and S_2 are not independent variables. Therefore, mutual information $I(S_1; S_2)$ between S_1 and S_2 is non-zero. Hence, information provided by S_1 about the visual world is redundant with information provided by S_2, although the latter is not exactly the same as the former. Since, by equation (3.18),

$$I(S_1; S_2) = H(S_1) + H(S_2) - H(S_1, S_2),$$ (3.27)

a non-zero $I(S_1; S_2)$ means $H(S_1) + H(S_2) > H(S_1, S_2)$. In general, given n signals S_1, S_2, ..., S_n,

$$\sum_i H(S_i) \geq H(S_1, S_2, ..., S_n),$$ (3.28)

with equality when all S's are mutually independent, i.e., when there is no redundancy. The degree of redundancy can be quantified by

$$\text{Redundancy} = \frac{\sum_{i=1}^{n} H(S_i)}{H(S_1, S_2, ..., S_n)} - 1,$$ (3.29)

which is non-negative, and is equal to zero when the variables are independent. Combining equation (3.27) with the definition $I(S_1; S_2) = H(S_2) - H(S_2|S_1)$ gives us

$$H(S_2|S_1) = H(S_1, S_2) - H(S_1), \tag{3.30}$$

which characterizes the amount of entropy in S_2 that is not present in S_1.

In order to practice what we learned above, it is useful to make a couple of simplifying assumptions or approximations about the statistics of natural images. First let us assume that, in natural images, the probability distribution $P(S_i)$ of the gray level S_i of an image pixel i over an ensemble of many images is independent of the location of this pixel. Second, we assume *the ergodicity condition*, which is

> the probability of a pixel, or of (a spatial configuration) a few pixels,
> sampled across the ensemble of images is the same as the probability   (3.31)
> obtained by sampling the pixel(s) across a single large image.

Hence, for example, the probability distribution $P(S_i)$ of S_i across many images at a particular pixel location i is the same as the probability distribution $P(S)$ of pixel value S across pixels in a sufficiently large image. Similarly, the joint probability distribution $P(S_1, S_2)$ of two pixel values S_1 and S_2 for two horizontally neighboring pixels can be approximated by sampling S_1 and S_2 over many pairs of horizontally neighboring pixels over a large image.

Now, let us examine information redundancy in natural images by using the ergodicity condition in a large image shown in Fig. 3.3 to calculate $H(S_1) = H(S_2) = H(S)$ and $H(S_1, S_2)$. In Fig. 3.3 A, the image pixel values are digitized to only two possible gray levels, denoted by $S = 0$ or $S = 1$ for dark or bright pixels. Two horizontally neighboring pixels are more likely to have the same gray level, both bright or dark, than different gray levels, one dark and one bright. This can be seen in the joint probability $P(S_1, S_2)$ written out as a 2×2 matrix whose row number and column number correspond to the S_1 and S_2 values, respectively,

$$P(S_1, S_2) = \begin{pmatrix} P(S_1 = 0, S_2 = 0) & P(S_1 = 0, S_2 = 1) \\ P(S_1 = 1, S_2 = 0) & P(S_1 = 1, S_2 = 1) \end{pmatrix} = \begin{pmatrix} 0.5247 & 0.0105 \\ 0.0105 & 0.4543 \end{pmatrix}. \tag{3.32}$$

The marginal probability $P(S_1) = \sum_{S_2} P(S_1, S_2)$ can then be obtained as $P(S_1 = 0) = 0.5352$, and $P(S_1 = 1) = 0.4648$, and similarly for $P(S_2)$. Then, the conditional probability $P(S_2|S_1) = P(S_1, S_2)/P(S_1)$ in a 2×2 matrix is

$$P(S_2|S_1) = \begin{pmatrix} P(S_2 = 0|S_1 = 0) & P(S_2 = 1|S_1 = 0) \\ P(S_2 = 0|S_1 = 1) & P(S_2 = 1|S_1 = 1) \end{pmatrix} = \begin{pmatrix} 0.9804 & 0.0196 \\ 0.0226 & 0.9774 \end{pmatrix}. \tag{3.33}$$

Hence, $P(S_2 = S_1|S_1) > 0.97$, i.e., given one pixel's value, the neighboring pixel has more than a 97% chance to be as bright or as dark. Figure 3.3 C shows that indeed $P(S_1, S_2)$ has its highest density at $S_1 \approx S_2$. Hence, two adjacent pixels should carry very redundant information. Using the probability values, we indeed obtain

$$H(S) = H(S_1) = H(S_2) = 0.9964 \text{ bits}, \qquad H(S_1, S_2) = 1.1434 \text{ bits},$$

$$\text{Redundancy} = \frac{2H(S)}{H(S_1, S_2)} - 1$$

$$= 0.74 \text{ for an image discretized to 2 gray levels (black and white)}.$$

We see that $H(S) \approx 1$ bit, the maximum entropy possible with two gray levels; see Fig. 3.2, because $P(S = 1) \approx P(S = 0)$; i.e., the pixel has a roughly equal chance to be either bright or dark. However, from equation (3.30),

$$H(S_2|S_1) = H(S_1, S_2) - H(S) = 0.1470 \ll 1 \text{ bit}, \tag{3.34}$$

because once pixel value S_1 is given, the probability $P(S_2|S_1)$ is very biased to one gray level $S_1 = S_2$. If $P(S_2 \neq S_1|S_1) = 0$ for each S_1 value, then S_1 always equals S_2, and $H(S_1, S_2) = H(S)$. In such a case, redundancy $= 1$, meaning that the two pixels are 100% redundant.

However, when S can take 256 different pixel values $S = 0, 1, 2, ..., 255$, as in Fig. 3.3 B, we have instead

$$H(S) = 7.63 \text{ bits}, \qquad H(S_1, S_2) = 11.31 \text{ bits},$$
$$\text{Redundancy} = 0.35 \quad \text{for an image discretized to 256 gray levels.} \tag{3.35}$$

So the redundancy is much reduced from 0.74 in the case of two gray levels. This means that at a finer resolution of the gray scale, the two pixels are less redundant. This is partly because, when S is described by more than 64 gray levels (6 bits of resolution), the exact pixel value (e.g., whether a particular pixel's gray value is $S = 105$ or $S = 106$ out of 256 gray levels) is roughly dictated by independent noise unrelated to visual objects in the scene (Petrov and Zhaoping 2003). For comparison, when we discretize images at a resolution of 6 bits using 64 gray levels, we have

$$H(S) = 5.64 \text{ bits}, \qquad H(S_1, S_2) = 7.58 \text{ bits},$$
$$\text{Redundancy} = 0.49 \quad \text{for an image discretized to 64 gray levels.} \tag{3.36}$$

A   B   C $P(S_1, S_2)$

Fig. 3.3: A photograph is displayed in two ways. In A, each image pixel takes one of two gray levels $S = 0$ or 1, as either darker or otherwise than the mean gray level. In B, each image pixel takes 256 gray levels $S = 0, 1, 2, ...255$. With two gray levels, the entropy of a single pixel gray value is $H(S) = 0.9964$ bits, the joint entropy of two horizontally neighboring pixels, S_1 and S_2, is $H(S_1, S_2) = 1.1434$ bits, and redundancy $\frac{2H(S)}{H(S_1,S_2)} - 1 = 0.74$. With 256 gray levels, $H(S) = 7.6267$ bits, $H(S_1, S_2) = 11.3124$ bits, and redundancy $= 0.35$. C plots the probability distribution $P(S_1, S_2)$, with higher probability values being rendered more brightly.

When information is represented redundantly, whether in natural visual images (Field 1987, Kersten 1987, Ruderman and Bialek 1994) or in other signals such as speech in a natural language such as English, the representation is not efficient. For example, if $\sum_{i=1}^{n} H(S_i) = 100$ bits $> H(S_1, S_2, ..., S_n) = 50$ bits, one uses 100 bits in the signal representation $\mathbf{S} \equiv (S_1, S_2, ..., S_n)$ to represent 50 bits of information. Sending the signals $\mathbf{S} \equiv (S_1, S_2, ..., S_n)$ (per unit time) through an information channel in this representation would require a channel capacity of at least 100 bits per unit time. Shannon and Weaver

(1949) showed that, theoretically, all the information about $\mathbf{S} \equiv (S_1, S_2, ..., S_n)$ could be faithfully transmitted through a channel having a capacity of only $H(S_1, S_2, ..., S_n)$ (e.g., 50 bits) per unit time. This is done by encoding $\mathbf{S}$ into some other form $f(\mathbf{S})$, where $f(.)$ is an encoding transform. In such a case, $f(\mathbf{S})$ would be a more efficient representation of the original information in $\mathbf{S}$, and the information channel could be most efficiently used (although the cost of encoding transform is ignored here). In reality, this Shannon limit is difficult to reach, but efficiency can still be improved by other encoding transforms.

Conversely, redundancy is useful for error correction. Although an efficient representation of signals can save information storage space or information channel capacity, it also reduces or removes the ability to recover information in the face of error, which can be introduced during signal transmission through a noisy channel. For instance, given a sentence conveyed noisily as "I w..ld lik. .o invite y.u f.r din.er" (in which each "." indicates some missing letter(s)), one can recover the actual sentence "I would like to invite you for dinner" using knowledge of English. This ability is licensed by the redundancy in English, in which there is substantial mutual information between different letters, words, or sentences, so that one can predict or guess some letters, words, or sentences from others. This can be stated in terms of probability and information in the following example. Without any neighboring letters or context, one can guess a missing letter S as one of any 26 letters in the alphabet based on a prior probability distribution $P(S)$ (in which some letters are more likely than other letters). One requires an amount $H(S) = -\sum_S P(S) \log_2 P(S)$ of information to ascertain this letter. However, the redundancy between neighboring letters is such that S can be narrowed down to fewer choices given the context. That is, the conditional probability $P(S|\text{contextual letters})$ typically has a substantial value only for a very few among the 26 letters in the alphabet, so that the amount of information needed to recover the letter S, the conditional entropy $H(S|\text{contextual letters})$, is less than $H(S)$. If there is so much redundancy that S can be almost completely determined by contextual letters, then a missing letter S in a word or sentence will not prevent language comprehension. Redundancy in natural language enables us to communicate effectively under conditions of noisy telephone lines, imperfect grammar, and unfamiliar accents. If everybody spoke clearly with a standard accent and perfect grammar, redundancy in language would be less necessary. How much redundancy is optimal in a representation depends on the level of noise, or tendency for errors, in the system. The end purpose or task that utilizes the transmitted information should also favor a particular type of redundancy over others.

3.2 Formulation of the efficient coding principle

The efficient coding principle was proposed by Barlow and others half a century ago (Barlow 1961). It states that one of the goals of early visual processing is to encode the visual input in an efficient form, taking advantage of its statistical regularities, such that the maximum amout of information can be extracted and transmitted under the constraint of limited neural resources. A mathematical formulation of this principle should facilitate its applications and predictions.

3.2.1 An optimization problem

One formulation (Laughlin 1981, Srinivasan, Laughlin and Dubs 1982, Linsker 1990, Atick and Redlich 1990, van Hateren 1992, Atick 1992) of this principle goes as follows. Let sensory input signal $\mathbf{S} \equiv (S_1, S_2, ..., S_i, ..., S_n)$ occur with probability $P(\mathbf{S})$. When $\mathbf{S}$ is a continuous variable, $P(\mathbf{S})$ is the probability density, such that $\mathbf{S} \in (\mathbf{S}, \mathbf{S} + d\mathbf{S})$ occurs with probability $P(\mathbf{S})d\mathbf{S}$. This book uses a bold font to denote or emphasize that a variable is a vector, although occasionally for notational simplicity the bold face is not used when the vector nature is evident from the context, or unimportant.

Due to input noise $\mathbf{N}$, which can come from sampling noise, the signals actually received by the sensory receptors are

$$\mathbf{S}' = \mathbf{S} + \mathbf{N}. \tag{3.37}$$

The amount of sensory information in this $\mathbf{S}'$ about $\mathbf{S}$ is $I(\mathbf{S}'; \mathbf{S})$. The information associated with receptor i is $I(S_i', S_i)$. However, due to information redundancy between S_i and S_j from different channels i and j, the sum $\sum_i I(S_i'; S_i)$ of information transmitted by all the receptors is more than $I(\mathbf{S}'; \mathbf{S})$. When referring to information transmitted per unit time, we will also refer to expressions like $I(S_i', S_i)$ and $I(\mathbf{S}'; \mathbf{S})$ as information rate or data rate, apologizing for the mild abuse of notation.

Consider an encoding transform K that transforms the variable $\mathbf{S}'$ to downstream neural responses or outputs (see Fig. 3.1)

$$\mathbf{O} = K(\mathbf{S}') + \mathbf{N_o}, \tag{3.38}$$

where $\mathbf{N_o}$ is the intrinsic output noise introduced in the encoding process and not attributable to input noise $\mathbf{N}$, and K can be a linear or nonlinear function. For instance, in a blowfly's compound eye, $\mathbf{S}$ is the input contrast, a scalar or a vector with only one component, and hence it could be written as simply S; S' is the actual signal, including the sampling noise, received by the receptor, and $K(\mathbf{S}')$ or $K(S')$ describes the sigmoid-like gain control of S' by large monopolar cells (LMC) in a fly's compound eye. In another example, $\mathbf{S} = (S_1, S_2, ..., S_n)$ could be a vector describing inputs to n photoreceptors (strictly speaking, $\mathbf{S}$ is the responses from the receptors when receptor sampling noise is treated as zero), $\mathbf{O}$ another vector for net inputs to many retinal ganglion cells, and the receptor-to-ganglion transform $K(\mathbf{S}')$ may be approximated linearly by a matrix K as

$$O_i = \left[\sum_j \mathsf{K}_{ij}(S_j + N_j)\right] + (N_o)_i. \tag{3.39}$$

This book uses a sans-serif font to denote matrices and italic fonts to denote functions; e.g., K in $K(\mathbf{S}')$. Hence, e.g., $[K(\mathbf{S}')]_i = \sum_j \mathsf{K}_{ij}S_j'$ describes a linear coding transform, using K for $K(\mathbf{S}')$ and K for K_{ij}. Here, K_{ij} is the effective neural connection from the jth receptor to the ith ganglion cell via the retinal interneurons, and $(N_o)_i$ is the ith component of $\mathbf{N_o}$.

The output $\mathbf{O}$ extracts and transmits $I(\mathbf{O}; \mathbf{S})$ amount of information about the sensory input $\mathbf{S}$. It also costs neural resources, in terms of, e.g., metabolic energy to fire action potentials or the space occupied by the number of neurons and axons concerned. More action potentials and more neurons can increase $I(\mathbf{O}; \mathbf{S})$, while also increasing the cost. Efficient coding requires us to find the most efficient K to minimize the quantity

$$E(K) = \text{neural cost} - \lambda \times I(\mathbf{O}; \mathbf{S}), \tag{3.40}$$

where the parameter λ balances the needs to extract information $I(\mathbf{O}; \mathbf{S})$ and to reduce or restrict cost. The value of λ may be determined by various requirements and constraints of the sensory system. If the subject, say a human, requires a large amount of information $I(\mathbf{O}; \mathbf{S})$ to see the world clearly in order to, e.g., read, λ should be large so that minimizing E is mainly or substantially influenced by maximizing $I(\mathbf{O}; \mathbf{S})$. Conversely, subjects, such as frogs, that do not require as much information can afford to sacrifice a large amount of $I(\mathbf{O}; \mathbf{S})$ in order to save the neural cost of representing and transmitting the output $\mathbf{O}$. Thus, the λ for these animals can be smaller than that for humans, and minimizing E is largely influenced by minimizing the neural cost. Both the neural cost and the information $I(\mathbf{O}; \mathbf{S})$ extracted should depend on K. The optimal code K is the solution(s) to equation $\partial E(\mathsf{K})/\partial K = 0$, assuming that E is differentiable with respect to K.

3.2.2 Exposition

While $\mathbf{O}$ extracts $I(\mathbf{O};\mathbf{S})$ amount of information about sensory input $\mathbf{S}$, it does so via $K(\mathbf{S}')$ and requires for each output O_i a data rate $H(O_i) - H((N_o)_i)$. This data rate consumes the transmission capacity resource needed to transmit the encoded signal $(K(\mathbf{S}'))_i$, since the information in O_i about $(K(\mathbf{S}'))_i$ is

$$I\left(O_i; (K(\mathbf{S}'))_i\right) = H(O_i) - H\left((N_o)_i\right). \tag{3.41}$$

This resource requires a sufficiently large $H(O_i)$; hence it requires a sufficient dynamic range in O_i, which is defined as the number of response levels of O_i. It also requires a sufficiently small noise $(N_o)_i$ or noise entropy $H\left((N_o)_i\right)$. However, this resource may not be most efficiently used, since the transmitted information includes some information about the sensory noise $\mathbf{N}$, and the information transmitted by output O_i may be redundant with the information transmitted by output O_j, for $j \neq i$.

In $\mathbf{O}$,

$$\text{the total output noise} \equiv K(\mathbf{S}') - K(\mathbf{S}) + \mathbf{N_o} \tag{3.42}$$

contains

$$\text{the transmitted input noise} = K(\mathbf{S}') - K(\mathbf{S}). \tag{3.43}$$

Of the total output information $I(\mathbf{O}; K(\mathbf{S}'))$ in $\mathbf{O}$ about $K(\mathbf{S}')$, the useful part about the sensory input $\mathbf{S}$ is

$$\begin{aligned} \text{extracted sensory information } I(\mathbf{O};\mathbf{S}) &= H(\mathbf{O}) - H(\mathbf{O}|\mathbf{S}) \\ &= H(\mathbf{S}) - H(\mathbf{S}|\mathbf{O}), \end{aligned} \tag{3.44}$$

where, e.g., $H(\mathbf{O}|\mathbf{S}) = -\int d\mathbf{O}d\mathbf{S}P(\mathbf{O},\mathbf{S})\log_2 P(\mathbf{O}|\mathbf{S})$, in which $P(\mathbf{O}|\mathbf{S})$ is the conditional probability of $\mathbf{O}$ given $\mathbf{S}$, and $P(\mathbf{O},\mathbf{S}) = P(\mathbf{O}|\mathbf{S})P(\mathbf{S})$ is the joint probability distribution. Note that $P(\mathbf{O}|\mathbf{S})$ depends on the probability distribution of $P_N(\mathbf{N})$ of the input noise $\mathbf{N}$ and $P_{N_o}(\mathbf{N_o})$ of the output intrinsic noise $\mathbf{N_o}$. The probability distribution of output $\mathbf{O}$ alone is $P(\mathbf{O}) = \int d\mathbf{S}P(\mathbf{O}|\mathbf{S})P(\mathbf{S})$.

An efficient coding transform K should minimize the use of resources such as the data rate $\sum_i [H(O_i) - H((N_o)_i)]$, while transmitting as much information $I(\mathbf{O};\mathbf{S})$ about the sensory signal $\mathbf{S}$ as possible.

Due to the addition of noise $\mathbf{N_o}$ in the encoding process, the extracted information $I(\mathbf{O};\mathbf{S})$ at the output $\mathbf{O}$ cannot exceed the amount of information $I(\mathbf{S}';\mathbf{S})$ received at the input stage. This is the data processing inequality

$$I(\mathbf{O};\mathbf{S}) \leq I(\mathbf{S}';\mathbf{S}). \tag{3.45}$$

To make $I(\mathbf{O};\mathbf{S})$ approach $I(\mathbf{S}';\mathbf{S})$ as closely as possible, the output noise $\mathbf{N_o}$ should be negligible compared to the transmitted input noise $K(\mathbf{S}') - K(\mathbf{S})$. For instance, this could be achieved by a large amplification factor F, making

$$\mathbf{O} = F \cdot (\mathbf{S} + \mathbf{N}) + \mathbf{N_o}. \tag{3.46}$$

Then, the total output noise, $F \cdot \mathbf{N} + \mathbf{N_o}$, would be dominated by the amplified input noise $F \cdot \mathbf{N}$. However, this makes the output dynamic range very large, costing a total output channel capacity of $\sum_i [H(O_i) - H((N_o)_i)]$.

Significant channel capacity cost can be saved by reducing the information redundancy between the output channels, which is inherited from the redundancy between the input channels. In particular, the amount of redundant information at the input stage is

$$\sum_i I(S'_i; S_i) - I(\mathbf{S}'; \mathbf{S}). \tag{3.47}$$

For instance, the input information rate $I(\mathbf{S}'; \mathbf{S})$ may be one megabyte/second, while using a data rate $\sum_i I(S'_i; S_i)$ of 10 megabyte/second. A suitable encoding K that reduces this redundancy could reduce the output dynamic range, thus saving neural cost, while maintaining the encoding of input information by making $I(\mathbf{O}; \mathbf{S})$ approach $I(\mathbf{S}'; \mathbf{S})$. In the example above, this means transmitting $I(\mathbf{O}; \mathbf{S})$ at a rate of nearly one megabyte/second, but using a data rate or channel capacity $\sum_i [H(O_i) - H((N_o)_i)]$ of much less than 10 megabyte/second.

3.2.2.1 Maximum entropy code, factorial code, and decorrelation in the high S/N limit

Since $I(\mathbf{O}; \mathbf{S}) = H(\mathbf{O}) - H(\mathbf{O}|\mathbf{S})$, the information extracted, $I(\mathbf{O}; \mathbf{S})$, is the difference between output entropy $H(\mathbf{O})$ and the conditional entropy $H(\mathbf{O}|\mathbf{S})$ of $\mathbf{O}$ given $\mathbf{S}$. This conditional entropy $H(\mathbf{O}|\mathbf{S})$ is the same as the entropy of the total output noise $K(\mathbf{S}') - K(\mathbf{S}) + \mathbf{N_o}$. This leads immediately to the following conclusion:

> When the input noise $\mathbf{N} \to 0$, $H(\mathbf{O}|\mathbf{S}) \to H(\mathbf{N_o}) = \text{constant}$; therefore, maximizing $I(\mathbf{O}; \mathbf{S})$ requires maximizing $H(\mathbf{O})$: *maximum entropy encoding.* $\tag{3.48}$

Here, $x \to y$ means that the value of x approaches the value of y. Hence, $\mathbf{N} \to 0$ implies high input S/N. When the total output channel capacity, $\sum_i H(O_i)$, is fixed or constrained by the available neural resources, the inequality $H(\mathbf{O}) \leq \sum_i H(O_i)$ implies that $H(\mathbf{O})$ is maximized when the equality $H(\mathbf{O}) = \sum_i H(O_i)$ is achieved. Mathematically, the equality holds when different output channels i (e.g., neurons) convey different information. If one neuron always responds exactly the same as another, information from the second neuron's response is redundant, and the total information conveyed by one neuron is the same as that by both. Thus, $H(\mathbf{O})$ is maximized when neurons respond independently, i.e.,

$$P(O_1, O_2, ...O_n) = P(O_1)P(O_2) \cdots P(O_n), \tag{3.49}$$

so the joint probability factorizes into a product of marginal probabilities. Such a coding scheme is said to be an *independent component code* (or *factorial code*) for input $\mathbf{S}$. This is why in the noiseless limit, $I(\mathbf{O}; \mathbf{S})$ is maximized when responses O_i and O_j are independent of each other or are decorrelated to all orders. In practice, independence is approximated by decorrelation to the second order, defined as

$$\int (O_i - \bar{O}_i)(O_j - \bar{O}_j)P(O_i, O_j)dO_i dO_j = 0, \quad \text{for } i \neq j, \tag{3.50}$$

in which $\bar{O}_i \equiv \int dO_i P(O_i)$ is the mean of O_i (similarly for $\bar{O}_j$). One should note, however, that such second order decorrelation often does not achieve complete independence.

If making different O_i's independent, i.e., making $H(\mathbf{O}) = \sum_i H(O_i)$, does not make $I(\mathbf{O}; \mathbf{S})$ provide sufficient information about $\mathbf{S}$, then increasing $\sum_i H(O_i)$, if this sum is not constrained, can further increase $I(\mathbf{O}; \mathbf{S})$. There are two ways to increase $\sum_i H(O_i)$. First, one can increase the dynamic range, i.e., the number of distinguishable response levels, for each O_i. Second, for a given dynamic range of O_i, $H(O_i)$ can be increased by equalizing the probabilities of different output response levels O_i. This is sometimes known as *histogram equalization.* For instance, if neuron i has only $m = 2$ possible response values O_i (per second), $H(O_i)$ can be maximized to $H(O_i) = \log_2 m = 1$ bit/second (when $m = 2$) of information when each response value is utilized equally often, while $m = 2$ is fixed. In this case

$$P(O_i = \text{a particular response}) = \frac{1}{m} \quad \text{for each response value.} \tag{3.51}$$

So n such (independent) neurons can jointly transmit n bits/second when $m = 2$. Furthermore, more information can be transmitted if m is larger, i.e., if the neuron has a larger dynamic range or a larger number of possible response levels.

Typically, natural scene signals $\mathbf{S}$, with probability $P(\mathbf{S})$, obey statistical regularities such that (1) different signal values do not occur equally often, and (2) different input channels S_i and S_j, e.g., inputs to neighboring photoreceptors, convey redundant information. Both (1) and (2) are for instance manifested in the observation that two neighboring pixels in images are more likely to have similar rather than very different intensity levels; see Fig. 3.3 C. Hence, responses from two neighboring photoreceptors give redundant information: given one response, the other one is largely predictable, and only the difference between the actual and the predicted response conveys additional, non-redundant, information. If n such photoreceptors (input channels) contain 8 bits/second of information in each channel i, and, say, 7 out of the 8 bits/second of information in each channel is redundant information already present in other channels, the total amount of joint information $H(\mathbf{S})$ is only about n bits/second (for large n), much less than the apparent $8 \times n$ bits/second. Transmitting the raw input directly to the brain using $\mathbf{O} = \mathbf{S}$ would be inefficient, or even impossible if, e.g., each of the n output channels $\mathbf{O} = (O_1, O_2, ..., O_n)$ has a limited capacity of only $H(O_i) = 1$ bit/second. The transform or coding $\mathbf{S} \to \mathbf{O} \approx K(\mathbf{S})$ could maximize efficiency such that (1) neurons O_i and O_j respond independently and (2) each response value of $\mathbf{O}$ is equally utilized. Then, (almost) all input information could be faithfully transmitted through responses $\mathbf{O}$ even though each output channel conveys only 1 bit/second. Accordingly, e.g., the connections from the photoreceptors to the retinal ganglion cells are such that, in bright illumination (i.e., when signal-to-noise is high), ganglion cells are tuned to respond to differences between nearby photoreceptor signals, making their responses more independent from each other. These ganglion cells are called feature detectors (Barlow 1961) as they respond to informative (rather than redundant) image contrast features.

3.2.2.2 Smoothing in the low S/N condition

However, when the input $S/N \ll 1$ is so poor that each input channel has no more than, say, $I(S'_i; S_i) = 0.1$ bits/second of useful information, optimal encoding no longer specifies that different channels should be mutually independent. In this case, from equation (3.26) for zero mean Gaussian signals $S'_i = S_i + N_i$, the S/N ratio is $\langle S_i^2 \rangle / \langle N_i^2 \rangle = (2^{0.1})^2 - 1 = 0.149$. (The notation $\langle ... \rangle$ denotes ensemble average, e.g., $\langle S_i^2 \rangle = \int dS_i P(S_i) S_i^2$.) In n such input channels, the total data rate is only $\sum_i I(S'_i; S_i) = 0.1n$ bits/second, and the total information rate $I(\mathbf{S}'; \mathbf{S})$ is even smaller when considering input redundancy. Such a small data rate is sufficient to fit into n output channels of 1 bit/second each even without an encoding transform, i.e., even when $\mathbf{O} = \mathbf{S}' + \mathbf{N_o}$ (at least when output intrinsic noise $\mathbf{N_o}$ is not too large). The output channel entropy $H(\mathbf{O}) = I(\mathbf{O}; \mathbf{S}) + H(\mathbf{O}|\mathbf{S})$ wastes a significant or possibly dominant fraction $H(\mathbf{O}|\mathbf{S})$ on transmitting input noise $\mathbf{N}$ which (compared to $\mathbf{S}$) is typically much less redundant between input channels. In fact, most or much of the output variability is caused by input noise rather than signal, costing metabolic energy to fire action potentials (Levy and Baxter 1996). To minimize this waste, a different transform K is desirable to average out input noise. Take the case that the input involves two channels with highly correlated inputs $S_1 \approx S_2$ but independent and identically distributed noise N_1 and N_2. Let the coding transform take the average of the inputs (assuming the output noise N_o is negligible compared with the input noise)

$$O_1 = (S_1' + S_2')/2 = (S_1 + S_2)/2 + (N_1 + N_2)/2.$$

This output O_1 has a signal $(S_1 + S_2)/2 \approx S_1$ and a noise $(N_1 + N_2)/2$. The variances, or average powers, of its signal and noise are

$$\left\langle \left(\frac{S_1 + S_2}{2}\right)^2 \right\rangle \approx \langle S_1^2 \rangle \quad \text{and} \quad \left\langle \left(\frac{N_1 + N_2}{2}\right)^2 \right\rangle = \frac{\langle N_1^2 \rangle}{4} + \frac{\langle N_2^2 \rangle}{4} + \frac{\langle N_1 N_2 \rangle}{2} = \frac{\langle N_1^2 \rangle}{2}.$$

In the above, we used $\langle N_1 N_2 \rangle = 0$ and $\langle N_1^2 \rangle = \langle N_2^2 \rangle$ because N_1 and N_2 are uncorrelated and have identical distributions. Hence, the averaging transform $O_1 = (S_1' + S_2')/2$ preserves the original signal power but halves the noise power. When there are many input channels $S_1', S_2', ...,$ and each output channel O_i calculates a weighted average of the correlated input channels (e.g., $O_i = (S_{i-1}' + S_i' + S_{i+1}')/3$), the different output channels O_i and O_j will be correlated and so carry redundant information. With a low input data rate, the output channel capacity (e.g., of n bits/second) is often not fully utilized, and the different output response levels are not equally utilized. This output redundancy, both by correlation between channels and by unequal utilization of the response levels in each channel, should help to recover the original signal S from O. Meanwhile, the averaging or smoothing transform K helps to utilize the output power $\langle O_i^2 \rangle$ better to transmit signal S rather than noise N.

Hence, efficient coding K requires different strategies in different conditions of input S/N. It employs decorrelation and/or output histogram equalization in the high S/N condition but input smoothing (or averaging) in the low S/N condition. In general, the efficient code is the solution K that minimizes $E(K) = $ neural cost $- \lambda \times I(O; S)$ as in equation (3.40), balancing the requirements of extracting information $I(O; S)$ and reducing or restricting cost. Especially in high S/N regime, this efficient coding transform K compresses the raw sensory data with minimum loss, such that maximum information $I(O; S)$ can be transmitted faithfully to higher visual areas despite information bottlenecks such as the optic nerve.

3.2.2.3 Neural cost

The neural cost is often $= \begin{cases} \text{output channel capacity } \sum_i H(O_i), & \text{or} \\ \text{output power } \sum_i \langle O_i^2 \rangle. \end{cases}$ (3.52)

This is because it costs a neuron energy (Laughlin 2001) to increase its response O_i or to generate a spike; it also costs to maintain a channel transmission capacity. For instance, the axon should be thick enough so that O_i can be sufficiently variable so that a given $H(O_i)$ can be attained. If one considers the cost as the metabolic energy used for signaling, then modeling this cost as $\sum_i \langle O_i^2 \rangle$ gives a relationship between information rate (for Gaussian signals) and cost that is similar to the relationship between information rate and ATP usage physiologically (Laughlin 2001), as will be elaborated later in this chapter.

3.2.2.4 Terms related or equivalent to efficient code and efficient coding

As mentioned in Section 3.2.2.1, in the high S/N limit, the efficient code is also referred to as the maximum entropy code (to maximize $H(O)$) and factorial or independent component code (to make $P(O) = \Pi_i P(O_i)$). The independence between the responses from different units in an efficient code has been suggested as enjoying additional cognitive advantages such as revealing the different perceptual entities that are responsible for the sensory data S—perhaps even (albeit not straightforwardly) visual objects. The efficient coding principle is sometimes also called *infomax* (i.e., maximizing $I(\mathbf{O}; \mathbf{S})$), *sparse coding* (i.e., minimizing $\sum_i H(O_i)$ or $\sum_i \langle O_i^2 \rangle$), *independent component analysis*, and (in low noise cases) *redundancy reduction* (Nadal and Parga 1993).

We now apply this principle to understand input sampling by the retinal cells and transformations by the receptive fields (RFs) of retinal and V1 cells. For better illustration, most examples below are simplified to focus only on the relevant sensory dimension(s). For example, when focusing on input contrast levels to the blowfly's eye, dimensions of space and time are ignored.

3.3 Efficient neural sampling in the retina

3.3.1 Contrast sampling in a fly's compound eye

In a fly's compound eye, the signal $\mathbf{S}$ is a scalar value S; it denotes the input contrast (the ratio between the input intensity increment at a location and the mean input intensity of the scene) to the photoreceptor. The encoding transform $K(S)$ is the contrast response function of the secondary neuron, which receives inputs from the photoreceptor and is called the large monopolar cell (LMC). Let us consider the condition when the input noise $N \to 0$; then

$$O = K(S) + N_o, \tag{3.53}$$

with a (scalar) intrinsic noise N_o in the LMC. Let us also consider neural cost to be of negligible importance. This is the situation when $\lambda \to \infty$ in equation (3.40). Hence, the goal is to find a function K that maximizes $I(O; S)$, the information extracted in O about S. The encoding $K(S)$ should be a monotonic function, e.g., to map a larger contrast input S to a larger response; otherwise, $K(S)$ is not uniquely invertible, and hence some information about S would be lost.

We saw in Section 3.2 (equation (3.48)) that, when the input S/N is large enough, $I(O; S)$ is maximized when the output entropy $H(O)$ is maximized. Meanwhile, $H(O)$ is maximized when the probability $P(O)$ = constant is independent of O within the range of allowable response levels $O \in [0, O_{\max}]$, where 0 and $O_{\max}$ are, respectively, the minimum and maximum responses possible. Since $O = K(S) + N_o$, a flat probability distribution $P(O)$ requires a flat probability distribution $P_K[K(S)]$.

The probability density $P_K[K(S)]$ depends on $P(S)$, the probability density of input S. A small interval $(S, S + dS)$ in the input value S gives the corresponding small interval $(K(S), K(S + dS))$ in the value of $K(S)$. Let

$$dK(S) \equiv K(S + dS) - K(S); \quad \text{then} \quad P(S)dS = P_K[K(S)]dK(S).$$

The equation above means that the probability for $S \in (S, S + dS)$ is the same as the probability for $K(S) \in (K(S), K(S) + dK(S))$. Consequently,

$$P_K[K(S)] = P(S) \left(\frac{dK(S)}{dS} \right)^{-1}. \tag{3.54}$$

Therefore, if the density $P_K[K(S)]$ is constant, then

$$\frac{dK(S)}{dS} \propto P(S), \quad \text{or} \quad dK(S) \propto P(S)dS; \quad \text{or,} \tag{3.55}$$

$$K(S) \propto \int_0^S dS' P(S'), \quad \text{which is the cumulative distribution of } S. \tag{3.56}$$

The $K(S)$ in the above equation is indeed consistent with the contrast response function in the LMC of the flies (Laughlin 1981). This encoding is intuitively illustrated in Fig. 3.4.

Note that $\frac{dK(S)}{dS}$ is also the sensitivity of the LMC response O to input S. Hence, the optimal $K(S)$, which obeys $\frac{dK(S)}{dS} \propto P(S)$, ensures that the responses are more sensitive to the more probable input contrasts. In other words, if one is to recover input contrast S from the responses O, the resolution for discriminating between different inputs S is better in the more probable input regions. As long as the input S/N is high enough, $K(S)$ should always follow the cumulative distribution of S, as expressed in equation (3.56), regardless of the background adaptation or light level—this has been empirically observed (Laughlin, Howard and Blakeslee 1987).

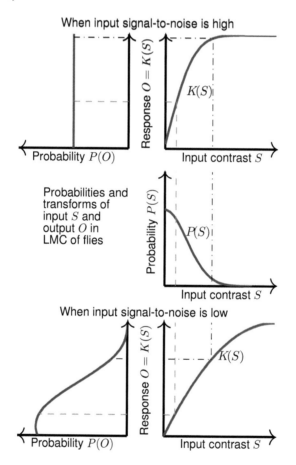

Fig. 3.4: Schematic illustration of input contrast encoding in the fly's LMC. The input contrast S starts at $S = 0$ on its axis, in arbitrary contrast units. When the input S/N is high, the contrast transform $K(S)$ closely follows the cumulative distribution of $P(S)$, such that the slope of $K(S)$ scales with input density $P(S)$, and the output distribution $P(O)$ is uniform. This makes all the LMC response levels equally utilized (qualitatively similar to the plots in Laughlin (1981), except that the negative contrast S, when the input intensity is lower than the adaptation background level, is omitted in the plot). When the input S/N is too low, $K(S)$ no longer follows the cumulative distribution of $P(S)$, and the $P(O)$ peaks near zero. This is qualitatively similar to the observations by Laughlin et al. (1987). The plots omit output noise.

However, when input noise N is substantial, the LMC response

$$O = K(S+N) + N_o \equiv K(S') + N_o, \quad \text{where } S' \equiv S + N, \tag{3.57}$$

should include a contribution from input noise N. Let S and N have standard deviations σ_S and σ_N respectively. Then, the input information rate before the encoding K transform is

$$I(S'; S) = H(S') - H(N) \approx \frac{1}{2} \log_2 \left(1 + \frac{\sigma_S^2}{\sigma_N^2} \right), \tag{3.58}$$

where the approximation, which is valid when the variables are Gaussian distributed, serves to illustrate that the input data rate $I(S'; S)$ increases with the input signal-to-noise σ_S^2/σ_N^2. The output data rate is

$$I(O; K(S')) = H(O) - H(N_o), \tag{3.59}$$

where $H(O)$ and $H(N_o)$ are the entropies of O and N_o, respectively. While $H(N_o)$ could be an intrinsic property of the output neuron, the entropy $H(O)$ of O can be increased by increasing the strength of the input signal S' or by increasing the gain associated with the transform $K(S')$. We note that $I(O; S') = I(O; K(S'))$ holds, since $I(O; S') = H(O) - H(N_o)$ is also valid. Since

$$I(O; S') = [I(O; S') - I(O; S)] + I(O; S), \tag{3.60}$$

the output data rate $I(O; K(S')) = I(O; S')$ is partly spent on $I(O; S') - I(O; S)$, which is the positive amount wasted on transmitting information about the noise N within S'. If K is made to maximize $H(O)$ by making $P(O)$ constant, this output data rate $I(O; K(S'))$ could be much greater than the input information rate $I(S'; S) \geq I(O; S)$, when there is too great a deal of input noise N. Imagine a situation in which the resource $I(O; K(S'))$ is 10 bits/second while the input data rate $I(S'; S)$ is only 1 bit/second; then a resource of 9 bits/second would be wasted. This is not desirable if neural cost is not negligible. To reduce cost, $P(O)$ should favor low response O levels (see Fig. 3.4), thereby reducing $H(O)$. Reducing the gain of $K(S')$ for typical S' levels can serve this purpose, so as not to amplify the input noise too greatly. An output data rate roughly matching the low input data rate (when S/N is low) is likely adequate; accordingly:

$$H(O) - H(N_o) \approx I(S'; S) \quad \text{when input S/N is low.} \tag{3.61}$$

Changes in the form of optimal encoding transforms K according to the input S/N (which increases with background light levels) suggest that the photoreceptors and/or LMC cells should be able to adapt to different input environments. These adaptive changes have indeed been observed experimentally (Laughlin et al. 1987)—adaptation can occur within seconds. As we will see later, this is just one example out of many in which sensory adaptation can be explained or predicted by the efficient coding principle.

3.3.2 Spatial sampling by receptor distribution on the retina

Just as more output response levels should be allocated to input contrast regions which are more likely (in inputs for the LMC of the fly), the human retina should allocate more cones to the fovea, where it is more likely to find images of relevant objects which are brought there by eye movements. Generalizing from equation (3.55), one therefore expects that cone density should match the distribution of the images of relevant objects on the retina, so that the limited resource of the finite number of cones (10^7 for primates) can be best utilized. (Here we are ignoring the underlying chicken-and-egg relationship between eye movements and cone allocation.[4] Thus we find an efficient allocation of cones assuming that eye movement properties are already given.) To derive meaningful results which can be compared with experimental data, we formulate this problem in the following way (Lewis, Garcia and Zhaoping 2003).

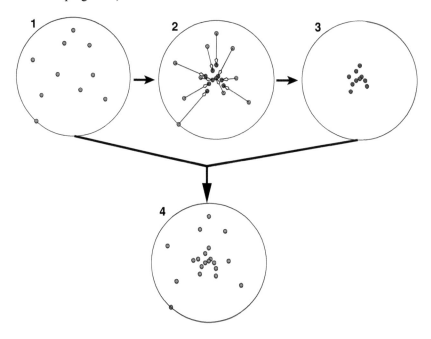

Fig. 3.5: In a scene in which several objects are present, one of them calls the attention of the observer. (1) In the beginning, the object could be anywhere in the visual field. This is represented by an initial uniform distribution $P_0(x)$. (2) Then, the attended object will, with some probability, elicit a saccade which brings it closer to the fovea. (3) This happens to many objects in an observer's visual life, so that the distribution of targets after saccades is concentrated around the center of the eye. (4) Combining the distributions of targets before and after saccades, we obtain the average distribution of relevant objects in visual space. In the full model, we allow for zero, one, or more than one saccade to be made to each target. Reproduced with permission from Lewis, A., Garcia, R., and Zhaoping, L., The distribution of visual objects on the retina: connecting eye movements and cone distributions, *Journal of Vision*, 3 (11), article 21, doi:10.1167/3.11.21, Fig. 2, copyright © 2003, Association for Research in Vision and Ophthalmology (ARVO).

Here, the input **S** is the image location x of a relevant visual object. This location x is sampled by photoreceptors. Hence, the output **O** is simplified to be just the identity or index ξ of the cone most excited by the object. (This simplification treats an object as being small and shapeless, which may limit the quality of the derived results.) The amount of information in the cone index ξ about the object's location x is $I(\mathbf{O}; \mathbf{S})$. By analogy with contrast sampling in blowfly's eye (in the high signal-to-noise limit), let $P(x)$ be the probability density of x; then the photoreceptors should be placed such that their density $D(x)$ scales with $P(x)$.

---

[4] The chicken-and-egg problem is the following. Eye movements bring objects of interest to central vision because there are more cones near the fovea, while more cones are allocated to the fovea because eye movements bring objects of interest to the fovea. A more complete design is to optimize the eye movement dynamics and the cone allocation simultaneously. Here, we assume the eye movement dynamics to be given and only optimize the cone allocation.

Meanwhile, $P(x)$ is shaped not only by the statistical properties of our visual world but also by our active eye movements, which bring objects closer to the center of vision. Accordingly, $P(x)$ is peaked near the fovea, consistent with experimental observations (see Fig. 2.17). This is illustrated schematically in Fig. 3.5.

This intuitive understanding can be confirmed by a detailed analysis, which we show next using the simplified case of a one-dimensional retina with a linear array of receptor cones. Further, we assume that, at any time, there is only one object in the scene, since typically attention is paid to only one object at a time (Cave and Bichot 1999). Hence, this formulation also means that we only consider attended objects in the scene. The object is treated as shapeless, as we only consider its position. Hence, one can either treat the object to have zero size, or consider x as merely the location of its center-of-mass. Let the object excite only one receptor cone, the one which is closest to the image location x of the object. We consider each cone to have an index, so that cone number ξ is at location $x(\xi)$. Given density $D(x)$ of cones at x, the distance between neighboring cones at x is $\delta x = \frac{1}{D(x)}$. So, if cone ξ at $x(\xi)$ is excited by the object, the object is considered to be positioned within the spatial range

$$x \in \left(x(\xi) - \frac{\delta x}{2}, \quad x(\xi) + \frac{\delta x}{2} \right]. \tag{3.62}$$

This means that the probability density of object position x given ξ is

$$P(x|\xi) = \begin{cases} 0, & \text{if } x \notin \left(x(\xi) - \frac{\delta x}{2}, \quad x(\xi) + \frac{\delta x}{2} \right], \\ \frac{1}{\delta x} = D[x(\xi)] & \text{otherwise.} \end{cases} \tag{3.63}$$

Meanwhile, given the prior probability density $P(x)$ of objects in x space, the probability that cone ξ is excited is, in the zero noise limit,

$$P_\xi = P[x(\xi)]\delta x. \tag{3.64}$$

Thus the information in ξ about x is

$$I(\xi; x) = H(x) - H(x|\xi) = H(x) + \sum_\xi P_\xi \int dx P(x|\xi) \log_2 P(x|\xi)$$

$$= H(x) + \sum_\xi P_\xi \int_{x(\xi)-\delta x/2}^{x(\xi)+\delta x/2} dx \frac{1}{\delta x} \log_2 D[x(\xi)]$$

$$= H(x) + \sum_\xi P_\xi \log_2 D[x(\xi)]$$

$$= H(x) + \sum_\xi P[x(\xi)]\delta x \log_2 D[x(\xi)]$$

$$= H(x) + \int dx P(x) \log_2 D(x). \tag{3.65}$$

Here, the neural cost is the total number of cones, which is $\int dx D(x)$. Hence, according to equation (3.40), $D(x)$ should minimize

$$E \equiv \int dx D(x) - \lambda \cdot I(\xi; x) = \int dx D(x) - \lambda \left(H(x) + \int dx P(x) \log_2 D(x) \right).$$

The optimal $D(x)$ is the solution to

$$\frac{\partial E}{\partial D(x)} = 1 - \lambda \frac{P(x)}{\log_2 e} \cdot \frac{1}{D(x)} = 0,$$

which gives our expected solution for an optimal density of cones as

$$D(x) \propto P(x). \tag{3.66}$$

To derive the predicted density $D(x)$, for comparison with experimental data, we need to calculate $P(x)$, the probability density of relevant objects on the retina. This density $P(x)$ is the outcome of a stationary Markov process, described below, involving two sources of statistical information. One is the a priori probability distribution $P_0(x)$ of attended objects on the retina if no active eye movement could affect it; the other is the statistical properties of visual behavior and eye movements to bring $P_0(x)$ to $P(x)$. The latter is described by two components. The first component is

$$K(x', x), \quad \begin{array}{l} \text{probability distribution of an object's retinal location } x' \text{ as a result} \\ \text{of making a saccade to the object's original location, } x. \end{array} \tag{3.67}$$

Hence, x' is the saccadic error. When saccades are error free, $x' = 0$ is defined as the center of fovea. However, due to imprecise eye movements, particularly to very peripheral targets, $K(x', x)$ is not a delta function $\delta(x')$ centered at the fovea but rather has a finite spread centered near the fovea. This distribution $K(x', x)$ depends on the original target location (or eccentricity) x, and it has been experimentally measured (Becker 1991). Sometimes, imprecise saccades are corrected by further saccades that bring objects closer to the fovea.

The second component of the visual behavior that brings $P_0(x)$ to $P(x)$ is:

$$\alpha^{(n)}(x), \quad \begin{array}{l} \text{the probability of saccading to an object currently at } x \\ \text{after } n \text{ saccadic attempts to the same object.} \end{array} \tag{3.68}$$

The $\alpha^{(n)}(x)$ describes the statistical properties of visual behavior. When $n = 0$, $\alpha^{(n)}(x)$ describes the probability of making a saccade to an object for the first time. Let us say that the first saccade brings the object from x to x'; then $\alpha^{(1)}(x')$ is the probability of making another saccade to it at x'. While the quantitative values of $\alpha^{(n)}(x)$ are unknown, they are related to the experimentally measurable probability distribution $f(x)$ of natural saccades made to a target at location x (Bahill, Adler and Stark 1975, Andrews and Coppola 1999). If $P(x, n)$ describes the probability that an object is at location x after n saccadic attempts to it, then

$$P(x, n+1) = \int dy \alpha^{(n)}(y) K(x, y) P(y, n). \tag{3.69}$$

Note that $\int dx P(x, n+1) \leq 1$ since $\alpha^{(n)}(y) \leq 1$, although $\int dx K(x, y) = 1$ is normalized. Hence, the overall probability distribution of an object on the retina is

$$P(x) = \sum_{n=0}^{\infty} P(x, n) = P(x, 0) + \sum_{n>0}^{\infty} P(x, n). \tag{3.70}$$

Meanwhile, $1 - \alpha^{(n)}(y)$ is the probability that no saccade is made to an object currently at location y after n saccades have been made to it. In such a case, attention is switched to an object in the background distribution $P_0(x)$. The probability distribution $P(x, 0)$ is made from the outcomes of such attention shifts to new objects; hence,

$$P(x,0) = \left[\int dy \sum_{n=0}^{\infty} \left(1 - \alpha^{(n)}(y)\right) P(y,n)\right] P_0(x)$$

$$= \left[\int dy \left(\sum_{n=0}^{\infty} P(y,n)\right) - \int dy \left(\sum_{n=0}^{\infty} \alpha^{(n)}(y) P(y,n)\right)\right] P_0(x)$$

$$= \left[\int dy P(y) - \int dy \left(\sum_{n=0}^{\infty} \alpha^{(n)}(y) P(y,n)\right)\right] P_0(x)$$

$$= \left[1 - \int dy \left(\sum_{n=0}^{\infty} \alpha^{(n)}(y) P(y,n)\right)\right] P_0(x).$$

Let

$$\omega f(x) \equiv \sum_{n=0}^{\infty} \alpha^{(n)}(x) P(x,n), \quad \text{in which} \quad \int dx f(x) = 1; \tag{3.71}$$

then $f(x)$ is the experimentally measured probability distribution of natural saccades made to target at location x (Bahill et al. 1975, Andrews and Coppola 1999), and ω is a parameter characterizing the probability that the attended object elicits a saccade. Then,

$$P(x,0) = (1 - \omega) P_0(x). \tag{3.72}$$

Also, combining equation (3.71) and equation (3.69), we have

$$\sum_{n>0}^{\infty} P(x,n) = \omega \int dy f(y) K(x,y). \tag{3.73}$$

Combining equations (3.70), (3.72), and (3.73),

$$P(x) = (1 - \omega) P_0(x) + \omega \int dy f(y) K(x,y). \tag{3.74}$$

On a two-dimensional retina, we approximately treat various probabilities as depending only on the retinal eccentricity and not on the azimuth angle. Let x now denote eccentricity on a two-dimensional retina, the above equation becomes (Lewis et al. 2003)

$$P(x) = (1 - \omega) P_0(x) + \frac{\omega}{2\pi \sin(x)} \int f(y) K(x,y) dy. \tag{3.75}$$

In the above equation, both $f(y)$ and $K(x,y)$ are experimental data available in the literature (Bahill et al. 1975, Andrews and Coppola 1999, Becker 1991), and both are peaked at $x = 0$. Consequently, $P(x)$ will also peak at the center of the fovea $x = 0$, since $P_0(x)$ should be more or less a constant over x. By varying ω in equation (3.75) and assuming that $P_0(x)$ is a constant over x, one can fit $P(x)$ to the experimentally observed cone density $D(x)$, achieving a reasonable agreement within the eccentricity range $5-30°$ (Lewis et al. 2003).

3.3.3 Optimal color sampling by the cones

Equation (3.40) has also been applied to understanding color sampling by three cones at a single spatial location. Here, the input is the visual surface color reflectance $\mathbf{S} = S(\lambda)$ as a function of light wavelength λ, and the output $\mathbf{O} = (O_r, O_g, O_b)^T$ models light absorption by the red (r), green (g), and blue (b) cones. For simplicity, one models the wavelength sensitivity of the different cones with the function $f(\lambda - \lambda_i)$ for $i = r, g, b$ with peak sensitivity occurring at optimal wavelength λ_i. Given illumination $I(\lambda)$ from sunlight, the light reflected from objects is then $S(\lambda)I(\lambda)$, and the cone absorption is $O_i = \int d\lambda f(\lambda - \lambda_i)S(\lambda)I(\lambda) + N_i$, where N_i is the photoreceptor noise. Assuming that the form of the function $f(.)$ is fixed for simplicity, one can adjust the values λ_i of the preferred wavelengths for the three cones to obtain an efficient coding of sensory signals $S(\lambda)$. Accordingly, the design of the cones can be seen as an optimization problem in which λ_i is chosen to maximize the information $I(\mathbf{O}; \mathbf{S})$ extracted about signals $S(\lambda)$. This optimization largely explains the sensitivities of cones in humans (Lewis and Zhaoping 2006), considering factors such as chromatic aberration and low signal-to-noise conditions in dim light. Naturally, this optimal solution is influenced by the statistics $P(\mathbf{S})$ of the surface reflectance $\mathbf{S}$.

Given the spectrum sensitivity functions of the three cones, one can also ask: what should be the relative proportions of the red, green, and blue cones in the cone mosaic that tiles the surface of the retina in order to extract the maximum visual information from the input? Now the input $\mathbf{S}$ is not only a function of light wavelength λ but also a function of space x. Its statistics $P(\mathbf{S})$, together with other sampling factors such as chromatic aberration and retinal light absorption, should then determine the optimal cone proportions to maximize sensory information extraction. It turns out that (Garrigan, Ratliff, Klein, Sterling, Brainard and Balasubramanian 2010), in the optimal composition, the blue cones should be relatively rare, while the total amount of information extracted is insensitive to the relative proportion of red and green cones. This optimal composition agrees with the experimental data.

3.4 Efficient coding by visual receptive field transforms

The efficient coding principle has been more extensively applied to understand the RF transforms of photoreceptor responses by retinal ganglion cells (or LGN cells) and V1 neurons. We denote the receptor responses by $\mathbf{S} + \mathbf{N}$, including both signal $\mathbf{S}$ and noise $\mathbf{N}$, and the output of the encoding by $\mathbf{O}$. The problem is simplified by approximating the receptive field transformations as being linear

$$\mathbf{O} = K(\mathbf{S} + \mathbf{N}) + \mathbf{N_o}, \quad \text{or, in component form,}$$
$$O_i = \sum_j K_{ij}(S_j + N_j) + (N_o)_i, \tag{3.76}$$

where $\mathbf{N_o}$ is the neural noise introduced by the transform. Hence, $K\mathbf{N} + \mathbf{N_o}$ is the total noise at the output. We will ignore any additional nonlinear transform of O_i, such as that of input contrast by the LMC cell in the blowfly discussed above. Although this can further improve the coding efficiency (Nadal and Parga 1994), our main concern is the features of the input of which the neurons are tuned, and these are mainly captured by the linear component of the transformation as in equation (3.76).

For simplicity, the variables $\mathbf{S}, \mathbf{N}, \mathbf{N_o}$ are assumed to have zero means

$$\langle S_i \rangle = \langle N_i \rangle = \langle (N_o)_i \rangle = 0.$$

This is equivalent to shifting the origins of coordinate systems for these variables. We remind ourselves that $\langle a \rangle$ means the average of a over the ensemble of patterns or signals.

In general, S_i denotes a visual input at a particular spatial location x, time t, eye of origin e, and cone c. As we saw in Section 2.3.2, the RF transform for a V1 cell, for instance, can reflect selectivities to all these input dimensions, so that a cell can be tuned to orientation (involving only x), motion direction (involving x, t), spatial scale (x), eye of origin (e), color (c), disparity (x, e), or combinations thereof. For simplicity, when we focus on only one particular dimension, e.g., space x, we may omit other dimensions when constructing $\mathbf{S}$ and the corresponding (e.g., spatial) receptive field, K.

To derive the optimal K according to the efficient coding principle, we need to know the probability $P(\mathbf{S})$, i.e., the joint probability distribution over all its components $(S_1, S_2, ..., S_n)$. Unfortunately, we do not know $P(\mathbf{S})$ precisely for large n, e.g., when $(S_1, ..., S_n)$ refers to the image pixel values in n image pixels. However, an $n \times n$ matrix R^S of second order correlation can be obtained for natural scene inputs; this has the matrix elements

$$\mathsf{R}^S_{ij} \equiv \langle S_i S_j \rangle \equiv \int dS_i dS_j S_i S_j P(S_i, S_j).$$

It can be used to specify a Gaussian distribution

$$P(\mathbf{S}) \propto \exp\left[-\frac{1}{2} \sum_{ij} S_i S_j (\mathsf{R}^S)^{-1}_{ij} \right], \tag{3.77}$$

where $(\mathsf{R}^S)^{-1}$ is the inverse matrix of R^S. This distribution has the same correlation $\langle S_i S_j \rangle = \mathsf{R}^S_{ij}$ as in the natural scenes. Furthermore, this Gaussian distribution has the largest entropy $H(\mathbf{S})$ among all probability distributions over the same variables which have the same correlation matrix R^S (this is proven later in Section 4.1.2). We can therefore use this Gaussian distribution as an approximation for the true $P(\mathbf{S})$. This approximation has the advantage of leading to analytical solutions for the optimal K (Linsker 1990, Atick and Redlich 1990, Atick, Li and Redlich 1992, Li and Atick 1994b, Li and Atick 1994a, Dong and Atick 1995, Li 1996), and of capturing our ignorance about the higher order statistics in the actual $P(\mathbf{S})$. In Chapter 4, we will show that this Gaussian approximation is reasonable and also discuss issues related to higher order statistics in natural visual inputs.

3.4.1 The general analytical solution for efficient coding of Gaussian signals

This section presents a summary of analytical results about the optimal encoding K when

$$\text{the neural cost} = \sum_i \langle O_i^2 \rangle \quad \text{and hence} \quad E(\mathsf{K}) = \sum_i \langle O_i^2 \rangle - \lambda I(\mathbf{O}; \mathbf{S}), \tag{3.78}$$

and when all signals and noises are assumed to be Gaussian. As these results will be illustrated step by step later in a more intuitive manner, it is not essential to digest the mathematical derivations here completely in order to follow the rest of this chapter.

For simplicity, it is assumed that the input noise N_i in different input channels is independently and identically distributed, and we denote $\langle N^2 \rangle \equiv \langle N_i^2 \rangle$ for all i. The same assumptions are made for $\mathbf{N_o}$, with $\langle N_o{}^2 \rangle \equiv \langle (N_o)_i^2 \rangle$ for all i.

The output $\mathbf{O} = \mathsf{K}(\mathbf{S} + \mathbf{N}) + \mathbf{N_o}$ contains

a transformed signal KS, and

the total output noise $\mathbf{N}_{total} \equiv \mathsf{K}\mathbf{N} + \mathbf{N_o}$.

Here, $\mathbf{N}_{total}$ comprises the intrinsic output noise $\mathbf{N_o}$ and the input noise $\mathbf{N}$ transformed by K. The output has correlation matrix R^O with elements

$$\begin{aligned}
\mathsf{R}^O_{ij} &= \langle O_i O_j \rangle = \langle ((\mathsf{K}(\mathbf{S}+\mathbf{N}) + \mathbf{N_o})_i (\mathsf{K}(\mathbf{S}+\mathbf{N}) + \mathbf{N_o})_j \rangle \\
&= \sum_{a,b} \mathsf{K}_{ia} \mathsf{K}_{jb} \langle S_a S_b \rangle + \sum_{a,b} \mathsf{K}_{ia} \mathsf{K}_{jb} \langle N_a N_b \rangle \\
&\quad + 2 \sum_{a,b} \mathsf{K}_{ia} \mathsf{K}_{jb} \langle S_a N_b \rangle + \langle (N_o)_i (N_o)_j \rangle \\
&= (\mathsf{K}\mathsf{R}^S \mathsf{K}^T)_{ij} + \langle N^2 \rangle (\mathsf{K}\mathsf{K}^T)_{ij} + \langle N_o{}^2 \rangle \delta_{ij}.
\end{aligned}$$

In the above, we assumed that $\mathbf{S}$, $\mathbf{N}$, and $\mathbf{N_o}$ are independent of each other. Thus,

$$R^O = KR^S K^T + \langle N^2 \rangle KK^T + \langle N_o{}^2 \rangle \mathbb{1} \equiv KR^S K^T + R^{N_{total}} \tag{3.79}$$

where $\mathbb{1}$ is an identity matrix such that $\mathbb{1}_{ij} = \delta_{ij}$, and $R^{N_{total}}$ is the correlation matrix of the total output noise $\mathbf{N}_{total}$

$$R^{N_{total}} \equiv \langle N^2 \rangle KK^T + \langle N_o{}^2 \rangle \mathbb{1} = \langle N^2 \rangle KK^T + \langle N_o{}^2 \rangle. \tag{3.80}$$

Sometimes, we simply write $\mathbb{1}$ as 1, or, as in the second equality above, omit it altogether. Denoting Tr(.) and det(.) as respectively the trace and determinant functions, we have

$$\text{neural cost} = \sum_i \langle O_i^2 \rangle = \text{Tr}(R^O) = \text{Tr}\left(KR^S K^T + R^{N_{total}}\right), \quad \text{and} \tag{3.81}$$

$$I(\mathbf{O}; \mathbf{S}) = \frac{1}{2} \log_2 \frac{\det(R^O)}{\det(R^{N_{total}})}. \tag{3.82}$$

The expression for $I(\mathbf{O}; \mathbf{S})$ holds for Gaussian variables $\mathbf{S}$ and $\mathbf{O}$. It is a generalization of $I(O; S)$ in equation (3.26), in which O and S were both scalars. In that case, $\det R^O = \sigma_o^2$ and $\det R^{N_{total}} = \sigma_n^2$. In particular, when R^o and $R^{N_{total}}$ are both diagonal matrices (we will show later how to diagonalize R^o),

$$I(\mathbf{O}; \mathbf{S}) = \frac{1}{2} \log_2 \frac{\det R^O}{\det R^{N_{total}}} = \frac{1}{2} \log_2 \left[\prod_i \frac{R_{ii}^O}{R_{ii}^{N_{total}}} \right] \tag{3.83}$$

$$= \frac{1}{2} \sum_i \log_2 \frac{\langle O_i^2 \rangle}{\langle (N_{total})_i^2 \rangle}. \tag{3.84}$$

It is known that for any matrix M, $\text{Tr}(M)$ and $\det(M)$ do not change when M is transformed to UMU^T by any orthonormal matrix U (here U^T denotes the transpose of U such that $(U^T)_{ij} = U_{ji}$). For an orthonormal matrix U, the equation $UU^T = \mathbb{1}$ holds. More generally, U could also be an unitary matrix (with complex elements), in which case, M is transformed to $UMU^\dagger$, leaving $\text{Tr}(M)$ and $\det(M)$ unchanged. Here, $U^\dagger = (U^T)^*$ is the conjugate transpose of U, i.e., $(U^\dagger)_{ij} = U_{ji}^*$, and $UU^\dagger = 1$ holds for unitary matrices.

Although R^O and $R^{N_{total}}$ depend on K through equations (3.79–3.80), results from linear algebra implies that $\text{Tr}(R^O)$, $\det(R^O)$, and $\det(R^{N_{total}})$ are all invariant to a change of the encoding matrix $K \rightarrow UK$ by any unitary matrix U (when the components of noise $\mathbf{N_o}$ are independent and identically distributed). In other words, the optimal encoding solutions K that minimize

$$E(K) = \sum_i \langle O_i^2 \rangle - \lambda I(\mathbf{O}; \mathbf{S}) = \text{Tr}(R^O) - \frac{\lambda}{2} \log_2 \frac{\det(R^O)}{\det(R^{N_{total}})} \tag{3.85}$$

are degenerate in that if K is a solution that minimizes $E(K)$, then $K' = UK$ is also a solution for any unitary U. In particular, consider a solution K; this leads to a transformed correlation matrix $KR^S K^\dagger$. It is known that a unitary matrix U will exist that diagonalizes $KR^S K^\dagger$, i.e., such that $U(KR^S K^\dagger)U^\dagger$ is diagonal. Then, since $U(KR^S K^\dagger)U^\dagger = (UK)R^S(UK)^\dagger$, this means $K' = UK$ diagonalizes R^S. Hence, there is a special solution K among all the solutions within this degenerate class such that $KR^S K^T$ is diagonal. The solution for this special K should be of the form

$\mathsf{K} = g\mathsf{K}_o$, such that

K_o is the unitary matrix to make $\mathsf{K}_o R^S \mathsf{K}_o^T$ diagonal, and $\qquad$ (3.86)

g, to be derived, is a diagonal matrix to keep $\mathsf{K} R^S \mathsf{K}^\dagger$ diagonal. $\qquad$ (3.87)

Now, the remaining problem is to choose the strength of the diagonal elements in g. Once this special K is found, other solutions $\mathsf{K}' = U\mathsf{K}$ can be generated by any U desired.

Let R^S have eigenvectors $V^{(1)}, V^{(2)}, ..., V^{(k)}, ...$; each normalized to unit length. Further, let the projection of $\mathbf{S}$ onto $V^{(k)}$ be $\mathcal{S}_k \equiv \sum_i S_i V_i^{(k)}$, and similarly, let $\mathcal{N}_k$ and $(\mathcal{N}_o)_k$ be the projections of $\mathbf{N}$ and $\mathbf{N}_o$ onto $V^{(k)}$, respectively.

Then the power $\langle \mathcal{S}_k^2 \rangle$ of $\mathcal{S}_k$ is the eigenvalue of R^S for the eigenvector $V^{(k)}$, since

$$\left\langle \left(\sum_i S_i V_i^{(k)} \right)^2 \right\rangle = \left\langle \sum_{ij} V_i^{(k)} S_i S_j V_j^{(k)} \right\rangle = (V^{(k)})^T R^S V^{(k)}.$$

Let the diagonal elements g_{kk} of matrix g be denoted by $g_k \equiv g_{kk}$ for all k. Then, the output $\mathbf{O} = \mathsf{K}(\mathbf{S} + \mathbf{N}) + \mathbf{N}_o$ using this special $\mathsf{K} = g\mathsf{K}_o$ should have components

$$\mathcal{O}_k = g_k(\mathcal{S}_k + \mathcal{N}_k) + (\mathcal{N}_o)_k, \quad \text{for} \quad k = 1, 2, ... \qquad (3.88)$$

These components $\mathcal{O}_k$ have variances

$$\langle \mathcal{O}_k^2 \rangle = g_k^2 \left(\langle \mathcal{S}_k^2 \rangle + \langle N^2 \rangle \right) + \langle N_o^2 \rangle. \qquad (3.89)$$

Different output channels $\mathcal{O}_k$ and $\mathcal{O}_{k'}$ are decorrelated, making R^O a diagonal matrix with elements

$$R^O_{kk'} = \langle \mathcal{O}_k \mathcal{O}_{k'} \rangle = [\mathsf{K}(R^S + \langle N^2 \rangle)\mathsf{K}^T + \langle N_o^2 \rangle]_{kk'}$$
$$= [g_k^2 (\langle \mathcal{S}_k^2 \rangle + \langle N^2 \rangle) + \langle N_o^2 \rangle] \delta_{kk'} = \langle \mathcal{O}_k^2 \rangle \delta_{kk'}.$$

Similarly, the correlation matrix $R^{N_{total}}$ of the total output noise $\mathbf{N}_{total}$ is also diagonal, with elements

$$(R^{N_{total}})_{kk'} = \left(g_k^2 \langle N^2 \rangle + \langle N_o^2 \rangle \right) \delta_{kk'}. \qquad (3.90)$$

Consequently,

$$E(\mathsf{K}) = \sum_k E_k(g_k), \quad \text{where} \qquad (3.91)$$

$$E_k(g_k) \equiv \langle \mathcal{O}_k^2 \rangle - \lambda I(\mathcal{O}_k; \mathcal{S}_k), \qquad (3.92)$$

$$I(\mathcal{O}_k; \mathcal{S}_k) = \frac{1}{2} \log_2 \frac{g_k^2 (\langle \mathcal{S}_k^2 \rangle + \langle N^2 \rangle) + \langle N_o^2 \rangle}{g_k^2 \langle N^2 \rangle + \langle N_o^2 \rangle}. \qquad (3.93)$$

Minimizing $E(\mathsf{K})$ requires finding the optimal gain g_k that minimizes each individual $E_k(g_k)$. This is the solution to $\partial E_k(g_k)/\partial g_k = 0$, which turns out to be a quadratic equation, leaving

$$g_k^2 \propto \max \left\{ \left[\frac{1}{1 + \frac{\langle N^2 \rangle}{\langle \mathcal{S}_k^2 \rangle}} \left(\frac{1}{2} + \frac{1}{2} \sqrt{1 + \frac{2\lambda}{(\ln 2)\langle N_o^2 \rangle} \frac{\langle N^2 \rangle}{\langle \mathcal{S}_k^2 \rangle}} \right) - 1 \right], 0 \right\}, \qquad (3.94)$$

in which $\max(x, y)$ equals the larger value between x and y. This g_k^2 value depends only on the S/N $\langle \mathcal{S}_k^2 \rangle / \langle N^2 \rangle$ once $\langle N_o^2 \rangle$ is given. In particular,

$$g_k^2 \propto \max\left(F_{\text{smoothing}} \cdot F_{\text{decorrelation}} - 1, 0\right), \quad \text{where}$$

$$F_{\text{smoothing}} \equiv \left(1 + \frac{\langle N^2 \rangle}{\langle S_k^2 \rangle}\right)^{-1}, \quad \text{and} \tag{3.95}$$

$$F_{\text{decorrelation}} \equiv \frac{1}{2} + \frac{1}{2}\sqrt{1 + \frac{2\lambda}{(\ln 2)\langle N_o^2 \rangle} \frac{\langle N^2 \rangle}{\langle S_k^2 \rangle}}.$$

The factor $F_{\text{smoothing}}$ increases with $\langle S_k^2 \rangle / \langle N^2 \rangle$ and saturates at $F_{\text{smoothing}} = 1$. In contrast, $F_{\text{decorrelation}}$ decreases with increasing $\langle S_k^2 \rangle / \langle N^2 \rangle$ toward a value $F_{\text{decorrelation}} = 1$. We will see later that $F_{\text{smoothing}}$ helps to smooth out input noise, and $F_{\text{decorrelation}}$ helps to decorrelate **S** by granting smaller g_k^2 to components S_k that have larger $\langle S_k^2 \rangle$ when $\langle S_k^2 \rangle / \langle N^2 \rangle \gg 1$.

In the high S/N regime, when $\langle S_k^2 \rangle / \langle N^2 \rangle \gg 1$, $F_{\text{smoothing}} \approx 1$ is almost independent of $\langle S_k^2 \rangle / \langle N^2 \rangle$, making g_k^2 depend on $\langle S_k^2 \rangle / \langle N^2 \rangle$ only through

$$F_{\text{decorrelation}} \approx 1 + \frac{\lambda}{(2 \ln 2)\langle N_o^2 \rangle} \frac{\langle N^2 \rangle}{\langle S_k^2 \rangle}, \quad \text{when } \langle S_k^2 \rangle / \langle N^2 \rangle \gg 1. \tag{3.96}$$

The approximation for $F_{\text{decorrelation}}$ above used a Taylor expansion $\sqrt{1+x} \approx 1 + x/2$ for small x. Hence, combining with $F_{\text{smoothing}} \approx 1$ for high S/N, equation (3.95) gives

$$g_k^2 \propto \langle S_k^2 \rangle^{-1} \quad \text{approximately,} \quad \text{when } \langle S_k^2 \rangle / \langle N^2 \rangle \gg 1. \tag{3.97}$$

This makes g_k^2 decrease with increasing $\langle S_k^2 \rangle$ at high S/N regions.

In contrast, in the low S/N regime, when $\langle S_k^2 \rangle / \langle N^2 \rangle \ll 1$,

$$F_{\text{smoothing}} \approx \langle S_k^2 \rangle / \langle N^2 \rangle, \quad F_{\text{decorrelation}} \approx \frac{1}{2}\sqrt{\frac{2\lambda}{(\ln 2)\langle N_o^2 \rangle} \frac{\langle N^2 \rangle}{\langle S_k^2 \rangle}},$$

$$\rightarrow g_k^2 \propto \max\left\{\alpha \langle S_k^2 \rangle^{1/2} - 1, 0\right\},$$

$$\text{where } \alpha = \left(\frac{\lambda}{2(\ln 2)\langle N^2 \rangle \langle N_o^2 \rangle}\right)^{1/2}. \tag{3.98}$$

This makes g_k^2 increase with increasing $\langle S_k^2 \rangle$. Summarizing the results above,

$$g_k^2 \propto \begin{cases} \langle S_k^2 \rangle^{-1}, & \text{if } \frac{\langle S_k^2 \rangle}{\langle N^2 \rangle} \gg 1, \\ \max\left\{\alpha \langle S_k^2 \rangle^{1/2} - 1, 0\right\}, & \text{if } \frac{\langle S_k^2 \rangle}{\langle N^2 \rangle} \ll 1, \text{ where } \alpha = \left(\frac{\lambda}{2(\ln 2)\langle N^2 \rangle \langle N_o^2 \rangle}\right)^{1/2}. \end{cases} \tag{3.99}$$

Hence, when the input S/N is high, g_k decreases with increasing signal power; conversely, when input S/N is low, g_k decreases with decreasing signal power. This makes g_k^2 a unimodal function of $\langle S_k^2 \rangle / \langle N^2 \rangle$, peaking around $\langle S_k^2 \rangle / \langle N^2 \rangle = 1$, as shown in Fig. 3.6.

Hence, when variables are Gaussian and when the neural cost is $\sum_i \langle O_i^2 \rangle$, the general optimal encoding transform $K = U g K_o$ can be decomposed into the following three conceptual components. First is the principal component decomposition of inputs by the unitary matrix K_o that diagonalizes R^S, thereby transforming the raw input $(S_1 + N_1, S_2 + N_2, ...)$ into their projections $(S_1 + \mathcal{N}_1, S_2 + \mathcal{N}_2, ..., S_k + \mathcal{N}_k, ...)$ onto the principal components of the signal **S**. Second is the gain control g_k for each component $S_k + \mathcal{N}_k$ according to its S/N. The third involves multiplexing the resulting components $\{\mathcal{O}_k\}$ by another unitary matrix U, giving new outputs $O_i = \sum_k U_{ik} \mathcal{O}_k$ as a linear combination of the original $\mathcal{O}_k$'s. These three components for constructing K are illustrated in Fig. 3.7.

In our formulation, the U component arises from the (mathematical) fact that $E(K)$ has many equally optimal encoding solutions K which are related to each other by unitary transforms. We will later discuss additional desiderata for the encoding that favor some optimal K over others.

Gain-squared g_k^2 varies with $\langle S_k^2\rangle/\langle N^2\rangle$ by the two factors $F_{\text{smoothing}}$ and $F_{\text{decorrelation}}$

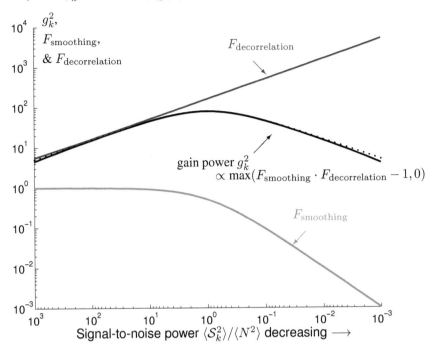

Fig. 3.6: Illustration of how the sensitivity-squared g_k^2 depends on the S/N $\langle S_k^2\rangle/\langle N^2\rangle$. The black solid curve shows a case for which $\frac{2\lambda}{(\ln 2)\langle N_o^2\rangle} = 10^5$, see equations (3.94–3.99). The horizontal axis has decreasing $\langle S_k^2\rangle/\langle N^2\rangle$, to match the decreasing signal power in the eigenvalue spectrum of the input correlation matrices R^S we will see later. The red curve shows $F_{\text{smoothing}} = (1 + \langle N^2\rangle/\langle S_k^2\rangle)^{-1}$; the blue curve shows $F_{\text{decorrelation}} = \frac{1}{2} + \frac{1}{2}\sqrt{1 + \frac{2\lambda}{(\ln 2)\langle N_o^2\rangle}\frac{\langle N^2\rangle}{\langle S_k^2\rangle}}$, $g_k^2 \propto \max(F_{\text{smoothing}} \cdot F_{\text{decorrelation}} - 1, 0)$. Note that when $\langle S_k^2\rangle/\langle N^2\rangle \gg 1$, $g_k^2 \propto (\langle S_k^2\rangle)^{-1}$, and when $\langle S_k^2\rangle/\langle N^2\rangle \ll 1$, g_k^2 decreases with decreasing $\langle S_k^2\rangle$.

The above formulation can be applied to signals in different domains, space, time, color, and stereo. Different domains have different dimensionalities in the input signals $\mathbf{S}$, different statistics $P(\mathbf{S})$, and different S/Ns. In the rest of the chapter, we will illustrate that these differences, and the interactions between the encodings in different domains, will lead to a diversity of transforms K which resemble the receptive field transforms evident in the population of neurons in the retina or V1.

In the brain, the coding transform K is not implemented by the three separate steps K_o, $\mathbf{g}$, and U. These just represent our mathematical understanding as to how K arises from the analytical structure of $E(\mathsf{K})$. The brain simply has an optimal K. The neural correlates of this K are the effective receptive fields of the neurons, and could be implemented via a whole cascade of transformations $\mathsf{K} = \mathsf{K}_m...\mathsf{K}_3\mathsf{K}_2\mathsf{K}_1$. Here K_1 does not correspond to our K_o, nor K_2 to our $\mathbf{g}$, etc., so one is unlikely to find the neural correlates of K_o, $\mathbf{g}$, and U except in exceptional circumstances. For example, if the overall K is the effective transform from the signals in the photoreceptors to those in the retinal ganglion cells, the intermediate layers or stages between the receptors and the ganglion cells include the layers of the bipolar cells, the horizontal cells, and the amacrine cells. For V1 simple cells, additional intermediate transforms go

Efficient coding K of Gaussian signals decomposed into three components: K_o, g, U.

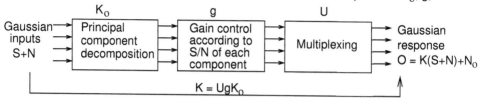

Fig. 3.7: Three conceptual components, K_o, g, and U, in the efficient coding $K = UgK_o$ of Gaussian signals. Adapted with permission from Zhaoping, L., Theoretical understanding of the early visual processes by data compression and data selection, *Network: Computation in Neural Systems*, 17 (4): 301–34, Fig. 3, copyright ©2006, Informa Healthcare.

through the retinal ganglion cells and the LGN cells. These intermediate processes work together to make the net transform K. They may be dictated by the hardware constraints of the neural mechanisms, as well as by the need to adapt or modify the net transform K according to changes in the input statistics $P(\mathbf{S})$. Furthermore, the actual K_i can be nonlinear; it is necessary to keep in mind that our K focuses only on the linear part of the overall K.

3.5 Case study: stereo coding in V1 as an efficient transform of inputs in the dimension of ocularity

For illustration (Fig. 3.8), we first focus on the input dimension of ocularity, or eye of origin, $e = L$ and $e = R$ for left and right eyes, with two-dimensional (2D) input signal $\mathbf{S}$ (Li and Atick 1994a). Recall from Section 2.3.6 that V1 neurons can be differentially sensitive to inputs from the left and right eyes, i.e., to the feature of ocularity. For convenience, we ignore their tuning to spatial location, which is equivalent to assuming that the input comes from only one location, or a particular weighted sum of locations, in each eye. Omitting the spatial dimension prevents us from studying disparity tuning, something we cover later.

In the ocular dimension, the receptive field of a neuron is merely described by its sensitivities to the two eyes to characterize whether the neuron prefers inputs from the left or right eye or is evenly driven by both eyes. We can ask whether these ocular tuning properties of the V1 neurons can be understood from the efficient coding principle.

We have

$$\text{input signal } \mathbf{S} = \begin{pmatrix} S_L \\ S_R \end{pmatrix}, \quad \text{and input noise } \mathbf{N} = \begin{pmatrix} N_L \\ N_R \end{pmatrix}. \quad (3.100)$$

The signal $\mathbf{S}$ only has two dimensions since its spatial, temporal, and color aspects are omitted. The signal S_e (for $e = L$ or $e = R$) may represent, e.g., the pixel value at a particular image location x, the average luminance in the image or a local image patch, or a Fourier component (at a particular spatiotemporal frequency; see Box 2.1) of the image. For simplicity, assume that they have zero mean and equal variance (or power) $\langle S_L^2 \rangle = \langle S_R^2 \rangle$. Analogous properties hold for input noise N_L and N_R.

Let us say that the inputs are recoded by the responses O_1 and O_2 of two V1 neurons. Applying equation (3.76), we get a coding transform K which is a 2×2 matrix

$$K = \begin{pmatrix} K_{1L} & K_{1R} \\ K_{2L} & K_{2R} \end{pmatrix}, \quad (3.101)$$

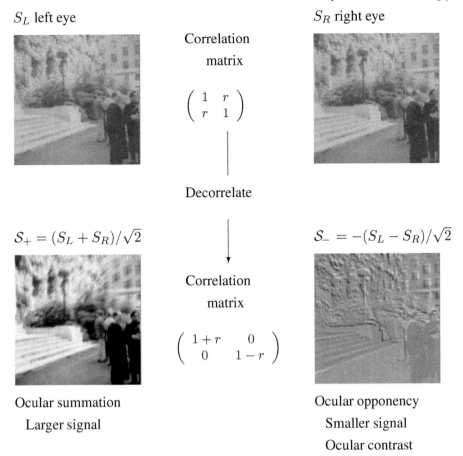

S_L left eye

S_R right eye

Correlation

matrix

$$\begin{pmatrix} 1 & r \\ r & 1 \end{pmatrix}$$

Decorrelate

$\mathcal{S}_+ = (S_L + S_R)/\sqrt{2}$

$\mathcal{S}_- = -(S_L - S_R)/\sqrt{2}$

Correlation

matrix

$$\begin{pmatrix} 1+r & 0 \\ 0 & 1-r \end{pmatrix}$$

Ocular summation

Larger signal

Ocular opponency

Smaller signal

Ocular contrast

Fig. 3.8: First step K_o of the efficient coding illustrated by stereo coding. Correlated inputs (S_L, S_R) from the two eyes are transformed to two decorrelated (by second order) signals $\mathcal{S}_\pm \propto S_R \pm S_L$. These two signals are ocular summation and opponency, respectively; they have different powers $\langle \mathcal{S}_+^2 \rangle > \langle \mathcal{S}_-^2 \rangle$. Adapted with permission from Zhaoping, L., Theoretical understanding of the early visual processes by data compression and data selection, *Network: Computation in Neural Systems*, 17(4): 301–34, Fig. 4, copyright © 2006, Informa Healthcare.

and the output responses as

$$\begin{pmatrix} O_1 \\ O_2 \end{pmatrix} = \begin{pmatrix} \mathsf{K}_{1L} & \mathsf{K}_{1R} \\ \mathsf{K}_{2L} & \mathsf{K}_{2R} \end{pmatrix} \begin{pmatrix} S_L + N_L \\ S_R + N_R \end{pmatrix} + \begin{pmatrix} (N_o)_1 \\ (N_o)_2 \end{pmatrix}, \tag{3.102}$$

in which $(N_o)_1$ and $(N_o)_2$ are the additional noise introduced during the encoding process. The coding transform K is linear, approximating the effective transform associated with the two receptive fields of the two neurons. Hence $(\mathsf{K}_{iL}, \mathsf{K}_{iR})$ is the receptive field of the ith neuron, implying that this neuron has sensitivities K_{iL} and K_{iR} for the inputs from the left and right eyes, respectively, such that its response is

$$O_i = \mathsf{K}_{iL}(S_L + N_L) + \mathsf{K}_{iR}(S_R + N_R) + (N_o)_i. \tag{3.103}$$

Let $\langle S_*^2 \rangle \equiv \langle S_L^2 \rangle = \langle S_R^2 \rangle$. Binocular input redundancy is represented by the correlation matrix

$$R^S \equiv \begin{pmatrix} \langle S_L^2 \rangle & \langle S_L S_R \rangle \\ \langle S_R S_L \rangle & \langle S_R^2 \rangle \end{pmatrix} \equiv \langle S_*^2 \rangle \begin{pmatrix} 1 & r \\ r & 1 \end{pmatrix},$$

$$r \equiv \frac{\langle S_L S_R \rangle}{(\langle S_L^2 \rangle \langle S_R^2 \rangle)^{1/2}} \text{ is the correlation coefficient, and } 0 \le r \le 1. \tag{3.104}$$

The correlation is manifest in the shape of the probability distribution of $\mathbf{S} = (S_L, S_R)^T$ in the input space shown in Fig. 3.9. In this distribution, each sample point (S_L, S_R) tends to have similar S_L and S_R values, since we tend to look at attended objects with both eyes. Hence, the distribution is shaped like an ellipse whose major and minor axes are not along the coordinate directions. From equation (3.77), the Gaussian approximation to the input distribution $P(\mathbf{S}) = P(S_L, S_R)$ is then

$$P(S_L, S_R) \propto \exp\left(-\frac{1}{2}S_L^2 (R^S)_{LL}^{-1} - \frac{1}{2}S_R^2 (R^S)_{RR}^{-1} - \frac{1}{2}S_L S_R \left[(R^S)_{LR}^{-1} + (R^S)_{RL}^{-1}\right]\right)$$

$$= \exp\left[-\frac{S_L^2 + S_R^2 - 2rS_L S_R}{2\langle S_*^2 \rangle (1 - r^2)}\right]. \tag{3.105}$$

According to Fig. 3.7, the encoding K that is most efficient according to the objective objective $E(\mathsf{K})$ is

$$\mathsf{K} = \begin{pmatrix} \cos(\theta) & \sin(\theta) \\ -\sin(\theta) & \cos(\theta) \end{pmatrix} \begin{pmatrix} g_+ & 0 \\ 0 & g_- \end{pmatrix} \begin{pmatrix} \cos(45^\circ) & \sin(45^\circ) \\ -\sin(45^\circ) & \cos(45^\circ) \end{pmatrix}. \tag{3.106}$$

That is, K can be decomposed into three conceptual steps—a 45° rotation, a gain control matrix g, and another rotation by angle θ—each corresponding to a matrix in equation (3.106), for didactic convenience. In the above, θ is an arbitrary parameter, and g_+ and g_- are calculated from equation (3.94) using the eigenvalues of R^S as $\langle S_+^2 \rangle$ and $\langle S_-^2 \rangle$, respectively. In a 2D space (as for $\mathbf{S}$ here), a generic unitary matrix has the form of the first matrix factor in K; it rotates the two axes of the space through angle θ.

In the following sections, we will explain the optimal coding K in the above equation step by step.

3.5.1 Principal component analysis K_o

An n-dimensional vector V is an eigenvector of an $n \times n$ matrix M if $MV = \lambda V$, where λ is a scalar value called the eigenvalue of this eigenvector (not to be confused with the Lagrange multiplier in our objective function $E(\mathsf{K})$). One can verify that the two eigenvectors, $V^{(+)}$ and $V^{(-)}$, of the correlation matrix R^S are

$$V^{(+)} \equiv \frac{1}{\sqrt{2}} \begin{pmatrix} 1 \\ 1 \end{pmatrix} \quad \text{and} \quad V^{(-)} \equiv \frac{1}{\sqrt{2}} \begin{pmatrix} -1 \\ 1 \end{pmatrix},$$

with eigenvalues $\lambda = \langle S_*^2 \rangle (1 \pm r)$, respectively. Different eigenvectors of a correlation matrix are orthogonal to each other, and we usually normalize them to have a unit length. These unit vectors can then define two new axes to span the signal space. We can describe the signals $\mathbf{S} = (S_L, S_R)^T$ by their projections $\mathcal{S}_+$ and $\mathcal{S}_-$ on the $V^{(+)}$ and $V^{(-)}$ axes, respectively (note the change of font for variables denoting projections to axes represented by the eigenvectors). These projections, $\mathcal{S}_+$ and $\mathcal{S}_-$, are called the principal components of the corresponding signals.

One can verify that the projections from $\mathbf{S} = (S_L, S_R)^T$ to $(\mathcal{S}_+, \mathcal{S}_-)^T$ are obtained by a 45° rotation of the coordinate system defined by the axes S_L and S_R (see equation (3.86)).

$$
K_o = \begin{pmatrix} (V^{(+)})^T \\ (V^{(-)})^T \end{pmatrix} = \begin{pmatrix} \cos(45^o) & \sin(45^o) \\ -\sin(45^o) & \cos(45^o) \end{pmatrix}, \quad \text{and} \tag{3.107}
$$

$$
\begin{pmatrix} S_+ \\ S_- \end{pmatrix} \equiv K_o \begin{pmatrix} S_L \\ S_R \end{pmatrix} = \frac{1}{\sqrt{2}} \begin{pmatrix} S_R + S_L \\ S_R - S_L \end{pmatrix}. \tag{3.108}
$$

Furthermore, the signal powers $\langle S_\pm^2 \rangle$ are the respective eigenvalues

$$
\begin{aligned}
\langle S_\pm^2 \rangle &= \frac{1}{2} \langle (S_R \pm S_L)^2 \rangle = \frac{1}{2} \langle S_R^2 + S_L^2 \pm 2 S_R S_L \rangle \\
&= \frac{\langle S_R^2 \rangle + \langle S_L^2 \rangle}{2} \pm \langle S_R S_L \rangle \\
&= \langle S_*^2 \rangle \pm r \left(\langle S_R^2 \rangle \langle S_L^2 \rangle \right)^{1/2} = \langle S_*^2 \rangle (1 \pm r). \tag{3.109}
\end{aligned}
$$

Here, we have used the fact that mean of a sum of terms is equal to the sum of the individual means, i.e., for any A and B, $\langle A + B \rangle = \langle A \rangle + \langle B \rangle$.

Since $r > 0$, the ocular summation signal S_+ is stronger than S_-. It conveys information about the average input from the two eyes, and thus has no stereo information. In comparison (Fig. 3.8), the weaker signal S_- conveys information about the ocular contrast (an "ocular edge") which underlies the perception of depth disparity. We note that S_+ and S_- are not correlated with each other:

$$
\langle S_+ S_- \rangle \propto \langle (S_R + S_L)(S_R - S_L) \rangle = \langle S_R^2 - S_L^2 \rangle = \langle S_R^2 \rangle - \langle S_L^2 \rangle = 0.
$$

This is a generic property of principal components. Since $S_\pm^2 = (S_L^2 + S_R^2 \pm 2 S_L S_R)/2$ and $\langle S_\pm^2 \rangle = (1 \pm r) \langle S_*^2 \rangle$, we have

$$
\frac{S_+^2}{2 \langle S_+^2 \rangle} + \frac{S_-^2}{2 \langle S_-^2 \rangle} = \frac{S_L^2 + S_R^2 - 2 r S_L S_R}{2 \langle S_*^2 \rangle (1 - r^2)}. \tag{3.110}
$$

Using this equality in equation (3.105), we can factorize $P(\mathbf{S})$ into component probabilities $P(S_+)$ and $P(S_-)$ as

$$
P(\mathbf{S}) \equiv P(S_+)P(S_-), \quad \text{in which} \quad P(S_\pm) \propto \exp \left(-\frac{S_\pm^2}{2 \langle S_\pm^2 \rangle} \right).
$$

This is expected, since S_+ and S_- are uncorrelated Gaussian signals, or are independent of each other.

The transform $(S_L, S_R)^T \to (S_+, S_-)^T \equiv K_o (S_L, S_R)^T$ is merely a 45° rotation of the coordinates by a rotational matrix K_o in the 2D space of the input signal, as indicated in Fig. 3.9. The directions for S_+ and S_- in the input signal space are exactly the major and minor axes of the probability distribution of input signals. As with any coordinate rotation, K_o preserves the total signal power

$$
\begin{aligned}
\sum_{i=+,-} \langle S_i^2 \rangle &= \left\langle \frac{1}{2}(S_R + S_L)^2 \right\rangle + \left\langle \frac{1}{2}(S_R - S_L)^2 \right\rangle \\
&= \frac{1}{2} \langle (S_R + S_L)^2 + (S_R - S_L)^2 \rangle \\
&= \frac{1}{2} \langle 2 S_R^2 + 2 S_L^2 \rangle = \langle S_R^2 \rangle + \langle S_L^2 \rangle = \sum_{i=L,R} \langle S_i^2 \rangle.
\end{aligned}
$$

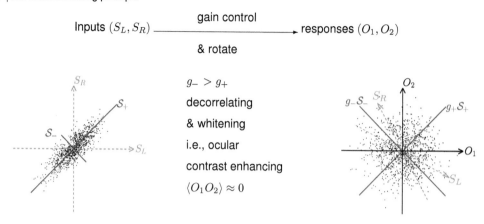

Fig. 3.9: Schematics of data **S** (in the left plot) and their transformation to responses **O** (in the right plot) by efficient coding in the noiseless condition. Each dot in either plot is a sample datum from the distribution $P(\mathbf{S})$ or $P(\mathbf{O})$ in the 2D space of **S** or **O**, respectively. Correlation $\langle S_L S_R \rangle > 0$ is manifested in the elliptical shape of the data distribution in the left plot. Gain control, $\mathcal{S}_\pm \to g_\pm \mathcal{S}_\pm$, produces, under high or low input S/N, decorrelated or correlated responses O_1 and O_2. This figure depicts the condition when S/N$\to \infty$. In this condition, the weaker signal $\mathcal{S}_-$ is relatively amplified for (ocular) contrast enhancement, $g_- > g_+$, leading to whitening or equal power responses $g_+^2 \langle \mathcal{S}_+^2 \rangle \approx g_-^2 \langle \mathcal{S}_-^2 \rangle$. Both O_1 and O_2 are the results of multiplexing $g_+\mathcal{S}_+$ and $g_-\mathcal{S}_-$ (by the U transform). Each O_i is excited by input from one eye (left or right, respectively) and inhibited by input from the other. Adapted with permission from Zhaoping, L., Theoretical understanding of the early visual processes by data compression and data selection, *Network: Computation in Neural Systems*, 17 (4): 301–34, Fig. 4, copyright © 2006, Informa Healthcare.

With sensory noise $\mathbf{N} = (N_L, N_R)^T$, the input signals become $O_{L,R} = S_{L,R} + N_{L,R}$. The rotational transform simply gives

$$\mathcal{O}_\pm = \mathcal{S}_\pm + \mathcal{N}_\pm,$$

where $\mathcal{N}_\pm \equiv (N_R \pm N_L)/\sqrt{2}$. Since S_i and N_i are independent Gaussian random variables for any $i = L, R$, then $\mathcal{S}_k$ and $\mathcal{N}_k$ are also independent Gaussian random variables for $k = +, -$. From equation (3.21), O_i or $\mathcal{O}_k$ is also a Gaussian random variable, and each has a variance $\langle O_i^2 \rangle = \langle S_i^2 \rangle + \langle N_i^2 \rangle$ or $\langle \mathcal{O}_k^2 \rangle = \langle \mathcal{S}_k^2 \rangle + \langle \mathcal{N}_k^2 \rangle$, which is the sum of the variances of its independent contributors.

Since N_L and N_R are independent and identically distributed Gaussian noise, and both have variance $\langle N^2 \rangle \equiv \langle N_i^2 \rangle$ for $i = L, R$, then

$$\langle \mathcal{N}_\pm^2 \rangle = \left\langle \frac{(N_R \pm N_L)^2}{2} \right\rangle = \frac{\langle N_R^2 \rangle + \langle N_L^2 \rangle}{2} = \langle N^2 \rangle, \quad \text{and}$$

$$\langle \mathcal{N}_+ \mathcal{N}_- \rangle = \left\langle \frac{(N_R + N_L)(N_R - N_L)}{2} \right\rangle = \frac{\langle N_R^2 \rangle - \langle N_L^2 \rangle}{2} = 0. \qquad (3.111)$$

Hence, $\mathcal{N}_+$ and $\mathcal{N}_-$ are also independent and identically distributed Gaussian noise with variance $\langle N^2 \rangle$. Therefore $\mathcal{O}_+$ and $\mathcal{O}_-$ are also decorrelated:

$$\langle \mathcal{O}_+ \mathcal{O}_- \rangle = \langle (\mathcal{S}_+ + \mathcal{N}_+)(\mathcal{S}_- + \mathcal{N}_-) \rangle = \langle \mathcal{S}_+ \mathcal{S}_- \rangle = 0.$$

They have factorized probability distribution

$$P(\mathbf{O}) = P(\mathcal{O}_+)P(\mathcal{O}_-) \propto \exp\left(-\frac{\mathcal{O}_+^2}{2\langle\mathcal{O}_+^2\rangle}\right) \exp\left(-\frac{\mathcal{O}_-^2}{2\langle\mathcal{O}_-^2\rangle}\right).$$

A neuron that receives $\mathcal{O}_+$ is a binocular cell, summing inputs from both eyes with equal weights, while a neuron receiving $\mathcal{O}_-$ is ocularly opponent or unbalanced.

Since the transform $(O_L, O_R) \rightarrow (\mathcal{O}_+, \mathcal{O}_-)$

$$\begin{pmatrix} \mathcal{O}_+ \\ \mathcal{O}_- \end{pmatrix} \equiv \begin{pmatrix} \cos(45^o) & \sin(45^o) \\ -\sin(45^o) & \cos(45^o) \end{pmatrix} \begin{pmatrix} O_L \\ O_R \end{pmatrix}$$

is merely a coordinate rotation, (O_L, O_R) and $(\mathcal{O}_+, \mathcal{O}_-)$ consume the same amount of total output power $\langle\mathcal{O}_+^2\rangle + \langle\mathcal{O}_-^2\rangle = \langle\mathcal{O}_L^2\rangle + \langle\mathcal{O}_R^2\rangle$. They also contain the same amount of information $I(\mathbf{O}; \mathbf{S})$ about input signal $\mathbf{S}$. This is because $(\mathcal{O}_+, \mathcal{O}_-)$ can be unambiguously derived from (O_L, O_R) and vice versa, so any information about $\mathbf{S}$ derived from knowing (O_L, O_R) can also be derived from the corresponding $(\mathcal{O}_+, \mathcal{O}_-)$ and vice versa. More mathematically, knowing either $(\mathcal{O}_+, \mathcal{O}_-)$ or (O_L, O_R) gives the same conditional probability distribution $P(\mathbf{S}|\mathbf{O})$ about $\mathbf{S}$, whether $\mathbf{O}$ is represented as $(\mathcal{O}_+, \mathcal{O}_-)$ or (O_L, O_R). In other words, knowing $(\mathcal{O}_+, \mathcal{O}_-)$ or (O_L, O_R) enables us to recover the original signal $\mathbf{S}$ to exactly the same precision.[5]

The equivalence between $(\mathcal{O}_+, \mathcal{O}_-)$ and (O_L, O_R) is the same as that between $(\mathcal{S}_+, \mathcal{S}_-)$ and $(\mathcal{S}_L, \mathcal{S}_R)$, which describe the original signal $\mathbf{S}$ in two different coordinate systems. Hence, the quantity $I(\mathbf{O}; \mathbf{S}) = H(\mathbf{O}) - H(\mathbf{O}|\mathbf{S})$ is the same whether $\mathbf{O}$ is represented by $(O_L, O_R)^T$ or $(\mathcal{O}_+, \mathcal{O}_-)^T$, and whether $\mathbf{S}$ is represented as $(\mathcal{S}_L, \mathcal{S}_R)^T$ or $(\mathcal{S}_+, \mathcal{S}_-)^T$. However, when $\mathbf{O}$ is represented by $(\mathcal{O}_+, \mathcal{O}_-)^T$, it is more natural to calculate $I(\mathbf{O}; \mathbf{S})$ using the representation $\mathbf{S} = (\mathcal{S}_+, \mathcal{S}_-)^T$.

From equation (3.26), we know that for Gaussian signals, the information in each channel $\mathcal{O}_k = \mathcal{S}_k + \mathcal{N}_k$ about the original signal $\mathcal{S}_k$, for $k = +$ or $-$, is

$$I(\mathcal{O}_k; \mathcal{S}_k) = \frac{1}{2}\log_2\frac{\langle\mathcal{O}_k^2\rangle}{\langle\mathcal{N}_k^2\rangle} = \frac{1}{2}\log_2\frac{\langle\mathcal{S}_k^2\rangle + \langle\mathcal{N}_k^2\rangle}{\langle\mathcal{N}_k^2\rangle} = \frac{1}{2}\log_2\left(1 + \frac{\langle\mathcal{S}_k^2\rangle}{\langle\mathcal{N}_k^2\rangle}\right), \quad (3.112)$$

which depends only on $\langle\mathcal{S}_k^2\rangle/\langle\mathcal{N}_k^2\rangle$. Since $\mathcal{O}_k$ is linked with the whole signal $\mathbf{S}$ only through component $\mathcal{S}_k$, $I(\mathcal{O}_k; \mathbf{S}) = I(\mathcal{O}_k; \mathcal{S}_k)$. Hence, from $\mathcal{O}_k$, the amount of information extracted about $\mathbf{S}$ is the same as that extracted about $\mathcal{S}_k$.

Because $\mathcal{O}_+$ and $\mathcal{O}_-$ are independent, the information extracted by $\mathcal{O}_+$ and $\mathcal{O}_-$ about $\mathbf{S}$, i.e., $I(\mathcal{O}_+; \mathcal{S}_+) = I(\mathcal{O}_+; \mathbf{S})$ and $I(\mathcal{O}_-; \mathcal{S}_-) = I(\mathcal{O}_-; \mathbf{S})$, respectively, is non-redundant. Consequently,

$$I(\mathbf{O}; \mathbf{S}) = I(\mathcal{O}_+; \mathcal{S}_+) + I(\mathcal{O}_-; \mathcal{S}_-).$$

In other words, the total information $I(\mathbf{O}; \mathbf{S})$ transmitted by $(\mathcal{O}_+, \mathcal{O}_-)$ is simply the sum $I(\mathcal{O}_+; \mathcal{S}_+) + I(\mathcal{O}_-; \mathcal{S}_-)$ of the information contributed by individual channels, since any information extracted by $\mathcal{O}_+$ is not repeated by $\mathcal{O}_-$. The sum $I(\mathcal{O}_+; \mathcal{S}_+) + I(\mathcal{O}_-; \mathcal{S}_-)$ is also the total data rate of the two channels.

---

[5]Because the transformation from (O_L, O_R) to $(\mathcal{O}_+, \mathcal{O}_-)$ preserves the volume in $\mathbf{O}$ space, $P(\mathcal{O}_+, \mathcal{O}_-) = P(O_L, O_R)$, and $P(\mathcal{O}_+, \mathcal{O}_-|\mathbf{S}) = P(O_L, O_R|\mathbf{S})$ for the corresponding $(\mathcal{O}_+, \mathcal{O}_-)$ and (O_L, O_R). Hence $H(\mathcal{O}_+, \mathcal{O}_-) = H(O_L, O_R)$, and $H(\mathcal{O}_+, \mathcal{O}_-|\mathbf{S}) = H(O_L, O_R|\mathbf{S})$, and the quantity $I(\mathbf{O}; \mathbf{S})$ is equal to the quantity $H(\mathbf{O}) - H(\mathbf{O}|\mathbf{S})$ regardless of whether $\mathbf{O}$ is represented by (O_L, O_R) or $(\mathcal{O}_+, \mathcal{O}_-)$.

In contrast, a non-zero correlation $\langle S_L S_R \rangle$ gives a non-zero $\langle O_L O_R \rangle$. Hence some of the information extracted by O_L and O_R, i.e., $I(O_L; S_L) = I(O_L; \mathbf{S})$ and $I(O_R; S_R) = I(O_R; \mathbf{S})$, respectively, is redundant. This means

$$I(\mathbf{O}; \mathbf{S}) < I(O_L; S_L) + I(O_R; S_R),$$

i.e., the total information transmitted is less than the sum of the information transmitted by the individual channels (or less than the total data rate of the output channels).

From the last two paragraphs, we have

$$I(\mathbf{O}; \mathbf{S}) = I(\mathcal{O}_+; \mathcal{S}_+) + I(\mathcal{O}_-; \mathcal{S}_-) < I(O_L; S_L) + I(O_R; S_R). \tag{3.113}$$

For example, let the original signal-to-noise power in each input channel O_L or O_R be $\langle S_L^2 \rangle / \langle N^2 \rangle = \langle S_R^2 \rangle / \langle N^2 \rangle = 10$, and let the binocular correlation be $r = 0.9$. Then

$$I(\mathcal{O}_\pm; \mathcal{S}_\pm) = \frac{1}{2} \log_2 \left[1 + \frac{(1 \pm r)\langle S_L^2 \rangle}{\langle N^2 \rangle} \right] = 2.16 \text{ or } 0.5 \text{ bits}$$

for the $\mathcal{O}_+$ or $\mathcal{O}_-$ channel, respectively;

$$I(O_{L,R}; S_{L,R}) = \frac{1}{2} \log_2 \left(1 + \frac{\langle S_L^2 \rangle}{\langle N^2 \rangle} \right) = 1.73 \text{ bits}$$

for either the O_L or O_R channel.

Therefore, comparing the two data representations $(\mathcal{O}_+, \mathcal{O}_-)$ and (O_L, O_R),

$$\text{the total data rate } I(\mathcal{O}_+; \mathcal{S}_+) + I(\mathcal{O}_-; \mathcal{S}_-) = I(\mathbf{O}; \mathbf{S}) = 2.66 \text{ bits};$$
$$\text{the total data rate } I(O_L; S_L) + I(O_R; S_R) = 3.46 \text{ bits} > I(\mathbf{O}; \mathbf{S}) = 2.66 \text{ bits}.$$
$$\tag{3.114}$$

The difference $I(O_L; S_L) + I(O_R; S_R) - I(\mathbf{O}; \mathbf{S}) = 3.46 - 2.66 = 0.8$ bits is the amount of redundant information between the two channels O_L and O_R, wasting the total data rate when using representation (O_L, O_R). Hence, the representation $(\mathcal{O}_+, \mathcal{O}_-)$ requires less total channel capacity to transmit the information, while the two representations $(\mathcal{O}_+, \mathcal{O}_-)$ and (O_L, O_R) transmit exactly the same 2.66 bits of information about the original signal $\mathbf{S}$. Hence, we say that the representation $\mathcal{O}_\pm$ is more efficient than $O_{L,R}$.

The quantity

$$\frac{\sum_{i=L,R} I(O_i; S_i)}{I(\mathbf{O}; \mathbf{S})} - 1$$

measures the degree of redundancy in the code $\mathbf{O} = (O_L, O_R)$. It is this redundancy that causes unequal signal powers $\langle \mathcal{O}_+^2 \rangle > \langle \mathcal{O}_-^2 \rangle$, because the non-zero correlation $\langle S_L S_R \rangle$ makes the summation $\mathcal{S}_+$ typically larger than the difference $\mathcal{S}_-$. Unequal information rates $I(\mathcal{O}_+; \mathcal{S}_+) > I(\mathcal{O}_-; \mathcal{S}_-)$ follow, consequently.

3.5.2 Gain control

In reality, the coding transform $\mathbf{O} = \mathbf{K}(\mathbf{S} + \mathbf{N}) + \mathbf{N}_o$ brings additional noise $\mathbf{N}_o$ to each of the output channels. Hence, when $\mathbf{K} = \mathbf{K}_o$ is the principal component transform,

$$\mathcal{O}_\pm = \mathcal{S}_\pm + \mathcal{N}_\pm + (\mathcal{N}_o)_\pm,$$

in which $\mathbf{N}_o$ is represented by its projections $(\mathcal{N}_o)_\pm$ onto the $V^{(\pm)}$ axes. Hence, the total output noise is now $\mathcal{N}_\pm + (\mathcal{N}_o)_\pm$. Assume $\langle (\mathcal{N}_o)_+^2 \rangle = \langle (\mathcal{N}_o)_-^2 \rangle$, and denote them both by

$\langle N_o{}^2 \rangle \equiv \langle (N_o)_\pm^2 \rangle$ for simplicity[6]. The output powers, which are the variances of the outputs, are now

$$\text{output power } \langle \mathcal{O}_\pm^2 \rangle = \langle \mathcal{S}_\pm^2 \rangle + \langle N^2 \rangle + \langle N_o^2 \rangle, \tag{3.115}$$

$$\text{output noise power} = \langle N^2 \rangle + \langle N_o^2 \rangle. \tag{3.116}$$

Here, again, we used the fact that the sum of independent Gaussian random variables is itself a Gaussian random variable, whose variance is equal to the sum of the individual variances (see equation (3.21)). This makes the output signal-to-noise ratio:

$$\text{output S/N}_\pm = \frac{\langle \mathcal{S}_\pm^2 \rangle}{\langle N^2 \rangle + \langle N_o{}^2 \rangle}.$$

Hence, adding noise N_o decreases output S/N from the original output $\text{S/N}_\pm = \langle \mathcal{S}_\pm^2 \rangle / \langle N^2 \rangle$. The extracted information is

$$I(\mathcal{O}_\pm; \mathcal{S}_\pm) = I_\pm \equiv \frac{1}{2}\log_2[1 + \text{output S/N}_\pm] = \frac{1}{2}\log_2 \frac{\langle \mathcal{O}_\pm^2 \rangle}{\langle N^2 \rangle + \langle N_o^2 \rangle}. \tag{3.117}$$

It is also reduced from its original amount $\frac{1}{2}\log_2[1 + \langle \mathcal{S}_\pm^2 \rangle / \langle N^2 \rangle]$. To reduce this loss of information, one can amplify the original $(\mathbf{S} + \mathbf{N})$ part of the output $\mathbf{O}$. In particular, amplifying $\mathcal{S}_\pm + N_\pm$ by a gain $g_\pm$ gives

$$\begin{pmatrix} \mathcal{O}_+ \\ \mathcal{O}_- \end{pmatrix} = \begin{pmatrix} g_+ & 0 \\ 0 & g_- \end{pmatrix} \begin{pmatrix} \mathcal{S}_+ + N_+ \\ \mathcal{S}_- + N_- \end{pmatrix} + \begin{pmatrix} (N_o)_+ \\ (N_o)_- \end{pmatrix}. \tag{3.118}$$

The above equation implies replacing the encoding matrix $\mathsf{K} = \mathsf{K}_o$ by (see equation (3.107) for K_o)

$$\mathsf{K} = g\mathsf{K}_o = \begin{pmatrix} g_+ & 0 \\ 0 & g_- \end{pmatrix} \begin{pmatrix} \cos(45^o) & \sin(45^o) \\ -\sin(45^o) & \cos(45^o) \end{pmatrix}. \tag{3.119}$$

This gives output power $\langle \mathcal{O}_\pm^2 \rangle = g_\pm^2 \left(\langle \mathcal{S}_\pm^2 \rangle + \langle N^2 \rangle \right) + \langle N_o{}^2 \rangle$, output noise power $g_\pm^2 \langle N^2 \rangle + \langle N_o{}^2 \rangle$, and extracted information

$$I_\pm = \frac{1}{2}\log_2 \left(1 + \frac{g_\pm^2 \langle \mathcal{S}_\pm^2 \rangle}{g_\pm^2 \langle N^2 \rangle + \langle N_o{}^2 \rangle} \right) \xrightarrow{\quad g_\pm \to \infty \quad} \frac{1}{2}\log_2 \left(1 + \frac{\langle \mathcal{S}_\pm^2 \rangle}{\langle N^2 \rangle} \right). \tag{3.120}$$

However, increasing the gains $g_\pm$ also increases the output power $\langle \mathcal{O}_\pm^2 \rangle$, which we are considering as the cost of coding. Critically, the extracted information,

$$I_\pm \equiv I(\mathcal{O}_\pm; \mathcal{S}_\pm) = \frac{1}{2}\log_2(\langle \mathcal{O}_\pm^2 \rangle) - \frac{1}{2}\log_2(\text{output noise power})$$

$$\to \frac{1}{2}\log_2(\text{ cost}) - \frac{1}{2}\log_2(\text{output noise power})$$

increases at most logarithmically with this cost (which remains roughly true if the cost is measured in metabolic terms (Laughlin 2001)). In particular, when the input noise $N_\pm \to 0$ is almost zero, $\langle \mathcal{O}_\pm^2 \rangle \to g_\pm^2 \langle \mathcal{S}_\pm^2 \rangle + \langle N_o{}^2 \rangle$, and output noise power approaches $\langle N_o{}^2 \rangle$, which is independent of the gain $g_\pm$. Hence, one can increase the output power $\langle \mathcal{O}_\pm^2 \rangle$ by increasing $g_\pm$ and thereby increase $I_\pm$ logarithmically with the power. This logarithmic return means

---

[6] Note that since its distribution is a spherical Gaussian, the average power in any component of $\mathbf{N_o}$ is the same no matter onto which orthogonal coordinate systems the noise is projected.

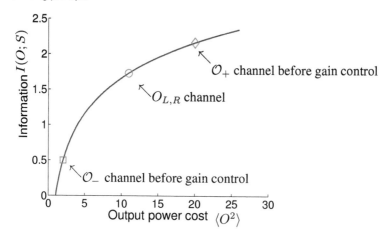

Fig. 3.10: The diminishing return of information $I = \frac{1}{2}\log_2\langle O_\pm^2 \rangle$ in the simplest case when the output noise power = 1 is fixed. In the plot, $\langle O_\pm^2 \rangle = (1 \pm r)\langle S_*^2 \rangle + 1$, $r = 0.9$, $\langle S_*^2 \rangle = 10$. Comparing the O_+ and O_- channels, the former consumes about 10 times as much power as the latter, but extracts only about 4–5 times as much information as the latter. If the O_+ channel reduces its power expenditure from $\langle O_+^2 \rangle = 20$ to $\langle O_+^2 \rangle = 11$, its extracted information $I(O_+; S_+)$ is reduced from 2.16 to 1.73 bits, a difference of only 0.43 bits. Putting the saved expenditure to the O_- channel, increasing $\langle O_-^2 \rangle$ from $\langle O_-^2 \rangle = 2$ to $\langle O_-^2 \rangle = 11$, increases extracted information $I(O_-; S_-)$ from 0.5 to 1.73, by the larger amount of 1.23 bits.

a diminishing information gain as the output power becomes larger. Hence, the return is greater if more of the power budget is devoted to the weaker channel O_- than the stronger one O_+. Figure 3.10 illustrates that shifting some power expense from the O_+ channel to the O_- channel should increase the total extracted information $I(O_+; S_+) + I(O_-; S_-)$. This motivates awarding different gains $g_\pm$ to the two channels, with $g_+ < g_-$. This implies amplifying the ocular contrast channel $S_- + N_-$ relative to the ocular summation channel $S_+ + N_+$. A key constraint on this is that the signal power in the contrast channel O_- is lower, and it is necessary to avoid wasting output power on over-amplifying the input noise N_-.

Balancing the need to reduce the cost in terms of the total output power cost $\langle O_+^2 \rangle + \langle O_-^2 \rangle$ against extracting more information $I(O; S) = I(O_+; S_+) + I(O_-; S_-)$, the optimal encoding is thus to find the gains $g_\pm$ that minimize (see equation (3.40))

$$E(K) = \text{cost} - \lambda \cdot I(O; S) \tag{3.121}$$
$$= \langle O_+^2 \rangle + \langle O_-^2 \rangle - \lambda\left[I(O_+; S_+) + I(O_-; S_-)\right] \tag{3.122}$$
$$= \sum_{k=+,-} \left[\langle O_k^2 \rangle - \lambda I(O_k; S_k)\right]. \tag{3.123}$$

Define

$$E_k(g_k) \equiv \langle O_k^2 \rangle - \lambda I(O_k; S_k)$$
$$= g_k^2\left(\langle S_k^2 \rangle + \langle N^2 \rangle\right) + \langle N_o^2 \rangle - \frac{\lambda}{2}\log_2 \frac{g_k^2\left(\langle S_k^2 \rangle + \langle N^2 \rangle\right) + \langle N_o^2 \rangle}{g_k^2\langle N^2 \rangle + \langle N_o^2 \rangle}.$$

We have

$$E(\mathsf{K}) = E_+(g_+) + E_-(g_-).$$

Here $E(\mathsf{K})$ as a function of K is now written as a function of g_+ and g_- through $E_+(g_+) + E_-(g_-)$. These gains $g_\pm$ are essential parameters that help characterize the full encoding transform $\mathsf{K} = \mathsf{g}\mathsf{K}_o$ via equation (3.119).

The optimal gain g_k can be obtained by $\frac{\partial E_k(g_k)}{\partial g_k} = 0$, or $\frac{\partial E_k(g_k)}{\partial (g_k^2)} = 0$, giving the same g_k^2 as in equation (3.95), which is repeated below:

$$g_k^2 \propto \max \left(F_{\text{smoothing}} \cdot F_{\text{decorrelation}} - 1, 0\right), \quad \text{where}$$
$$F_{\text{smoothing}} \equiv \left(1 + \frac{\langle N^2 \rangle}{\langle S_k^2 \rangle}\right)^{-1}, \quad F_{\text{decorrelation}} \equiv \frac{1}{2} + \frac{1}{2}\sqrt{1 + \frac{2\lambda}{(\ln 2)\langle N_o{}^2 \rangle} \frac{\langle N^2 \rangle}{\langle S_k^2 \rangle}}, \tag{3.124}$$

where $\max(x, y)$ takes the maximum among the two variables x and y. Hence, this optimal gain g_k depends on the input signal-to-noise ratio, $\langle S_k^2 \rangle / \langle N^2 \rangle$, through two factors, $F_{\text{smoothing}}$ and $F_{\text{decorrelation}}$. As the value of $\langle S_k^2 \rangle / \langle N^2 \rangle$ increases from $\langle S_k^2 \rangle / \langle N^2 \rangle \ll 1$ to $\langle S_k^2 \rangle / \langle N^2 \rangle \gg 1$, the factor $F_{\text{smoothing}}$ increases from 0 to 1, while the factor $F_{\text{decorrelation}}$ decreases from a very large value to 1, making g_k^2 increase from $g_k^2 = 0$ to a peak value when $\langle S_k^2 \rangle / \langle N^2 \rangle \approx 1$ before decreasing to $g_k^2 = 0$ again. In particular, when $\frac{\langle S_k^2 \rangle}{\langle N^2 \rangle} \ll 1$, $g_k^2 \propto \max \left\{\alpha \langle S_k^2 \rangle^{1/2} - 1, 0\right\}$ (where $\alpha = \left(\frac{\lambda}{2(\ln 2)\langle N^2 \rangle \langle N_o{}^2 \rangle}\right)^{1/2}$), and, when $\langle S_k^2 \rangle / \langle N^2 \rangle \gg 1$, $g_k^2 \propto \langle S_k^2 \rangle^{-1}$, see equation (3.99) and Fig. 3.6.

3.5.3 Contrast enhancement, decorrelation, and whitening in the high S/N regime

Let us apply what we learned so far to an example with a high S/N. In this example, $\langle N^2 \rangle = 10$, $\langle S_-^2 \rangle = 30$, $\langle S_+^2 \rangle = 310$, and $\langle N_o{}^2 \rangle = 1$. The total information at input is

$$I(\mathbf{S} + \mathbf{N}; \mathbf{S}) = \sum_{k=+,-} I(\mathcal{S}_k + \mathcal{N}_k; \mathcal{S}_k)$$
$$= \frac{1}{2} \log_2 \left(1 + \frac{\langle S_+^2 \rangle}{\langle N^2 \rangle}\right) + \frac{1}{2} \log_2 \left(1 + \frac{\langle S_-^2 \rangle}{\langle N^2 \rangle}\right) = 2.5 + 1 = 3.5 \text{ bits.}$$

If $\mathcal{S}_\pm + \mathcal{N}_\pm$ are directly sent to the output without extra gain control, $\mathcal{O}_\pm = \mathcal{S}_\pm + \mathcal{N}_\pm + \mathcal{N}_{o,\pm}$, the total information at the output is

$$I(\mathbf{O}; \mathbf{S}) = \frac{1}{2} \log_2 \left(1 + \frac{\langle S_+^2 \rangle}{\langle N^2 \rangle + \langle N_o^2 \rangle}\right) + \frac{1}{2} \log_2 \left(1 + \frac{\langle S_-^2 \rangle}{\langle N^2 \rangle + \langle N_o^2 \rangle}\right)$$
$$= 2.43 + 0.95 = 3.38 \text{ bits,}$$

which is less than $I(\mathbf{S} + \mathbf{N}; \mathbf{S}) = 3.5$ bits because of the extra noise $\mathbf{N}_o$. The total output power is

$$\sum_{k=+,-} \langle \mathcal{O}_k^2 \rangle = \left(\langle S_+^2 \rangle + \langle N^2 \rangle + \langle N_o{}^2 \rangle\right) + \left(\langle S_-^2 \rangle + \langle N^2 \rangle + \langle N_o{}^2 \rangle\right)$$
$$= 321 + 41 = 362,$$

of which a large amount, 321, is consumed by the $\mathcal{O}_+$ channel.

If we weaken the $\mathcal{S}_+$ channel by a gain $g_+ = 0.5$ and meanwhile amplify the $\mathcal{S}_-$ channel by a gain of $g_- = 1.6 = g_+\sqrt{\frac{\langle \mathcal{S}_+^2\rangle}{\langle \mathcal{S}_-^2\rangle}}$ according to equation (3.99) for the high S/N situation, the total extracted information $I(\mathbf{O}; \mathbf{S})$ at the output $\mathcal{O}_\pm = g_\pm(\mathcal{S}_\pm + \mathcal{N}_\pm) + (\mathcal{N}_o)_\pm$ is

$$\sum_{k=+,-} I_k = \frac{1}{2}\log_2\left(1 + \frac{g_+^2\langle \mathcal{S}_+^2\rangle}{g_+^2\langle N^2\rangle + \langle N_o^2\rangle}\right) + \frac{1}{2}\log_2\left(1 + \frac{g_-^2\langle \mathcal{S}_-^2\rangle}{g_-^2\langle N^2\rangle + \langle N_o^2\rangle}\right)$$

$$= 2.27 + 0.98 = 3.25 \text{ bits},$$

which is roughly as much as before the gain control. This is achieved by slightly increasing the information transmitted in the $\mathcal{S}_-$ channel while slightly reducing that in the $\mathcal{S}_+$ channel. Meanwhile, the total output power,

$$\sum_{k=+,-} \langle \mathcal{O}_k^2\rangle = \left[g_+^2\left(\langle \mathcal{S}_+^2\rangle + \langle N^2\rangle\right) + \langle N_o{}^2\rangle\right] + \left[g_-^2\left(\langle \mathcal{S}_-^2\rangle + \langle N^2\rangle\right) + \langle N_o{}^2\rangle\right]$$

$$= 81 + 104.3 = 185.3,$$

is reduced substantially from 362, which was the output power required without the gain control. Thus, including gain control makes $\mathsf{g}\mathsf{K}_o$ better than K_o at reducing $E(\mathsf{K}) = $ output power cost $-\lambda \cdot I(\mathbf{O}; \mathbf{S})$.

Since $g_- > g_+$, this encoding emphasizes the binocular difference, or contrast, channel $\mathcal{S}_-$ relative to the ocular summation channel $\mathcal{S}_+$, which conveys the common aspects of inputs to the two eyes. The gain control therefore achieves relative contrast enhancement.

This gain $g_\pm \propto \langle \mathcal{S}_\pm^2\rangle^{-1/2}$ also equalizes output power $\langle \mathcal{O}_+^2\rangle \approx \langle \mathcal{O}_-^2\rangle$, since $\langle \mathcal{O}_\pm^2\rangle = g_\pm^2\langle \mathcal{S}_\pm^2\rangle$ + noise power (in the above equation, this equality in output power is not accurate since the noise power is still substantial). Combining $\langle \mathcal{O}_+^2\rangle \approx \langle \mathcal{O}_-^2\rangle$ with decorrelation $\langle \mathcal{O}_+\mathcal{O}_-\rangle = 0$ gives an output correlation matrix R^O, which is proportional to an identity matrix, with elements

$$R_{ab}^O = \langle \mathcal{O}_a\mathcal{O}_b\rangle = \delta_{ab} \cdot \text{constant}.$$

Such a transform $\mathbf{S} \rightarrow \mathbf{O}$, which leaves output channels decorrelated and with equal power, is called whitening. The term "whitening" comes from the observation that the output signals $g_\pm\mathcal{S}_\pm$ are like white noise, for which different (typically spatial or temporal) frequency channels are uncorrelated and have equal power.

3.5.4 Many equivalent solutions of optimal encoding

Any coordinate rotation $\mathbf{O} \rightarrow \mathsf{U}\mathbf{O}$ by angle θ in the 2D space $\mathbf{O}$ multiplexes the channels $\mathcal{O}_+$ and $\mathcal{O}_-$ to give two alternative channels O_1 and O_2. Let us use $\mathsf{U}(\theta)$ to indicate that the matrix U is parameterized by θ. Then

$$\begin{pmatrix} O_1 \\ O_2 \end{pmatrix} = \mathsf{U}(\theta)\begin{pmatrix} \mathcal{O}_+ \\ \mathcal{O}_- \end{pmatrix} \equiv \begin{pmatrix} \cos(\theta) & \sin(\theta) \\ -\sin(\theta) & \cos(\theta) \end{pmatrix}\begin{pmatrix} \mathcal{O}_+ \\ \mathcal{O}_- \end{pmatrix} \qquad (3.125)$$

$$= \begin{pmatrix} \cos(\theta)\mathcal{O}_+ + \sin(\theta)\mathcal{O}_- \\ -\sin(\theta)\mathcal{O}_+ + \cos(\theta)\mathcal{O}_- \end{pmatrix}. \qquad (3.126)$$

Since $(O_1, O_2)^T$ can be uniquely obtained from $(\mathcal{O}_+, \mathcal{O}_-)^T$ and vice versa, the amount of extracted information $I(\mathbf{O}; \mathbf{S})$ is unchanged whether $\mathbf{O}$ is represented by $(O_1, O_2)^T$ or $(\mathcal{O}_+, \mathcal{O}_-)^T$. Meanwhile, since $\langle \mathcal{O}_+\mathcal{O}_-\rangle = 0$,

$$\langle O_1^2 \rangle = \left\langle (\cos(\theta)\mathcal{O}_+ + \sin(\theta)\mathcal{O}_-)^2 \right\rangle$$
$$= \cos^2(\theta)\langle \mathcal{O}_+^2 \rangle + \sin^2(\theta)\langle \mathcal{O}_-^2 \rangle + 2\cos(\theta)\sin(\theta)\langle \mathcal{O}_+\mathcal{O}_- \rangle$$
$$= \cos^2(\theta)\langle \mathcal{O}_+^2 \rangle + \sin^2(\theta)\langle \mathcal{O}_-^2 \rangle.$$

Similarly, $\langle O_2^2 \rangle = \sin^2(\theta)\langle \mathcal{O}_+^2 \rangle + \cos^2(\theta)\langle \mathcal{O}_-^2 \rangle$. Hence,

$$\langle O_1^2 \rangle + \langle O_2^2 \rangle = \langle \mathcal{O}_+^2 \rangle + \langle \mathcal{O}_-^2 \rangle,$$

i.e., the total output power cost is also invariant to the rotation from $(\mathcal{O}_+, \mathcal{O}_-)^T$ to $(O_1, O_2)^T$. Hence, both encoding schemes $S_{L,R} \to \mathcal{O}_\pm$ and $S_{L,R} \to O_{1,2}$, with the former a special case of the latter when $\theta = 0$, are equally optimal in minimizing $E(K) =$ output power cost $-\lambda I(\mathbf{O}; \mathbf{S})$. This is a particular manifestation of a degeneracy in the optimal encoding solutions, when more than one solution is available, as discussed in Section 3.4.1.

Since $\mathsf{K} = \mathsf{g}\mathsf{K}_o$ (see equation (3.119)) is the encoding transform for $S_{L,R} \to \mathcal{O}_\pm$, then $\mathsf{K} = \mathsf{U}(\theta)\mathsf{g}\mathsf{K}_o$ is the encoding transform for $S_{L,R} \to O_{1,2}$. The general coding transform $\mathsf{K} = \mathsf{U}(\theta)\mathsf{g}\mathsf{K}_o$ is then

$$\mathsf{K} = \begin{pmatrix} \cos(\theta) & \sin(\theta) \\ -\sin(\theta) & \cos(\theta) \end{pmatrix} \begin{pmatrix} g_+ & 0 \\ 0 & g_- \end{pmatrix} \begin{pmatrix} \cos(45^o) & \sin(45^o) \\ -\sin(45^o) & \cos(45^o) \end{pmatrix}. \tag{3.127}$$

Focusing on the signals (and thus omitting noise, such that $\mathcal{O}_\pm = g_\pm(\pm S_L + S_R)$),

$$\begin{pmatrix} O_1 \\ O_2 \end{pmatrix} = \begin{pmatrix} S_L[\ \cos(\theta)g_+ - \sin(\theta)g_-] + S_R[\ \cos(\theta)g_+ + \sin(\theta)g_-] \\ S_L[-\sin(\theta)g_+ - \cos(\theta)g_-] + S_R[-\sin(\theta)g_+ + \cos(\theta)g_-] \end{pmatrix}. \tag{3.128}$$

The neuron receiving the signal O_1 has a sensitivity $\cos(\theta)g_+ - \sin(\theta)g_-$ for input S_L from the left eye but a sensitivity $\cos(\theta)g_+ + \sin(\theta)g_-$ for input S_R from the right eye. Hence, in general, the two neurons coding O_1 and O_2 are differentially sensitive to inputs from different eyes. Varying $\mathsf{U}(\theta)$ (by varying θ) leads to a whole spectrum of possible neural ocularities from very binocular to very monocular. Indeed, the population of V1 neurons have a whole spectrum of ocularities.

Let us examine this rotation $(O_1, O_2)^T = \mathsf{U}(\theta)(\mathcal{O}_+, \mathcal{O}_-)^T$ when the input S/N is very high, and so whitening $\langle \mathcal{O}_+^2 \rangle = \langle \mathcal{O}_-^2 \rangle$ is achieved. Then, since $\langle \mathcal{O}_+\mathcal{O}_- \rangle = 0$, O_1 and O_2 are still decorrelated:

$$\langle O_1 O_2 \rangle = \langle (\cos(\theta)\mathcal{O}_+ + \sin(\theta)\mathcal{O}_-)(-\sin(\theta)\mathcal{O}_+ + \cos(\theta)\mathcal{O}_-) \rangle$$
$$= -\cos(\theta)\sin(\theta)(\langle \mathcal{O}_+^2 \rangle - \langle \mathcal{O}_-^2 \rangle) + \langle \mathcal{O}_+\mathcal{O}_- \rangle (\cos^2(\theta) - \sin^2(\theta))$$
$$= -\cos(\theta)\sin(\theta)(\langle \mathcal{O}_+^2 \rangle - \langle \mathcal{O}_-^2 \rangle) \tag{3.129}$$
$$= 0, \quad \text{when } \langle \mathcal{O}_+^2 \rangle = \langle \mathcal{O}_-^2 \rangle.$$

Using the same assumptions, one can verify that $\langle O_1^2 \rangle = \langle O_2^2 \rangle$, and thus whitening is also still maintained. Figure 3.9 shows intuitively that reading out responses from any two orthogonal axes would be equivalent to using the ones depicted by (O_1, O_2).
In particular,

when $\theta = -45^o$, the responses
$$O_1 \propto (g_+ + g_-)S_L + (g_+ - g_-)S_R \quad \text{and} \tag{3.130}$$
$$O_2 \propto (g_+ - g_-)S_L + (g_+ + g_-)S_R$$

are those shown in Fig. 3.9. The neuron O_1 is excited by the left-eye signal S_L and inhibited by the right-eye signal S_R when $g_- > g_+$ in the high S/N region; O_2 has the opposite ocular preference. Both neurons emphasize the ocular contrast signal.

3.5.5 Smoothing and output correlation in the low S/N region

Even though $\langle S_+^2 \rangle - \langle S_-^2 \rangle$ is substantial when the ocular correlation coefficient r is close to 1, when input S/N is too low, $\langle S_+^2 \rangle - \langle S_-^2 \rangle$ can be overwhelmed by noise. We can then have

$$\langle S_+^2 \rangle - \langle S_-^2 \rangle < \langle N^2 \rangle \quad \text{or even} \quad \langle S_+^2 \rangle - \langle S_-^2 \rangle \ll \langle N^2 \rangle.$$

Since $\langle S_+^2 \rangle - \langle S_-^2 \rangle = 2\langle S_L S_R \rangle$, the inequalities above mean that the binocular correlation $\langle S_L S_R \rangle$ is swamped by the independent input noise in the two channels. When $\langle S_+^2 \rangle - \langle S_-^2 \rangle \ll \langle N^2 \rangle$, we can have

$$\langle S_+^2 \rangle + \langle N^2 \rangle \approx \langle S_-^2 \rangle + \langle N^2 \rangle,$$

so the shape of the probability distribution of the noisy inputs $\mathbf{S} + \mathbf{N}$ looks more isotropic than ellipsoidal; see Fig. 3.11. This is in contrast to the situation in which the input noise is negligible (e.g., in Fig. 3.9). When the input $S_\pm + N_\pm$ is dominated by noise $N_\pm$, its contribution to the output $\mathcal{O}_\pm = g_\pm(S_\pm + N_\pm) + $ encoding noise will also be dominated by the noise contribution $g_\pm N_\pm$. Hence, according to $g_k^2 \propto \max\{a\langle S_k^2 \rangle^{1/2} - 1, 0\}$ in equation (3.99), the gain g_k should decrease with decreasing signal power $\langle S_k^2 \rangle$. Since $\langle S_-^2 \rangle < \langle S_+^2 \rangle$ when $\langle S_L S_R \rangle > 0$, this leads to

$$g_- < g_+, \quad \text{to reduce wasting output power on transmitted input noise } g_- N_-.$$

Although the two channels $S_+ + N_+$ and $S_- + N_-$ have comparable average power $\langle S_+^2 \rangle + \langle N^2 \rangle \approx \langle S_-^2 \rangle + \langle N^2 \rangle$ (when S/N is small) before the gain control, the gain g_+ can be very different from g_- since $g_\pm$ is determined by $\langle S_\pm^2 \rangle$ rather than $\langle S_\pm^2 \rangle + \langle N^2 \rangle$.

With $g_- < g_+$, the weaker binocular contrast signal is downplayed or totally suppressed, as illustrated in Fig. 3.11. This is called smoothing, i.e., smoothing out the difference between inputs from the two eyes by averaging over them (which is the effect of the ocular sum channel). Smoothing is thus the opposite of contrast enhancement. It helps to average out the noise in the inputs, since the noise from the left and right eyes are mutually independent.

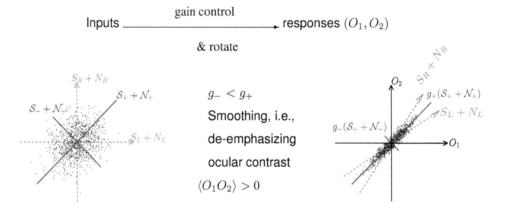

Fig. 3.11: Stereo coding when S/N$\ll$ 1; the weaker signal S_- is de-emphasized, suppressed, or abandoned to avoid transmitting too much input noise. In the right plot, the two red-dashed arrow lines come from the two axes $S_R + N_R$ and $S_L + N_L$ in the left plot after application of the differential gains $g_\pm$ to the two dimensions $S_\pm + N_\pm$. Both O_1 and O_2 integrate the left and right inputs to smooth out noise, while (slightly) preferring the left and right eye, respectively. Adapted with permission from Zhaoping, L., Theoretical understanding of the early visual processes by data compression and data selection, *Network: Computation in Neural Systems*, 17(4): 301–34, Fig. 4, copyright © 2006, Informa Healthcare.

The output channels $\mathcal{O}_+$ and $\mathcal{O}_-$ are still decorrelated, although they are no longer equally powered. Since $\langle S_+^2\rangle \gg \langle S_-^2\rangle$, then with $g_+ > g_-$, we have

$$\langle \mathcal{O}_+^2\rangle = g_+^2\langle S_+^2\rangle + \text{its noise power} > \langle \mathcal{O}_-^2\rangle = g_-^2\langle S_-^2\rangle + \text{its noise power.}$$

However, when $\mathcal{O}_+$ and $\mathcal{O}_-$ are multiplexed by a rotation matrix U to give general O_1 and O_2 output channels, we have according to equation (3.126)

$$O_1 = \cos(\theta)\mathcal{O}_+ + \sin(\theta)\mathcal{O}_-, \quad O_2 = -\sin(\theta)\mathcal{O}_+ + \cos(\theta)\mathcal{O}_-.$$

Both O_1 and O_2 will be dominated by inputs from the S_+ channel when g_+ is sufficiently larger than g_-. In particular,

$$\text{when } g_+ \gg g_-, \text{ then } O_{1,2} \propto g_+(S_L + S_R) + \text{noise.}$$

Then both output channels model binocular cell responses that integrate the correlated inputs S_L and S_R to smooth out noise. Consequently, the two output channels are correlated with each other. Indeed, equation (3.129) indicates that

$$\langle O_1 O_2\rangle \propto \langle \mathcal{O}_+^2\rangle - \langle \mathcal{O}_-^2\rangle \neq 0 \quad \text{when } \langle \mathcal{O}_+^2\rangle > \langle \mathcal{O}_-^2\rangle.$$

Consider the example from Section 3.5.3, in which $\langle S_+^2\rangle = 310$, $\langle S_-^2\rangle = 30$, $\langle N^2\rangle = 10$, and output noise power is $\langle N_o^2\rangle = 1$. Let us reduce the input signal and noise power such that $\langle S_+^2\rangle = 0.31$, $\langle S_-^2\rangle = 0.03$, and $\langle N^2\rangle = 1$. Consequently, the total input information rate is reduced to

$$I(\mathbf{S}+\mathbf{N};\mathbf{S}) = \frac{1}{2}\sum_{k=+,-}\log_2\left(1 + \frac{\langle S_i^2\rangle}{\langle N^2\rangle}\right) = 0.195 + 0.021 = 0.216 \text{ bits,}$$

most of which (0.195 bits) is from the S_+ channel. Sending $S_\pm + N_\pm$ directly to $\mathcal{O}_\pm = S_\pm + N_\pm + N_{o,\pm}$ gives output information

$$I(\mathbf{O};\mathbf{S}) = \sum_{k=+,-}I(\mathcal{O}_k;S_k) = \frac{1}{2}\sum_{k=+,-}\log_2\frac{\langle S_k^2\rangle + \langle N^2\rangle + \langle N_o^2\rangle}{\langle N^2\rangle + \langle N_o^2\rangle}$$
$$= 0.1039 + 0.0107 = 0.115 \text{ bits.}$$

This $I(\mathbf{O};\mathbf{S})$ is much reduced from $I(\mathbf{S}+\mathbf{N};\mathbf{S})$ due to the extra noise $\mathbf{N}_o$, whose power is substantial compared with that of the weak inputs $\mathbf{S}+\mathbf{N}$. The total output power cost is

$$\langle \mathcal{O}_+^2\rangle + \langle \mathcal{O}_-^2\rangle = 2.31 + 2.03 = 4.34,$$

of which 2.03 is spent on sending $I(\mathcal{O}_-;S_-) = 0.0107$ bits, a tiny fraction of the total $I(\mathbf{O};\mathbf{S}) = 0.115$ bits of information. To balance the need to cut down $\sum_k\langle \mathcal{O}_k^2\rangle$ and the need to extract $\sum_k I(\mathcal{O}_k;S_k)$ through minimizing $E(K) = \sum_k\langle \mathcal{O}_k^2\rangle - \lambda\sum_k I(\mathcal{O}_k;S_k)$, $g_+ > g_-$ is required, according to $g_k^2 \propto \max\{\alpha\langle S_k^2\rangle^{1/2} - 1, 0\}$ by equation (3.99). Let us say that this makes $g_+ = 1.0$ and $g_- = 0$; then the contrast channel is abandoned, losing $I(\mathcal{O}_-;S_-) = 0.0107$ bits of information. Meanwhile, the total output cost $\langle \mathcal{O}_+^2\rangle + \langle N_o^2\rangle = 2.31 + 1 = 3.31$ is reduced by almost a quarter (the baseline output power cost from the encoding noise $\mathbf{N}_o$ cannot be saved).

Multiplexing $\mathcal{O}_+$ and $\mathcal{O}_-$ by a $-45°$ rotation, as in Section 3.5.4, will spread the same total power cost in the two channels O_1 and O_2. However, the total information extracted,

$I(\mathbf{O};\mathbf{S})$, is unchanged. When $g_- = 0$, we get $O_i = g_+ S_+/\sqrt{2} + \text{noise}$ for both $i = 1, 2$. Then,

$$I(\mathbf{O};\mathbf{S}) = I_+ = 0.1039 \text{ bits},$$

$$I(O_i; S_+) = \frac{1}{2}\log_2\left(1 + \frac{g_+^2\langle S_+^2\rangle/2}{g_+^2\langle N^2\rangle/2 + \langle N_o^2\rangle}\right) = 0.0709, \quad \text{for both } i = 1, 2.$$

Each output channel extracts more than half of the total output information of $I(\mathbf{O};\mathbf{S}) = 0.1039$ bits, giving a redundancy of $2 \times 0.0709/0.1039 - 1 = 0.37$. This is expected, since the two output channels are correlated, and the redundancy should help the input signal recovery. If the channel capacity of a neuron is determined by the higher data rate in the high S/N condition, then this capacity is not fully used in the low S/N condition. The low S/N regime requires a smaller dynamic range than available, consumes less output cost, has less information to transmit, and leaves the responses of the output neurons more strongly correlated than in the higher S/N regime.

3.5.6   A special, most local, class of optimal coding

A special case of the general efficient stereo coding transform $\mathsf{K} = \mathsf{U}(\theta)g\mathsf{K}_o$ is when $\theta = -45°$. In this case, $\mathsf{U}(\theta) = \mathsf{K}_o^{-1}$ is the inverse transform of the particular K_o that transformed $(S_L, S_R)^T$ to $(S_+, S_-)^T$. In this case, if $g_+ = g_- = g$ and noise is ignored, this special U matrix will simply make $(O_1, O_2) = g(S_L, S_R)$, such that there is no mixing of the left and right eye signals. According to equation (3.124), $g_+ = g_-$ occurs when $\langle S_+^2\rangle = \langle S_-^2\rangle$, which in turn occurs when there is zero binocular correlation $\langle S_L S_R\rangle$. This is intuitively correct as a coding scheme, since, if there is no redundancy in the original signals, there is no need to decorelate the channels by mixing them.

In general, among the degenerate class of optimal encoding transforms $\mathsf{K} = \mathsf{U}g\mathsf{K}_o$, the particular one $\mathsf{K} = \mathsf{K}_o^{-1}g\mathsf{K}_o$ is special in the following sense. Define $e_i \equiv O_i - S_i$ as the signal change caused by encoding $\mathsf{K} = \mathsf{U}g\mathsf{K}_o$. Then, the summed squared change $\sum_i e_i^2$ is the smallest when $\mathsf{U} = \mathsf{K}_o^{-1}$, making

$$\mathsf{K} = \mathsf{K}_o^{-1}g\mathsf{K}_o. \tag{3.131}$$

This above encoding transform is the most local among all $\mathsf{K} = \mathsf{U}g\mathsf{K}_o$, in the sense that it distorts the original signal $\mathbf{S}$ the least (Atick, Li and Redlich 1993), as shown in Box 3.1. For instance,[7] this most local sense could entail the least volume of neural wiring to create the encoding K, or the smallest receptive fields for the filters (these are the row vectors of K). We will see an example of this for the spatial encoding in the retina in Section 3.6.1.

3.5.7   Adaptation of the optimal code to the statistics of the input environment

When input environment changes, the correlation matrix R^S and the input noise characteristics also change. They can be manifest as changes in signal-to-noise, ocular correlation r, or in the balance or symmetry between the two eyes. It leads to the changes in the eigenvectors of R^S, the principal components of $\mathbf{S}$, and the S/N of the principal components. According to the efficient coding principle, the encoding transform K should adapt to these changes.

---

[7] These various examples of "most local sense" could be formulated as an additional term in $E(\mathsf{K})$. We do not delve into these formulations in this book.

Box 3.1: $K = K_o^{-1} g K_o$ as the most local encoding

We seek the optimal $U = M$ that makes the encoding $K = M g K_o$ produce the least change $|S - M g K_o \cdot S|$ to the input signal S (Atick et al. 1993). This can be formulated as finding the M that minimizes

$$E'\{M\} = \langle |S - M g K_o \cdot S|^2 \rangle - \mathrm{Tr}\left[\rho(M \cdot M^T - \mathbb{1})\right],$$

where $\rho = \rho^T$ is a symmetric matrix. The first term is $\sum_i e_i^2$, with $e_i \equiv [S_i - (M g K_o \cdot S)_i]$, the difference between the original signal S_i and the encoded outcome $(M g K_o \cdot S)_i$. The second term is to constrain M as an unitary matrix with a Lagrange multiplier matrix ρ. This desired M is the solution to the variational equation $\frac{\delta E\{M\}}{\delta M} = 0$. This gives solution $M = K_o^{-1}$.

Changes in input statistics can be caused by short-term environmental adaptation, such as going from daytime to night vision (which leads to a huge change in the S/N ratio), or long-term differences such as in different visual development conditions (Li 1995). Different animal species also experience different input statistics, since they have different habitats and also different interocular distances.

Within a single animal, changes in ocular input statistics can also be caused by changes in receptive field (RF) properties in the dimensions of space, time, and color. For simplicity, we have so far omitted these to focus on ocularity. Simplistically, these omissions mean that we are examining the efficient encoding K in the ocular dimension while holding fixed the encoding properties (e.g., the size or orientation of the RF) in the other dimensions. However, the other tuning properties do affect the input statistics in the ocular dimension. For instance, the size and orientation of the RF affect the power of the input signals (S_L, S_R) and ocular correlation coefficient r (as will be shown later). Hence, as far as the encoding in the ocular dimension is concerned, V1 can be viewed as containing many subpopulations of neurons, each population being defined by its non-ocular RF properties which determine the input statistics R^S in the ocular dimension.

The efficient coding principle should predict how ocular encoding changes with the non-ocular RF properties, in terms of adaptation to input statistics. For ease of exposition, we have postponed consideration of efficient coding in the other dimensions to later parts of the chapter. For the moment we simply treat the non-ocular RF properties as given according to the V1 data.

Thus, we examine whether changes in ocular encoding K due to changes in input environments, developmental conditions, animal species, and RF properties in non-ocular dimensions can be understood as the adaptation of efficient coding to input statistics. To relate the coding properties to experimental data, let us first examine how the 2×2 matrix K is manifest in the ocularities of the neurons.

3.5.7.1 Binocular cells, monocular cells, and ocular dominance columns

Omitting noise, the stereo coding $O = KS$ gives the response in neuron $i = 1$ or $i = 2$ as

$$O_i = K_{iL} S_L + K_{iR} S_R. \tag{3.132}$$

When $K_{iL} \approx K_{iR}$, inputs from the two eyes add together to affect the response. Physiologically, such a neuron would be about equally sensitive to monocular stimulation in either eye alone and respond maximally to simultaneous stimulation in both eyes. Such a neuron is called a binocular neuron. When K_{iL} and K_{iR} have opposite signs, $K_{iL} K_{iR} < 0$, input from one eye excites the neuron whereas input from the other eye inhibits it. Real V1 neurons are nonlinear, see equations (2.42–2.43), with a near zero spontaneous level of response for zero input S.

Hence, inputs to the inhibitory eye alone should cause zero rather than negative responses. Such a neuron, which appears physiologically to respond to input from one eye only, would be called a monocular cell. Its responses to binocular stimulation should be weaker than to monocular stimulation to the preferred eye. (A linear cell in the linear encoding model could be represented by a complementary pair of such nonlinear cells such that input S to one neuron has the same effect as input $-S$ to the other neuron.) In the population of V1 neurons, there is a whole spectrum of ocular preferences from extreme monocularity to extreme binocularity. Many neurons prefer inputs from one eye to a degree, while still responding to monocular inputs from either eye.

The efficient stereo coding framework predicts that the balance of binocular versus monocular neurons depends on the relative gains g_+ and g_-. Without loss of generality, we focus on one of our usual two neurons (whose responses have been denoted as O_1 and O_2) derived on the basis of the efficient coding principle, denoting its response as

$$O = g_L S_L + g_R S_R, \tag{3.133}$$

where g_e is the sensitivity to input S_e for $e = L$ or R. According to equation (3.128), one can always find some angle ϕ to write,

$$g_L = \cos(\phi)g_+ - \sin(\phi)g_-, \quad \text{and} \quad g_R = \cos(\phi)g_+ + \sin(\phi)g_-. \tag{3.134}$$

This can be achieved for O_1 in equation (3.128) by making $\theta = \phi$ and for O_2 by $\theta = \phi - \pi/2$. We can define an ocularity index as

$$OI = \frac{2g_L g_R}{g_L^2 + g_R^2} = \frac{\cos^2(\phi)g_+^2 - \sin^2(\phi)g_-^2}{\cos^2(\phi)g_+^2 + \sin^2(\phi)g_-^2}. \tag{3.135}$$

This ocularity index OI is related to, but not quite the same as, the ocular dominance index used in physiological studies. If $OI = 1$ (i.e., $\phi = 0$), then $g_L = g_R$, and the cell is extremely binocular; if $OI = -1$ (i.e., $\phi = \pi/2$), then $g_L = -g_R$ and the cell is extremely monocular and ocularly opponent. One may qualitatively state that cells with $OI \leq 0$ are monocular, and cells with $OI > 0$ are binocular. If ϕ is evenly distributed across all angles (making $\cos^2(\phi)$ vary from 1 to 0 and $\sin^2(\phi)$ simultaneously vary from 0 to 1), then the distribution of positive or negative OI depends on the relative values of g_+ and g_-. Hence

$$\begin{aligned} &\text{when } g_+ > g_-, \text{ cells are more likely binocular;} \\ &\text{when } g_+ < g_-, \text{ cells are more likely monocular.} \end{aligned} \tag{3.136}$$

In V1, neurons tuned to the same eye of origin tend to cluster together, forming ocular dominance columns visualized by anatomical studies or optical imaging, as shown in Fig. 2.23 C. Binocular neurons cluster at the boundaries between neighboring columns tuned to different eyes, making imaged ocular dominance columns appear fuzzy there. Hence ocular dominance columns should appear weaker when there are more binocular neurons. In contrast, an abundance of monocular cells makes stark-looking ocular dominance columns.

3.5.7.2 Correlation between ocular dominance and the size of the receptive field

In V1, different neurons have differently sized receptive fields (RFs), with the RF size being inversely correlated with the preferred spatial frequency of the neuron. Change of stereo coding K with the RF size, or the preferred spatial frequency, is one example of how efficient coding is affected by input statistics. Two aspects of the input correlation matrix R^S changes

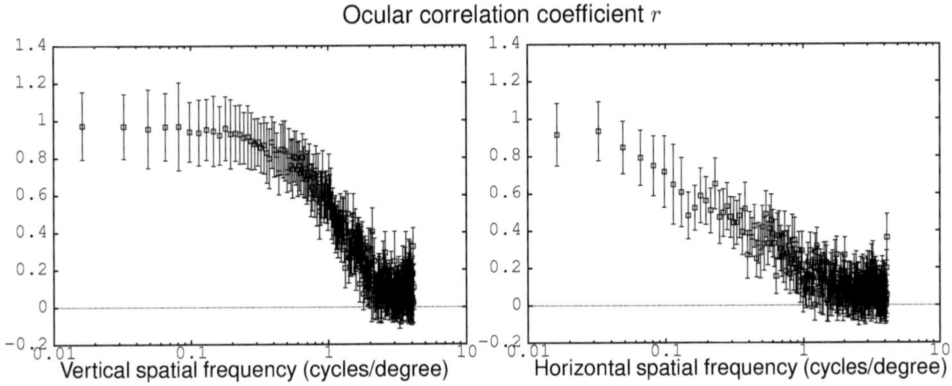

Fig. 3.12: The ocular correlation coefficient $r = \frac{\langle S_L S_R^* \rangle}{\langle |S_*|^2 \rangle}$ with $\langle |S_*|^2 \rangle = \langle |S_L|^2 \rangle = \langle |S_R|^2 \rangle$. The values of r were estimated from 127 stereo photographs, mostly taken from scenes in Central Park in Manhattan in the 1990s. S_L and S_R are the Fourier frequency components of the left and right images, respectively. Here, r value versus frequency along two slices in 2D frequency space are shown: one for vertical frequency (in the left plot) and the other for horizontal frequency (right plot). The r decreases with spatial frequency; it is also larger for vertical than horizontal frequencies. Adapted with permission from Li, Z. and Atick, J. J., Efficient stereo coding in the multiscale representation, *Network: Computation in Neural Systems*, 5(2): 157–74, Fig. 1, copyright © 1994, Informa Healthcare.

with RF size: the ocular correlation coefficient r and the signal strength $\langle S_*^2 \rangle$. As the RF becomes smaller, the ocular correlation coefficient r decreases—as is shown in Fig. 3.12 (via the correlation between small RF sizes and the high spatial frequency preference). This occurs since left eye and right eye inputs are only similar at a coarse scale. In the extreme case, when $r \to 0$, the two monocular channels are independent, ($\langle S_+^2 \rangle = \langle S_-^2 \rangle$, making $g_+ = g_-$), making it unnecessary to transform the S_L and S_R channels at all. Assuming that the cortex adopts the most local among the efficient codes, as described in Section 3.5.6, such that no binocular multiplexing (i.e., $O_1 \propto S_L$ and $O_2 \propto S_R$) occurs when there is no input redundancy, the neurons will then be monocular (Li and Atick 1994a, Li 1995).

Meanwhile, the signal strength $\langle S_*^2 \rangle$ also decreases as the RFs get smaller, since the input signal arises from integrating visual signals (and smoothing over noise) over an image area comparable to the size of the RF. In natural images, the signal power in the spatial frequency (component) f is proportional to f^{-2} (Field 1987). Hence, for neurons tuned to spatial frequency f, with RF size $\sim 1/f$, the input signal power $\langle S_*^2 \rangle$ scales with f^{-2} (this will be elaborated later in the chapter); see Fig. 3.13 A. Therefore, neurons with a large RF size and tuned to small spatial frequencies f enjoy a large input S/N. According to Section 3.5.3, they should exhibit stereo encoding with $g_- > g_+$, emphasizing the ocular contrast S_-. These neurons are therefore more likely monocular according to equation (3.136). Neurons with smaller RFs experience lower S/N ratios, such that when S/N ratios are so small $g_- < g_+$ coding should be employed, emphasizing binocular cells. Figure 3.13 illustrates how g_-/g_+ evolves as the RF size becomes smaller (or as the preferred spatial frequency becomes larger), stemming from the decrease in both the S/N ratio and the ocular correlation r with decreasing RF size.

Combining the effects of changing signal-to-noise and changing r with receptive field sizes, we have:

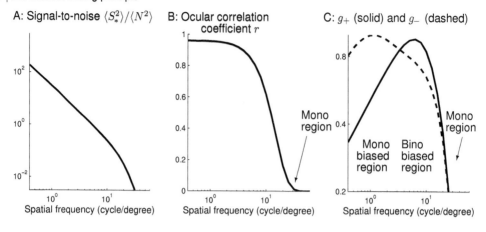

Fig. 3.13: Illustration of how g_-/g_+, and neural ocularity, should vary with the RF size or the preferred spatial frequency f of the neurons. This arises since the input $\langle S_*^2 \rangle / \langle N^2 \rangle$ and ocular correlation coefficient r both decrease with decreasing RF size (in A and B). The contrast sensitivity g_+ and g_- (in arbitrary units) for the ocular summation S_+ and ocular contrast S_- signals are calculated according to the efficient coding prescription (equation (3.124)) using $\langle S_\pm^2 \rangle / \langle N^2 \rangle = \langle S_*^2 \rangle / \langle N^2 \rangle (1 - r)$ and a parameter $\frac{2\lambda}{(\ln 2)\langle N_o^2 \rangle} = 6000$. The $g_- > g_+$ and $g_- < g_+$ regimes lead to neurons that are more likely to be monocular and binocular, respectively. The highest spatial frequency regime gives rise to monocular neurons because of a lack of ocular redundancy, as $r \to 0$. Consider an adaptation of the data in Fig. 3.12 to the case of active vision, for which two eyes are more aligned than is true for the passive stereo camera used to collect that data. For these data, for spatial frequency $f \neq 0$ (and $f = 0$ is not plotted), we have $\langle S_*^2 \rangle / \langle N^2 \rangle = (30/f^2)(M(f))^2$, where $M(f) = \exp(-(f/f_\mu)^2)$, with $f_\mu = 30$ cycles/degree, is the optical modulation transfer function of the eye, $r = 0.96 \exp(-(f/f_o)^2)$ (with $f_o = 15$ cycles/degree). The actual curves plotted in C are the overall effective filters $g_\pm M(f)$, which correspond to what should be observed experimentally.

As RF size decreases,
    (1) the S/N ratio decreases,
        so $\frac{g_+}{g_-}$ increases, and more neurons are binocular;
    (2) r decreases—when $r \to 0$, neurons stay monocular. (3.137)

The two effects act in opposite directions. However, since r is not zero until the RF size is very small, we should have (see Fig. 3.13 C):

As RF size decreases,
        RFs first become more binocular, (3.138)
        then become more monocular.

This predicted correlation can be tested physiologically. In such a test, one should consider the possibility that high spatial frequency cut-off of the population of cortical neurons may occur while the ocular correlation r is still substantial. This is especially the case for animals or animal species with poor spatial acuity and short interocular distance (which makes r larger). In such a case, one should only observe a trend of increasing binocularity of the neurons with decreasing RF size. This interaction between spatial coding, associated with the sizes of the RFs, and stereo coding, associated with ocularity, is an example of coupling between various input dimensions that efficient coding predicts. More such examples will be presented later.

Comparing Fig. 3.13 C with Fig. 2.15, we can make an analogy between stereo coding and color coding. Luminance and chromatic signals are made by summing and differentiating, respectively, input signals from different cones. Similarly, ocular summation and ocular contrast signals $\mathcal{S}_+$ and $\mathcal{S}_-$ are made by summing and differentiating, respectively, input signals from different eyes. At low spatial frequencies, contrast sensitivity is higher for the ocular contrast or chromatic signals than it is for the ocular summation or luminance signals. At higher spatial frequencies, contrast sensitivity is higher for the ocular summation or luminance signals (excluding the very high spatial frequency in stereo coding). At very high spatial frequencies, color and stereo coding differ because the ocular correlation $r \to 0$ whereas the correlation between cone inputs remains. The analogy between the color and stereo coding will be elaborated later in this chapter.

3.5.7.3 Strabismus

In strabismus, the two eyes are not properly aligned, reducing the ocular correlation r. Since $\langle \mathcal{S}_\pm^2 \rangle \propto 1 \pm r$, strabismus reduces the difference between the S/Ns of the $\mathcal{S}_+$ and $\mathcal{S}_-$ channels. Figure 3.13 C indicates that stereo coding switches from the $g_- > g_+$ regime at lower spatial frequencies to the $g_- < g_+$ regime at higher spatial frequencies. This switch occurs at the spatial frequency when the S/N associated with $\mathcal{S}_-$ becomes so weak that it is not worth amplifying $\mathcal{S}_-$. Since strabismus leads to $\langle \mathcal{S}_-^2 \rangle$ being stronger, this switch should occur at a higher spatial frequency or a smaller RF size. This means that the coding regime $g_- > g_+$ should cover a larger range of RF sizes within the overall multiscale representation afforded by the whole population of V1 neurons. Furthermore, because strabismus reduces ocular correlation r, the monocular regime at the highest spatial frequency region emerges at a smaller spatial frequency, further squeezing out the binocular regime in the regime involving intermediate spatial frequencies. Hence, both the expanded regime where $g_- > g_+$ and the encroaching regime of zero ocular correlations tend to boost monocularity in V1. Consequently, the ocular dominance columns should appear stronger or have sharper boundaries, since fewer binocular cells are present to make the boundaries fuzzy. This has indeed been observed in animals whose eyes are misaligned surgically or optically during development (Hubel and Wiesel 1965).

Although the $\mathcal{S}_-$ channel has stronger signals in strabismus, and depth information is derived from the $\mathcal{S}_-$ channel, individuals with strabismus tend to have poorer depth perception. This is a case of failure of the decoding stage to properly decode the depth implicitly encoded in O; we do not discuss it here in detail.

3.5.7.4 Correlation between ocular dominance and preferred orientation of V1 neurons

Since the two eyes are displaced from each other horizontally,

the ocular correlation r is greater for horizontally than vertically oriented inputs,　(3.139)

as has been measured in natural scenes; see Fig. 3.12 (note that horizontally or vertically oriented inputs are characterized by vertical or horizontal Fourier frequencies, respectively). This statistical fact arises as follows. Figure 2.23 shows that the projections of an object onto the two retinas are typically at slightly different locations relative to the fovea, implying that the object has a disparity. Larger disparity inputs make the ocular correlation r smaller. Disparity has both horizontal and vertical components; however, because the two eyes are horizontally displaced, horizontal disparities tend to be larger than vertical disparities. A horizontal disparity is most apparent between the two images of a vertical bar in the 3D scene, whereas a vertical disparity is most apparent in a horizontally oriented input. Statement (3.139) above follows by combining the above arguments.

Since a smaller r makes the input statistics more like those for strabismic animals, and since more cells are monocular in strabismus, one can predict the following (Li and Atick 1994a):

> V1 neurons tuned to orientations around horizontal are more
> likely to be binocular than those tuned to orientations around vertical. (3.140)

Figure 3.14 shows support for this prediction in the physiological, single unit data recorded from V1 of paralyzed and anesthetized cats (Zhaoping and Anzai, in preparation). The preferred orientation and spatial frequency were measured by drifting gratings. Using gratings of the optimal spatial frequency and orientation, responses O_L and O_R to left eye and right eye stimulation at 10% contrast were obtained. Since O_L and O_R are proportional to g_L and g_R respectively at a constant input contrast, they were used to obtain an ocular balance index (OBI) defined as

$$\text{OBI} = 1 - \left| \frac{O_L - O_R}{O_L + O_R} \right|. \qquad (3.141)$$

This index thus approaches 1 for binocular cells and 0 for monocular cells. Data from 126 recorded neurons indeed show that the average OBI for cells preferring horizontal orientations is significantly higher than that for cells preferring vertical orientations.

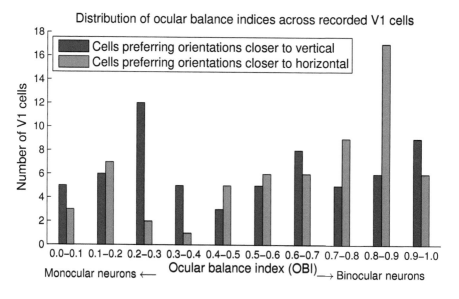

Fig. 3.14: Physiological data supporting the prediction that neurons tuned to near-horizontal orientations are more likely to be binocular than those preferring near-vertical orientations. From 136 recorded neurons, 62 neurons (26 simple cells, 36 complex cells) are tuned to within 44° from horizontal and 64 neurons (43 simple cells and 21 complex cells) are tuned to within 44° from vertical orientation. Their OBI distributions are plotted. The average OBIs within each group are OBI (horizontal) $= 0.6192 \pm 0.036$ and OBI (vertical) $= 0.51 \pm 0.038$. A two-tail t-test of the OBI values show that the two groups have significantly different OBIs ($p = 0.0345$). These data on OBIs and preferred orientations of individual cells were collected by Aki Anzai and his colleagues in experiments for a publication (Anzai et al. 1995), but the relationship between the OBIs and the preferred orientations was not published in that paper.

3.5.7.5 Adaptation of stereo coding to light levels

In dimmer environments, the S/N is lowered for cells of all RF sizes. According to the argument in equation (3.137), we should then expect more V1 neurons to be binocular. This causes weaker sensitivity to depth information which is derived from the S_- channel. Indeed, the disparity threshold measured behaviorally increases when image contrast is reduced (Legge and Yuanchao 1989).

3.5.7.6 Adaptation of stereo coding with animal species

Animals with short interocular distances, such as squirrel monkeys, enjoy larger binocular correlation (i.e., greater r) than other primates, including humans. This is the opposite situation from that of strabismus; now the S_- channel has weaker signals. Consequently, more cells should be binocular, and the ocular dominance columns should be weaker. Indeed, ocular dominance columns are lacking in squirrel monkeys. Since these columns can emerge in squirrel monkeys made strabismic within a few weeks of birth (Livingstone 1996), the weakness or absence of the ocular dominance columns in normal squirrel monkeys is unlikely to be caused by a mere lack of clustering of neurons preferring the same eye. Developmental situations involving strong ocular correlation can also be simulated by artificially synchronous inputs to the two eyes, leading to similar consequences (Stryker 1986). One could also expect that animals with a larger distance between the two eyes should have fewer binocular cells.

3.5.7.7 Monocular deprivation

If one eye is deprived of inputs during development, the correlation matrix R^S becomes asymmetric, $R^S_{LL} = \langle S^2_L \rangle \neq R^S_{RR} = \langle S^2_R \rangle$. Consequently, the eigenvectors and eigenvalues of R^S are different from their usual form (Li 1995). The ocular summation channel S_+ should be dominated by the strong eye, whereas the ocular contrast channel S_- should be dominated by the weak eye and be much weaker overall (i.e., has a much smaller corresponding eigenvalue) than normal. Under severe forms of monocular deprivation, the signal power of S_- at most spatial scales will be negligible compared to the noise power. In such a case, it should be suppressed via zero gain g_-. Consequently, a majority of V1 cells should be dominated by the strong eye input. Assuming neurons preferring the same eye cluster together, this should give rise to thicker ocular dominance columns for the stronger eye, as observed physiologically (Hubel, Wiesel and LeVay 1977).

3.5.8 A psychophysical test of the adaptation of the efficient stereo coding

One can test the efficient coding principle in the case of stereo psychophysically, by examining the behavioral effects of adapting the visual system to particular input environment. In particular, the most important binocular input statistic is the ocular correlation between the inputs to the two eyes. One can increase or decrease this correlation by exposing viewers to left eye inputs and right eye inputs which are more or less similar to each other than usual. Here, we report one such test.

Behavioral experiments inevitably involve interrogating the perception of observers, which in turn involves stages of visual selection and decoding beyond visual encoding (see Fig. 1.5). Thus, in designing experiments, it is important to ensure that selection and decoding proceed such that the experimental observations reflect the encoding process which is of interest.

Probing the sensitivity g_+ relative to g_- from perceived direction of drifting gratings

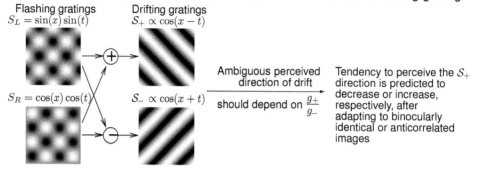

Fig. 3.15: Dichoptic stimulus (see equations (3.143–3.144)) produces a percept of a grating whose direction of drift is ambiguous. The chance of perceiving the drift in the direction of the S_+ grating should increase with g_+/g_-. Efficient stereo coding thus predicts that adaptation to strong $\langle S_+^2 \rangle$ or $\langle S_-^2 \rangle$ signals should change the motion perception (May et al. 2012).

3.5.8.1 Motion perception of dichoptic stimuli is altered by adaptation to increased or decreased ocular correlation

Efficient coding prescribes that when signal-to-noise is high enough, so whitening dominates, the gains g_+ and g_- associated with the ocular summation S_+ and ocular contrast S_- signals, respectively, should be $g_\pm^2 \propto \langle S_\pm^2 \rangle^{-1}$. Since, given an ocular correlation coefficient r,

$$\frac{\langle S_+^2 \rangle}{\langle S_-^2 \rangle} = \frac{1+r}{1-r} \qquad \xrightarrow{\text{whitening}} \qquad \frac{g_+^2}{g_-^2} = \frac{1-r}{1+r}, \qquad (3.142)$$

adaptation to an environment with a higher r should reduce the ratio g_+/g_- between the gains. One extreme example of this environment is when inputs to the two eyes are identical, i.e., $S_L = S_R$; then $\langle S_+^2 \rangle \gg \langle S_-^2 \rangle$, and consequently, g_+ should be much weaker than g_-. Similarly, overexposure to an unnatural environment in which the inputs to the two eyes are anticorrelated, i.e., $r < 0$ (in the extreme case $S_L = -S_R$) and $\langle S_+^2 \rangle \ll \langle S_-^2 \rangle$, should make $g_+ > g_-$.

In a recent psychophysical experiment, we tested this prediction, using the following inputs to probe the relative sensitivities g_+ and g_- (May et al. 2012). Let S_+ and S_- each contain a drifting grating, drifting in opposite directions:

$$\begin{aligned} S_+ &= A_+ \cos(kx + D \cdot \omega t + \phi_+), \\ S_- &= A_- \cos(kx - D \cdot \omega t + \phi_-), \end{aligned} \qquad (3.143)$$

in which $A_\pm$ and $\phi_\pm$ are amplitudes and phases of the gratings and $D = 1$ or $D = -1$ is a parameter that indicates the direction of the drift in the S_+ input. In the special case that $A_+ = A_-$, the input to each eye is a flashing grating:

$$\begin{aligned} S_L &\propto S_+ - S_- \propto \sin(kx + (\phi_+ + \phi_-)/2) \sin(\omega t + (\phi_+ - \phi_-)/2), \\ S_R &\propto S_+ + S_- \propto \cos(kx + (\phi_+ + \phi_-)/2) \cos(\omega t + (\phi_+ - \phi_-)/2). \end{aligned}$$

This is illustrated in Fig. 3.15. When human observers are presented with this stimulus, their percept is ambiguous such that when answering the question "in which direction do you see the grating drift?", they reply with the direction of the S_+ and S_- gratings with probabilities P_+ and P_- respectively (Shadlen and Carney 1986). We can expect that the ratio P_+/P_-

Psychophysical experimental test confirmed the predictions

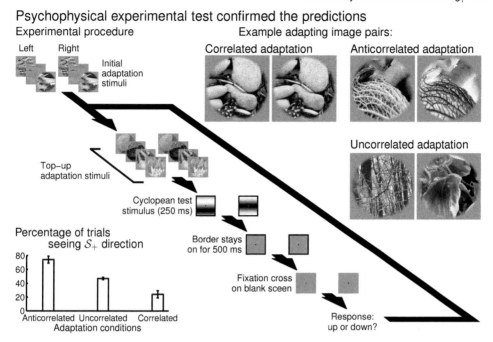

Fig. 3.16: Psychophysical confirmation of the prediction in Fig. 3.15. The experimental procedure is illustrated for the correlated adaptation condition. Example image pairs are shown (upper right) for the three adaptation conditions: correlated, anticorrelated, and uncorrelated. Testing trials immediately followed initial adaptation; these always started with four pairs of top-up adapting stimuli, shown one after another for 1.5 seconds each, followed immediately by a 250 ms display of a test grating which was horizontal. The observer had to report whether they perceived the grating as drifting up or down. Observers fixated on the central fixation point (present on each image) throughout the adaptation and testing. The percentage of trials in which the $\mathcal{S}_+$ direction was seen in the test trials after each adaptation condition is plotted lower left. Data were averaged over six observers. The test grating was $4^o \times 4^o$ wide, had a low contrast with $A_+/A_- \approx 1/1.45$, $k/(2\pi) = 0.25$ cycles/degree, and $\omega/(2\pi) = 6$ Hz. The grating phases $\phi_\pm$ and the $\mathcal{S}_+$ direction (up or down) were random in each trial. The circular adapting images were about 11° in diameter. In the anticorrelated adaptation, a picture and its negative were allocated to left and right eyes at random.

should increase with the sensitivity ratio g_+/g_-. In particular, the efficient coding principle predicts that overexposure to strong $\langle\mathcal{S}_+^2\rangle$ or $\langle\mathcal{S}_-^2\rangle$ signals should decrease or increase P_+/P_-, respectively.

This prediction was confirmed; see Fig. 3.16. In the experimental test, adaptation was done by having observers watch for 48 seconds a sequence of pairs of photographs. Each pair of photographs was shown for 1.5 second, with its two images designated for the two respective eyes. Each photograph (for the initial and top-up adaptation) was taken from natural scenes and suitably processed. The three most relevant adaptation conditions were termed binocularly correlated, uncorrelated, and anticorrelated, respectively. In the correlated adaptation, the two photographs in each pair were identical to each other; in the anticorrelated adaptation, the two photographs were negative images of each other; and in the uncorrelated adaptation, the two photographs were unrelated to each other. Hence, the signal power $\langle\mathcal{S}_+^2\rangle$ was highest in the correlated adaptation, lower in the uncorrelated adaptation, and lowest in the anticorrelated adaptation. Similarly, the power $\langle S_-^2 \rangle$ was lowest in the correlated and highest in the anticorrelated adaptation. The amplitude ratio A_+/A_- was adjusted such that $P_+/P_- = 1$ after the uncorrelated adaptation. Indeed, after binocularly correlated adaptation, observers were more likely to perceive the grating drift in the S_- than in the S_+ direction. The reverse was observed after anticorrelated adaptation.

3.5.9 How might one test the predictions physiologically?

Figure 3.14 showed one physiological test of the efficient stereo coding. Let us explore other possible tests. According to equation (3.134), one can always find some angle ϕ such that

$$g_L = \cos(\phi)g_+ - \sin(\phi)g_- \quad \text{and} \quad g_R = \cos(\phi)g_+ + \sin(\phi)g_-. \quad (3.144)$$

Therefore, one can test the adaptation that is predicted for $g_\pm$ by examining the changes to the sensitivities g_L and g_R for neurons. For example, take an extremely monocular cell which prefers inputs from the right eye, and let us say that we have identified an optimal spatiotemporal input that excites this neuron most effectively when presented to the right eye alone. Let us say that presenting the same spatiotemporal input to the left eye additionally will neither facilitate nor suppress the neuron's response to inputs from the right eye. According to the equation above, since $g_L = 0$ for this neuron, we expect that $\cos(\phi)g_+ = \sin(\phi)g_-$. Furthermore, since $g_R > 0$ for the optimal input, both these terms should be positive: $\cos(\phi)g_+ = \sin(\phi)g_- > 0$. According to the efficient coding principle, in a dimmer environment, g_- should be reduced in favor of g_+. This would imply, at least for this particular neuron and input, that $\cos(\phi)g_+ = \sin(\phi)g_-$ should be replaced by $\cos(\phi)g_+ > \sin(\phi)g_-$, thereby increasing g_L from $g_L = 0$ to $g_L > 0$. Consequently, presenting the same spatiotemporal input (which is optimal to the right eye) to the left eye should now have a facilitatory effect onto this neuron's response in a dimmer environment. Thus, input from the non-dominant eye that a monocular neuron normally ignores could facilitate responses to inputs from the dominant eye in a dimmer environment. Conversely, brightening the environment should make the input from the non-dominant eye suppressive.

As another example, Fig. 3.13 C shows that for low spatial frequencies, $g_- > g_+$, while for the intermediate or higher spatial frequencies, $g_+ > g_-$. Consequently, we can predict that ocular dominance columns will be stronger, or more easily seen, using gratings with low spatial frequencies than those with high spatial frequencies. This is because the sensitivity g_- to the ocular contrast signal is what leads to the appearance of ocular dominance.

3.6 The efficient receptive field transforms in space, color, time, and scale in the retina and V1

Stereo coding illustrates a general recipe, shown in Fig. 3.7, for an optimally efficient linear coding transformation $\mathbf{O} = \mathsf{K}(\mathbf{S} + \mathbf{N}) + \mathbf{N_o}$. This is for Gaussian signals $\mathbf{S}$, with correlation matrix R^S, given independent Gaussian input noise $\mathbf{N}$ and additional coding noise $\mathbf{N_o}$. Our interest is in the encoding transform K. Thus, for convenience and when it does not cause confusion, we typically leave the noise out of our notations and write $\mathbf{O} = \mathsf{K}\mathbf{S}$, or perhaps $\mathbf{O} = \mathsf{K}\mathbf{S} + \text{noise}$, instead of $\mathbf{O} = \mathsf{K}(\mathbf{S} + \mathbf{N}) + \mathbf{N_o}$. Of course, in calculating K or g, the noise is never omitted, since its power influences the S/Ns that in turn determines the gain control values in g. The recipe contains three conceptual (though not neural) ingredients: $\mathsf{K_o}$, g, and U, as follows.

    1. $\mathbf{S} \to \mathcal{S} = \mathsf{K_o}\mathbf{S}$ —find principal components $\mathcal{S} = (\mathcal{S}_1, \mathcal{S}_2, ..., \mathcal{S}_k, ...)$ by using unitary transform $\mathsf{K_o}$,

Recipe for efficient coding of Gaussian signals illustrated

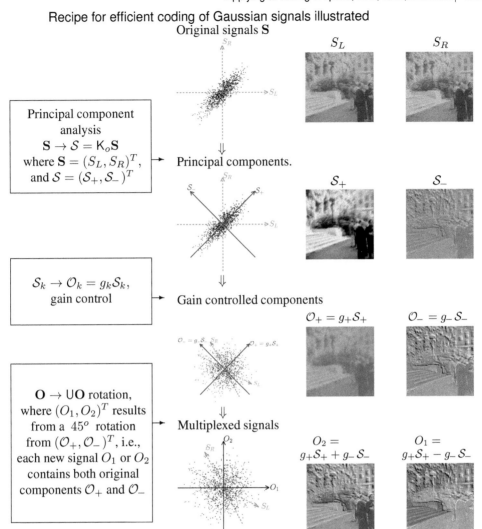

Fig. 3.17: Illustration of the three conceptual ingredients in the recipe for efficient coding, using the example of stereo coding. Each scatter plot contains random samples from distribution, (top to bottom) $P(\mathbf{S} = (S_L, S_R))$, $P(S_+, S_-)$, $P(\mathcal{O}_+, \mathcal{O}_-)$, or $P(O_1, O_2)$. The relationship between the transformed signals $S_\pm$, $\mathcal{O}_\pm$, $O_{1,2}$, and the original signals $S_{L,R}$ are evident in the directions of the axes for the transformed and the original signals in each plot. Examples of stereo images and their transforms are shown in the two rightmost columns.

2. $S_k \to \mathcal{O}_k = g_k S_k$—apply gain control g_k (a function of $\langle S_k^2 \rangle / \langle N^2 \rangle$) to each principal component S_k according to equation (3.94),

3. $\mathbf{O} \to \mathbf{U}\mathbf{O}$—rotate the resulting efficient code by a suitable unitary transform $\mathbf{U}$ to suit additional purpose.[8]

---

[8] Note that the invariance to $\mathbf{U}$ that underpins the freedom in the last step holds when the cost is $\sum_i \langle O_i^2 \rangle$ or $H(\mathbf{O})$, but, except in the noiseless case, not when the cost is $\sum_i H(O_i)$. Given finite noise, the cost of $\sum_i H(O_i)$ would break the invariance to $\mathbf{U}$, as there would be a preferred rotation $\mathbf{U} = \mathbb{1}$ that eliminates the second order correlation between output channels. The observation that early vision does not usually have the identity $\mathbf{U} = \mathbb{1}$

This recipe, with its three conceptual components, is illustrated in Fig. 3.17 using the example of stereo coding in the high signal-to-noise limit. Although the recipe was originally outlined in Section 3.4.1, we reiterate it below in order to consolidate the concepts, after having familiarized ourselves with them for the case of stereo coding.

The overall effective transform is

$$K = UgK_o, \tag{3.145}$$

where g is a diagonal matrix with elements $g_{kk} = g_k$. All such solutions, for any unitary U, retain the maximum information about S for a given output cost $\sum_k \langle O_k^2 \rangle$. When $U = 1$, the optimal coding transform is $K = gK_o$. The resulting $O = (O_1, O_2, ...) = (\mathcal{O}_1, \mathcal{O}_2, ...)$ has decorrelated components. Other unitary transforms U lead to outputs O with the same information $I(O, S)$ and cost, and, in the zero noise limit, the same decorrelation. The operation including all the three conceptual steps is equivalent to the single matrix operation K which is a solution to $\frac{\partial E}{\partial K} = 0$ where $E(K) = \text{cost} - \lambda \cdot I(O; S)$. The solution is degenerate, i.e., there are many equally good solutions obtained by arbitrary choices of the unitary transform (or rotation) U. The input statistics, which are manifest in the correlation matrix R^S, determine the optimal coding K through at least the first two conceptual steps. In particular, the S/N controls g_k, giving contrast enhancement and decorrelation for high S/Ns and input smoothing and response correlation for low S/Ns.

The actual implementation in the neural system of this encoding transform K does not have to involve the separate stages corresponding to the three conceptual transforms K_o, g, and U. For instance, in the retina, the coding transform from the receptors to the retinal ganglion cells is implemented through various neural mechanisms involving interneurons such as bipolar cells, horizontal cells, and amacrine cells. The computations associated with these interneurons do not correspond to the particular transforms K_o, g, and U, but the net effect of the signal transforms from the receptors to these interneurons, from the interneurons to each other, and eventually to the ganglion cells is the overall transform K. One might wonder why there should be multiple levels of interneurons just for a single overall transform K which could be achieved by a direct set of linear connections from the receptors to the ganglion cells without the interneurons. However, the transform K has to be adaptable to various changes in the input statistics $P(S)$ and S/Ns, and the interneurons may be instrumental in realizing these changes.

The correlation between the inputs are described by R^S. Ignoring the contributions from noise, the correlation between the outputs caused by these inputs are

$$\langle O_i O_j \rangle = \sum_{a,b} K_{ia} K_{jb} \langle S_a S_b \rangle = (KR^S K^T)_{ij} = (UgK_o R^S K^T gU^T)_{ij}. \tag{3.146}$$

In the more general case, K_o and U can be complex matrices. Although O represents real vectors that represent outputs of the encoding, the above formula can be modified in general to treat all the variables, including O, as if they are complex:

$$\langle O_i O_j \rangle \to \langle O_i O_j^* \rangle = \left[U(gK_o R^S K_o^\dagger g)U^\dagger \right]_{ij}, \tag{3.147}$$

where the superscript $\dagger$ denotes the conjugate transpose of a matrix, e.g., $K_{ij} = (K^\dagger)_{ji}^*$ and the superscript $*$ indicates the complex conjugate.

---

suggests that the cost is more likely output power $\sum_i \langle O_i^2 \rangle$ than the summed output data rate $\sum_i H(O_i)$. For instance, from our Gaussian approximation, retinal coding actually maximizes the second order output correlation given $\sum_i \langle O_i^2 \rangle$ and $I(O; S)$. This output correlation (without increasing the output cost) might aid in the signal recovery.

Since $UU^\dagger = 1$, i.e., $(UU^\dagger)_{ij} = \delta_{ij}$, we will have $\langle O_i O_j^* \rangle \propto \delta_{ij}$ when $gK_oR^SK_o^\dagger g$ is proportional to the identity matrix. The definition of K_o means that

$$(K_oR^SK_o^\dagger)_{kk'} = \delta_{kk'}\lambda_k \equiv \delta_{kk'}\langle|\mathcal{S}_k|^2\rangle, \tag{3.148}$$

where $\lambda_k \equiv \langle|\mathcal{S}_k|^2\rangle$ is the kth eigenvalue of R^S. In matrix form

$$K_oR^SK_o^\dagger = \Lambda \tag{3.149}$$

where Λ is a diagonal matrix with diagonal elements $\Lambda_{kk} = \lambda_k = \langle|\mathcal{S}_k|^2\rangle$. Note that $\mathcal{S}_k$ is also treated as complex (in fact, it could be, since it is an auxiliary variable that may never be realized during the encoding process). Thus $gK_oR^SK_o^\dagger g$ is proportional to the identity when $g_k^2 \propto \langle|\mathcal{S}_k|^2\rangle^{-1}$, which is the prescription from efficient coding when the S/N ratio is sufficiently high for all principal components. Hence, as expected, output channels are decorrelated:

$$\langle O_i O_j^* \rangle \propto \delta_{ij} \quad \text{when the S/N is sufficiently high for all input components.} \tag{3.150}$$

By contrast, since $g_k^2 \not\propto \langle|\mathcal{S}_k|^2\rangle^{-1}$ when the S/N is low, output channels are correlated:

$$\langle O_i O_j^* \rangle \not\propto \delta_{ij} \quad \text{when the S/N is low for some} \atop \text{input principal components } \mathcal{S}_k. \tag{3.151}$$

Hence, efficient coding in general does not lead to decorrelated outputs, except when all channels have sufficiently high signal-to-noise.

A special encoding is $K = K_o^{-1}gK_o$. As for the case of stereo, this is the efficient code that distorts the original signal $\mathbf{S}$ the least.

We now apply our understanding to visual coding in space, time, and color, always approximating signals and noise as Gaussian.

3.6.1 Efficient spatial coding in the retina

Efficient spatial coding has been studied by various researchers (Srinivasan et al. 1982, Linsker 1990, Atick and Redlich 1990). A visual image $S_{x_i} = S(x_i)$ is a set of signals, one for each of a collection of image locations x_1, x_2, etc. It can also be written as a vector

$$\mathbf{S} \equiv \begin{pmatrix} S_{x_1} \\ S_{x_2} \\ ... \\ S_{x_n} \end{pmatrix}. \tag{3.152}$$

We follow the recipe in Section 3.6 to construct an efficient spatial coding scheme for a collection of such images. We simplify the notation by writing either S_x or $S(x)$ to denote the same entity, and similarly S_k or $S(k)$; g_k or $g(k)$; $\mathcal{O}_k$ or $\mathcal{O}(k)$, and O_x or $O(x)$. Although we use these alternative notations interchangeably, we favor the S_x form when considering an element in a vector and the $S(x)$ form for a value in a function.

3.6.1.1 First step in the spatial coding: decorrelation transform $\mathbf{S} \rightarrow \mathcal{S} = K_o\mathbf{S}$

The input correlation is

$$R^S_{x_ix_j} = \langle S_{x_i}S_{x_j}\rangle.$$

Here, to make the notation more meaningful, the ith element in $\mathbf{S}$ of equation (3.152) is written as S_{x_i} instead of S_i. Readers should understand them as meaning the same thing.

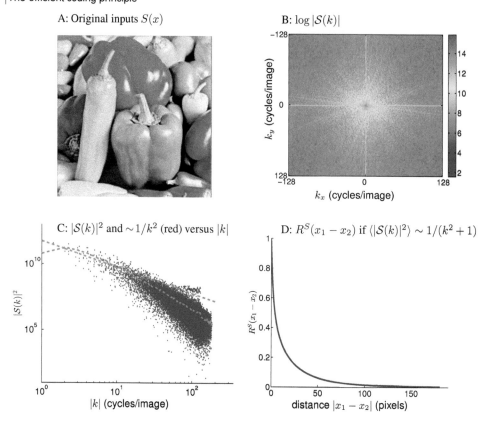

Fig. 3.18: A: An example visual input $S(x)$ (also denoted as S_x) in space x. The image has 256×256 pixels. B: Fourier transform $S(k)$ of image $S(x)$, visualized as $\log |S(k)|$ versus the 2D spatial frequency $k = (k_x, k_y)$. In the text, $S(k)$ is also denoted as S_k. C: Power $|S(k)|^2$ versus $|k|$ for input $S(x)$ in A; red line: $\sim 1/|k|^2$; green curve: the average of $|S(k)|^2$ over a local range of k for the input image in A. D: Spatial correlation $R^S(x_1 - x_2)$ assuming $\langle |S(k)|^2 \rangle = 1/(k^2 + k_o^2)$ for low cut-off frequency $k_o = 1$ cycle/image.

Similarly, $R^S_{x_i x_j}$ means the $(i, j)^{\text{th}}$ element of matrix R^S. Although this is a slight abuse of notations, such a meaningful notation system will be analogously used for other matrices and vectors in this book.

As one can see in Fig. 3.18 A, nearby image pixels tend to have similar input intensities, in just the same way for stereo that inputs to the two eyes tend to be similar. Furthermore, this similarity decreases with increasing distance between the two pixels (Fig. 3.18 D). This means $R^S_{xx'}$ decreases with increasing distance $|x - x'|$. One can also expect $R^S_{xx'}$ to be translation invariant, depending only on $x - x'$. Hence, we denote $R^S(x - x') \equiv R^S_{xx'}$ as the autocorrelation function of the spatial inputs. We remind ourselves that sans-serif fonts denote matrices, as in matrix R^S, while italic font is for a function, as in the correlation function $R^S(x)$ of spatial displacement x.

For images, location x is generally two-dimensional (2D). However, the formulation of spatial coding applies to images of any dimension (e.g., one-dimensional (1D)). Thus, we denote x using a non-bold font.

To see an explicit input correlation matrix R^S, we take the simple case of a 1D retina sampled at evenly spaced locations $x_1 = \delta x$, $x_2 = 2\delta x$, $x_3 = 3\delta x$, ..., $x_n = n\delta x$. This models a small patch of the retina where the spatial sampling is relatively even. Then the $n \times n$ matrix R^S takes the form

$$R^S = \begin{pmatrix} R^S_{x_1 x_1} & R^S_{x_1 x_2} & R^S_{x_1 x_3} & \cdots & R^S_{x_1 x_n} \\ R^S_{x_2 x_1} & R^S_{x_2 x_2} & R^S_{x_2 x_3} & \cdots & R^S_{x_2 x_n} \\ \vdots & & & & \\ R^S_{x_n x_1} & R^S_{x_n x_2} & R^S_{x_n x_3} & \cdots & R^S_{x_n x_n} \end{pmatrix}. \tag{3.153}$$

Due to the translation invariance property, $R^S_{x_1 x_1} = R^S_{x_2 x_2} = R^S_{x_i x_i}$ for all i. Additionally, consider space to be periodic, such that locations x_1 and x_n are neighbors. This boundary condition is obviously not realistic, but it is often used mathematically to ignore the breakdown of translation invariance at the spatial boundary and it is acceptable as long as we do not try to apply the encoding solution to locations close to the spatial boundary. Then, translation invariance also gives $R^S_{x_1 x_2} = R^S_{x_n x_1} = R^S_{x_i x_{i+1}}$ for all i, and in general, $R^S_{x_i, x_{i+m}} = R^S_{x_j, x_{j+m}}$ for all i, j, and m (if $i + m > n$, it is replaced by $i + m - n$, and similarly for $j + m$). Then, R^S is of the form

$$R^S = R^S_{x_1 x_1} \begin{pmatrix} 1 & r_1 & r_2 & \cdots & & r_1 \\ r_1 & 1 & r_1 & r_2 \cdots & & r_2 \\ r_2 & r_1 & 1 & r_1 & r_2 & \\ \vdots & & & \vdots & & \cdots \\ \vdots & & & \vdots & & \\ r_1 & \cdots & & & r_2 & r_1 & 1 \end{pmatrix}, \tag{3.154}$$

where we have denoted $r_1 \equiv R^S_{x_i x_{i+1}} / R^S_{x_i x_i}$, $r_2 \equiv R^S_{x_i x_{i+2}} / R^S_{x_i x_i}$ as correlation coefficients. The matrix is symmetric by the definition of correlation. A matrix like R^S, in which the ij^{th} element only depends on the distance $i - j$ (or in our case the $x_i x_j^{th}$ element only depends on the distance $x_i - x_j$), is called a Toeplitz matrix. The eigenvectors of a Toeplitz matrix have the form

$$V \equiv \frac{1}{\sqrt{n}} \begin{pmatrix} e^{ikx_1} \\ e^{ikx_2} \\ \vdots \\ e^{ikx_n} \end{pmatrix} \tag{3.155}$$

in which $k = \frac{2\pi}{x_n} \cdot (m - 1)$, and m is an integer, so that x_n is a period of the spatial wave e^{ikx} (i.e., this wave is periodic in spatial extent x_n). The integer m can take values $m = 1, 2, ..., n$, so that there can be n different eigenvectors, each denoted by the value of this integer m, or equivalently, the value of k. Let us denote the eigenvector with a particular k value by $V^{(k)}$. One can verify that $V^{(k)}$ is an eigenvector of R^S by noting that $R^S V^{(k)} = \lambda_k V^{(k)}$, with an eigenvalue λ_k. Explicitly

$$\sum_j R^S_{x_i, x_j} V^{(k)}_{x_j} = \sum_j R^S_{x_i, x_j} \frac{1}{\sqrt{n}} e^{ikx_j} = \frac{1}{\sqrt{n}} e^{ikx_i} \left[\sum_j R^S(x_i - x_j) e^{-ik(x_i - x_j)} \right]$$

$$\equiv \lambda_k^S \frac{1}{\sqrt{n}} \cdot e^{ikx_i}, \tag{3.156}$$

where

$$\lambda_k^S \equiv \sum_j R^S(x_i - x_j) e^{-ik(x_i - x_j)} \tag{3.157}$$

is a constant independent of i or x_i because $R^S(x_i - x_j)$ depends only on $x_i - x_j$. Hence, the eigenvector $V^{(k)}$ is a Fourier wave with wave number k, which is used here as the index for the eigenvector. Meanwhile, the corresponding eigenvalue $\lambda_k = \sum_x R^S(x)e^{-ikx}$ is proportional to the corresponding Fourier component of the correlation function $R^S(x)$.

Hence, the principal component transform matrix K_o is the Fourier transform (writing $k_m = 2\pi(m-1)/x_n$):

$$\mathsf{K}_o = \begin{pmatrix} (V^{(k_1)})^\dagger \\ (V^{(k_2)})^\dagger \\ \vdots \\ (V^{(k_n)})^\dagger \end{pmatrix} = \frac{1}{\sqrt{n}} \begin{pmatrix} e^{-ik_1 x_1} & e^{-ik_1 x_2} & \dots & e^{-ik_1 x_n} \\ e^{-ik_2 x_1} & e^{-ik_2 x_2} & \dots & e^{-ik_2 x_n} \\ & & \vdots & \\ e^{-ik_n x_1} & e^{-ik_n x_2} & \dots & e^{-ik_n x_n} \end{pmatrix}, \qquad (3.158)$$

where $(V^{(k_i)})^\dagger$ is a row vector whose complex conjugate is the transpose of the eigenvector $V^{(k_i)}$, i.e., $\left[(V^{(k_i)})^\dagger\right]_a = \left[V_a^{(k_i)}\right]^*$. Now, applying K_o to $\mathbf{S}$ to obtain $\mathcal{S} = \mathsf{K}_o \mathbf{S}$, we derive

$$\begin{pmatrix} \mathcal{S}_{k_1} \\ \mathcal{S}_{k_2} \\ \vdots \\ \mathcal{S}_{k_n} \end{pmatrix} \propto \begin{pmatrix} e^{-ik_1 x_1} & e^{-ik_1 x_2} & \dots & e^{-ik_1 x_n} \\ e^{-ik_2 x_1} & e^{-ik_2 x_2} & \dots & e^{-ik_2 x_n} \\ & & \vdots & \\ e^{-ik_n x_1} & e^{-ik_n x_2} & \dots & e^{-ik_n x_n} \end{pmatrix} \begin{pmatrix} S_{x_1} \\ S_{x_2} \\ \vdots \\ S_{x_n} \end{pmatrix},$$

hence

$$\mathcal{S}_{k_a} = \sum_b (\mathsf{K}_o)_{k_a x_b} S_{x_b} \sim \sum_x e^{-ik_a x} S_x \propto \int dx e^{-ik_a x} S(x). \qquad (3.159)$$

In the last step, we used notation $S(x)$ for S_x. Hence $\mathcal{S}_{k_m}$ is the mth Fourier component of the input image $S(x)$.

When $m = 1$ and $k_m = \frac{2\pi}{x_n}(m-1) = 0$, the Fourier wave has zero frequency. The signal $\mathcal{S}_{k_m}$ at $m = 1$ is thus analogous to the $\mathcal{S}_+$ mode in stereo vision, signaling the average input across different input channels or image pixel locations. When $k \neq 0$, the signal $\mathcal{S}_k$ senses the input difference or contrast between different locations x; hence, it is analogous to the mode $\mathcal{S}_-$ in stereo vision. For example, when $m = 2$ and $k_m = \frac{2\pi}{x_n}$, the $\mathcal{S}_{k_m}$ senses the first non-zero Fourier mode, with a Fourier wave covering the whole image in one period. Hence, this $\mathcal{S}_k$ senses roughly the difference between one half side of the image and the other half side of the image. Since there are more spatial dimensions (one per location) than ocular dimensions (two, one for each eye), there are more forms of input contrast for spatial inputs $\mathbf{S}$. Each $\mathcal{S}_k$ measures the degree of image variation in the shape of a Fourier wave of frequency k.

Figure 3.18 AB give an example input $S(x)$ and the amplitudes of its Fourier components $|\mathcal{S}_k|$. It shows a trend toward higher signal power $\langle|\mathcal{S}_k^2|\rangle$ for modes with lower spatial frequencies $|k|$. This is again analogous to stereo vision, for which $\langle\mathcal{S}_+^2\rangle > \langle\mathcal{S}_-^2\rangle$. Correlations between intensities in different input pixels make signal power stronger in input modes that smooth or average over input pixels.

As expected, the average powers of the Fourier modes are the eigenvalues λ_k^S of R^S:

$$\langle|\mathcal{S}_k|^2\rangle \propto \int dx dx' e^{-ik(x-x')} \langle S_x S_{x'}\rangle$$

$$= \int dx dx' e^{-ik(x-x')} R^S(x-x') \propto \int dx e^{-ikx} R^S(x) \propto \lambda_k^S.$$

Hence, $\langle|\mathcal{S}_k|^2\rangle$ is also a Fourier component of $R^S(x)$.

In 2D images of natural scenes (Field 1987), the power spectrum is roughly $\langle |S_k|^2 \rangle \sim 1/|k|^2$. Meanwhile, the general variation of signal power with frequency k in any specific example (such as Fig. 3.18 AB) will be similar, but not identical, to $1/|k|^2$. Although $R^S(x)$ can be measured directly, it can also be obtained from $\langle |S_k|^2 \rangle$ as

$$R^S(x) = \langle S(x')S(x'+x) \rangle \propto \int dk \langle |S(k)|^2 \rangle e^{ikx} \qquad (3.160)$$

as the inverse Fourier transform of $\langle |S(k)|^2 \rangle$ (here $S(k)$ is used to denote S_k).

Figure 3.18 D shows that this correlation $R^S(x)$ can be non-zero for long distances x if $\langle |S(k)|^2 \rangle \sim (|k|^2 + k_o^2)^{-1}$ for a low cut-off frequency of $k_o = 1$ cycle/image.

Now let us consider input noise, $\mathbf{N} = (N_{x_1}, N_{x_2}, ..., N_{x_n})$, which we show added to an image in Fig. 3.22. The noise terms at different locations are assumed to be uncorrelated, and thus

$$\langle N_x N_{x'} \rangle \equiv \langle N^2 \rangle \delta_{xx'}.$$

Let $\mathcal{N}_k \propto \sum_x e^{-ikx} N_x$ be the Fourier component of the noise corresponding to wave number k, i.e., the vector $(\mathcal{N}_{k_1}, \mathcal{N}_{k_2}, ..., \mathcal{N}_{k_m}, ...)^T = \mathsf{K}_o \mathbf{N}$. Uncorrelated noise at different image locations means that the input noise is white noise. Hence, the power spectrum of the noise is independent of k,

$$\langle N^2 \rangle \equiv \langle |\mathcal{N}_k|^2 \rangle = \langle |\mathcal{N}_{k'}|^2 \rangle.$$

This is analogous to $\langle N_+^2 \rangle = \langle N_-^2 \rangle$ in stereo vision.

3.6.1.2 Second step in the spatial coding: gain control on the spatial Fourier component $S_k \to O_k = g_k S_k$

According to the second step in the recipe for efficient coding, gain control arises from the diagonal matrix

$$\mathsf{g} = \begin{pmatrix} g_{k_1} & 0 & 0 & \cdots & & 0 \\ 0 & g_{k_2} & 0 & 0 & \cdots & \\ \vdots & & & & & \\ \cdots & \cdots & \cdots & g_{k_i} & & \cdots \\ \vdots & & & & & \\ \cdots & \cdots & \cdots & 0 & g_{k_n} \end{pmatrix}, \qquad (3.161)$$

in which the value of g_{k_i} should be determined by the signal-to-noise ratio $\langle |S_{k_i}|^2 \rangle / \langle N^2 \rangle$ in the principal component corresponding to frequency k_i.

As we noted above, in natural visual images, the signal power $\langle |S_k^2| \rangle$ decreases very quickly with increasing spatial frequency $|k|$. Thus, for small $|k|$, when the noise is comparatively negligible, the gain $g_k \propto \langle |S_k|^2 \rangle^{-1/2}$ should approximate a whitening solution and rise quickly with increasing $|k|$. In particular, $g_k \propto k$ when $\langle |S_k|^2 \rangle \propto |k|^{-2}$. This whitening gain emphasizes higher spatial frequencies, extracting information about image contrast. However, as $|k|$ increases further, the signal is less dominating over the noise. Let k_p be the spatial frequency at which $\langle |S_k|^2 \rangle = \langle N^2 \rangle$, i.e., when the S/N is one. The signal-to-noise ratio $\langle |S_k|^2 \rangle / \langle N^2 \rangle < 1$ will be low for frequency $|k| > k_p$, and so, from equation (3.94), g_k should also be low. This prevents image contrast noise from being over-amplified. Using notation $g(k)$ for g_k, we note that $g(k)$ as a function of k should peak at around k_p; see Fig. 3.19.

3.6.1.3 Third step in the spatial coding: multiplexing by an inverse Fourier transform $\mathcal{O} \to \mathbf{O} = \mathsf{K}_o^{-1} \mathcal{O}$

Consider the special case in which there are as many output neurons as input spatial sampling locations. This is approximately true around fovea, where the densities of cones and retinal

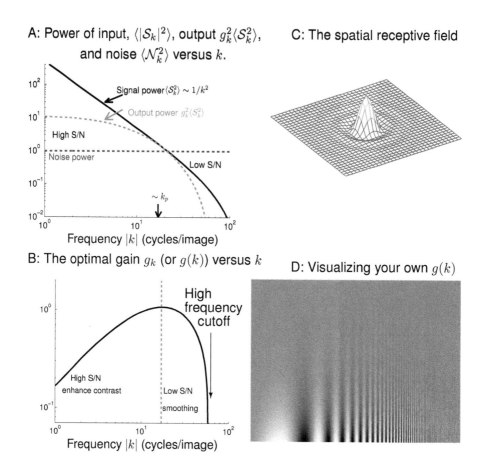

A: Power of input, $\langle|\mathcal{S}_k|^2\rangle$, output $g_k^2\langle\mathcal{S}_k^2\rangle$, and noise $\langle\mathcal{N}_k^2\rangle$ versus k.

C: The spatial receptive field

B: The optimal gain g_k (or $g(k)$) versus k

D: Visualizing your own $g(k)$

Fig. 3.19: Illustration of how input statistics determine optimal encoding in space. A: The power spectra of the input signal $\langle\mathcal{S}_k^2\rangle = 500/(|k|^2 + k_o^2) \cdot (M(k))^2$, the output $g^2(k)\langle\mathcal{S}_k^2\rangle$, and white noise $\langle\mathcal{N}_k^2\rangle = 1$, with $k_o = 0.1$ and $M(k) = \exp[-(k/100)^2]$ is the modulation transfer function of the eye. At $k = k_p$, $\langle\mathcal{S}_k^2\rangle = \langle\mathcal{N}_k^2\rangle$. B: The optimal $g(k)$ (in arbitrary units) by equation (3.94), given input $\langle\mathcal{S}_k^2\rangle/\langle\mathcal{N}_k^2\rangle$ in A, when $\frac{2\lambda}{(\ln 2)\langle N_o^2\rangle} = 60$. The actual curve plotted in B is the effective gain $\tilde{g}(k) = g(k)M(k)$ (which should be proportional to the contrast sensitivity curves observed in experiments). Note that $g(k)$ peaks around $k = k_p$, and $g(k) \propto k$ for small k. C: The shape of the receptive field $K(x) \sim \int \tilde{g}(k)e^{ikx}dk$, determined by $\tilde{g}(k)$. The size of the receptive field should be roughly $1/k_p$. All frequencies k are in units of cycles/image. D: An image of gratings with spatial frequency k increasing from left to right, and contrast increasing from top to bottom. The boundary between invisible and visible contrast regions demonstrates the human contrast sensitivity function $g(k)$. Picture courtesy of Mark Georgeson, 2013.

ganglion cells are comparable. If $\mathsf{U} = \mathsf{K}_o^{-1} = \mathsf{K}_o^\dagger$, the whole transform $\mathsf{K} = \mathsf{U}g\mathsf{K}_o$ takes the form

$$
\mathbf{K} \propto
\begin{pmatrix}
e^{ik_1 x_1} & e^{ik_2 x_1} & \cdots & e^{ik_n x_1} \\
e^{ik_1 x_2} & e^{ik_2 x_2} & \cdots & e^{ik_n x_2} \\
& \vdots & & \\
e^{ik_1 x_n} & e^{ik_2 x_n} & \cdots & e^{ik_n x_n}
\end{pmatrix}
\cdot
\begin{pmatrix}
g_{k_1} & 0 & \cdots & 0 \\
0 & g_{k_2} & \cdots & 0 \\
& & \vdots & \\
0 & & \cdots & 0 \quad g_{k_n}
\end{pmatrix}
$$

$$
\begin{pmatrix}
e^{-ik_1 x_1} & e^{-ik_1 x_2} & \cdots & e^{-ik_1 x_n} \\
e^{-ik_2 x_1} & e^{-ik_2 x_2} & \cdots & e^{-ik_2 x_n} \\
& \vdots & & \\
e^{-ik_n x_1} & e^{-ik_n x_2} & \cdots & e^{-ik_n x_n}
\end{pmatrix}.
\tag{3.162}
$$

Element $\mathbf{K}_{x_i x_j} = (\mathbf{U}g\mathbf{K}_o)_{x_i x_j}$ of this matrix is

$$
\mathbf{K}_{x_i x_j} = \sum_k \mathbf{U}_{x_i k} g_{kk} (\mathbf{K}_o)_{k x_j} \propto \sum_k g_k e^{ik(x_i - x_j)}.
\tag{3.163}
$$

The output neural responses $\mathbf{O} = \mathbf{K}\mathbf{S}$ should be

$$
\begin{pmatrix}
O_{x_1} \\
O_{x_2} \\
\vdots \\
O_{x_i} \\
\vdots
\end{pmatrix}
= \mathbf{K} \cdot
\begin{pmatrix}
S_{x_1} \\
S_{x_2} \\
\vdots \\
S_{x_i} \\
\vdots
\end{pmatrix}.
\tag{3.164}
$$

Hence,

$$
\begin{aligned}
O_{x_i} &= \sum_{x_j} \mathbf{K}_{x_i x_j} S_{x_j} \propto \sum_{x_j} \sum_k g_k e^{ik(x_i - x_j)} S_{x_j} \\
&= \sum_k g_k e^{ikx_i} \sum_{x_j} e^{-ikx_j} S_{x_j} \propto \sum_k (g_k S_k) e^{ikx_i}.
\end{aligned}
\tag{3.165}
$$

Equation (3.163) shows that $\mathbf{K}_{x_i x_j}$ depends only on $x_i - x_j$. When $K(x_i - x_j) = \mathbf{K}_{x_i x_j}$ is written as a spatial function (analogous to function $R^S(x)$ for matrix $\mathbf{R}^S$), it can be seen as a spatial filter

$$
K(x) \propto \sum_k g_k e^{ikx} = \sum_k g(k) e^{ikx}
\tag{3.166}
$$

whose frequency sensitivity function is $g(k)$. Since $g(k)$ peaks at around $k = k_p$, the spatial filter $K(x)$ is a band-pass filter. Using $S(x)$ to substitute for S_x, equation (3.165) can then be written as

$$
O_x = \sum_{x'} K(x - x') S(x').
\tag{3.167}
$$

Hence, O_x is the response from a neuron whose spatial receptive field $K(x - x')$ is centered at location x, as illustrated in Fig. 3.20. In the output vector $\mathbf{O} = (O_{x_1}, O_{x_2}, ..., O_{x_n})^T$, with $O_{x_i} = \sum_x K(x_i - x) S(x)$, different output neurons have the same receptive field shape but different RF centers at x_i for $i = 1, 2, ..., n$. Let $O(x)$ denote O_x to express output $\mathbf{O}$ as a function of space x. Then $O(x) = \sum_{x'} K(x - x') S(x')$ is the result of convolving the visual input image $S(x)$ with filter $K(x)$.

In natural scenes, the signal power $\langle |S_k|^2 \rangle \sim 1/|k|^2$ is approximately independent of the direction of the spatial frequency k. Consequently, the gain $g(k)$ is also independent of the direction of k. This makes the filter $K(x) = \sum_k g(k) e^{ikx}$ spatially isotropic, or radially symmetric. It should be the filter for the retinal output (ganglion) cells. It performs a center-surround transform on the input image $S(x)$ and emphasizes the intermediate frequency band around $k = k_p$ for which the S/N is of order 1. That is, this filter enhances image contrast up to an appropriate level of spatial detail without amplifying contrast noise too greatly. The size of the receptive field is of order $1/k_p$. The gain $g(k)$ is manifest in the contrast sensitivity function of the neurons (or indeed of the animal) to input gratings; see Fig. 3.19 BD.

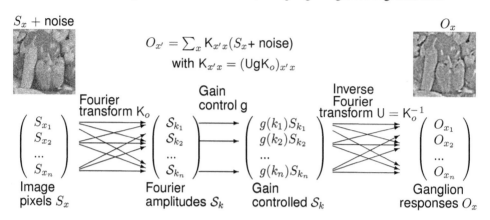

Fig. 3.20: Illustration of the three mathematical, but not neural, ingredients, K_o, g, and $\mathsf{U} = \mathsf{K}_o^{-1}$ that combine to achieve the retinal spatial coding $\mathsf{K} = \mathsf{U}\mathsf{g}\mathsf{K}_o$. Here K_o is the Fourier transform, and U is the inverse Fourier transform. For a retinal neuron whose receptive field is centered at location x', the effective connection strength to this neuron from retinal input S_x at location x is $\mathsf{K}_{x'x} = \sum_k \mathsf{U}_{x'k}\mathsf{g}_{kk}(\mathsf{K}_o)_{kx} \propto \sum_k e^{ikx'}g(k)e^{-ikx} = \sum_k g(k)e^{ik(x'-x)}$. The filter $K(x' - x) \equiv \mathsf{K}_{x'x}$ filters out the high frequency inputs where the noise dominates and emphasizes the intermediate frequency inputs. In the brain, O_x should also include additional neural coding noise $(N_o)_x$.

Our choice $\mathsf{U} = \mathsf{K}_o^{-1}$ is an inverse Fourier transform (since K_o is the Fourier transform). This makes all neurons share the same shape for their receptive fields, but with their center locations x translated. It also makes the RF shape small or localized. As argued in Section 3.5.6, $\mathsf{U} = \mathsf{K}_o^{-1}$ makes a special efficient code for which the signal S is minimally distorted; see Fig. 3.22. If U is an identity matrix, the resulting $\mathsf{K} = \mathsf{g}\mathsf{K}_o$ would have many different shapes of receptive field, each being characterized by a row vector in K. For example, the kth row vector would be

$$(\mathsf{K}_{kx_1}, \mathsf{K}_{kx_2}, ..., \mathsf{K}_{kx_n}) \propto g_k(e^{-ikx_1}, e^{-ikx_2}, ..., e^{-ikx_n}). \tag{3.168}$$

It would be associated with a receptive field function as $K(x) = g(k)e^{-ikx}$, or in real values, $K(x) = g(k)\cos(kx)$. This receptive field is as formally big as the whole visual field, shaped as a sinusoidal wave of a particular spatial frequency k. The output neuron's activity would be $O = \sum_x K(x)S(x) = \sum_x g(k)\cos(kx)S(x)$. Other neurons would have receptive fields as $g(k)\sin(kx)$, $g(k')\cos(k'x)$, and $g(k')\sin(k'x)$ for different frequencies k'. Each output neuron would have to be connected with all input receptors at all locations x with a connection strength, e.g., $g(k)\cos(kx)$. This would make the eye ball truly huge, filled with neural wiring. By contrast, for the actual retinal code, $\mathsf{K} = \mathsf{K}_o^{-1}\mathsf{g}\mathsf{K}_o$, the wiring in the net transform K is local and can be realized by retinal processes involving the bipolar, horizontal, and amacrine cells.

From equation (3.165), we see that $O_x \propto \sum_k (g_k \mathcal{S}_k) e^{ikx}$ is the outcome of an inverse Fourier transform of $g_k \mathcal{S}_k$. This is expected, since U is the inverse Fourier transform. Denoting $\mathcal{S}_k$ by $\mathcal{S}(k)$, analogous to $g(k)$ for g_k, we should then have the Fourier transform of $O(x)$ as $\mathcal{O}(k) \equiv g(k)\mathcal{S}(k)$. Since, for low spatial frequency $k < k_p$, $g^2(k) \propto \langle |\mathcal{S}(k)|^2 \rangle^{-1}$ at least approximately, the mean output power

$$\langle |\mathcal{O}(k)|^2 \rangle = g^2(k) \langle |\mathcal{S}(k)|^2 \rangle \approx \text{constant for } |k| < k_p, \tag{3.169}$$

as illustrated in Fig. 3.19. This means that the output image $O(x)$ has the form of spatial white noise up to spatial frequency k_p. This can also be seen from the output correlation between two neurons O_x and $O_{x'}$ at locations x and x':

$$\langle O_x O_{x'} \rangle = \sum_{ab} \mathsf{K}_{xa} \mathsf{K}_{x'b} \langle \mathcal{S}_a \mathcal{S}_b \rangle = (\mathsf{K} \mathsf{R}^S \mathsf{K}^\dagger)_{xx'} \tag{3.170}$$

$$= (\mathsf{U}g\mathsf{K}_o \mathsf{R}^S \mathsf{K}_o^\dagger g\mathsf{U}^\dagger)_{xx'} = \int dk e^{ik(x-x')} g^2(k) \langle |\mathcal{S}(k)|^2 \rangle.$$

If $g^2(k)\langle |\mathcal{S}(k)|^2 \rangle$ is independent of k, then, since $\int dk e^{ik(x-x')} \propto \delta_{xx'}$, we have that

when $S/N \to \infty$, $\langle O_x O_{x'} \rangle \propto \delta_{xx'}$, and
$\mathsf{K} = \mathsf{K}_{\text{whitening}}$

$$\equiv \mathsf{K}_o^{-1} \begin{pmatrix} (\langle |\mathcal{S}_{k_1}|^2 \rangle)^{-1/2} & 0 & \cdots & 0 \\ 0 & (\langle |\mathcal{S}_{k_2}|^2 \rangle)^{-1/2} & 0 & \cdots \\ \vdots & & & \\ \cdots & \cdots & 0 & (\langle |\mathcal{S}_{k_n}|^2 \rangle)^{-1/2} \end{pmatrix} \mathsf{K}_o. \tag{3.171}$$

This is just as expected for white noise. However, for $|k| > k_p$ the output power decays quickly with increasing k to cut down the amplification of input noise, and so the output is not white at higher spatial frequencies. In general, when $g^2(k)\langle |\mathcal{S}(k)|^2 \rangle$ is approximately a constant only for $|k| < k_p$,

$$\langle O_x O_{x'} \rangle \approx 0, \quad \text{when } |x - x'| > \tfrac{1}{k_p}, \tag{3.172}$$

$$\langle O_x O_{x'} \rangle \neq 0, \quad \text{when } |x - x'| < \tfrac{1}{k_p}. \tag{3.173}$$

Thus, output correlation is particularly significant when the S/N is low, when output power $g^2(k)\langle |\mathcal{S}(k)|^2 \rangle$ decays with k for a large range of k. Large output correlations do indeed occur physiologically (Puchalla, Schneidman, Harris and Berry 2005).

3.6.1.4 Spatial shape of the receptive fields adapts to signal-to-noise levels

In a dimmer environment, inputs are weakened. For example,

$$\frac{\langle \mathcal{S}_k^2 \rangle}{\langle N^2 \rangle} \sim 500/k^2 \quad \text{becomes} \quad \frac{\langle \mathcal{S}_k^2 \rangle}{\langle N^2 \rangle} \sim 5/k^2$$

in a lower level of light. Then, the peak sensitivity of $g(k)$, at which $\frac{\langle \mathcal{S}_k^2 \rangle}{\langle N^2 \rangle} \approx 1$, occurs at a lower frequency, i.e.,

$$k_p \quad \text{is reduced by a factor of 10.}$$

Effectively this reduction of k_p makes $g(k)$ (almost) a low-pass function of k, as shown in Fig. 3.21. Accordingly, rather than enhancing the spatial contrast of the image, the filter

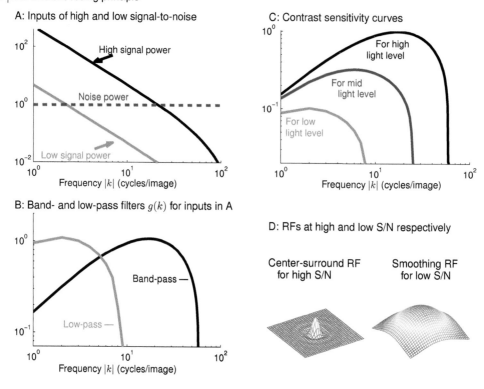

Fig. 3.21: Spatial receptive fields (RFs) adapt to the input signal-to-noise ratio (S/N). A: Input power spectra $\langle S_k^2 \rangle = \hat{S}^2/(|k|^2 + k_o^2) \cdot (M(k))^2$ (where $M(k)$ is the modulation transfer function of the eye) for high and low S/Ns, $\hat{S}^2 = 500$ and $\hat{S}^2 = 5$, respectively. B: For the inputs in A, the optimal effective gains $M(k)g(k)$ lead to band- and low-pass filters, respectively. The two curves have similar peak heights, since the peaks arise when $\frac{\langle S_k^2 \rangle}{\langle N^2 \rangle} \approx 1$. C: (Normalized) contrast sensitivity curves $g(k)\hat{S}$, made by scaling each $g(k)$ by the mean light level $\hat{S}$. The red and black curves correspond to those in B; the blue curve is for $\langle S_k^2 \rangle = \hat{S}^2/(|k|^2 + k_o^2) \cdot (M(k))^2$ when $\hat{S}^2 = 50$. D: Receptive fields $K(x)$ for high (left) and low (right) S/N conditions in A and B, plotted with the same scaling for space, but the amplitude for the smoothing RF is relatively amplified for visualization. The center-surround RF is the same as in Fig. 3.19 C. Other parameters, including λ, k_o, and those for $M(k)$, are the same as in Fig. 3.19.

$K(x)$ integrates over space and smooths the image to boost signal-to-noise at the expense of spatial resolution. This explains the dark adaptation of the RFs of the retinal ganglion cells or LGN cells (Barlow, Fitzhugh and Kuffler 1957, Enroth-Cugell and Robson 1966, Kaplan, Marcus and So 1979), in which the neural receptive fields go from being center-surround contrast enhancing (band-pass) to Gaussian-like smoothing (low-pass) filters. When $k_p \neq 0$, the gain $g(k = 0)$ for the zero-frequency component of spatial input is smaller than the peak $g(k)$ value (when $k \approx k_p$). In this case, the receptive field should have an inhibitory surround beyond the expanded excitatory central region (if the neuron is an on-center cell). This inhibitory surround can only be noticeable by examining a much larger visual field. Despite this antagonistic surround, the filter could still been seen as performing smoothing (although not a strictly low-pass filter), with the antagonism occurring after smoothing inputs in a larger central region and a larger surrounding region. The smoothing filters lead to the

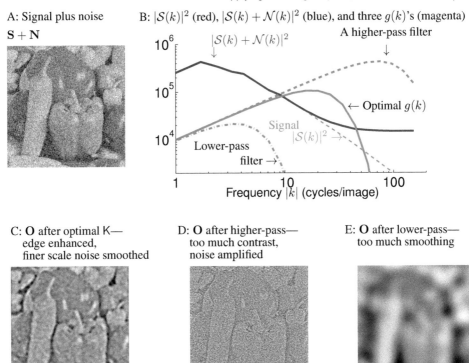

A: Signal plus noise
$S + N$

B: $|\mathcal{S}(k)|^2$ (red), $|\mathcal{S}(k) + \mathcal{N}(k)|^2$ (blue), and three $g(k)$'s (magenta)

A higher-pass filter

$|\mathcal{S}(k) + \mathcal{N}(k)|^2$

$\leftarrow$ Optimal $g(k)$

Signal
$|\mathcal{S}(k)|^2 \rightarrow$

Lower-pass
filter $\rightarrow$

Frequency $|k|$ (cycles/image)

C: **O** after optimal K—
edge enhanced,
finer scale noise smoothed

D: **O** after higher-pass—
too much contrast,
noise amplified

E: **O** after lower-pass—
too much smoothing

Fig. 3.22: Signal transform by optimal and non-optimal codings of spatial visual inputs. A: Image **S** with added white noise **N**. B: Signal power $|\mathcal{S}(k)|^2$ (red-dashed, partly covered by the blue curve), total input power $|\mathcal{S}(k) + \mathcal{N}(k)|^2$ (blue), and three filter sensitivity functions $g(k)$ (magenta) as functions of frequency k. (The vertical axis has an arbitrary scale.) Solid magenta curve: optimal sensitivity $g(k)$, which peaks near k_p where $|\mathcal{S}(k) + \mathcal{N}(k)|^2$ starts to deviate from $|\mathcal{S}(k)|^2$ significantly. For comparison, sensitivity curves with higher and lower-pass characteristics are shown as magenta dashed and magenta dash dotted, respectively. C: Response $\mathbf{O} = \mathsf{K}(\mathbf{S} + \mathbf{N})$ after optimal filtering K with the optimal sensitivity curve $g(k)$. Here, contrast in the image (i.e., edges) is enhanced for low k, where $g(k)$ increases with k; but the image is smoothed for high k, where $g(k)$ decreases with k, to avoid transmitting too much input noise at finer spatial scales. D and E: Outputs **O** when the filters are higher and lower pass, respectively, as depicted in B. Gray scale values shown in A, C, D, and E are normalized to the same range. Adapted with permission from Zhaoping, L., Theoretical understanding of the early visual processes by data compression and data selection, *Network: Computation in Neural Systems*, 17(4): 301–34, Fig. 5, copyright © 2006, Informa Healthcare.

responses of neighboring output neurons being highly correlated, especially when the filter diameters are larger than the distances between the centers of the RFs.

An input $\mathcal{S}(k)$ has contrast $\mathcal{S}(k)/$mean light level; see equation (2.13). Since $g(k)$ is the sensitivity to input $\mathcal{S}(k)$, this implies that $g(k) \times$ mean light level is the contrast sensitivity function. Assuming a mean light level proportional to $\sqrt{\langle|\mathcal{S}(0)|^2\rangle}$, Fig. 3.21 C shows the contrast sensitivity curves that the efficient coding predicts at different levels of adaptation. These can be seen to be in a qualitative agreement with the data in Fig. 2.8.

Consider suddenly leaving a bright outdoor environment and entering a dim room. Before it fully adapts to the lower S/N, the filter K is more sensitive to higher spatial frequencies than is optimal. This non-optimal filter thus passes too much high frequency input noise to the brain, giving rise to something that would have an image content rather like Fig. 3.22 D, in which the salt-and-pepper noise is overwhelming. Conversely, if one goes from a dim environment to a brighter one, the filter would continue to smooth the input images more than necessary until it is fully adapted to the new signal-to-noise regime. The effect of this is demonstrated in Fig. 3.22 E. Of course, in reality, photoreceptors also adapt their dynamic ranges to the mean light level of the environment; hence, vision will be affected by both the adaptation of the photoreceptors *and* the adaptation of the transformation K to the retinal ganglion cells (and of other processes including the pupil size).

3.6.2 Efficient coding in time

Temporal redundancy in natural input implies that much of the information in input signals S arriving at time t has already been communicated by the input at previous time $t' < t$. The efficient coding principle requires that the temporal receptive fields, or the temporal filters, should be optimally designed based on the temporal redundancy (Srinivasan et al. 1982, van Hateren 1992, Li 1992, Dong and Atick 1995, Li 1996). When the S/N is high, it is better to save cost by making the response **O** at time t convey mostly non-redundant information that has not yet been received. Intuitively, this should lead to temporal filters which are more sensitive to temporal contrasts (or temporal derivatives) in the inputs. When the S/N is low, temporal filters should smooth out temporal noise while extracting temporally correlated weak signal.

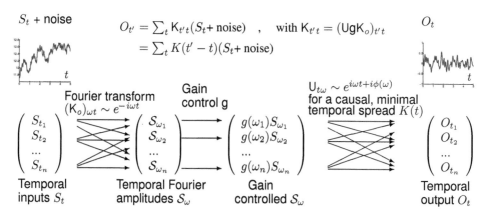

Fig. 3.23: Illustration of the three ingredients, K_o, g, and U, of the transform $K = UgK_o$ that achieves efficient temporal coding, analogous to the efficient spatial coding in the retina (see Fig. 3.20). An additional constraint on the temporal filter $K(t)$ is causality, such that the temporal output $O(t)$ depends only on the previous inputs $S(t')$ at $t' \leq t$.

Coding in time $O_t = \sum_{t'} K_{tt'} S_{t'}$ + noise is mostly analogous to coding in space, with input S_x, indexed by space x, being replaced by input S_t, indexed by time t. Just as for space, the temporal correlation $R_{tt'}^S = \langle S_t S_t' \rangle$ is expected to depend only on $t - t'$ and can be described by a correlation function $R^S(t - t')$. Hence, R^S can be characterized by the power spectrum in time

$$\int dt R^S(t) e^{-i\omega t} \propto \langle |S_\omega|^2 \rangle$$

where $\mathcal{S}_\omega = \sum_t (K_o)_{\omega,t} S_t \sim \int dt e^{-i\omega t} S(t)$ is the temporal Fourier component of $S(t)$ at temporal frequency ω, representing the amplitude of an input principal component. One can expect that $R^S(t)$ will decay monotonically and smoothly with t, and thus that $\langle |\mathcal{S}(\omega)|^2 \rangle$ will decay with ω. Indeed, it has been measured that (Dong and Atick 1995),

$$\langle |\mathcal{S}(\omega)|^2 \rangle \propto \omega^{-2}. \tag{3.174}$$

Given $\langle |\mathcal{S}(\omega)|^2 \rangle$ and the power spectrum of the noise, the temporal frequency sensitivity

$$g(\omega) \sim \left| \int dt K(t) e^{-i\omega t} \right|$$

A: $g(\omega)$ in a V1 cell

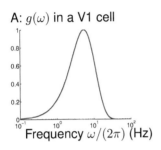

B: The phase spectrum $\phi(\omega)$ for A

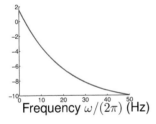

C: The $K(t)$ from A and B

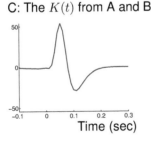

D: $g(\omega)$ for the $K(t)$ in E

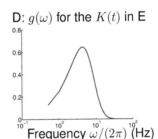

E: $K(t)$ inferred from behavior

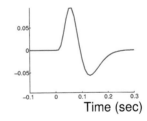

Fig. 3.24: Examples of temporal frequency sensitivity functions $g(\omega)$, impulse response functions $K(t)$, and phase spectra $\phi(\omega)$. A: $g(\omega) = \exp(-(0.1f)^{1.5}) - \exp(-(0.3f)^{1.5})$ (where $f = \omega/(2\pi)$ is the frequency), as typically observed by Holub and Morton-Gibson (1981) in neurons from cat visual area 17 (V1) in response to drifting gratings. B: A particular choice of phase spectrum, $\phi(\omega) = -2\pi\tau_p \cdot [20(1 - e^{-f/20})] + \phi_o$ for $\tau_p = 0.1$ second and $\phi_o = \pi/2$, that makes $K(t)$ causal and of small temporal spread. Note that $\phi(\omega)$ varies approximately linearly with ω except for very large ω. C: The impulse response function $K(t) = \int d\omega g(\omega) e^{i[\omega t + \phi(\omega)]}$ from A and B. D: $g(\omega)$, the amplitude of Fourier transform of the $K(t)$ in E inferred from behavior. E: An impulse response function $K(t) = e^{-\alpha t}[(\alpha t)^5/5! - (\alpha t)^7/7!]$, inferred by Adelson and Bergen (1985) from human visual behavior. Here, $\alpha = 70$/second.

is determined by equation (3.94) according to $\langle|S_\omega|^2\rangle/\langle|N_\omega|^2\rangle$ at this frequency ω. Since $\langle|S(\omega)|^2\rangle$ decays with ω, then, just as for the case of spatial coding, $g(\omega)$ should increase with ω until $\omega = \omega_p$, when the S/N is of order 1. In the example of Fig. 3.24 AD, $\omega_p/(2\pi) \sim 5$ Hz.

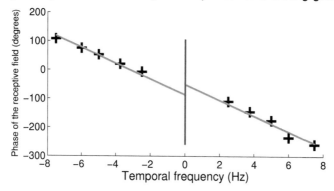

Response phase of a monkey V1 simple cell to a drifting grating

Fig. 3.25: The phase of the temporal filter in a monkey V1 simple cell is roughly a linear function $-\omega\tau_p + \text{sign}(\omega)\phi_o$ of the temporal frequency ω, as in equation (3.176). The data (black crosses) were obtained from the neuron's response to a drifting grating with spatial frequency $k/(2\pi) = 1.42$ cycles/degree. The two red lines, constrained to have the same slope, were fit to the data for positive and negative ω's, respectively. Their vertical offsets on the blue line (at $\omega = 0$) should be at $\phi(k) + \text{sign}(\omega)\phi_o$, where $\phi(k)$ is the phase of the neural spatial filter not considered here when we focus on temporal coding. This ϕ_o, which determines the shape of the temporal filter, is nearly zero here, as is typical in V1 neurons. Data from figure 2b of Hamilton, D. B. and Albrecht, D.G. and Geisler, W.S., Visual cortical receptive fields in monkey and cat: spatial and temporal phase transfer function, *Vision Research*, 29(10): 1285–1308, 1989, Elsevier.

There is, however, a major difference between the spatial and temporal coding. The temporal transform $K_{tt'}$ has to be translation invariant, such that $K_{tt'}$ can only depend on $t - t'$ (described by a function $K(t - t')$), and has to be causal, such that $K_{tt'} = 0$ for $t < t'$, or $K(t) = 0$ for $t < 0$. The function $K(t)$ is also the impulse response function of a neuron. In implementation, it is also desirable for it to have a minimum temporal spread and a short latency. This means that $K(t)$ should be non-zero just for $0 < t < t^*$ for a small t^*. This can be arranged by an appropriate (Dong and Atick 1995, Li 1996) choice of the U matrix in the third step of the efficient coding. In particular

$$U_{t,\omega} \propto e^{i\omega t + i\phi(\omega)}, \tag{3.175}$$

should have the appropriate choice of $\phi(\omega)$, such that the temporal filter

$$K_{tt'} = (UgK_o)_{tt'} = \sum_\omega U_{t,\omega} g_{\omega\omega}(K_o)_{\omega t'} \propto \sum_\omega g(\omega)e^{i\omega(t-t')+i\phi(\omega)}$$

has the desired property. Recall that, in efficient spatial coding, we chose a zero phase for all spatial frequencies k for the U, and this choice helped to make the spatial RF small in space. Such a choice ($\phi(\omega) = 0$ for all ω) for temporal coding would make the temporal filter acausal. The expression for $K_{tt'}$ above is analogous to that for $K_{x_i x_j}$ for the spatial filter in equation (3.163), via the correspondence between (x_i, x_j, k) and (t, t', ω); the difference is in the additional phase factor $e^{i\phi(\omega)}$. The above expression makes it clear that $K_{tt'}$ depends only on $t - t'$, as designed. Hence, we replace $K_{tt'}$ by $K(t - t')$ to get

$$K(t) \sim \int_{-\infty}^{\infty} d\omega g(\omega) e^{i[\omega t + \phi(\omega)]} \propto \int_0^{\infty} d\omega g(\omega) \cos[\omega t + \phi(\omega)].$$

Note that we applied the condition $\phi(\omega) = -\phi(-\omega)$ in the last step above in order that the filter $K(t)$ be real valued. Arranging for $K(t)$ to have a minimal temporal spread means that the individual cosine waves $g(\omega) \cos[\omega t + \phi(\omega)]$ for various frequencies ω should superpose constructively around a particular time τ_p when $|K(t)|$ is large and substantial, and destructively at other times t when $K(t) \approx 0$. Meanwhile, $\tau_p > 0$ is implied by causality. The constructive superposition can be achieved when all waves $g(\omega) \cos[\omega t + \phi(\omega)]$ of various frequencies ω have similar phases (i.e., have temporal coherence) at $t \approx \tau_p$. Thus

$$\omega \tau_p + \phi(\omega) \quad \text{is almost independent of } \omega \text{ for } \omega > 0.$$

Therefore,

$$\phi(\omega) \approx -\omega \tau_p + \text{sign}(\omega)\phi_o, \tag{3.176}$$
$$\text{where sign}(\omega) = 1 \text{ or } -1 \text{ for positive and negative } \omega, \tag{3.177}$$
$$\text{and } \phi_o \text{ is a constant that determines the shape of } K(t). \tag{3.178}$$

Consequently

$$K(t) \stackrel{\propto}{\approx} \int d\omega g(\omega) \cos[\omega(t - \tau_p) + \phi_o].$$

Figure 3.24 illustrates such an example. Since $\tau_p > 0$ is at or near the time when the temporal filter $K(t = \tau_p)$ is at its peak, τ_p is thus effectively the latency of the impulse response function. Physiologically (Hamilton et al. 1989), neurons' temporal filters indeed have phase spectra $\phi(\omega)$ that are roughly linear functions of ω, as in equation (3.176); see Fig. 3.25. Note that $\tau_p = 0$ and $\phi_o = 0$ correspond to the situation in efficient spatial coding that has no causality constraint and has a spatially isotropic RF shape.

3.6.2.1 Predictive coding

A typical band-pass temporal filter $K(t)$, as shown in Fig. 3.24 CE, has the properties of integrating over short time t to average out the noise but having an opponency between $K(t)$ and $K(t')$ over longer times $t - t' > 0$ to boost temporal changes. This is because, typically, $g(\omega \approx 0) \approx 0$ by efficient coding prescription to an overwhelming signal power $\langle |S(\omega = 0)|^2 \rangle$, making the filter relatively insensitive to stationary or slowly changing inputs. Such a filter is often called a predictive filter (in the sense of predictive coding), responding minimally except when inputs change substantially with time. This is illustrated in Fig. 3.26 BC. Since inputs are correlated in time, stationary or slowly changing inputs are expected or predicted. This input predictability implies that the non-zero responses $O(t)$ of the optimal filter $K(t)$ are mainly caused by unpredictable or faster changing inputs, thus minimizing the response amplitudes or the output cost.

The filter responses to natural inputs should have a flat power spectrum, like that of white noise, i.e., $\langle O^2(\omega) \rangle = $ constant, at least for frequencies ω up to an ω_p. This makes outputs $O(t_1)$ and $O(t_2)$ roughly decorrelated for time differences $|t_1 - t_2|$ larger than $1/\omega_p$. This is illustrated in Fig. 3.26 DE, and has been confirmed physiologically for (LGN) neurons which receive inputs from retinal ganglion cells (Dan, Atick and Reid 1996).

Given a stationary (temporally non-changing) input $S(t)$ over time t, the output $O(t) = \int K(t - t')S(t')dt'$ may be more sustained or transient depending on whether the filter

A: A neural filter $K(t)$

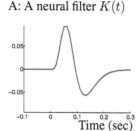

B: An input $S(t)$ with onset and offset

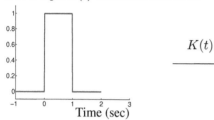

$K(t)$

C: Response $O(t)$ to input $S(t)$ in B

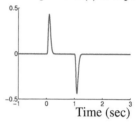

D: "Natural"-looking input $S(t)$

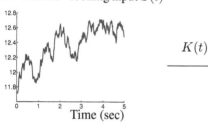

$K(t)$

E: Response $O(t)$ to input $S(t)$ in D

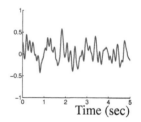

Fig. 3.26: Predictive filtering. A: A neural filter $K(t)$ as in Fig. 3.24 E. B: An input $S(t)$. C: Response $O(t)$ to the input $S(t)$ in B by the filter $K(t)$ in A. Note that responses are mainly to changes in input. D: An input $S(t)$ which has Fourier amplitude $|S(\omega)| \sim 1/(\omega + \omega_o)$ with $\omega_o/(2\pi) = 0.0017$ Hz and random phases. Note that signal $S(t_1)$ and $S(t_2)$ are correlated even for long interval $t_2 - t_1$. E: Response $O(t)$ to the input $S(t)$ in D by the filter in A (the scale on the vertical axis is arbitrary). Note that the response $O(t_1)$ and $O(t_2)$ are not correlated for $t_2 - t_1$ much longer than the width of $K(t)$, and that in the shorter time scale (within 0.1 seconds) input fluctuations are smoothed out in the response.

$g(\omega)$ is more low pass (performing temporal smoothing) or band pass (enhancing temporal contrast) (Srinivasan et al. 1982, Li 1992, Dong and Atick 1995, Li 1996, van Hateren and Ruderman 1998). As in spatial coding, dark adaptation makes the temporal filter of the neurons lower pass, making the responses more sustained (Kaplan et al. 1979).

3.6.3 Efficient coding in color

Color coding (Buchsbaum and Gottschalk 1983, Atick et al. 1992) is analogous to stereo coding, especially if we consider the simple case when there are only two cone types, red and green. This is true of the fovea, which contains few blue cones. The input signal is then

$$\mathbf{S} = (S_r, S_g)^T, \tag{3.179}$$

representing the responses of the red and green cones. The 2×2 input correlation matrix is

$$R^S = \begin{pmatrix} R_{rr}^S & R_{rg}^S \\ R_{gr}^S & R_{gg}^S \end{pmatrix}, \tag{3.180}$$

in which $R_{ij}^S \equiv \langle S_i S_j \rangle$ for $i = r, g$ and $j = r, g$. A substantial correlation R_{rg}^S between the responses of red and green cones is expected, since the cones' spectral sensitivity curves overlap (Fig. 2.16 A) and since inputs at different wavelengths are correlated.

The red and green cones have comparable signal power (Ruderman, Cronin and Chiao 1998, Garrigan et al. 2010), i.e., $R_{rr}^S \approx R_{gg}^S$. Therefore, the situation is almost identical to the stereo coding with a symmetry between the left and right eyes when one identifies (S_r, S_g) with (S_L, S_R). Analogously, the decorrelated channels are

$$\text{Achromatic luminance signal } \mathcal{S}_{LUM} = (S_r + S_g)/\sqrt{2} \text{ and}$$
$$\text{Chromatic red-green opponent signal } \mathcal{S}_{RG} = (S_r - S_g)/\sqrt{2} \tag{3.181}$$

using

$$\begin{pmatrix} \mathcal{S}_{LUM} \\ \mathcal{S}_{RG} \end{pmatrix} = \mathsf{K}_o \begin{pmatrix} S_r \\ S_g \end{pmatrix} \quad \text{with } \mathsf{K}_o = \frac{1}{\sqrt{2}} \begin{pmatrix} 1 & 1 \\ 1 & -1 \end{pmatrix}. \tag{3.182}$$

Like $\mathcal{S}_+$ in stereo coding, the cone summation signal $\mathcal{S}_{LUM}$ (luminance) has a higher signal power than the cone opponent signal $\mathcal{S}_{RG}$ (chromatic). Optimal coding awards appropriate gains, g_{LUM} and g_{RG}, to them according to their respective signal-to-noise ratios. Analogous to the diminishing sensitivity to the ocular opponent signal $\mathcal{S}_-$ in dim light, the gain g_{RG} should diminish in dim light. This is manifest as the loss of color vision in dim light, when the luminance channel $\mathcal{S}_{LUM}$ dominates perception.

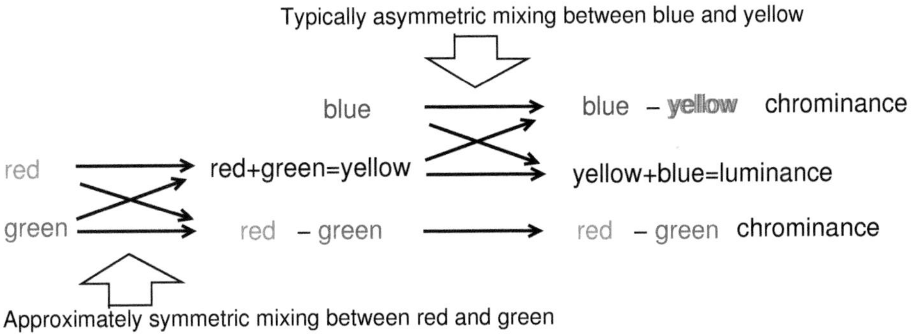

Typically asymmetric mixing between blue and yellow

Approximately symmetric mixing between red and green

Fig. 3.27: Color decorrelation understood by two separate decorrelation steps.

Outside the fovea, three cone types are available, and so the input **S** is

$$\mathbf{S} = (S_r, S_g, S_b)^T, \tag{3.183}$$

with blue signal S_b. The resulting input correlation matrix R^S is a 3×3 matrix:

$$R^S = \begin{pmatrix} R_{rr}^S & R_{rg}^S & R_{rb}^S \\ R_{gr}^S & R_{gg}^S & R_{gb}^S \\ R_{br}^S & R_{bg}^S & R_{bb}^S \end{pmatrix}. \tag{3.184}$$

If we make the approximation (motivated by observations (Ruderman et al. 1998, Garrigan et al. 2010)) of symmetry between the red and green cones, such that $R_{rr}^S = R_{gg}^S$ and $R_{rb}^S = R_{gb}^S$

(noting also that $R_{ij}^S = R_{ji}^S$ for any i, j by definition), then the three decorrelated channels will comprise a luminance channel and two chromatic channels:

$$
\begin{aligned}
\mathcal{S}_{LUM} &= \cos(\alpha)\frac{S_r+S_g}{\sqrt{2}} + \sin(\alpha)S_b, \\
\mathcal{S}_{RG} &= \frac{S_r-S_g}{\sqrt{2}}, \\
\mathcal{S}_{BY} &= -\sin(\alpha)\frac{S_r+S_g}{\sqrt{2}} + \cos(\alpha)S_b, \\
\text{in which} \quad \alpha &\equiv \tfrac{1}{2}\tan^{-1}\frac{4R_{rb}^S}{\sqrt{2}(R_{rr}^S+R_{rg}^S-R_{bb}^S)}.
\end{aligned}
\tag{3.185}
$$

The decorrelating matrix K_o for $(\mathcal{S}_{LUM}, \mathcal{S}_{RG}, \mathcal{S}_{BY})^T = K_o(S_r, S_g, S_b)^T$ is thus

$$
K_o = \begin{pmatrix} \cos(\alpha)/\sqrt{2} & \cos(\alpha)/\sqrt{2} & \sin(\alpha) \\ 1/\sqrt{2} & -1/\sqrt{2} & 0 \\ -\sin(\alpha)/\sqrt{2} & -\sin(\alpha)/\sqrt{2} & \cos(\alpha) \end{pmatrix}.
\tag{3.186}
$$

The luminance signal $\mathcal{S}_{LUM}$ is a weighted sum of the three cone inputs, the chromatic channel $\mathcal{S}_{RG}$ is the same red-green opponency channel in the two-cone simplification. The additional chromatic channel $\mathcal{S}_{BY}$ is an opponent channel between blue and yellow inputs.

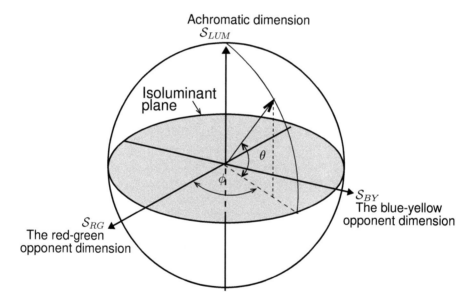

Fig. 3.28: Three-dimensional color space (in the style of Fig. 1A of Lennie et al. (1990)) contains the achromatic dimension $\mathcal{S}_{LUM}$ and the two chromatic dimensions, $\mathcal{S}_{RG}$ and $\mathcal{S}_{BY}$, for red-green and blue-yellow opponencies, respectively. These dimensions have been called the *cardinal dimensions* in color space. Any vector **S** in this space has a color-hue angle ϕ from the red-green axis and an elevation angle θ from the isoluminant plane.

In this red-green symmetric simplification, the color decorrelation from $\mathbf{S} = (S_r, S_g, S_b)^T$ to $\mathbf{S} = (\mathcal{S}_{LUM}, \mathcal{S}_{RG}, \mathcal{S}_{BY})^T$ can be treated as a cascade of two decorrelation steps, each between two input channels. In the first step, the red and green channels are transformed into a summation or yellow channel and a red-green opponency channel; in the second step, the yellow and blue channels are transformed into a yellow-blue weighted summation channel and a yellow-blue weighted opponency channel. This is illustrated in Fig. 3.27. In this simplification, we can borrow many intuitions from efficient stereo coding. In particular, decorrelation between blue and yellow is similar to the atypical stereo coding when there is monocular deprivation, when inputs in one eye dominate inputs in the other eye. This is why typically different weights, $\cos(\alpha)$ and $\sin(\alpha)$, respectively, in equation (3.185) are applied to the yellow and blue channels. In particular, when $\sin(\alpha)$ is very small, as happens when R_{rr}^S and R_{rg}^S are, respectively, much larger than R_{bb}^S and R_{rb}^S (i.e., in a normal environment in which the input powers from red and green cones are much stronger than that from blue cones), the sensitivity $\sin(\alpha)$ of the luminance channel to blue cone input can be quite small. Indeed, it is known empirically that blue cone's input to luminance perception is weak (Vos and Walraven 1971).

Small departures from exact red-green symmetry will modify the weights for the cones in each of the three channels but leave the qualitative composition the same. These three channels, as depicted in Fig. 3.28, have been termed the *three cardinal directions*. However, given adaptation to an environment with different input statistics, for instance in which R_{rr}^S and R_{gg}^S are substantially different from each other or R_{rb}^S and R_{gb}^S are substantially different from each other, then the composition of cones in each decorrelated dimension should change qualitatively. This is because the eigenvectors of the R^S matrix will be structurally different.

Since each of the chromatic channels S_{RG} and S_{BY} arise from a difference between correlated cone signals, their signals will be weaker than that associated with luminance S_{LUM}, which arises as a sum of the cone signals. However, if the input noise is negligible, the gains g_{LUM}, g_{RG}, and g_{BY} for the three channels should specify whitening, so that the ensemble averaged output powers in different decorrelated channels will be equal to each other. In contrast, in high noise conditions, the chromatic channels should desensitize. Explicitly,

$$g_{LUM} < g_{RG} \text{ and } g_{LUM} < g_{BY}, \text{ when signal-to-noise ratio} \gg 1, \quad (3.187)$$

$$g_{LUM} > g_{RG} \text{ and } g_{LUM} > g_{BY}, \text{ when signal-to-noise ratio is small.} \quad (3.188)$$

3.6.3.1 Color appearance change due to adaptation of color coding to input statistics

A purple color becomes more reddish after adapting to strong input variations along the blue-yellow axis

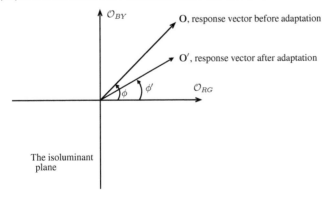

Fig. 3.29: The response vector $\mathbf{O} \equiv (\mathcal{O}_{RG}, \mathcal{O}_{BY})$ in the isoluminant plane provides information about the hue of the input color. Adapting to inputs with strong power $\langle |S_{BY}|^2 \rangle$ in the blue-yellow hue direction reduces g_{BY}. Since $(\mathcal{O}_{RG}, \mathcal{O}_{BY}) = (g_{RG} S_{RG}, g_{BY} S_{BY})$, this adaptation can make a purple input color appear more reddish than usual.

Perceptual color distortions after color adaptation can also be understood from encoding changes in both the compositions and gains g_k of the luminance and chromatic channels that are induced by changes in input statistics. For example, let a purple input color $\mathbf{S} = (\mathcal{S}_{LUM}, \mathcal{S}_{RG}, \mathcal{S}_{BY})^T$ evoke equal output strength in the red-green and blue-yellow channels, i.e.,

$$\mathcal{O}_{RG} = g_{RG}\mathcal{S}_{RG} = \mathcal{O}_{BY} = g_{BY}\mathcal{S}_{BY}; \qquad (3.189)$$

see Fig. 3.29. Suppose that the visual system decodes the hue of colors on the basis of the ratio $\mathcal{O}_{BY} : \mathcal{O}_{RG}$ of the responses of the chromatic channels from the response vector $\mathbf{O} = (\mathcal{O}_{RG}, \mathcal{O}_{BY})$ in the isoluminant plane. This ratio is related to the hue angle ϕ of this vector. The larger (smaller) the ratio associated with an input color, the more bluish (reddish) the color appears. If the visual system adapts to a strong input power in the $\mathcal{S}_{BY}$ channel, then in the near noiseless condition, the efficient coding principle requires that the gain g_{BY} should be reduced. This reduction in gain implies that, in response to the original color, the ratio $\mathcal{O}_{BY} : \mathcal{O}_{RG}$ will be smaller after the adaptation, and so the color will appear redder than usual. This account (Atick et al. 1993) of color adaptation can indeed explain the psychophysical data on apparent hue shift due to adaptation (Webster and Mollon 1991).

3.6.4 Coupling space and color coding in the retina

Physiologically, the encoding of color and space are coupled in, e.g., the red-center-green-surround opponent RFs (Fig. 3.30 and Fig. 2.16 B) of the retinal ganglion cells. This can be understood as follows (Atick, Li and Redlich 1990). When the temporal and stereo input dimensions are ignored, visual inputs $\mathbf{S}$ only depends on space x and cone type $c = (r, g, b)$. Hence,

$$\mathbf{S} = [S_r(x), S_g(x), S_b(x)]^T,$$

in which $S_c(x)$ describes a spatial image in cone $c = r, g, b$.

Meanwhile, the output $\mathbf{O}$ can also be described by three spatial images $O_i(x)$ for $i = 1, 2, 3$. They should be

$$\begin{pmatrix} O_1(x) \\ O_2(x) \\ O_3(x) \end{pmatrix} = \sum_{x'} \begin{pmatrix} K_{1r}(x, x') & K_{1g}(x, x') & K_{1b}(x, x') \\ K_{2r}(x, x') & K_{2g}(x, x') & K_{2b}(x, x') \\ K_{3r}(x, x') & K_{3g}(x, x') & K_{3b}(x, x') \end{pmatrix} \begin{pmatrix} S_r(x') \\ S_g(x') \\ S_b(x') \end{pmatrix}, \qquad (3.190)$$

in which each $K_{ic}(x, x')$ characterizes the effective connection from input $S_c(x')$ in the c cone at location x' to the ith output neuron at location x.

The input correlation matrix R^S is

$$R^S = \begin{pmatrix} R^S_{rr}(x_1, x_2) & R^S_{rg}(x_1, x_2) & R^S_{rb}(x_1, x_2) \\ R^S_{gr}(x_1, x_2) & R^S_{gg}(x_1, x_2) & R^S_{gb}(x_1, x_2) \\ R^S_{br}(x_1, x_2) & R^S_{bg}(x_1, x_2) & R^S_{bb}(x_1, x_2) \end{pmatrix}, \qquad (3.191)$$

in which

$$R^S_{cc'}(x_1, x_2) = \langle S_c(x_1) S_{c'}(x_2) \rangle$$

is the correlation between $S_c(x_1)$ and $S_{c'}(x_2)$ for cone types $c, c' = r, g, b$. A simple assumption (Atick et al. 1990, Atick et al. 1992), confirmed by subsequent measurements (Ruderman et al. 1998), is that the correlation matrix R^S is separable into a cross product of correlations in spatial and chromatic dimensions:

A: Retinal space-color selectivity from multiplexing the luminance and chromatic channels

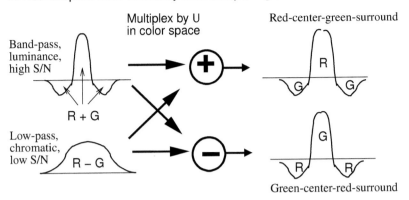

B: Human contrast sensitivity to luminance and red-green chromatic signals

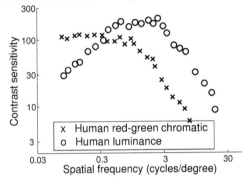

Fig. 3.30: Coupling coding in space and color (restricted, for simplicity to red (R) and green (G)). A: Multiplexing the center-surround, contrast enhancing, achromatic (R + G) filter with the input smoothing chromatic (R − G) filter gives, e.g., a red-center-green-surround RF, as observed in the retina. One-dimensional profiles of the filters $K(x)$ are shown, with the horizontal axis representing 1D space x, and the vertical axis representing the magnitude $K(x)$ at location x. The markings R, G, $R + G$, and $R − G$ indicate the selectivity of the filter $K(x)$ to cones at particular spatial locations x. Adapted with permission from Zhaoping, L., Theoretical understanding of the early visual processes by data compression and data selection, *Network: Computation in Neural Systems*, 2006; 17 (4): 301–34, Fig. 6, copyright © 2006, Informa Healthcare. B: human contrast sensitivity to luminance and red-green signals. Data from Mullen, K. T., The contrast sensitivity of human color vision to red-green and blue-yellow chromatic gratings, *The Journal of Physiology*, 359(1): 381–400, 1985.

$$R^S = R^{S(x)} \otimes R^{S(c)}$$

$$\equiv \begin{pmatrix} R^S_{x_1 x_1} & R^S_{x_1 x_2} & R^S_{x_1 x_3} & \cdots & R^S_{x_1 x_n} \\ R^S_{x_2 x_1} & R^S_{x_2 x_2} & R^S_{x_2 x_3} & \cdots & R^S_{x_2 x_n} \\ \vdots & & & & \\ R^S_{x_n x_1} & R^S_{x_n x_2} & R^S_{x_n x_3} & \cdots & R^S_{x_n x_n} \end{pmatrix} \otimes \begin{pmatrix} R^S_{rr} & R^S_{rg} & R^S_{rb} \\ R^S_{gr} & R^S_{gg} & R^S_{gb} \\ R^S_{br} & R^S_{bg} & R^S_{bb} \end{pmatrix}. \qquad (3.192)$$

This means $R^S_{cc'}(x_1, x_2)$ can be factorized into $R^S_{cc'}(x_1, x_2) = R^{S(x)}_{x_1 x_2} \cdot R^{S(c)}_{cc'}$. Here $R^{S(x)}$ describes the spatial correlation as in Section 3.6.1 between the same type of cones at two spatial locations. So if there are n spatial locations, $\mathsf{R}^{S(x)}$ is an $n \times n$ matrix with elements $\mathsf{R}^{S}_{x_1 x_2}$. Meanwhile $\mathsf{R}^{S(c)}$ is a 3×3 matrix, with elements $\mathsf{R}^{S}_{cc'}$, describing the correlation between different cone signals at the same spatial location, as in Section 3.6.3. So the whole matrix $\mathsf{R}^S = \mathsf{R}^{S(x)} \otimes \mathsf{R}^{S(c)}$ is a $(3n) \times (3n)$ matrix. This book uses a superscript in brackets, e.g., (x) and (c), to denote matrices (e.g., $\mathsf{R}^{S(x)}$ and $\mathsf{R}^{S(c)}$) for a particular input dimension like space x or color c.

Consequently, we may think of input signal $\mathbf{S}$ as being composed of three parallel channels of spatial inputs

$$\mathcal{S}_{LUM}(x), \mathcal{S}_{RG}(x), \mathcal{S}_{BY}(x)$$

for three decorrelated channels, LUM, RG, and BY, in the color dimension. Each of these channels of spatial inputs is an n-dimensional vector for n spatial locations, and should have its own efficient spatial coding, as described in Section 3.6.1. Accordingly, the stronger luminance channel $\mathcal{S}_{LUM}(x)$ requires a center-surround or band-pass spatial filter $K_{LUM}(x)$ to enhance image contrast, while the weaker chromatic channels $\mathcal{S}_{RG}(x)$ and $\mathcal{S}_{BY}(x)$ require spatial smoothing filters $K_{RG}(x)$ and $K_{BY}(x)$ to average out noise. Hence, $K_{RG}(x)$ and $K_{BY}(x)$ should be relatively lower-pass filters compared to $K_{LUM}(x)$, as observed behaviorally and physiologically (Fig. 3.30 and Fig. 2.15). Thus, color vision has a lower spatial resolution than achromatic vision. The output from the spatial encoding is

$$\mathcal{O}_\kappa(x) = \sum_{x'} K_\kappa(x - x') \mathcal{S}_\kappa(x') = (K_\kappa * \mathcal{S}_\kappa)(x)$$

for $\kappa = LUM$, RG, or BY, remembering that $f * g$ denotes convolution of functions f and g.

We can multiplex $\mathcal{O}_{LUM}(x)$, $\mathcal{O}_{RG}(x)$, and $\mathcal{O}_{BY}(x)$ using a 3×3 matrix $\mathsf{U}^{(c)}$ in the color dimension, making

$$\begin{pmatrix} O_1(x) \\ O_2(x) \\ O_3(x) \end{pmatrix} = \mathsf{U}^{(c)} \begin{pmatrix} (K_{LUM} * \mathcal{S}_{LUM})(x) \\ (K_{RG} * \mathcal{S}_{RG})(x) \\ (K_{BY} * \mathcal{S}_{BY})(x) \end{pmatrix}. \tag{3.193}$$

The response $O_i(x)$ is associated with a neuron at location x in the ith output channel, for $i = 1, 2, 3$. Thus

$$O_i = \mathsf{U}^{(c)}_{i,LUM} \cdot K_{LUM} * \mathcal{S}_{LUM} + \mathsf{U}^{(c)}_{i,RG} \cdot K_{RG} * \mathcal{S}_{RG} + \mathsf{U}^{(c)}_{i,BY} \cdot K_{BY} * \mathcal{S}_{BY}. \tag{3.194}$$

This expression means that $O_i(x)$ has three spatial receptive field components, $\mathsf{U}^{(c)}_{i,LUM} K_{LUM}(x)$ for input $\mathcal{S}_{LUM}(x)$, $\mathsf{U}^{(c)}_{i,RG} K_{RG}(x)$ for input $\mathcal{S}_{RG}$, and $\mathsf{U}^{(c)}_{i,BY} K_{BY}(x)$ for input $\mathcal{S}_{BY}$. Since

$$(\mathcal{S}_{LUM}, \mathcal{S}_{RG}, \mathcal{S}_{BY})^T = \mathsf{K}^{(c)}_o (S_r, S_g, S_b)^T,$$

we have

$$O_i = \left[\sum_\kappa \mathsf{U}^{(c)}_{i\kappa} \left(\mathsf{K}^{(c)}_o \right)_{\kappa r} K_\kappa \right] * S_r$$

$$+ \left[\sum_\kappa \mathsf{U}^{(c)}_{i\kappa} \left(\mathsf{K}^{(c)}_o \right)_{\kappa g} K_\kappa \right] * S_g$$

$$+ \left[\sum_\kappa \mathsf{U}^{(c)}_{i\kappa} \left(\mathsf{K}^{(c)}_o \right)_{\kappa b} K_\kappa \right] * S_b. \tag{3.195}$$

Hence, this neuron has a spatial receptive field $\sum_\kappa \mathsf{U}^{(c)}_{i\kappa} \left(\mathsf{K}^{(c)}_o \right)_{\kappa c} K_\kappa(x)$ for $c = r$, g, or b cone inputs, respectively.

In the simplified case when color vision has only red and green cone inputs with $\langle S_r^2 \rangle = \langle S_g^2 \rangle$, the 2×2 decorrelation matrix is

$$K_o^{(c)} = \frac{1}{\sqrt{2}} \begin{pmatrix} 1 & 1 \\ 1 & -1 \end{pmatrix}. \tag{3.196}$$

There are then two cardinal channels, representing luminance $\mathcal{S}_{LUM}(x)$ and cone opponency $\mathcal{S}_{RG}(x)$. The spatial filter $K_{LUM}(x)$ for the stronger $\mathcal{S}_{LUM}(x)$ is a band-pass filter, while $K_{RG}(x)$ for the weaker $\mathcal{S}_{RG}(x)$ should be a lower-pass or smoothing filter. Let $U^{(c)} = \left(K_o^{(c)}\right)^{-1}$, which in this case happen to be $K_o^{(c)}$ itself. Then, according to equation (3.194),

$$O_i = \frac{1}{\sqrt{2}} (K_{LUM} * \mathcal{S}_{LUM} \pm K_{RG} * \mathcal{S}_{RG}). \tag{3.197}$$

Therefore, the neuron associated with O_1 has a $K_{LUM}(x)$ filter for signal $\mathcal{S}_{LUM}(x)$ and the $K_{RG}(x)$ filter for signal $\mathcal{S}_{RG}(x)$; the other neuron associated with O_2 has also a $K_{LUM}(x)$ filter for signal $\mathcal{S}_{LUM}(x)$ but has a $-K_{RG}(x)$ filter for signal $\mathcal{S}_{RG}(x)$. In terms of cone signals $S_r(x)$ and $S_g(x)$,

$$O_i = \frac{1}{2} [(K_{LUM} \pm K_{RG}) * S_r + (K_{LUM} \mp K_{RG}) * S_g]. \tag{3.198}$$

Hence, the neuron associated with O_1 (or O_2) has a receptive field $\frac{1}{2}(K_{LUM} + K_{RG})$ for the red (green) cone image and another receptive field $\frac{1}{2}(K_{LUM} - K_{RG})$ for the green (red) cone image. $K_{LUM}(x)$ is a band-pass filter with a center-surround opponency; and $K_{RG}(x)$ is a low-pass smoothing filter with a larger center and a negligible surround. Adding them gives approximately a center spot; subtracting them gives approximately a surround which opposes the center spot. A red-center-green-surround receptive field, or a green-center-red-surround receptive field, arises accordingly, as illustrated in Fig. 3.30.

The intuitive solution above can be more formally obtained by applying our recipe for efficient coding to the $3n$-dimensional input vector $\mathbf{S}$ as follows (readers not interested in the mathematical details may skip the rest of this section). The $3n \times 3n$ correlation matrix $R^S = R^{S(x)} \otimes R^{S(c)}$ of the input $\mathbf{S}$ is a cross product of an $n \times n$ matrix $R^{S(x)}$, which specifies spatial correlation, and a 3×3 matrix $R^{S(c)}$, which specifies cone correlation. The kth eigenvector of $R^{S(x)}$ is the Fourier wave e^{-ikx}, whereas the κth eigenvector of $R^{S(c)}$ is a 3D vector along $\kappa = LUM$, RG, or BY channel as in Fig. 3.28. Hence, the decorrelating matrix for the $3n \times 3n$ matrix R^S is a cross product of the decorrelating matrix $K_o^{(x)}$ in space and $K_o^{(c)}$ in color

$$K_o = K_o^{(x)} \otimes K_o^{(c)}, \quad \text{such that} \tag{3.199}$$

$$[K_o]_{k,\kappa,x,c} = \left(K_o^{(x)}\right)_{kx} \left(K_o^{(c)}\right)_{\kappa c} = \frac{1}{\sqrt{n}} e^{-ikx} (K_o^{(c)})_{\kappa c}, \tag{3.200}$$

where $K_o^{(x)}$ is an $n \times n$ matrix as in equation (3.158), $K_o^{(c)}$ is a 3×3 matrix as in equation (3.186), and

the $(k,\kappa)^{\text{th}}$ eigenvector of $R^S = k^{\text{th}}$ row vector in $K_o^{(x)} \otimes \kappa^{\text{th}}$ row vector in $K_o^{(c)}$. (3.201)

Let $\lambda_k^{S(x)}$ and $\lambda_\kappa^{S(c)}$ be the corresponding eigenvalues for $R^{S(x)}$ and $R^{S(c)}$. The mean power of the principal component (k, κ) is also a product

$$\langle |\mathcal{S}_{k,\kappa}|^2 \rangle = \lambda_k^{S(x)} \lambda_\kappa^{S(c)},$$

where, typically, $\lambda_k^S \sim 1/k^2$, whereas $\lambda_\kappa^{S(c)}$ is larger for $\kappa = LUM$ than for the chrominance channels. Combined with the power of noise, the signal power $\langle |\mathcal{S}_{k,\kappa}|^2 \rangle$ determines the input signal-to-noise ratio, and so, from equation (3.94), the gain $g_{k,\kappa} \equiv g_\kappa(k)$. This should then determine the diagonal $(3n) \times (3n)$ gain control matrix in the second step of the efficient coding

$$\mathsf{g} \equiv \begin{pmatrix} g_{LUM}^{(x)} & 0 & 0 \\ 0 & g_{RG}^{(x)} & 0 \\ 0 & 0 & g_{BY}^{(x)} \end{pmatrix},$$

in which for $\kappa = LUM, RG,$ or BY,

$$g_\kappa^{(x)} = \begin{pmatrix} g_\kappa(k_1) & 0 & 0 \\ 0 & g_\kappa(k_2) & 0 \\ & \cdots & \\ 0 & 0 & g_\kappa(k_n) \end{pmatrix}. \tag{3.202}$$

The spatial contrast sensitivity to achromatic inputs of spatial frequency k is proportional to the gain $g_{LUM}(k)$; whereas those for chromatic inputs are proportional to $g_{RG}(k)$ and $g_{BY}(k)$. As shown in the behavioral data in Fig. 3.30, the achromatic contrast sensitivity function is band-pass, whereas the chromatic one is lower pass. This is because the achromatic channel has higher signal power, e.g., $\langle |S_{k,LUM}|^2 \rangle > \langle |S_{k,RG}|^2 \rangle$ for each k.

If, for the third step of the efficient coding, we choose an $(3n) \times (3n)$ matrix U as

$$\mathsf{U} = \mathsf{U}^{(c)} \otimes \left(\mathsf{K}_o^{(x)} \right)^{-1}, \tag{3.203}$$

which is a cross product between a 3×3 unitary transform in color dimension and the $n \times n$ matrix which is the inverse of $\mathsf{K}_o^{(x)}$, then the whole $(3n) \times (3n)$ coding matrix is

$$\mathsf{K} = \left[\mathsf{U}^{(c)} \otimes \left(\mathsf{K}_o^{(x)} \right)^{-1} \right] \times \mathsf{g} \times \left[\mathsf{K}_o^{(x)} \otimes \mathsf{K}_o^{(c)} \right]$$

$$= \mathsf{U}^{(c)} \begin{pmatrix} \left(\mathsf{K}_o^{(x)} \right)^{-1} g_{LUM}^{(x)} \mathsf{K}_o^{(x)} & 0 & 0 \\ 0 & \left(\mathsf{K}_o^{(x)} \right)^{-1} g_{RG}^{(x)} \mathsf{K}_o^{(x)} & 0 \\ 0 & 0 & \left(\mathsf{K}_o^{(x)} \right)^{-1} g_{BY}^{(x)} \mathsf{K}_o^{(x)} \end{pmatrix} \mathsf{K}_o^{(c)}$$

$$\equiv \mathsf{U}^{(c)} \begin{pmatrix} \mathsf{K}_{LUM}^{(x)} & 0 & 0 \\ 0 & \mathsf{K}_{RG}^{(x)} & 0 \\ 0 & 0 & \mathsf{K}_{BY}^{(x)} \end{pmatrix} \mathsf{K}_o^{(c)}. \tag{3.204}$$

In the 3×3 array within the (...) above, each element $\mathsf{K}_\kappa^{(x)}$ denotes a $n \times n$ matrix in the spatial dimension and is treated as a single matrix element in the three-dimensional color space by the two 3×3 flanking matrices $\mathsf{U}^{(c)}$ and $\mathsf{K}_o^{(c)}$. Each of the spatial matrices $\mathsf{K}_{LUM}^{(x)}$, $\mathsf{K}_{RG}^{(x)}$, and $\mathsf{K}_{BY}^{(x)}$ defined above is exactly analogous to the coding transform K in the efficient spatial coding, giving spatial receptive field filters $K_{LUM}(x)$, $K_{RG}(x)$, or $K_{BY}(x)$ defined by

$$K_\kappa(x) \propto \sum_k g_\kappa(k) e^{ikx} \tag{3.205}$$

for $\kappa = LUM, RG,$ or BY. We can thus write K as a 3×3 matrix of receptive field functions

$$\mathsf{K}(x - x') = \mathsf{U}^{(c)} \times \begin{pmatrix} K_{LUM}(x - x') & 0 & 0 \\ 0 & K_{RG}(x - x') & 0 \\ 0 & 0 & K_{BY}(x - x') \end{pmatrix} \times \mathsf{K}_o^{(c)} \tag{3.206}$$

in which

$$K_\kappa(x - x') = \left[\left(\mathsf{K}_o^{(x)} \right)^{-1} \times \mathsf{g}_\kappa^{(x)} \times \mathsf{K}_o^{(x)} \right]_{xx'}, \quad \text{for } \kappa = LUM, RG, \text{ and } BY.$$

Since

$$\begin{pmatrix} \mathcal{S}_{LUM}(x) \\ \mathcal{S}_{RG}(x) \\ \mathcal{S}_{BY}(x) \end{pmatrix} = \mathsf{K}_o^{(c)} \begin{pmatrix} S_r(x) \\ S_g(x) \\ S_g(x) \end{pmatrix},$$

the response from the ith (for $i = 1, 2$, or 3) output channel in a neuron whose RF is centered at x is (omitting noise)

$$O_i(x) = \sum_{\kappa = LUM, RG, BY} \mathsf{U}_{i\kappa}^{(c)} \sum_{x'} K_\kappa(x - x') \mathcal{S}_\kappa(x') \tag{3.207}$$

$$\equiv \sum_{c=r,g,b} \sum_{x'} K_{ic}(x - x') S_c(x'), \tag{3.208}$$

in which $K_{ic}(x - x') \equiv [\mathsf{K}(x - x')]_{ic}$ is the $(ic)^{\text{th}}$ element of the 3×3 matrix of functions $\mathsf{K}(x - x')$ in equation (3.206) and describes the receptive field of the neuron in terms of cone inputs $S_c(x)$ for $c = r, g, b$. The above equation is the same as equations (3.194–3.195).

3.6.5 Spatial coding in V1

| A: A filter preferring vertical bars | B: A filter preferring vertical edges | C: A filter preferring tilted bars | D: A filter preferring left tilted, smaller edges |

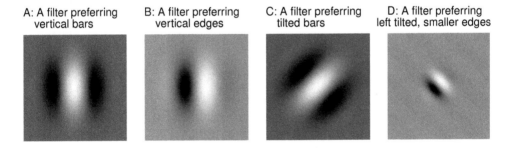

Fig. 3.31: Illustration of oriented, multiscale, spatial filters modeling V1 neurons.

V1 receives retinal outputs via the LGN. The receptive fields of V1 neurons are orientation selective and shaped like small (Gabor) bars or edges (equation (2.44)). Different RFs have different preferred orientations, and they collectively span the whole orientation range. RFs also have a range of different sizes but are roughly scaled versions of each other; this is called multiscale or wavelet coding (Daubechies 1992). Figure 3.31 shows examples of RFs that prefer vertically oriented bars, vertical edges, right tilted bars, and smaller, left tilted edges.

If we focus on spatial coding, can we understand V1 RFs in terms of the efficient coding principle? A second question is whether and why we need a different version of the efficient code in V1 from the one in the retina. We will postpone the discussion on the second question to Chapter 4. First, let us ask whether V1's spatial RFs can be seen as the components of an efficient code $\mathsf{K} = \mathsf{U}g\mathsf{K}_o$ (equation (3.145)) based on an appropriate U. This unitary transformation would be different from the one adopted in the retina (for which $\mathsf{U} = \mathsf{K}_o^{-1}$); so it could allow V1 to exhibit properties in addition to coding efficiency. One of the suggested goals (Li and Atick 1994b) for V1 is to build a translation and scale invariant representation such that a change in the position or scale of visual input will only shift the evoked neural activity pattern from one group of neurons to another, while preserving the same underlying form. To put this more explicitly:

> If two inputs $S(x)$ and $S'(x)$ are related by $S(x) = S'(\alpha x + \Delta x)$, i.e., by a scale change α and a position change Δx, then responses $\mathbf{O} = \mathsf{K}\mathbf{S}$ and $\mathbf{O}' = \mathsf{K}\mathbf{S}'$ should be related as $O_i = O'_{i+a}$ by a shift a of the index of the neurons in the neural population, where a depends only on α and Δx.

(3.209)

For example, when $\alpha = 1$ and $\Delta x \neq 0$, the two inputs $\mathbf{S}$ and $\mathbf{S}'$ are translated versions of each other. If, in addition, the K is as in equations (3.165–3.167), then the respective response patterns $\mathbf{O}$ and $\mathbf{O}'$ will be index-shifted versions of each other, e.g., $(O_1, O_2, ..., O_{10}) = (O'_{21}, O'_{22}, ..., O'_{30})$. However, if $\Delta x = 0$ and $\alpha \neq 1$, then $S(x) = S'(\alpha x)$, and the same code produces response patterns that cannot be so easily related. However, if the receptive fields of two subpopulations of neurons are scaled versions of each other according to factor α, then it is possible for the response pattern to shift from one neural subpopulation to the other as the input scales. The multiscale representation in V1 offers just such a code.

To see which U might lead to the V1 RFs, we first examine two extreme examples Us: one leads to a code with a pure scale invariance property, and the other to a code with a pure translation invariance property. The first one is when $\mathsf{U} = \mathbb{1}$, i.e., the identity matrix. Then, $\mathsf{K} = \mathsf{g}\mathsf{K}_o$ has Fourier wave RFs, as discussed in equation (3.168) and the accompanying text. Each output neuron can be indexed by k, representing the unique input frequency to which it is sensitive. This neuron does not respond to inputs of any other spatial frequency and is thus infinitely finely tuned to frequency. Different neurons, each tuned to a unique k, have different receptive field shapes defined by their respective frequency k. It has a desirable scale-invariance character, since a RF tuned to spatial frequency $\mathbf{k}$ can be seen as a spatially scaled version of another RF tuned to another spatial frequency $\mathbf{k}' \propto \mathbf{k}$. However, when the image of an object is scaled in size, the linear frequency bandwidth for the image changes, thereby changing the number of stimulated neurons by the object image. Hence, this coding is not object scale invariant. Meanwhile, this neuron has no spatial selectivity to inputs and requires very long and numerous neural connections to connect to inputs from all input locations x. Each neuron is tuned to an orientation perpendicular to the 2D wave vector $\mathbf{k} = (k_x, k_y)$, although this orientation tuning is much sharper than for the actual V1 neurons. Apparently, though, our visual system did not choose such a coding, as there is no evidence for global Fourier wave receptive fields with a zero-frequency tuning width.

We previously studied the second example, $\mathsf{U} = \mathsf{K}_o^{-1}$, in the context of retinal RFs (see Section 3.6.1). In detail, this U takes the form,

$$\mathsf{U} = \frac{1}{\sqrt{n}} \begin{pmatrix} e^{ik_1 x_1} & e^{ik_2 x_1} & \cdots & e^{ik_j x_1} & \cdots \\ e^{ik_1 x_2} & e^{ik_2 x_2} & \cdots & e^{ik_j x_2} & \cdots \\ \vdots & & & \vdots & \vdots \\ e^{ik_1 x_m} & e^{ik_2 x_m} & \cdots & e^{ik_j x_m} & \cdots \\ \vdots & & & \vdots & \vdots \end{pmatrix},$$

(3.210)

giving a K for which

$$K(x' - x) \equiv \mathsf{K}_{x', x} = (\mathsf{U}\mathsf{g}\mathsf{K}_o)_{x', x} = \frac{1}{\sqrt{n}} \sum_k e^{ikx'} \left(\frac{1}{\sqrt{n}} g_k e^{-ikx} \right)$$

(3.211)

$$= \frac{1}{n} \sum_k g_k e^{ik(x' - x)}.$$

(3.212)

The right-hand side of equation (3.211) indicates that the net filter is a weighted sum of Fourier waves $\frac{1}{\sqrt{n}} g_k e^{-ikx}$, according to the frequency specific weights $\frac{1}{\sqrt{n}} e^{ikx'}$. Consequently, the resulting filter $K_{x',x}$ is sensitive to all Fourier frequencies with a sensitivity function $g(k)$. Also, as shown in equation (3.212), the weights in the sum are such that the component Fourier waves add constructively at location x' and destructively at locations sufficiently far away from x'. Consequently, this filter is spatially localized around location x', which becomes the center of the resulting RF. Different output neurons, each indexed by their unique x', have the same RF shape and differ only by the center locations of their RFs. This provides a translation invariance property for the representation such that an object shifted in spatial location will activate the same response pattern, albeit in a different population of neurons. Meanwhile, since the RFs are not spatially scaled versions of each other, this representation is not scale invariant. Furthermore, no RF is orientation tuned.

3.6.5.1 The multiscale coding, quadrature phase relationship, and the 1.6 octave frequency bandwidth

The two example Us above represent two extremes: the first picks a single Fourier wave filter and gives no spatial selectivity, whereas the second multiplexes all Fourier wave filters and gives RFs with no diversity in their scale selectivities. In-between these two extremes is a (compromise) code which has multiscale, orientation tuned, and spatially selective RFs. It takes the form of a block diagonal matrix:

$$ \mathsf{U} = \mathsf{U}^{(multiscale)} \equiv \begin{bmatrix} \boxed{\mathsf{U}^{(0)}} & & & \\ & \boxed{\mathsf{U}^{(1)}} & & \\ & & \mathsf{U}^{(2)} & \\ & & & \ddots \end{bmatrix} $$

$$(3.213)$$

Each submatrix $\mathsf{U}^{(s)}$ is itself unitary, so that the whole U is unitary. Different submatrices $\mathsf{U}^{(s)}$ are concerned with different bands of spatial frequencies, such that different frequency bands do not overlap and collectively cover the whole frequency range. (Here, the superscript (s) denotes the sth band, not to be confused with a feature dimension marked as s.) The component $\mathsf{U}^{(s)}$, covering the sth frequency band, has a form like the U in equation (3.210), except that the frequency range for $\mathsf{U}^{(s)}$ is limited within the band. Hence, at the inter-block level, U is like an identity matrix that does not multiplex between different frequency bands. Within a band, each $\mathsf{U}^{(s)}$ multiplexes all frequency filters $\sim g(k)e^{-ikx}$, making RFs of a particular scale.

To see this more clearly, we consider a simplification of a 1D space. In this case, k is a scalar, and $\mathcal{S}(k)$ has power $\langle |\mathcal{S}(k)|^2 \rangle \sim 1/k$, analogous to $\langle |\mathcal{S}(k)|^2 \rangle \sim 1/k^2$ for 2D natural images. In the noiseless (whitening) limit, $g(k) \sim \sqrt{k}$. Visual input space is sampled at locations $x_1, x_2, ..., x_n$, such that $x_m = m\delta x$ for a particular sampling interval δx. The spatial frequencies are thus $k_j = \frac{2\pi}{x_n} j$ for $j = 0, \pm 1, \pm 2, ..., \pm(n/2 - 1), n/2$, whose magnitudes run from a minimum of $|k| = 0$ and a maximum of $|k| = (n\pi/x_n)$. This range of frequencies can be divided into separate bands, $s = 0, 1, ...,$ where $|k|$ increases with s, such that the sth band includes the following spatial frequencies

$$k_j = \pm \frac{2\pi}{x_n} j, \quad \text{for } j^s < j \le j^{s+1} \quad \text{and } j^{s+1}/j^s = \alpha > 1. \tag{3.214}$$

Note that j^s does not mean j to the power of s (similarly for j^{s+1}); the s and $s + 1$ are just superscript indices that denote the scale number. Meanwhile α defines the frequency bandwidth in octaves (see equation (2.48))

$$\text{Bandwidth} \equiv \log_2 \frac{\text{highest frequency in the band}}{\text{lowest frequency in the band}} = \log_2 \alpha. \tag{3.215}$$

In particular, $\alpha = 2$ implies the bandwidth is one octave, which is commonly used in image processing. Meanwhile, $\alpha = 3$ would give a bandwith of around 1.6 octaves, which is comparable to the frequency-tuning bandwidth of the V1 neurons. The sth band contains $n^{(s)} \equiv 2((j^{s+1} - j^s)$ frequencies, which come in pairs of positive and negative frequencies of the same magnitude. Consecutive bands s cover neighboring, but non-overlapping, frequencies, such that they collectively cover the whole frequency range. The bands are also called scales, since different frequencies correspond to different scales. Let us denote the frequencies in the sth band by $\pm k_1^s, \pm k_2^s, ..., \pm k_{n^{(s)}/2}^s$. If we list these frequencies as

$$(f_1, f_2, f_3, f_4, ..., f_{n^{(s)}-1}, f_{n^{(s)}}) = (k_1^s, -k_1^s, k_2^s, -k_2^s, ..., k_{n^{(s)}/2}^s, -k_{n^{(s)}/2}^s),$$

then the submatrix $U^{(s)}$ for scale s in the simplest case (we will see later why this is the simplest case) is an $n^{(s)} \times n^{(s)}$ matrix:

$$U^{(s)} = \frac{1}{\sqrt{n^{(s)}}} \begin{pmatrix} e^{if_1 x_1} & e^{if_2 x_1} & \cdots & e^{if_{n^{(s)}} x_1} \\ e^{if_1 x_2} & e^{if_2 x_2} & \cdots & e^{if_{n^{(s)}} x_2} \\ \vdots & & & \vdots \\ e^{if_1 x_{n^{(s)}}} & e^{if_2 x_{n^{(s)}}} & \cdots & e^{if_{n^{(s)}} x_{n^{(s)}}} \end{pmatrix}. \tag{3.216}$$

Note that each row vector in $U^{(s)}$ is devoted to a given location x_m and covers both the positive and negative frequencies for the band. The locations x_m take the values $x_m = (n/n^{(s)})m\delta x$, with $m = 1, 2, ..., n^{(s)}$, in order to tile the whole input spatial range. As in equations (3.211–3.212), albeit replacing the sum by integration, the RFs in this band are

$$K^{(s)}(x_m - x) = \frac{1}{\sqrt{n^{(s)}}} \int_{|k| \in (k_{j^s}, \alpha k_{j^s}]} dk \left(e^{ikx_m} \left[\frac{1}{\sqrt{n}} g(k) e^{-ikx} \right] \right)$$

$$= \frac{1}{\sqrt{n^{(s)}}n} \int_{k_{j^s}}^{k_{j^{s+1}}} dk \{ g(k) \cos[k(x_m - x)] \}. \tag{3.217}$$

Hence the mth neuron in this band has its receptive field centered at location x_m, while different neurons share the same receptive-field shape. Let us compare this RF filter with a 1D Gabor filter of the form (see equations (2.44) and (2.47))

$$\text{Gabor filter } K(x' - x) \propto \exp\left(-\frac{(x' - x)^2}{2\sigma^2} \right) \cos[k(x' - x)]$$

$$\propto \int dk' \exp\left(-\frac{(k' - k)^2 \sigma^2}{2} \right) \cos[k'(x' - x)].$$

We can see that the two filters have similar shapes. Both are weighted sums of coherent cosine waves in a restricted band of frequencies. The Gabor filter applies its restriction by specifying weights according to a Gaussian envelope centered on the preferred frequency.

By comparison, the restrictions associated with filter $K^{(s)}(x_m - x)$ involves hard frequency bounds k_{j^s} and k_{j^s+1}, with the weights $g(k)$ playing a lesser role.

Meanwhile, spatially downscaling this RF filter $K^{(s)}(x)$ by a factor α produces the RF for the next, $(s+1)^{\text{th}}$, scale. This can be seen by noting that $g(\alpha k) = \sqrt{\alpha} g(k)$ since $g(k) \propto \sqrt{k}$, and that $n^{(s+1)} = \alpha n^{(s)}$, and hence

$$
\begin{aligned}
K^{(s+1)}(x_m - x) &= \frac{1}{\sqrt{n^{(s+1)} n}} \int_{k_{j^s+1}}^{k_{j^s+2}} dk \left\{ g(k) \cos[k(x_m - x)] \right\} \\
&= \frac{\alpha}{\sqrt{n^{(s+1)} n}} \int_{k_{j^s+1}}^{k_{j^s+2}} \frac{dk}{\alpha} \left\{ g\left(\alpha \frac{k}{\alpha}\right) \cos\left[\frac{k}{\alpha} \alpha(x_m - x)\right] \right\} \\
&= \frac{\alpha}{\sqrt{n^{(s)} n}} \int_{k_{j^s}}^{k_{j^s+1}} dk \left\{ g(k) \cos[k\alpha(x_m - x)] \right\} \\
&= \alpha K^{(s)}[\alpha(x_m - x)].
\end{aligned}
\tag{3.218}
$$

Hence, the receptive fields in different scales s are spatially scaled versions of each other according to a scale factor α. (Given an α, we note that the scale invariance is limited, since only image scaling by a factor of α or its integer power can achieve invariance.) Meanwhile, the total spatial weight $\int dx K^{(s)}(x)$ of receptive fields $K^{(s)}(x)$ is independent of s. When $g(k)$ is not exactly a power of k in the case of substantial input noise, the scaling relationship between $K^{s+1}(x)$ and $K^s(x)$ holds only approximately. Note that the RFs are now much smaller than a full Fourier wave but still larger than the "retinal ganglion" RF, which involves multiplexing all the Fourier waves across the whole frequency range.

Although equation (3.217) suggests that the different RFs within a frequency band share the same RF shape, this is not generally the case. When the bandwidth scale factor $\alpha > 2$, it is more complex to ensure that the submatrix $\mathsf{U}^{(s)}$ is unitary when the frequency band does not cover the whole frequency range. Specifically, and without diving into the details in the footnote,[9] it turns out that α is restricted to particular discrete values. One discrete value is

[9] Here are some details (this is amplified in Li and Atick (1994b)). We have seen that to make the RFs in each scale s share the same RF shape, the submatrix $\mathsf{U}^{(s)}$ is forced to have the form $\left(\mathsf{U}^{(s)}\right)_{m,k_j^s} \propto \exp\left[ik_j^s x_m\right]$. However, to make $\mathsf{U}^{(s)}$ unitary for $\alpha > 2$, this form has to be compromised. If one allows several, say q, RF shapes for neurons within this scale band, unitary $\mathsf{U}^{(s)}$ is still possible. In this case, each cell has a RF which is identical to (or is the off-cell type of) the RF of another cell at a center location q lattice space away in the same scale (i.e., $x_m \to x_{m+q}$). Hence, $\left(\mathsf{U}^{(s)}\right)_{m,k_j^s} \propto \exp\left[ik_j^s x_m\right]$ should be modified to

$$
\left(\mathsf{U}^{(s)}\right)_{m,k_j^s} = \frac{1}{\sqrt{n^{(s)}}} \exp\left[ik_j^s x_m + i\text{sign}(k_j^s)(-m \cdot \psi + \phi)\right],
\tag{3.219}
$$

in which ϕ is an arbitrary phase, the sign function $\text{sign}(k_j^s) = 1$ or -1 for positive or negative k_j^s, respectively, and

$$
\psi = p\pi/q
\tag{3.220}
$$

for two relatively prime integers p and q. This means the number of RF shapes in any given scale will be q. The constraint that $\mathsf{U}^{(s)}$ be unitary is now

$$
\sum_{j=j^s+1}^{j^{s+1}} \left(\mathsf{U}^{(s)}\right)_{m,k_j^s} \left(\left(\mathsf{U}^{(s)}\right)_{m',k_j^s}\right)^* + c.c. = \delta_{m,m'}
\tag{3.221}
$$

where $c.c.$ means complex conjugate. The solution to the above equation gives $\psi = \frac{2j^s+1}{j^{s+1}-j^s}\frac{\pi}{2}$. Combining with $\psi = p\pi/q$ gives a non-trivial relationship

$$
j^{s+1} = \frac{(q+p)}{p} j^s + \frac{q}{2p}.
\tag{3.222}
$$

$\alpha = 3$, which gives a frequency bandwidth comparable to V1 RFs. It gives

$$K^{(s)}(x_m - x) \propto \int_{k_{js}}^{k_{js+1}} dk \left(g(k) \cos \left[k(x_m - x) - m\frac{\pi}{2} + \phi \right] \right), \qquad (3.226)$$

and bandwidth = $\log_2 \alpha = 1.6$ octaves. (3.227)

Now the phase $-m\frac{\pi}{2} + \phi$ of the receptive field centered at location x_m changes from location to location, i.e., from neuron to neuron, indexed by integer m. Neighboring RFs have a phase difference of $\pi/2$ and hence have a quadrature relationship with each other. This has been observed in physiological experiments (Pollen and Ronner 1981, DeAngelis, Ghose, Ohzawa and Freeman 1999) but remains somewhat controversial. When $\phi = 0$, the two RF shapes are even and odd functions of space, respectively, corresponding to bar and edge detectors. Meanwhile, the bandwidth of the RFs are 1.6 octaves, resembling the physiological values (De Valois et al. 1982). On a linear scale, the bandwidth $\Delta k \equiv k_{js+1} - k_{js}$ increases linearly with increasing center spatial frequency $(k_{js+1} + k_{js})/2$ of the receptive fields.

In the 2D space, the frequency is a 2D vector, $\mathbf{k} = (k_x, k_y)$. When the frequency range for submatrix $\mathsf{U}^{(s)}$ is non-isotropic, the resulting RF is spatially oriented. This is rather similar to the way that the preferred spatial frequencies of an oriented RF in Fig. 3.31 are not isotropically distributed. Thus, the orientation selectivity of the RFs arises from the multi-band decomposition.

3.6.5.2 An overcomplete representation

In principle, a complete multiscale representation, with as many V1 neurons as spatial samples on the retina, could be constructed by making each 2D RF filter a cross product of two 1D RF filters (i.e., the 2D filter is separable into the two spatial dimensions), each along one of the two orthogonal spatial dimensions (Li and Atick 1994b). If the V1 spatial code was constructed this way, there would only be two possible preferred orientations of the neurons corresponding to the two spatial dimensions chosen (e.g., one horizontal and one vertical). Such a coding would also lead to some complex shaped RFs, e.g., shaped like checker patterns, which have not been observed in V1 RFs in physiological experiments. However, in fact, the V1 code is overcomplete, with about 100 times as many V1 neurons as retinal ganglion cells. The preferred orientations of the V1 neurons span the whole range of $180°$. The spatial frequency bands sampled by different neurons overlap with each other extensively, and the frequency

In a discrete system, the only acceptable solutions have $q/2p$ be an integer, e.g., $q = 2$ and $p = 1$ give the scaling $j^{s+1} = 3j^s + 1$. This solution corresponds to the V1 coding. In the continuum limit, $n \to \infty$ and $j^s \to \infty$; then we simply have $j^{s+1} = \frac{(q+p)}{p} j^s$ for any q and p. Summarizing,

$$\left(\mathsf{U}^{(s)} \right)_{m,k_j^s} = \frac{1}{\sqrt{n^{(s)}}} \exp \left[ik_j^s x_m + i\frac{k_j^s}{|k_j^s|} \left(-m\frac{p\pi}{q} + \phi \right) \right], \qquad (3.223)$$

in which p and q are relative prime integers and ϕ is an arbitrary phase angle. This implies for the RFs and the bandwidth factor α:

$$K^{(s)}(x_m - x) \propto \int_{k_{js}}^{k_{js+1}} dk \left(g(k) \cos \left[k(x_m - x) - m\frac{p\pi}{q} + \phi \right] \right) \quad \text{and} \quad \alpha = \frac{p+q}{p}. \qquad (3.224)$$

For $p = q = 1$, different receptive fields (associated with different m) have the same shape (except for their polarity), and receptive fields between neighboring scales are scaled by a factor of $\alpha = 2$, giving a bandwidth of one octave. However, the situation in V1 resembles the case with $p = 1$ and $q = 2$, when

$$\left(\mathsf{U}^{(s)} \right)_{m,k_j^s} = \frac{1}{\sqrt{n^{(s)}}} \exp \left[ik_j^s x_m + i\frac{k_j^s}{|k_j^s|} \left(-m\frac{\pi}{2} + \phi \right) \right], \qquad (3.225)$$

leading to the results in equation (3.226).

space is thus redundantly sampled. The center locations of the RFs are more densely distributed than required for a complete tiling of the visual space. Such an overcomplete representation of scales and locations should improve scale and translation invariance, since it allows invariance for many possible changes of scales and positions of an object image.

Figure 3.32 schematically illustrates how different frequency subbands are carved up by the V1 coding scheme. The ranges of the different frequency bands s overlap extensively. Since the size of a RF is inversely proportional to its frequency bandwidth, the V1 RFs are smaller than the large RFs, which span the whole spatial range, when U is the identity matrix and picks single frequencies. However, the V1 RFs are larger than the retinal RFs which, with $U = K_o^{-1}$, multiplex all frequencies. The RFs in different bands s can be scaled and rotated versions of each other when the frequency bands are scaled as in Fig. 3.32, such that the linear bandwidth scales with the central frequency of the band. The lowest frequency band, which covers the zero frequency $k = 0$, is an exception. The RFs tuned to higher frequencies are smaller than those tuned to lower frequencies. The number of neurons in each band s scales with the linear frequency bandwidth, so that there are more neurons having smaller RFs than there are neurons having larger RFs, i.e., more neurons are tuned to higher spatial frequencies. (Physiologically, when the RF sizes are too small, the number of neurons observed decreases with decreasing RF size (De Valois et al. 1982). This is expected, since there should be a cut-off frequency for frequency sampling in the cortex. Additionally, when the signal-to-noise is poor at higher spatial frequencies, neurons with low sensitivities $g(k)$ at high spatial frequencies k are hard to observe, because they are less responsive to inputs.) This V1 coding scheme is similar to many multiscale image processing schemes, such as those involving wavelets.

Note that at the smallest frequencies in Fig. 3.32, the frequency range sampled is isotropic, leading to a non-oriented receptive field. The small bandwidth of this lowest frequency band means that the RFs are large, and the neurons in this band are far fewer than neurons tuned to orientation.

The overcompleteness of the V1 representation is difficult to understand based purely on the efficient coding principle. This is because overcompleteness is associated with increased redundancy and requires additional cost for the additional neurons; see Chapter 4 for a more detailed discussion. There have been proposals that the overcompleteness helps to build a tight frame, such that weighted linear sums of the responses of various V1 neurons can be used to construct arbitrary functional mappings from inputs to outcomes or to reconstruct visual input (Lee 1996, Salinas and Abbott 2000). Such goals are beyond the efficient coding framework and are likely to serve visual decoding.

Meanwhile, we can ask whether the efficient coding principle can provide insights into V1's representation of other feature dimensions—color, time, and stereo. The representation of these non-spatial features in V1 is typically manifest in the interaction between the spatial and non-spatial features. For example, a disparity selective receptive field involves both spatial and ocular dimensions. Hence, to examine V1's non-spatial feature coding, we simply model the V1 spatial coding using the form shown in Fig. 3.32 and employ this model to see whether we can understand V1 coding of non-spatial features. In particular, each spatial RF is modeled parametrically as

$$K(x) = A \exp\left(-\frac{x^2}{2\sigma^2}\right)\cos(\hat{k}x + \phi), \tag{3.228}$$

$$\text{in which} \quad A \propto g(\hat{k}) \quad \text{and} \quad \sigma \propto \frac{1}{\hat{k}},$$

in which $\hat{k}$ is the preferred spatial frequency of the neuron, ϕ is a phase parameter, σ

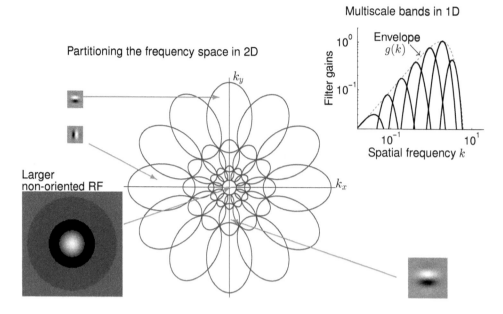

Fig. 3.32: Schematic illustration of how the frequency space $\mathbf{k} = (k_x, k_y)$ can be carved up in a multiscale representation. Each ellipse (and its mirror image with respect to the origin) in the $\mathbf{k} = (k_x, k_y)$ space marks the rough boundary of the frequency range for a particular group of neurons tuned to that range. Four examples of RFs are illustrated for four ellipses: three of them are oriented, but the one at the lowest frequency is isotropic. Note that the size of the RF decreases with the higher center frequency of the band. At the upper right corner is an illustration, in 1D frequency space, of how the partition of the frequency space into multiscale filters (the black solid curves) is enveloped by the gain function $g(k)$ (the red dashed curve) determined by efficient coding.

characterizes the RF size, and A is the amplitude of the Gabor filter, approximated (according to equation (3.226)) as being proportional to the contrast sensitivity $g(\hat{k})$ at the preferred frequency $\hat{k}$. As will be clear soon, in terms of interaction with the coding of non-spatial features, the two most relevant characteristics of the above model for $K(x)$ are the amplitude A and phase ϕ. The amplitude A is proportional to $g(\hat{k})$, the contrast sensitivity, which is determined by the input signal-to-noise according to the efficient coding principle. The phase ϕ is a free parameter corresponding to the free parameter ϕ in the $\mathsf{U}^{(s)}$ matrix. The $K(x)$ model above omits any explicit expression for the spatial dimension along the preferred orientation of the neuron, in order to avoid notation that is unnecessary for the purpose of understanding V1 coding in non-spatial features.

3.6.6 Coupling the spatial and color coding in V1

According to equation (3.194), an efficient code for the cone input images

$$(S_r(x), S_g(x), S_b(x))^T$$

gives neural response in the ith output channel

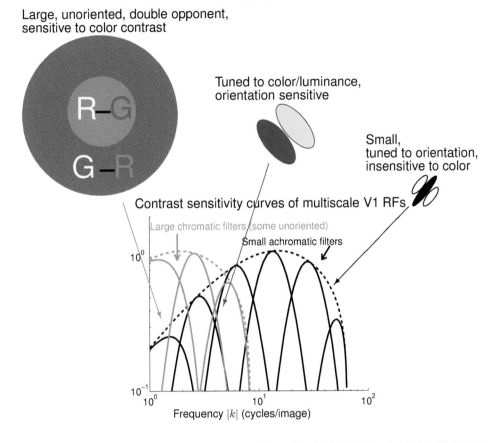

Large, unoriented, double opponent, sensitive to color contrast

Tuned to color/luminance, orientation sensitive

Small, tuned to orientation, insensitive to color

Contrast sensitivity curves of multiscale V1 RFs

Large chromatic filters (some unoriented)

Small achromatic filters

Frequency $|k|$ (cycles/image)

Fig. 3.33: Interaction between color and spatial coding in V1 (Li and Atick 1994b). V1 neurons combine chromatic and achromatic filters, which are sensitive to coarser and finer scales, respectively. V1 neurons tuned to color tend to have larger receptive fields and are often not tuned to orientation, since the chromatic channel has lower signal power. Orientation tuned neurons tuned to intermediate scales tend to be sensitive to both color and luminance. Neurons tuned to the highest spatial frequencies tend to be untuned to color.

$$O_i(x) = \mathsf{U}^{(c)}_{i,LUM} \cdot (K_{LUM} * \mathcal{S}_{LUM})(x) \tag{3.229}$$

$$+ \, \mathsf{U}^{(c)}_{i,RG} \cdot (K_{RG} * \mathcal{S}_{RG})(x) \tag{3.230}$$

$$+ \, \mathsf{U}^{(c)}_{i,BY} \cdot (K_{BY} * \mathcal{S}_{BY})(x), \tag{3.231}$$

in which

$$(\mathcal{S}_{LUM}(x), \mathcal{S}_{RG}(x), \mathcal{S}_{BY}(x))^T = \mathsf{K}^{(c)}_o \, (S_r(x), S_g(x), S_b(x))^T$$

are the three cardinal images after decorrelation transform $\mathsf{K}^{(c)}_o$ in the cone feature dimension, and $K_{LUM}(x)$, $K_{RG}(x)$, and $K_{BY}(x)$ are the three respective spatial receptive fields associated with the cardinal images based on their respective signal-to-noise power spectra.

In multiscale coding in V1, different groups of neurons are involved in representing different spatial frequency bands. Consequently, each spatial filter $K_\kappa(x)$ in the above equation should be replaced by its corresponding multiscale V1 version $K^{(s)}(x)$ involving neurons

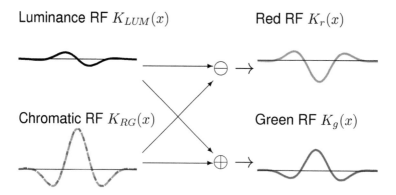

Fig. 3.34: Illustration of how to combine luminance (R + G) and chromatic (R − G) input channels (when blue cone inputs are omitted) $K_{LUM}^{(s)}(x)$ and $K_{RG}^{(s)}(x)$ to make receptive fields $K_r^{(s)}(x)$ and $K_g^{(s)}(x)$ for the red and green inputs, according to equation (3.235). In this example, the chromatic channel is most sensitive, leading to an opponency between red and green in the receptive fields. This typically occurs in low spatial frequency bands. Because this cell has opponencies in both color and space, it is called a double-opponent cell.

associated with the sth spatial frequency band. This leads to the neural response[10]

$$O_i^{(s)}(x) = \mathsf{U}_{i,LUM}^{(c)} \int dx' K_{LUM}^{(s)}(x - x') \mathcal{S}_{LUM}(x')$$
$$+ \mathsf{U}_{i,RG}^{(c)} \int dx' K_{RG}^{(s)}(x - x') \mathcal{S}_{RG}(x')$$
$$+ \mathsf{U}_{i,BY}^{(c)} \int dx' K_{BY}^{(s)}(x - x') \mathcal{S}_{RG}(x'). \tag{3.232}$$

This means that the signal $\mathcal{S}_\kappa(x)$ in each cardinal channel κ is spatially transformed by the corresponding multiscale filter $K_\kappa^{(s)}(x)$ before being multiplexed in color space by the matrix $\mathsf{U}^{(c)}$.

Taking the parametric model of the V1 spatial filter $K_\kappa^{(s)}(x)$ from equation (3.228), we have

$$K_{LUM}^{(s)} = g_{LUM}(\hat{k}) \exp\left(-\frac{x^2}{2\sigma_x^2}\right) \cos(kx + \phi_{LUM}),$$
$$K_{RG}^{(s)} = g_{RG}(\hat{k}) \exp\left(-\frac{x^2}{2\sigma_x^2}\right) \cos(kx + \phi_{RG}),$$
$$K_{BY}^{(s)} = g_{BY}(\hat{k}) \exp\left(-\frac{x^2}{2\sigma_x^2}\right) \cos(kx + \phi_{BY}),$$
$$\tag{3.233}$$

[10]More rigorously, equation (3.232) can be derived by replacing the U matrix in equation (3.203) by a form which is not a cross product of two separate U matrices in the two respective feature dimensions. The U matrix for the spatial dimension is no longer constrained to be $\mathsf{K}_o^{(x)}$ but takes the multiscale version in equation (3.213). Additionally, it can involve different parameters (e.g., the ϕ in equation (3.223)) for the different images $\mathcal{S}_{LUM}(x)$, $\mathcal{S}_{RG}(x)$, and $\mathcal{S}_{BY}(x)$. Consequently, the filters $K_{LUM}^{(s)}(x)$, $K_{RG}^{(s)}(x)$, and $K_{BY}^{(s)}(x)$ in equation (3.232) can have different phases or shapes.

in which $\hat{k}$ is the preferred spatial frequency of the filter for this scale. Importantly, the relative strengths of $K_{LUM}^{(s)}(x)$, $K_{RG}^{(s)}(x)$, and $K_{BY}^{(s)}(x)$ determine whether the response $O_i^{(s)}(x)$ will be more or less tuned to color (Li and Atick 1994b). They are themselves determined by the frequency sensitivities $g_{LUM}(\hat{k})$, $g_{RG}(\hat{k})$, and $g_{BY}(\hat{k})$ in the cardinal channels for the preferred spatial frequency $\hat{k}$. Since the achromatic channel has a higher signal-to-noise ratio than the chromatic channels, according to the efficient coding prescription illustrated in Fig. 3.21, the chromatic sensitivities $g_{RG}(\hat{k})$ and $g_{BY}(\hat{k})$ should dominate at lower spatial frequencies, whereas the achromatic sensitivity $g_{LUM}(\hat{k})$ should dominate at higher spatial frequencies. This is illustrated in Fig. 3.33. Consequently, the color selectivity of the neurons should be influenced by their preferred spatial frequencies or RF sizes. Neurons with larger receptive fields, especially those untuned to orientation since they are tuned to the lowest spatial frequencies (see Fig. 3.32), tend to be selective to color and less sensitive to luminance. In contrast, neurons with smaller receptive fields (and these neurons are typically orientation tuned since they are tuned to higher spatial frequencies) tend to be less sensitive to color.

3.6.6.1 Double-opponency to space and color

To examine joint tuning to space and color in more detail, consider the simple case for which there are only red and green inputs, as in Fig. 3.30. Then,

$$S_{LUM}(x) = \frac{1}{\sqrt{2}}[S_r(x) + S_g(x)], \quad S_{RG}(x) = \frac{1}{\sqrt{2}}[S_r(x) - S_g(x)],$$

$$U^{(c)} = \frac{1}{\sqrt{2}} \begin{pmatrix} 1 & 1 \\ 1 & -1 \end{pmatrix}.$$

The spatial receptive fields are Gabor filters in these two channels:

$$K_{LUM}^{(s)}(x) = A_{LUM} \exp\left(-\frac{x^2}{2\sigma_x^2}\right) \cos(\hat{k}x + \phi_{LUM}), \quad A_{LUM} \propto g_{LUM}(\hat{k}),$$

$$K_{RG}^{(s)}(x) = A_{RG} \exp\left(-\frac{x^2}{2\sigma_x^2}\right) \cos(\hat{k}x + \phi_{RG}), \quad A_{RG} \propto g_{RG}(\hat{k}).$$

After multiplexing by the $U^{(c)}$ matrix, the responses from two output channels ($i = 1, 2$) are (applying equation (3.232), but without the third channel BY)

$$O_i^{(s)}(x) = \frac{1}{\sqrt{2}} \int dx' K_{LUM}^{(s)}(x - x') S_{LUM}(x') \pm \frac{1}{\sqrt{2}} \int dx' K_{RG}^{(s)}(x - x') S_{RG}(x')$$

$$\propto \int dx' \left[K_{LUM}^{(s)}(x - x') \pm K_{RG}^{(s)}(x - x')\right] S_r(x')$$

$$+ \int dx' \left[K_{LUM}^{(s)}(x - x') \mp K_{RG}^{(s)}(x - x')\right] S_g(x'). \tag{3.234}$$

The two expressions in the square brackets [...] in equation (3.234) are thus the receptive fields for the red and green inputs, respectively. They are Gabor filters

$$K_{r,g}^{(s)}(x) \equiv A_{r,g} \exp\left(-\frac{x^2}{2\sigma_x^2}\right) \cos(kx + \phi_{r,g}),$$

whose amplitudes $A_{r,g}$ and phases $\phi_{r,g}$ satisfy

$$A_r e^{i\phi_r} \propto A_{LUM} e^{i\phi_{LUM}} \pm A_{RG} e^{i\phi_{RG}} \quad \text{and} \quad A_g e^{i\phi_g} \propto A_{LUM} e^{i\phi_{LUM}} \mp A_{RG} e^{i\phi_{RG}}. \tag{3.235}$$

The last line above is derived by viewing each Gabor filter $K(x) = A \exp\left(-\frac{x^2}{2\sigma_x^2}\right) \cos(\hat{k}x + \phi)$ as the real part of a complex number with amplitude $A \exp\left(-\frac{x^2}{2\sigma_x^2}\right)$ and phase $\hat{k}x + \phi$. Hence, summation or difference of filters gives the real part of the summation or difference of the corresponding complex numbers. The last equation arises from noting that the Gaussian factor $\exp\left(-\frac{x^2}{2\sigma_x^2}\right)$ is the same in all the filters, and so is the phase component kx. This equation says that, if a Gabor filter $K(x) = A\exp\left(-\frac{x^2}{2\sigma_x^2}\right)\cos(\hat{k}x + \phi)$ can be viewed as a vector with amplitude A and direction ϕ, then the Gabor filter for one cone image (e.g., red) is the vector summation, and that for the other cone image (e.g., green) is the vector subtraction, of the Gabor filters for luminance and chromatic images.

At low spatial frequencies $\hat{k}$, $g_{LUM}(\hat{k})$ can be much weaker than $g_{RG}(\hat{k})$; see Fig. 3.33. This should give $A_{LUM} \ll A_{RG}$, and according to equation (3.235), $A_r e^{i\phi_r} \approx -A_g e^{i\phi_g}$. Consequently, the RFs for the red and green inputs for the neurons should have similar shape and strength but opposite polarity, making the neurons strongly color opponent such that they can be excited by one cone input (e.g., red) while being inhibited by the other cone input (e.g., green) at the same spatial location. This is illustrated in Fig. 3.34. If the chromatic contrast sensitivity $g_{RG}(\hat{k} = 0)$ for the zero-frequency spatial mode is nearly zero, or if the frequency band of the Gabor filter does not include the zero frequency, then the neuron should be insensitive to zero-frequency inputs, making its RF spatially opponent. Because the receptive fields are quite large when the preferred spatial frequency is low, the spatially opponent regions may be large and far apart. In such neurons, different subregions within the receptive fields exhibit opposite color preferences, e.g., one subregion can be excited by red and inhibited by green while another subregion can be inhibited by red and excited by green, as in Fig. 3.34. In such neurons, color preference observed using small input spots can disappear when the input spot is enlarged, since opposite color preferences from different subregions can cancel each other out. Because such neurons have both color and spatial opponency, they are called *double-opponent cells*, and have been observed physiologically (Livingstone and Hubel 1984). In comparison, single-opponent cells in the retina (in Fig. 3.30) simply prefer one color without substantial suppression from the other color at a given subregion (center or surround) within a RF.

3.6.6.2   Correlation between selectivities to color, orientation, and scale

For intermediate frequencies $\hat{k}$ where the two dashed contrast sensitivity curves cross in Fig. 3.33, the sensitivities for chromatic and achromatic channels are similar: $g_{LUM}(\hat{k}) \approx g_{RG}(\hat{k})$. This implies that the chromatic and achromatic filters have similar amplitudes: $A_{LUM} \approx A_{RG}$. However, different phases ϕ_{LUM} and ϕ_{RG} can lead to a diversity of cone-specific receptive fields according to equation (3.235), as illustrated in Fig. 3.35.

At even higher spatial frequencies $\hat{k}$, the input power is overwhelmed by noise, and so the efficient code should attenuate the chromatic sensitivity. This gives $g_{LUM}(\hat{k}) \gg g_{RG}(\hat{k})$, and according to equation (3.235), $A_r e^{i\phi_r} \approx A_g e^{i\phi_g}$. The neurons should be insensitive to color since their receptive fields for red and green inputs will be very similar. These neurons are typically tuned to orientation, due to their high spatial frequency preferences.

The correlations between V1 selectivities for color, spatial frequency, and orientation are thus the consequence of the different contrast sensitivity curves $g_{LUM}(k)$ and $g_{RG}(k)$ for the achromatic and chromatic signals, respectively. Since, according to the efficient coding principle, these curves are affected by the input signal-to-noise ratios, the strengths of the correlations observed physiologically should depend on the adaptation conditions in the experiments concerned. For example, in a dim environment, the chromatic signals S_{RG} could be so weak that $g_{RG}(k)$ is only appreciable for very small spatial frequencies $k \approx 0$, when neurons tend to be orientation untuned. Then, a strong correlation between a presence of color tuning and an absence of orientation tuning should be easily observed. However, in bright

Receptive fields $K_{LUM}(x)$ (in black) and $K_{RG}(x)$ (colored) for luminance and chromatic channels —in these examples, they have the same amplitudes but may differ in phase.

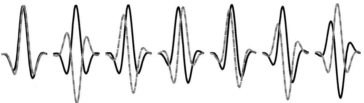

The resulting receptive fields $K_r(x) = K_{LUM}(x) - K_{RG}(x)$ and $K_g(x) = K_{LUM}(x) + K_{RG}(x)$ for the red and green cone inputs, respectively.

Fig. 3.35: Seven examples that demonstrate the various receptive-field shapes for red and green inputs as the result of multiplexing the luminance and chromatic channels with the same input sensitivities $A_{LUM} = A_{RG}$ but different RF phases ϕ_{LUM} and ϕ_{RG}. Each RF for red and green inputs in the bottom row is determined from equation (3.235) based on the RF above it in the top row.

conditions, $g_{RG}(k)$ is appreciable also for higher spatial frequencies when most or all neurons are orientation tuned. Consequently, the correlation between color selectivity and a lack of orientation tuning may not be apparent. The correlation between a lack of color selectivity and high frequency preference may also be absent unless the experimental conditions include very high spatial frequencies, where $g_{RG}(k) \ll g_{LUM}(k)$. In sum, different adaptation conditions in different experiments may cause apparent disagreements between experimental findings regarding these correlations.

3.6.6.3 Conjunctive tuning to color and orientation, and the McCollough effect

As long as the signal-to-noise ratios of the chromatic channels are not too weak, there should be neurons which are tuned to both orientation and color, as discussed above (see Fig. 3.33). Conjunctive tuning to both color and orientation is thought to be behind an illusion called the McCollough effect, which is a color adaptation aftereffect that is contingent on the orientation of the visual inputs. This is demonstrated in Fig. 3.36. A striking characteristic of this effect is its longevity—as long as months; the efficient coding principle does not explain this.

3.6.6.4 A diversity of preferred colors among V1 cells

Although the statistics of natural scenes and properties of the cones lead to the existence of just two chromatic axes for color space, namely red-green and blue-yellow opponency, monkey V1 cells have a diversity of color preferences, including purple, beige, and cyan (Friedman, Zhou and Von der Heydt 2003). This diversity can be understood from equation (3.232), which states that a single neuron has three receptive field components, $U_{i,LUM}^{(c)} K_{LUM}^{(s)}(x)$, $U_{i,RG}^{(c)} K_{RG}^{(s)}(x)$, and $U_{i,BY}^{(c)} K_{BY}^{(s)}(x)$, to inputs along the three cardinal axes respectively. Let us describe $U_{i,\kappa}^{(c)} K_\kappa^{(s)}(x)$ by a Gabor filter with amplitude A_κ and phase ϕ_κ. Using a procedure

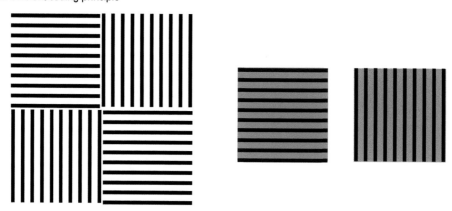

Fig. 3.36: McCollough effect, an aftereffect of color adaptation that depends on the orientation of the visual input (McCollough 1965). To see this, look at the red and green gratings on the right for about 1–2 minutes, alternating your gaze between the two colored gratings. Then look at the black and white grating on the left. The horizontal gratings should appear pinkish, and the vertical gratings should appear greenish.

analogous to that in equations (3.234–3.235), the Gabor filters $A_\kappa e^{i\phi_\kappa}$ for the cardinal signals $\kappa = LUM, RG, BY$ can be transformed into Gabor filters, $A_r e^{i\phi_r}$, $A_g e^{i\phi_g}$, and $A_b e^{i\phi_b}$ for red, green, and blue inputs, respectively. Hence, using $U^{(c)} = (K_o^{(c)})^{-1}$ (with the $K_o^{(c)}$ in equation (3.186) in the 3D color space),

$$O_1(x) = \frac{\cos(\alpha)}{\sqrt{2}}(K_{LUM} * S_{LUM})(x) + \frac{1}{\sqrt{2}}(K_{RG} * S_{RG})(x) - \frac{\sin(\alpha)}{\sqrt{2}}(K_{BY} * S_{BY})(x),$$
$$O_2(x) = \frac{\cos(\alpha)}{\sqrt{2}}(K_{LUM} * S_{LUM})(x) - \frac{1}{\sqrt{2}}(K_{RG} * S_{RG})(x) - \frac{\sin(\alpha)}{\sqrt{2}}(K_{BY} * S_{BY})(x),$$
$$O_3(x) = \sin(\alpha)(K_{LUM} * S_{LUM})(x) + \cos(\alpha)(K_{BY} * S_{BY})(x).$$

$$(3.236)$$

Since $(S_{LUM}, S_{RG}, S_{BY})^T = K_o^{(c)}(S_r, S_g, S_b)^T$, the receptive fields for various cones r, g, b of the response O_1 and O_2 are

$$A_r e^{i\phi_r} = \frac{\cos^2(\alpha)}{2}A_{LUM}e^{i\phi_{LUM}} \pm \frac{1}{2}A_{RG}e^{i\phi_{RG}} + \frac{\sin^2(\alpha)}{2}A_{BY}e^{i\phi_{BY}},$$
$$A_g e^{i\phi_g} = \frac{\cos^2(\alpha)}{2}A_{LUM}e^{i\phi_{LUM}} \mp \frac{1}{2}A_{RG}e^{i\phi_{RG}} + \frac{\sin^2(\alpha)}{2}A_{BY}e^{i\phi_{BY}},$$
$$A_b e^{i\phi_b} = \frac{\cos(\alpha)\sin(\alpha)}{\sqrt{2}}(A_{LUM}e^{i\phi_{LUM}} - A_{BY}e^{i\phi_{BY}}).$$

Note that the first two equations are the same as equation (3.235) when the blue cones are ignored by taking $\alpha = 0$. The receptive fields of the response O_3 are

$$A_{r,g}e^{i\phi_{r,g}} = \frac{\cos(\alpha)\sin(\alpha)}{\sqrt{2}}(A_{LUM}e^{i\phi_{LUM}} - A_{BY}e^{i\phi_{BY}}),$$
$$A_b e^{i\phi_b} = \sin^2(\alpha)A_{LUM}e^{i\phi_{LUM}} + \cos^2(\alpha)A_{BY}e^{i\phi_{BY}}.$$

Figure 3.35 demonstrates that a diversity of preferences for red or green can arise from merely mixing $K_{LUM}^{(s)}(x)$ and $K_{RG}^{(s)}(x)$ of various phases but comparable amplitudes $A_{LUM} \approx A_{RG}$. An even greater diversity of possibilities for $A_r e^{i\phi_r}$, $A_g e^{i\phi_g}$, and $A_b e^{i\phi_b}$ can arise from moderate variations in the relative strengths among A_{LUM}, A_{RG}, and A_{BY}. Variations in the contrast sensitivity curves ($g_{LUM}(k)$, $g_{RG}(k)$, and $g_{BY}(k)$), and variations in the preferred spatial frequency $\hat{k}$, can all cause such variations in color selectivity of the neurons. For example, if for the preferred frequency $\hat{k}$ of a neuron, $g_{LUM}(\hat{k})$ is much weaker than $g_{RG}(\hat{k})$ and $g_{BY}(\hat{k})$ such that $0 \approx A_{LUM} \ll A_{RG}$ and $A_{LUM} \ll A_{BY}$, and if $\phi \equiv \phi_{RG} = \phi_{BY}$ and $A_1 \equiv A_{RG}/2 > A_2 \equiv A_{BY}\sin^2(\alpha)/2$, then, while the O_3 neuron should have blue-yellow opponency, the $O_{1,2}$ neurons have their color selectivities (to r, g, b):

$$A_r e^{i\phi_r} \approx (\pm A_1 + A_2)e^{i\phi}, \quad A_g e^{i\phi_g} \approx (\mp A_1 + A_2)e^{i\phi}, \quad A_b e^{i\phi_b} \propto -A_2 e^{i\phi}. \quad (3.237)$$

One neuron should exhibit opponency between cyan (blue-green) and orange (red-yellow), and the other neuron between magenta (red-blue) and lime (green-yellow). Some parts of the receptive field are excited by one and inhibited by the other; other parts of its RF have the opposite preferences. If A_{LUM} becomes non-negligible, as for a neuron preferring a higher spatial frequency $\hat{k}$, its preferred colors should include contribution from luminance, e.g., favoring pink or brown.

In some physiological experiments (Lennie, Krauskopf and Sclar 1990), one searches for the input color and/or luminance modulation that most effectively modulates a neuron's response. For example, if a neuron's response is best modulated by an isoluminant input patch that oscillates between being red and green, then the neuron is described as having red-green opponency, with an optimal direction of input modulation being along the direction of $\phi = \theta = 0$ in the 3D color space in Fig. 3.28. If this neuron is also driven by luminance modulations, the optimal direction should have $\theta > 0$. According to our analysis above, the optimal modulation directions for V1 neurons should favor diverse ϕ angles, associated with a diversity of preferred colors. Meanwhile, the optimal θ values, which indicate the degree of sensitivity to luminance, should be greater for neurons preferring higher spatial frequencies, since $g_{LUM}(k)$ is relatively stronger for a larger k. This is consistent with physiological observations (Lennie et al. 1990), showing that the optimal θ values tend to be larger for neurons with oriented rather than non-oriented receptive fields (which are duly tuned to smaller spatial frequencies; see Fig. 3.33).

3.6.7 Coupling spatial coding with stereo coding in V1—coding disparity

The encoding of disparity can emerge from the coupling of spatial and ocular dimensions (Li and Atick 1994a). This is analogous to the coupling between spatial and color dimensions, treating the two eyes (left and right) like the two main cone types (red and green). Luminance and red-green opponent signals (S_{LUM}, S_{RG}) correspond to ocular summation and opponent signals $(S_+, S_-) \equiv \frac{1}{\sqrt{2}}(S_L + S_R, -S_L + S_R)$.

Analogous to equation (3.234), a V1 neuron, whose receptive field is centered at location $x = 0$, responds to the ocular summation $S_+(x)$ and opponent $S_-(x)$ signals according to

$$O = \frac{1}{\sqrt{2}}\left(\int dx K_+(x) S_+(x) + \int dx K_-(x) S_-(x) \right), \quad (3.238)$$

in which $K_\pm$ are Gabor filters:

$$K_\pm(x) = g_\pm \exp\left(-\frac{x^2}{2\sigma^2} \right) \cos(\hat{k}x + \phi_\pm) \quad (3.239)$$

with gains $g_\pm$, phases $\phi_\pm$, and the most preferred frequency $\hat{k}$. (Here, we omit the scale or subband index s to avoid notational clutter.) Analogous to equation (3.235), the neuron's

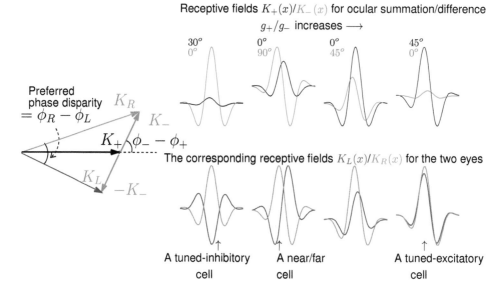

Receptive fields $K_+(x)/K_-(x)$ for ocular summation/difference

g_+/g_- increases $\longrightarrow$

| $30°$ $0°$ | $0°$ $90°$ | $0°$ $45°$ | $45°$ $0°$ |

Preferred phase disparity K_R $= \phi_R - \phi_L$

K_-

K_+ $\phi_- - \phi_+$

K_L

$-K_-$

The corresponding receptive fields $K_L(x)/K_R(x)$ for the two eyes

A tuned-inhibitory cell   A near/far cell   A tuned-excitatory cell

Fig. 3.37: Construction of the receptive fields $K_{L,R}(x)$ for individual eyes, and the preferred disparity of a neuron from the receptive fields $K_\pm(x)$ for the ocular summation and difference channels. On the left, $K_L = g_L e^{i\phi_L}$, $K_R = g_R e^{i\phi_R}$, $K_+ = g_+ e^{i\phi_+}$, and $K_- = g_- e^{i\phi_-}$ are shown as vectors with amplitude g and angle ϕ. The left plot visualizes the relationship in equation (3.240) (ignoring the factor of $1/\sqrt{2}$). On the right are four examples of applying this relationship to obtain the receptive fields. In each example, $K_+(x)$ and $K_-(x)$ are plotted in black and magenta, with their ϕ_+ and ϕ_- indicated by correspondingly colored text; the resulting $K_L(x)$ and $K_R(x)$ are plotted in blue and red. Note that, as g_+/g_- increases, the two receptive fields for the two eyes become more similar and the preferred disparity decreases. Note also that the neuron is more binocular when g_+/g_- is large or when $|\phi_+ - \phi_-|$ is closer to $\pi/2$. Note examples of tuned-excitatory, tuned-inhibitory, and near/far cells.

receptive fields for the left and right eyes are also Gabor filters $K_{L,R} \propto K_+(x) \pm K_-(x)$ with gains $g_{L,R}$ and phases $\phi_{L,R}$ satisfying

$$
\begin{aligned}
g_L \exp(i\phi_L) &= \tfrac{1}{\sqrt{2}} \left[g_+ \exp(i\phi_+) - g_- \exp(i\phi_-) \right], \\
g_R \exp(i\phi_R) &= \tfrac{1}{\sqrt{2}} \left[g_+ \exp(i\phi_+) + g_- \exp(i\phi_-) \right].
\end{aligned}
\tag{3.240}
$$

The above equation should be distinguished from equation (3.134). The angle ϕ in equation (3.134) was associated with the freedom to choose unitary matrix $U^{(e)}$ in the ocular dimension; here, we have removed this freedom by making $U^{(e)} = \left(K_o^{(e)} \right)^{-1}$ (as in color coding), leading to equation (3.238). Meanwhile, the phase angles $\phi_{L,R}$ and $\phi_\pm$ in the above equation arise from the freedom to choose the unitary submatrices $U^{(s)}$ that relate to the spatial dimension. This freedom can be expressed in terms of parameter ϕ in equation (3.225) or equation (3.226) and results in the phases $\phi_\pm$ of the receptive fields $K_\pm(x)$ being free parameters. Apparently, V1 coding uses this freedom for the receptive field phases in its overcomplete representation.

When $\phi_L = \phi_R$, the receptive fields for the two eyes have the same spatial shape and differ only in their overall gains. Such a neuron should prefer zero disparity inputs, when the location of an object in the left eye is the same as that in the right eye (cf. equation 2.63). (For simplicity, we assume that the contrast polarity and position of the object images match those of the receptive fields for individual eyes.) In general, the preferred phase disparity is

$$\phi_R - \phi_L = \arccos\left(\frac{g_+^2 - g_-^2}{\sqrt{(g_+^2 + g_-^2)^2 - 4g_+^2 g_-^2 \cos^2(\phi_+ - \phi_-)}}\right), \tag{3.241}$$

which can be derived from the illustration in Fig. 3.37. This phase disparity can be translated to preferred spatial disparity $= (\phi_R - \phi_L)/\hat{k}$ through the wavenumber $\hat{k}$. When $g_- = 0$, then $\phi_R = \phi_L$ always holds. Therefore, the ocular difference signal through $g_- \neq 0$ is necessary to encode non-zero disparity. In particular, neurons tend to be tuned to larger disparities when $g_- > g_+$, since, from Fig. 3.37 and equation (3.241), it follows that,

$$\begin{aligned} &\text{when } g_- < g_+, \text{ the preferred phase } |\phi_R - \phi_L| < \pi/2, \\ &\text{when } g_- > g_+, \text{ the preferred phase } |\phi_R - \phi_L| > \pi/2. \end{aligned} \tag{3.242}$$

The sensitivities $g_{L,R}$ for the receptive fields of the individual eyes are

$$\begin{aligned} g_R &\propto \left[g_+^2 + g_-^2 + 2g_+g_- \cos(\phi_+ - \phi_-)\right]^{1/2}, \\ g_L &\propto \left[g_+^2 + g_-^2 - 2g_+g_- \cos(\phi_+ - \phi_-)\right]^{1/2}. \end{aligned} \tag{3.243}$$

This makes a neuron equally sensitive to the two eyes when $|\phi_+ - \phi_-| = \pi/2$. However, according to equation (3.241), this in turn makes the two monocular receptive fields different, i.e., $\phi_L \neq \phi_R$, unless $g_- = 0$. By equations (3.241) and (3.243), the receptive fields for the two eyes have similar sensitivities *and* phases when $g_+ \gg g_-$. This is the generalization of equation (3.136) as to how ocular balance depends on the relative gains g_+/g_-.

Meanwhile, according to equation (3.137), the ratio g_+/g_- increases as the size of the V1 receptive field decreases, due to the dependence of signal-to-noise ratio and ocular correlation on the spatial scale of visual inputs. This combined with equation (3.242) leads to:

As receptive size decreases, the preferred disparity phase decreases. (3.244)

This prediction is consistent with physiological observation (Anzai, Ohzawa and Freeman 1999). It is the extension of the prediction from equation (3.137) about the dependence of stereo coding on the spatial scale of the receptive field. A neuron whose preferred disparity $|\phi_L - \phi_R|$ is near zero is called a *tuned excitatory cell*, meaning that the neuron is excited by a zero disparity. An example of such a cell is shown in Fig. 3.37. Conversely, a neuron whose preferred disparity $|\phi_L - \phi_R| \approx \pi$ is inhibited by zero-disparity inputs.

When both $K_L(x)$ and $K_R(x)$ are approximately even functions, $|\phi_L - \phi_R| \approx \pi$ gives rise to a *tuned inhibitory cell* (see an example in Fig. 3.37) that is less inhibited, or even excited, by both near and far inputs. Conversely, when both $K_L(x)$ and $K_R(x)$ are approximately odd functions, $|\phi_L - \phi_R| \approx \pi$ leads to a *near or far cell* (see an example in Fig. 3.37), which is excited by one sign of disparity but inhibited by the other.

Since $|\phi_R - \phi_L|$ decreases with increasing preferred spatial frequency $\hat{k}$, tuned excitatory cells tend to have smaller RFs. Note further that the preferred disparity in visual angle scales inversely with $\hat{k}$ for a given preferred phase disparity $\phi_R - \phi_L$. Hence, cells with smaller RFs prefer zero or small disparities not only because they prefer small phase disparities, but also because their RFs are small. The following observations on disparity tuning in V1 are however not accounted for by the efficient coding principle. Firstly, a V1 cell can also signal non-zero disparity if it has different center locations for the RFs from the two eyes. Also, complex cells can have a disparity tuning curve that is invariant to the input contrast polarity and insensitive to the position of inputs within its RF (Qian 1997, Ohzawa et al. 1990), as shown in equation (2.76).

3.6.8 Coupling space and time coding in the retina and V1

When considering both space x and time t, the visual input can be written as $S(x,t)$. It is assumed to have a translation invariant correlation function

$$\langle S(x_1,t_1)S(x_2,t_2)\rangle = R^S(x - x', t - t') \tag{3.245}$$

with independent components

$$S(k,\omega) \propto \int dx dt\, S(x,t) e^{-ikx-i\omega t}. \tag{3.246}$$

Hence, according to the general efficient coding scheme $\mathsf{K} = \mathsf{U}\mathsf{g}\mathsf{K}_o$, the matrix K_o has elements

$$(\mathsf{K}_o)_{k,\omega,x,t} \propto e^{-ikx-i\omega t}, \tag{3.247}$$

and can be seen as the cross product of two matrices

$$\mathsf{K}_o \equiv \mathsf{K}_o^{(x)} \otimes \mathsf{K}_o^{(t)}, \tag{3.248}$$

one each for spatial and temporal dimensions. Their individual matrix elements are

$$(\mathsf{K}_o^{(x)})_{k,x} \propto e^{-ikx}, \quad (\mathsf{K}_o^{(t)})_{\omega,t} \propto e^{-i\omega t}. \tag{3.249}$$

It has been observed (Dong and Atick 1995) that the signal power $\langle |S(k,\omega)|^2 \rangle$ decreases with increasing frequencies $|k|$ and $|\omega|$. This is expected since the input correlation $\langle S(x_1,t_1)S(x_2,t_2)\rangle$ is larger for smaller $|x_1 - x_2|$ and $|t_1 - t_2|$.

The diagonal elements $g(k,\omega)$ of the gain matrix g in the coding scheme $\mathsf{K} = \mathsf{U}\mathsf{g}\mathsf{K}_o$ are determined by the signal to noise ratios $\langle |S(k,\omega)|^2 \rangle / \mathcal{N}^2$ following equation (3.94).

If, in the final step, the U matrix is the identity, the filters $\mathsf{K} = \mathsf{g}\mathsf{K}_o$ would be acausal spatiotemporal Fourier filters. The filter denoted by frequencies (k,ω) would have an overall sensitivity of $g(k,\omega)$ and could be written:

$$\mathsf{K}_{k,\omega,x,t} \propto g(k,\omega) e^{-ikx-i\omega t}. \tag{3.250}$$

Its output would be $\mathcal{O}(k,\omega) = \sum_{x,t} \mathsf{K}_{k,\omega,x,t} S(x,t) \propto g(k,\omega)S(k,\omega)$. However, calculating this is unrealistic, since it would require inputs to summed over all space x and time t according to weights $\mathsf{K}_{k,\omega,x,t}$.

The simplest U that makes the coding filter causal, with minimum temporal spread (see equation (3.175) and the accompanying text), and also spatially local (see equation (3.162) and the accompanying text), is

$$\mathsf{U}_{x,t,k,\omega} = (\mathsf{U}^{(x)})_{x,k}\left(\mathsf{U}^{(t)}\right)_{t,\omega} \propto e^{ikx} \cdot e^{i\omega t + i\phi(k,\omega)}, \tag{3.251}$$

where $\mathsf{U}^{(x)}_{x,k} \propto e^{ikx}$, and $\mathsf{U}^{(t)}_{t,\omega} \propto e^{i\omega t + i\phi(k,\omega)}$. As in equation (3.175), the phase spectrum $\phi(k,\omega)$ (which satisfies $\phi(k,\omega) = \phi(-k,\omega) = -\phi(k,-\omega)$) is chosen to satisfy the requirements for causality and temporal compactness. This phase spectrum $\phi(k,\omega)$ as a function of ω depends on the spatial frequency k, since given k, $\phi(k,\omega)$ is determined by the sensitivity $g(k,\omega)$, which depends on both k and ω (because $\langle |S(k,\omega)|^2 \rangle$ is a function of both k and ω). (If $\phi(k,\omega)$ did not depend on k, then this U could be written as a cross product between a spatial matrix and a temporal matrix $U \equiv U^{(x)} \otimes U^{(t)}$.) The U in equation (3.251) leads to a spatiotemporal receptive field

$$K(x - x', t - t') \equiv K_{x,t,x',t'} = (UgK_o)_{x,t,x',t'}$$
$$= \int dkd\omega \left[U_{x,t,k,\omega} g(k, \omega)(K_o)_{k,\omega,x',t'} \right]$$
$$\propto \int dkd\omega \left[g(k, \omega) e^{ik(x-x')+i\omega(t-t')+i\phi(k,\omega)} \right]. \qquad (3.252)$$

This filter generalizes the spatial filter in equation (3.163). It is spatiotemporally translation invariant, and, if $g(k, \omega)$ is isotropic with respective to the direction of k, the filter is spatially isotropic, even though temporally causal. For a fixed k, the signal power $\langle |S(k, \omega)|^2 \rangle$ monotonically decreases with increasing ω, and thus the gain $g(k, \omega)$ is usually a unimodal function of ω, peaking at some non-zero ω value. Therefore, the filter $K(x - x', t - t')$ should be more sensitive to moving than static stimuli. However, due to its spatial isotropy, the filter is equally sensitive to all directions of motion. It therefore describes the spatiotemporal RFs of retina ganglion or LGN neurons (Dong and Atick 1995) in primates or cats.

Looked at another way, for a given temporal frequency ω, $g(k, \omega)$ is (proportional to) the spatial contrast sensitivity curve of the retinal or LGN cells (to a grating of spatial frequency k). In turn, this curve should depend on the temporal frequency ω; see Fig. 2.14 B.

3.6.8.1 A spectrum of directional selectivities of receptive fields in V1

The multiscale coding in V1 makes the $U^{(x)}$ matrix take a form like that in equation (3.213). Let us consider whether this can cause V1 cells to be directionally selective.

In the simplest departure from the LGN spatiotemporal coding, we replace the $U^{(x)}$ component used in equation (3.251) by the $U^{(s)}$ in equation (3.225) for the sth band in the V1 multiscale spatial coding (using 1D space for simplicity). Then, the receptive field in equation (3.252) becomes (for the mth neuron in the sth scale)

$$K^{(s)}(x_m - x', t - t') \propto \int_{k_{\text{low}}}^{k_{\text{high}}} dk \left\{ \cos\left[k(x_m - x') - m\frac{\pi}{2} + \phi_x \right] \cdot \right.$$
$$\left. \left[\int d\omega g(k, \omega) e^{i\omega(t-t')+i\phi(k,\omega)} \right] \right\}$$
$$= \int_{k_{\text{low}}}^{k_{\text{high}}} dk \int_0^\infty d\omega g(k, \omega) \left\{ \cos\left[k(x_m - x') - m\frac{\pi}{2} + \phi_x \right] \cdot \right.$$
$$\left. \cos[\omega(t - t') + \phi(k, \omega)] \right\}. \qquad (3.253)$$

Here, ϕ_x corresponds to the ϕ in equation (3.225) or equation (3.226), and the subscript x is added in order to distinguish ϕ_x from the phase $\phi(k, \omega)$ in the temporal dimension. The filter is now restricted to a spatial frequency range $k_{\text{low}} < k \leq k_{\text{high}}$ (with $k_{\text{low}} = k_{j_s}$ and $k_{\text{high}} = k_{j_s+1}$) defined for the multiscale code but is still equally sensitive to the two opposite directions of motion (note that it is almost spatiotemporally separable, and it would be exactly so were there no integration over k and ω). However, matrices $U^{(s)}$ and $U^{(t)}$ will remain unitary by the parameter change $\phi_x \to \phi_x + \alpha$ and $\phi(k, \omega) \to \phi(k, \omega) + \beta$ (in equation (3.253)) for any constants α and β. These freedoms can lead to other choices of the overall spatiotemporal unitary matrix U, and the consequent filters $K^{(s)}(x, t)$ within a spatial frequency band. For example, making $\alpha = \beta = \pi/2$ would lead to another filter

$$K^{(s)}(x_m - x', t - t') \propto \int_{k_{\text{low}}}^{k_{\text{high}}} dk \int_0^\infty d\omega g(k, \omega) \left\{ \sin\left[k(x_m - x') - m\frac{\pi}{2} + \phi_x \right] \cdot \right.$$
$$\left. \sin[\omega(t - t') + \phi(k, \omega)] \right\}, \tag{3.254}$$

which is identical to the one in equation (3.253) except for changing each $\cos(.)$ to $\sin(.)$. Adding or substracting the two filters in equations (3.253–3.254), which are in a spatiotemporal quadrature relationship with each other, leads to a new filter:

$$K^{(s)}(x_m - x', t - t') \propto \int_{k_{\text{low}}}^{k_{\text{high}}} dk \int_0^\infty d\omega g(k, \omega) \cos[k(x_m - x') \mp \omega(t - t')$$
$$-m\frac{\pi}{2} + \phi_x \mp \phi(k, \omega)]. \tag{3.255}$$

This new filter is fully direction selective, as should arise from adding the two filters in a quadrature pair (see equations (2.55–2.61)). However, to perform such operations without compromising the unitary nature of the overall U matrix, neighboring filters within the scale s must satisfy certain relationships, since they are jointly subject to $U^{(s)}$. After some derivation, sketched in the footnote,[11] the general solution for all the spatiotemporal filters in this scale in a complete representation can be written as

$$K^s(x_m - x, t - t') \propto \int_{k_{\text{low}}}^{k_{\text{high}}} dk \int_0^\infty d\omega g(k, \omega)\{A^+ \cos[\psi(x_m - x) + \phi(t - t')]$$
$$+ A^- \cos[\psi(x_m - x) - \phi(t - t')]\},$$

where

$$\psi(x_m - x) = k(x_m - x) - \frac{m\pi}{2} + \phi_x,$$
$$\phi(t - t') = \omega(t - t') + \phi(k, \omega) + \phi_t,$$

if m is odd, $(A^+, A^-, \phi_t) = (\cos(\gamma), \sin(\gamma), \phi_1)$,

otherwise, $(A^+, A^-, \phi_t) = (\sin(\gamma), \cos(\gamma), \phi_2)$,

$$\tag{3.257}$$

based on the parameters γ, ϕ_1, ϕ_2, and ϕ_x. The sensitivities of this RF to the two opposite directions of motion are A^+ and A^-. Its directional selectivity can thus be characterized by $|\cos(\gamma) - \sin(\gamma)|/|\cos(\gamma) + \sin(\gamma)|$, while parameters ϕ_x and ϕ_i further characterize the spatiotemporal shape of the receptive field. Different values of γ (and the values of ϕ_1, ϕ_2, and ϕ_x) correspond to choosing different U matrices in the efficient code. Apparently, with overcomplete coding in V1, multiple values of γ are employed in V1 to make a wide spectrum of directional selectivities among V1 neurons. The rational for exhibiting such a wide spectrum is likely coupled to the rational for having an overcomplete representation in the first place, about which the efficient coding principle is mute.

---

[11] The goal is to create a general spatiotemporal U which is unitary for scale s for the case of a complete (rather than overcomplete) efficient code. Let this submatrix (for scale s) have elements

$$U_{t,\omega,m,k_j} = \exp\{i[k_j x_m + \omega t + \phi(k_j, \omega)]\} \cdot F, \quad \text{where}$$

$$F \equiv \begin{cases} A^+(m) \exp\left[i(\Phi^+(m) - m\pi/2)\right], & \text{if } k_j > 0, \omega > 0, \\ A^+(m) \exp\left[-i(\Phi^+(m) - m\pi/2)\right], & \text{if } k_j < 0, \omega < 0, \\ A^-(m) \exp\left[i(\Phi^-(m) - m\pi/2)\right], & \text{if } k_j > 0, \omega < 0, \\ A^-(m) \exp\left[-i(\Phi^-(m) - m\pi/2)\right], & \text{if } k_j < 0, \omega > 0. \end{cases} \tag{3.256}$$

This form can lead to the receptive field in equation (3.257). Translation invariance between different coding units m within the scale requires $A^\pm(m) \exp[i\Phi^\pm(m)] = A^\pm(m+2) \exp[i\Phi^\pm(m+2)]$, reducing the free parameters to only two sets of values $A^\pm(m)$ and $\Phi^\pm(m)$ for m odd and even. Making U be unitary imposes further requirements on the relationship between $A^\pm(m)$ and $\Phi^\pm(m)$ for even and odd m. This relationship is reflected in the way that the filters $K^s(x, t)$ for all neurons in the scale band are parameterized using only four quantities γ, ϕ_1, and ϕ_2, and ϕ_x in equation (3.257). See Li (1996) for more details and a complete proof.

3.6.8.2 Correlation between preferred spatial frequency and temporal frequencies, and a weak speed tuning

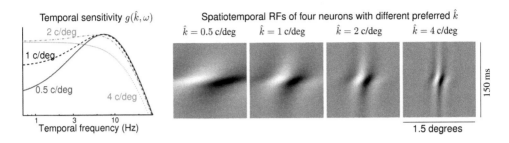

Temporal sensitivity $g(\hat{k}, \omega)$

2 c/deg

1 c/deg

0.5 c/deg

4 c/deg

Temporal frequency (Hz)

Spatiotemporal RFs of four neurons with different preferred $\hat{k}$

$\hat{k} = 0.5$ c/deg $\hat{k} = 1$ c/deg $\hat{k} = 2$ c/deg $\hat{k} = 4$ c/deg

150 ms

1.5 degrees

Fig. 3.38: Temporal coding depends on the spatial frequency band in a multiscale code. The preferred temporal frequency ω_p or speed v_p of a neuron should decrease as its preferred spatial frequency $\hat{k}$ increases (in the plots, "cycles/degree" is abbreviated to "c/deg"). Note that the temporal sensitivity curve $g(\hat{k}, \omega)$ changes from being band-pass for $\hat{k} = 0.5$ cycles/degree to being low pass for $\hat{k} = 4$ cycles/degree. All four example RFs are directionally selective. Adapted with permission from Li, Z., A theory of the visual motion coding in the primary visual cortex, *Neural Computation*, 8(4): 705–30, Fig. 3, copyright © 1996, The MIT Press.

Let $\hat{k} = \frac{1}{2}(k_{\text{low}} + k_{\text{high}})$ be the preferred spatial frequency of a V1 neuron. If the contrast sensitivity $g(k, \omega)$ changes slowly enough with k, we can approximately write $g(k, \omega) \approx g(\hat{k}, \omega)$ for all frequencies k within the narrow frequency band $(k_{\text{low}}, k_{\text{high}})$ for a given V1 neuron. This enables us to study the temporal frequency sensitivity of this neuron—i.e., how $g(\hat{k}, \omega)$ varies with ω given $\hat{k}$.

According to our analysis in Section 3.6.2, $g(\hat{k}, \omega)$ should be a unimodal function of ω, peaking at ω_p, where the input signal-to-noise ratio $\langle |\mathcal{S}(\hat{k}, \omega)|^2 \rangle / \langle N^2 \rangle \approx 1$. This comes from gain control which enhances temporal contrast when input noise is negligible for low $\omega < \omega_p$ (i.e., whitening), but suppresses the input noise at higher $\omega > \omega_p$. Since the signal power $\langle |\mathcal{S}(\hat{k}, \omega)|^2 \rangle$ also decreases with increasing $\hat{k}$, the efficient coding principle predicts:

$$\text{the optimal temporal frequency } \omega_p \text{ should be smaller} \atop \text{for neurons with higher preferred spatial frequencies } \hat{k}. \tag{3.258}$$

A neuron with preferred frequencies ω_p and $\hat{k}$ should prefer a grating drifting at velocity $v_p = \omega_p / \hat{k}$. Therefore, the optimal speed for the neuron also decreases with increasing $\hat{k}$, as observed physiologically (Holub and Morton-Gibson 1981) and illustrated in Fig. 3.38. However, since typically the curve $g(\hat{k}, \omega)$ as a function of ω is such that the frequency ω tuning width is quite wide, the neurons are typically not sharply tuned to temporal frequency or speed.

3.6.9 V1 neurons tuned simultaneously to multiple feature dimensions

Because natural images have input correlations between nearby spatial locations, between nearby temporal instants, between cones, and between eyes, signal power decreases with increasing signal differentiation in space, time, color, and ocular dimension; see Fig. 3.39. This is true for chromatic versus luminance channels, which differentiate signals between cones, for ocular opponent versus summation channels, which differentiate signals between the eyes, and

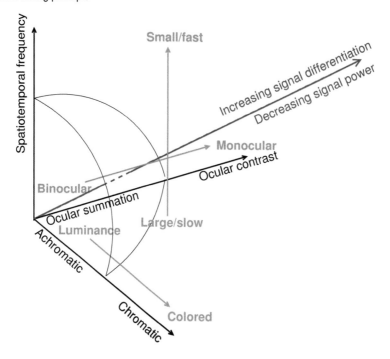

Fig. 3.39: The arrows point in the directions of increasing differentiation between input signals in spatiotemporal, ocular, and cone feature dimensions. Increasing differentiation of signals decreases signal power to a degree that is greater when differentiation arises for multiple dimensions. According to the efficient coding principle, encoding sensitivities g increase with moderate degrees of differentiation to enhance contrast but decrease with further differentiation.

for high versus low temporal and spatial frequencies. These forms of differentiation all weaken signal power. Hence, chromatic power is weaker than the luminance power, ocular opponent power is weaker than the ocular summation power, and powers in higher frequency channels are weaker than those in lower frequency channels. Differentiating inputs simultaneously in multiple feature dimensions makes the signal power even weaker than differentiation in just one feature dimension. For example, a chromatic channel tuned to non-zero spatial frequency is the outcome of differentiating in both color and space; its power is weaker than a luminance channel tuned to the same spatial frequency, since the luminance channel does not differentiate between cones, although it does the same differentiation as the chromatic channel between spatial locations.

According to the efficient coding principle, small or zero sensitivities g should be awarded to channels of weak signals. This is likely the reason why fewer V1 neurons are tuned simultaneously to multiple feature dimensions. For example, when input differences between cones are used to create color tuning, it helps to sum signals along space and time dimensions to boost the signal power. Hence, neurons tuned to color tend to have larger receptive fields in space and time, and thus they tend to be insensitive to fine spatial details or rapid motion. Similarly, when a neuron is tuned to spatial variations along the horizontal axis, it compensates by summing spatially along the vertical axis—this neuron is tuned to vertical orientation. If a neuron is tuned to higher spatial frequencies, it compensates by summing across time to boost signal power; hence the neuron is likely to be tuned to lower temporal frequencies. Neurons

Fig. 3.40: Color flicker becomes difficult to perceive when the color is isoluminant. Consider a version of this checkerboard of red and green squares in which squares flicker in color between red and green synchronously in time. When the flickering frequency is sufficiently high, the flicker becomes barely noticeable if the red and green squares have the same luminance. This is because, according to the efficient coding principle, the chromatic channel for red-green opponency is at best poorly sensitive to high temporal frequencies. This is because there is little signal power in a channel involving simultaneous differentiation in two feature dimensions—time and cone space—and so it has weak sensitivity g. The flicker becomes noticeable (unless the flicker frequency is too high) when the red and green squares do not have the same luminance, since the luminance channel is more sensitive to high temporal frequencies. The flickering frequency at which the flicker becomes imperceptible should be smaller for smaller sizes of the squares, which weaken the signal further by reducing spatial summation.

tuned to higher spatial frequencies tend to be binocular, to sum signals across the eyes after severe signal differentiation in space. For the same reason, neurons tuned to color are more likely to be binocular, and neurons tuned to ocular contrast are less likely to be color tuned. If a neuron is tuned to both color and ocular contrast (it should appear as a monocular color tuned neuron), it is likely to have a larger receptive field and be insensitive to higher temporal frequencies.

The property that neurons tuned to high temporal frequencies are less likely to be tuned to color is used behaviorally to find the isoluminant balance between red and green (or between blue and yellow). One can make a checkerboard pattern of red and green squares and swap the colors of the squares in time; see Fig. 3.40. If the red and green squares have the same luminance, the color flicker becomes unnoticeable when the frequency at which the colors are swapped is too high. At this same high frequency, if the luminances of the red and green squares are not balanced, the flicker should become noticeable again. A related phenomenon is that perceived motion slows down or stops for moving stimuli that become more isoluminant (Cavanagh, Tyler and Favreau 1984).

Similarly, fewer neurons should be tuned to motion in depth, since this signal requires simultaneous signal differentiation in the dimensions of space, time, and ocularity. When one does find a V1 cell tuned to motion in depth, its receptive field size in space and time should be relatively large, such that the spatial and temporal signal differentiation occur only after substantial spatiotemporal signal integration. Analogously, it should be even more difficult to find a V1 neuron jointly tuned to color and motion in depth. Arguments along these lines also predict that one is more likely to find V1 neurons tuned to multiple feature dimensions when adapted to brighter conditions, when input powers are stronger. The efficient coding principle can be used to guide our expectations about the kinds of neurons we are likely to find in V1. These expectations can be compared with, and tested by, experimental data.

3.7 The efficient code, and the related sparse code, in low noise limit by numerical simulations

In Section 3.4, we approximated the various variables $\mathbf{S}$, $\mathbf{N}$, and $\mathbf{N_o}$ as having Gaussian distributions in order to obtain analytical solutions of the optimal efficient coding K. This coding minimizes $E(\mathsf{K}) =$ neural cost $- \lambda I(\mathbf{O}; \mathbf{S})$, where, again, λ is the Lagrange multiplier that balances reducing the neural cost against extracting information $I(\mathbf{O}; \mathbf{S})$. The Gaussian approximations can be avoided if the optimal K is instead obtained through numerical simulation. In particular, one can start from an initial guess K, and improve it incrementally $\mathsf{K} \rightarrow \mathsf{K} + \Delta\mathsf{K}$ by gradient descent

$$\Delta\mathsf{K} \propto -\frac{\partial E}{\partial\mathsf{K}}, \tag{3.259}$$

such that the change K reduces $E(\mathsf{K})$. Since $E(\mathsf{K})$ depends on the statistics $P(\mathbf{S})$, $P(\mathbf{N})$, and $P(\mathbf{N_o})$, so will $\partial E/\partial\mathsf{K}$. Then, the question becomes one of how this dependency can be enforced during the simulation without the Gaussian approximation.

In practice, this simulation is typically carried out in the following special case (Bell and Sejnowski 1997), which makes the problem simpler. First, take the zero noise limit $\mathbf{N} \rightarrow 0$; then maximizing $I(\mathbf{O}; \mathbf{S})$ is equivalent to maximizing $H(\mathbf{O})$ (see equation (3.48)), and, since $\mathbf{O} = \mathsf{K}(\mathbf{S})$, one does not have to worry about the statistics of the noise. Second, constrain the neural cost by forcing $\mathbf{O}$ to live within a fixed output dynamic range. This can be achieved by having $\mathbf{u} = \mathsf{K}\mathbf{S}$, followed by a nonlinear function $O_i = g(u_i)$, e.g., $g(u) = (1 + e^{-u})^{-1}$, which is bounded within the range (e.g., $g(u) \in (0, 1)$). Having fixed the nonlinear transform $g(.)$, one can then proceed to obtain the optimal K. Third, note that when the neural cost is constrained in this way, minimizing $E =$ neural cost $- \lambda I(\mathbf{O}; \mathbf{S})$ requires maximizing $I(\mathbf{O}; \mathbf{S})$, or maximizing $H(\mathbf{O})$ in the zero noise limit. Hence:

> With zero noise, when $O_i = g((\mathsf{K}\mathbf{S})_i)$ for a fixed sigmoid function g, efficient coding finds K that maximizes $H(\mathbf{O})$. $\tag{3.260}$

In particular,

$$\frac{\partial E}{\partial\mathsf{K}} \rightarrow -\frac{\partial H(\mathbf{O})}{\partial\mathsf{K}}, \tag{3.261}$$

when the output dynamic range is constrained and noise is omitted.

Fourth, given $O_i = g((\mathsf{K}\mathbf{S})_i)$, when K is invertible, the probability $P(\mathbf{O})$ is related to $P(\mathbf{S})$ by $P(\mathbf{S})d\mathbf{S} = P(\mathbf{O})d\mathbf{O}$; hence,

$$P(\mathbf{O}) = \frac{P(\mathbf{S})}{J}, \quad \text{where}$$

$$J \equiv \det\left(\frac{\partial O_i}{\partial S_j}\right) \text{ is the determinant of the Jacobian matrix whose } (i, j)^{\text{th}} \text{ element is } \frac{\partial O_i}{\partial S_j}.$$

Thus,

$$H(\mathbf{O}) = -\int d\mathbf{O} P(\mathbf{O}) \log_2 P(\mathbf{O}) = -\int d\mathbf{S} P(\mathbf{S}) \log_2 \left(\frac{P(\mathbf{S})}{J} \right)$$

$$= -\int d\mathbf{S} P(\mathbf{S}) \log_2 P(\mathbf{S}) + \int d\mathbf{S} P(\mathbf{S}) \log_2 J$$

$$= H(\mathbf{S}) + \int d\mathbf{S} P(\mathbf{S}) \log_2 J. \tag{3.262}$$

The first term above, $H(\mathbf{S})$, does not depend on K, so maximizing $H(\mathbf{O})$ is equivalent to maximizing the second term $\int d\mathbf{S} P(\mathbf{S}) \log_2 J$. This second term can also be written as $\langle \log_2 J \rangle$, i.e., the average value of $\log_2 J$ over the visual input ensemble. Thus,

$$\Delta \mathsf{K} \propto -\frac{\partial E}{\partial \mathsf{K}}$$

$$= \frac{\partial H(\mathbf{O})}{\partial \mathsf{K}} \quad \text{(when output dynamic range is constrained and noise is zero)}$$

$$= \frac{\partial \langle \log_2 J \rangle}{\partial \mathsf{K}} = \left\langle \frac{\partial \log_2 J}{\partial \mathsf{K}} \right\rangle, \quad \text{when K is invertible.} \tag{3.263}$$

The average of $\dfrac{\partial \log_2 J}{\partial \mathsf{K}}$ over the input ensemble can be calculated by averaging $\dfrac{\partial \log_2 J}{\partial \mathsf{K}}$ over samples $\mathbf{S}$ of visual inputs. This can be done by an online algorithm called stochastic gradient descent in which K is modified in light of each sample $\mathbf{S}$ by an amount proportional to $\dfrac{\partial \log_2 J}{\partial \mathsf{K}}$ evaluated for that particular $\mathbf{S}$. This algorithm has an advantage of directly averaging over the ensemble defined by the probability distribution $P(\mathbf{S})$ of input samples $\mathbf{S}$, without having to know the explicit form of this distribution.

Since, in the noiseless limit, $O_i = g((\mathsf{KS})_i)$, the simulation algorithm can be seen as aiming to find the transformation K such that the vector $\mathbf{O} = (O_1, O_2, ...)$ is evenly and uniformly distributed in the output space $\mathbf{O}$ to maximize the output entropy $H(\mathbf{O})$.

When $\mathbf{S}$ is a visual image with its components denoting the pixel values of the image, the row vectors in the matrix K are spatial filters or receptive fields for the image. When the above online algorithm is applied to such images, the K obtained contains row vectors which qualitatively resemble the spatial receptive fields of the V1 neurons, and the responses $\mathbf{O}$ are presumed to model V1 neural responses. Since images $\mathbf{S}$ containing n image pixels require an $n \times n$ matrix K, in practice, simulations are often done using small image patches of, say, 12×12 pixels. This gives 144-dimensional vectors $\mathbf{S}$, and a 144×144 matrix K. This limitation to smaller image sizes for $\mathbf{S}$ means that the statistical property of position and scale invariance of natural visual scenes cannot be properly captured. For example, the image size 12×12 would allow a spatial frequency range of no more than 3 octaves (from one cycle every 2 pixels to one cycle every 12 pixels); this range is insufficient to make the resulting K properly manifest the scale invariance property in the statistics of visual input.

3.7.1 Sparse coding

Note that once $\mathbf{O}$ is obtained, if K is invertible (i.e., when $\mathbf{O}$ is a complete or overcomplete representation), $\mathbf{S}$ can be reconstructed by

$$\mathbf{S} = \mathbf{S}' + \text{reconstruction error},$$
$$\text{with } \mathbf{S}' \equiv \mathsf{K}^{-1}\mathbf{u}, \quad u_i = g^{-1}(O_i). \tag{3.264}$$

If $g(x) = x$, this gives

$$\mathbf{S}' = \mathsf{K}^{-1}\mathbf{O}, \quad \text{i.e.,} \quad S'_i = \sum_j O_j(\mathsf{K}^{-1})_{ij}.$$

This means that $\mathbf{S}'$ can be seen as a weighted sum of basis vectors, with weights given by coefficients O_j. Each basis vector is a column vector of matrix K^{-1}.

Although input reconstruction is not the goal of efficient coding, it is worth noting the link between efficient coding and another line of work often referred to as sparse coding, which is also aimed at providing an account for early visual processing (Olshausen and Field 1997, van Hateren and Ruderman 1998, Simoncelli and Olshausen 2001). Sparse coding proposes that visual input $\mathbf{S}$ is represented by $\mathbf{O}$, from which it can be reconstructed, as in equation (3.264), and that, given $P(\mathbf{S})$, these basis functions should be such that the coefficients O_i have a sparse distribution $P(O_i)$ for all i. Sparse coding is a term originally used to describe distributed binary representation of various entities, when the binary representation has few 1's and many 0's; thus the number of 1's is "sparse". Hence, a sparser representation has a smaller entropy and utilizes a smaller resource. In the use of sparse coding for understanding early visual processing, sparseness means that O_i tends to be small, so that $P(O_i)$ is peaked near $O_i = 0$—in such a case, a neuron's most likely response is zero response. Of course, a Gaussian distributed O_i also makes zero response the most likely. However, since natural scenes have non-Gaussian $\mathbf{S}$, the sparse coding aims to produce O_i's which are sparser than Gaussian.

In any case, a sparser O_i requires a smaller resource $H(O_i)$, and since a larger $I(\mathbf{O};\mathbf{S})$ means a better reconstruction $\mathbf{S}' = \mathsf{K}^{-1}\mathbf{O}$ of $\mathbf{S}$, i.e., $|\mathbf{S} - \mathsf{K}^{-1}\mathbf{O}|$ is small, sparse coding is an alternative formulation of the efficient coding principle. Indeed, in practice, the sparse coding algorithm finds $\mathbf{O}$ and K^{-1} by minimizing an objective function

$$\mathcal{E} = \lambda \sum_i \langle \text{Sp}(O_i) \rangle + \langle |\mathbf{S} - \mathsf{K}^{-1}\mathbf{O}|^2 \rangle \tag{3.265}$$

where $\text{Sp}(O_i)$, e.g., $\text{Sp}(O_i) = |O_i|$, describes a cost of non-sparseness. Reducing such a cost encourages the distribution $P(O_i)$ to be sharply peaked at 0, and thus, a low $H(O_i)$. The second term in $\mathcal{E}$ is the ensemble average of the squared reconstruction error $|\mathbf{S} - \mathbf{S}'|^2$. This objective function $\mathcal{E}$ is closely related to our efficient coding objective $E(\mathsf{K}) = $ neural cost $-\lambda I(\mathbf{O};\mathbf{S})$ in equation (3.40). The first term "neural cost" in $E(\mathsf{K})$ has an effect akin to the first term $\lambda \sum_i \langle \text{Sp}(O_i) \rangle$ in $\mathcal{E}$, since minimizing $\langle \text{Sp}(O_i) \rangle$ typically reduces the cost of the channel capacity $H(O_i)$. Further, the averaged squared reconstruction error $\langle |\mathbf{S} - \mathbf{S}'|^2 \rangle$ should be smaller when $\mathbf{O}$ provides more information about $\mathbf{S}$, i.e., when $I(\mathbf{O};\mathbf{S})$ is larger.[12] Hence minimizing $\langle |\mathbf{S} - \mathsf{K}^{-1}\mathbf{O}|^2 \rangle$ in $\mathcal{E}$ is closely related to maximizing the $I(\mathbf{O};\mathbf{S})$ in $E(\mathsf{K})$. In fact, when the reconstruction error $\mathbf{S} - \mathsf{K}^{-1}\mathbf{O}$ is forced to be zero, and when $\text{Sp}(O_i)$ takes such a form that for the optimal K, the distribution of $g(O_i = (\mathsf{K}\mathbf{S})_i)$ is uniform for

---

[12]More explicitly, the second term $\langle |\mathbf{S} - \mathsf{K}^{-1}\mathbf{O}|^2 \rangle = \langle |\mathbf{S} - \mathbf{S}'|^2 \rangle$ in $\mathcal{E}$ can be understood as follows. The mutual information between the original input $\mathbf{S}$ and the reconstructed input $\mathbf{S}'$ is $I(\mathbf{S};\mathbf{S}') = H(\mathbf{S}) - H(\mathbf{S}|\mathbf{S}')$, giving $H(\mathbf{S}|\mathbf{S}') = H(\mathbf{S}) - I(\mathbf{S};\mathbf{S}')$. Since $\mathbf{S}'$ is completely determined by $\mathbf{O}$ by an invertible matrix K, $I(\mathbf{S};\mathbf{S}') = I(\mathbf{S};\mathbf{O})$, i.e., the $\mathbf{O}$ and $\mathbf{S}'$ convey the same amount of information about $\mathbf{S}$. Thus $H(\mathbf{S}|\mathbf{S}') = H(\mathbf{S}) - I(\mathbf{O};\mathbf{S})$. We note that $H(\mathbf{S}|\mathbf{S}')$, the ignorance about $\mathbf{S}$ after knowing $\mathbf{S}'$, is the entropy of the reconstruction error $\mathbf{S} - \mathbf{S}'$. Approximating this error as Gaussian, we have $\langle |\mathbf{S} - \mathbf{S}'|^2 \rangle \approx 2^{H(\mathbf{S}|\mathbf{S}')}$ by equation (3.23). Since $H(\mathbf{S}|\mathbf{S}') = H(\mathbf{S}) - I(\mathbf{O};\mathbf{S})$, and $H(\mathbf{S})$ is a constant independent of K, we have, approximately,

$$\langle |\mathbf{S} - \mathsf{K}^{-1}\mathbf{O}|^2 \rangle \propto 2^{-I(\mathbf{O};\mathbf{S})}. \tag{3.266}$$

each i, sparse coding is exactly equivalent to the efficient coding in the noiseless limit as in equation (3.260). Hence, these sparse coding algorithms (Olshausen and Field 1997), which were mostly simulated from sampled images for the zero noise condition, produce results similar to those by simulation algorithms (Bell and Sejnowski 1997) (applied also in the noiseless limit) to minimize $E(\mathsf{K})$ for efficient coding.

These simulation algorithms have the advantage of being performed online while being exposed to individual natural images $\mathbf{S}$ presumably drawn from the true $P(\mathbf{S})$, allowing all orders of statistics in $P(\mathbf{S})$ potentially to influence the outcome. These algorithms have been mostly used to derive the V1 visual receptive fields in space and time. Particularly for V1 simple cell RFs, their resulting K (Bell and Sejnowski 1997, Olshausen and Field 1997, van Hateren and Ruderman 1998, Simoncelli and Olshausen 2001) are qualitatively similar to our analytical ones (Li and Atick 1994b, Li 1996), obtained by approximating $P(\mathbf{S})$ by a Gaussian distribution and imposing an additional requirement of multiscale coding. The similarity may arise from the observation (discussed in Chapter 4) that the Gaussian approximation is quite reasonable. The disadvantages of these simulation algorithms include: (1) tampering with translation and scale invariance in input statistics because of the small image patches used in the simulations, and (2) a difficulty in studying the influence of noise on the resulting receptive fields. These disadvantages limit the predictive power of the efficient coding theory. For example, it is not straightforward from these algorithms to predict how the receptive fields should adapt to changes in $P(\mathbf{S})$ and signal-to-noise ratios, which can arise from changes in animal species, developmental conditions, or local visual environment (such as brightness). This book will thus not delve further into the simulation algorithms for efficient coding. The next chapter, however, will provide some further discussions on higher order input statistics and their role in vision.

3.8 How to get efficient codes by developmental rules and unsupervised learning

So far, we have mainly talked about "why" the receptive fields should take certain forms or "what" these forms should be. Another question is "how" the brain forms such receptive fields, i.e., how does a neuron "know" which synaptic strengths should be employed to connect to which other neurons such that neurons in a population collectively achieve an efficient code? Except in the simulation algorithms for efficient coding, which could be exploited for developing or adapting neural connections to input environment, we have not looked into this question. Since the receptive fields should be able to adapt to changes in input statistics very quickly, the mature neural connections for the receptive fields are not pre-determined by genes. Instead, there should be developmental or adaptive rules (for which genes could play a role) that govern the way these receptive fields form. Since there is no plausible teaching signal or supervisor, the rules must specify how RFs change in an unsupervised or self-organized manner, in the light of the statistics of inputs. In this section, we briefly discuss some models of unsupervised learning for the receptive fields. As will be apparent, these models are very simple and do not closely model what might happen physiologically in development. However, they illustrate that efficient codes could possibly arise from rules that operate without an external teacher and only involve information that is purely local to synapses, such as pre- and postsynaptic activities and other signals close to the particular connections concerned. Many believe that the brain employs local development (or adaptive) or plasticity rules to produce encodings and representations that have globally attractive properties.

3.8.1 Learning for a single encoding neuron

One simple model was suggested by Linsker (1988) and Oja (1982) in the 1980s. Consider an output neuron O receiving inputs from zero mean Gaussian signals $\mathbf{S} = (S_1, S_2, ..., S_i, ..., S_n)$ via weights $\mathsf{K} = (K_1, K_2, ..., K_i, ..., K_n)$,

$$O = \sum_i K_i S_i. \tag{3.267}$$

Here we still use notation K for the weights even though this is now only a $1 \times n$-dimensional matrix, i.e., a vector. In response to each input, let K_i be adjusted according to the learning rule:

$$\dot{K}_i = \epsilon O (S_i - O K_i), \tag{3.268}$$

where ϵ is called the learning rate. This ϵ is so small that K_i does not change appreciably as the input signal $\mathbf{S}$ varies from one sample to another drawn from the ensemble with probability $P(\mathbf{S})$. Note that K_i adapts according to the signals O, S_i, and K_i, all of which are local to this synapse; there is no remote teacher to provide instruction. In particular, the first term OS_i on the right-hand side of the above equation is a Hebbian learning term (Hebb 1949) which increases the connection K_i when presynaptic and postsynaptic activities are correlated. As K_i changes sufficiently slowly, one may average the right-hand side of the equation above over the samples $\mathbf{S}$. Plugging equation (3.267) into the first term in equation (3.268), we get

$$\dot{K}_i = \epsilon \left(\sum_j K_j \langle S_j S_i \rangle - \langle O^2 \rangle K_i \right). \tag{3.269}$$

Multiplying both sides of equation (3.269) by $2K_i$ and summing over i gives

$$\frac{d}{dt} \sum_i K_i^2 = 2\epsilon \left(\sum_{ij} K_i R_{ij}^S K_j - \langle O^2 \rangle \sum_i K_i^2 \right) = 2\epsilon \langle O^2 \rangle \left(1 - \sum_i K_i^2 \right), \tag{3.270}$$

where $\sum_{ij} K_i R_{ij}^S K_j = \langle O^2 \rangle$ was used. Hence, at the end of learning, which is defined by $\dot{K}_i = 0$ for all i, when the left side of this equation vanishes, $\sum_i K_i^2$ converges to $\sum_i K_i^2 = 1$. Further, at that point, from equation (3.269)

$$\sum_j \langle S_i S_j \rangle K_j = \langle O^2 \rangle K_i, \tag{3.271}$$

making the weight vector K an eigenvector of the correlation matrix $R^S = \langle \mathbf{S} \mathbf{S}^T \rangle$.

One can see that even though each neural connection K_i adapts according to a local rule without a global teacher, the connections evolve collectively to achieve two global properties: making K an eigenvector of R^S and making $\sum_i K_i^2 = 1$. Noting that

$$\langle OS_i \rangle = \sum_j R_{ij}^S K_j = \frac{1}{2} \partial \left(\sum_{ab} K_a R_{ab}^S K_b \right) / \partial K_i, \tag{3.272}$$

one may also see the evolution of K by equation (3.269) as tending to minimize

$$E(\mathsf{K}) = -\sum_{ab} K_a R_{ab}^S K_b = -\langle O^2 \rangle, \tag{3.273}$$

or maximize the output variance $\langle O^2 \rangle$ subject to the constraint that $\sum_i K_i^2 = 1$.

Further, consider the case that the input signals S_i in each channel i come corrupted by zero mean Gaussian input noise N_i with variance $\langle N^2 \rangle$. Then, with S_i replaced by $S_i + N_i$, the output is $O = \sum_i K_i(S_i + N_i)$. Applying the same learning rule (equation (3.268)), but with this noise-corrupted input and output, maximizes the mutual information between O and $\mathbf{S} = (S_1, S_2, ...)$:

$$I(O; \mathbf{S}) = \frac{1}{2} \log \frac{\langle O^2 \rangle}{\langle N^2 \rangle \sum K_i^2} = \frac{1}{2} \log \frac{\langle O^2 \rangle}{\langle N^2 \rangle}, \tag{3.274}$$

which is another global property of the system.

3.8.2 Learning simultaneously for multiple encoding neurons

The learning rule applies when there is only a single output neuron, making O extract the largest principal component of the input $\mathbf{S}$. When there are multiple output neurons, the values O_1, O_2, and O_3, etc., need to be decorrelated (at least in the high signal-to-noise ratio limit) in order to avoid transmitting the same (redundant) information about the same principal component of the input $\mathbf{S}$.

Goodall (1960) proposed an algorithm for doing this which uses recurrent inhibitory connections between O_i and O_j. This method has been used to realize an efficient color coding network that adapts to changes in input color statistics (Atick et al. 1993). Consider the case that O_i receives direct excitatory input from S_i but inhibition from other O_j units in the form

$$\tau \dot{O}_i = S_i - \sum_j W_{ij} O_j, \tag{3.275}$$

where τ is the time constant of this dynamical system. When the above equation converges to $\dot{O}_i = 0$, we have $S_i = \sum_j W_{ij} O_j$ or $O_i = \sum_j (W^{-1})_{ij} S_j$. If there is no inhibition between output neurons, except for the self-inhibition (which is the tendency to decay to the resting state) $W_{ij} = \delta_{ij}$, this system is an identity mapping from $\mathbf{S}$ to $\mathbf{O}$. However, if it were the case that $W^{-1} = K_o^{-1} \mathbf{g} K_o$, then this network would have created an efficient code. (Note that this also means that a linear feedforward network can be equivalent to another linear recurrent network if we only consider signal transformation at the equilibrium point of the latter.) Goodall's algorithm suggests that connections W_{ij} should adapt according to

$$\tilde{\tau} \dot{W}_{ij} = S_i O_j - W_{ij}, \tag{3.276}$$

with a time constant $\tilde{\tau} \gg \tau$, so that W_{ij} evolves much more slowly than the neural activities $\mathbf{O}$. The learning of the synapse W_{ij} depends only on the information (S_i, O_j, and W_{ij}) that can be seen as local to the synapse concerned, since S_i is at least one of the other inputs to the postsynaptic cell O_i. Again, the slow dynamics of learning means that one may average $S_i O_j$ on the right-hand side of the above equation over the activity ensemble to get $\tilde{\tau} \dot{W}_{ij} = \langle S_i O_j \rangle - W_{ij}$. When the learning converges, $\dot{W}_{ij} = 0$, we will have $\langle S_i O_j \rangle = W_{ij}$. Using $S_i = \sum_k W_{ik} O_k$, we have

$$\langle S_i O_j \rangle = \sum_k W_{ik} \langle O_k O_j \rangle = W_{ij}. \quad \text{Hence,} \quad \langle O_k O_j \rangle = \delta_{kj}. \tag{3.277}$$

Thus the output neurons are decorrelated from each other, with each having unit variance. This implies that $W^{-1} = U \mathbf{g} K_o$ with $\mathbf{g}$ a diagonal matrix whose diagonal elements are the inverse of the eigenvalues of R^S. As noted above, this W^{-1} is the efficient coding transform in the noiseless limit.

The Goodall algorithm is therefore capable of creating an efficient code in the noiseless limit even when the input mapping is just the identity matrix. However, consider the case that the initial inputs S_i have already been transformed into an efficient code, so that $\langle S_i S_j \rangle = \delta_{ij}$. In this case, the Goodall algorithm will arrange for $\mathsf{W} = \mathbb{1}$ to be the identity matrix, with no recurrent inhibition between the output units. If the correlations in the sensory environment then change, for instance when the illumination changes hue (Atick et al. 1993), then W will adapt to compensate. The initial condition $\mathsf{W} = \mathbb{1}$ will lead W to a particular $\mathsf{W} = \mathsf{K}_o^{-1} \mathsf{g}^{-1} \mathsf{K}_o$ in which the output $\mathbf{O}$ is least distorted from the original input $\mathbf{S}$.

4   V1 and information coding

This chapter discusses the difficulties the efficient coding principle faces in accounting for V1 properties. Some research works relevant to this issue are summarized through this discussion. It is argued that this is not a failing of the application of efficient coding—resolvable, for instance, by reducing higher order redundancies in the statistics of the natural visual inputs. Rather, the difficulties motivate the search for another functional role for V1. This ushers in the hypothesis of the next chapter, that V1 creates a bottom-up saliency map to guide bottom-up selection. Readers not interested in these discussions may skip this chapter.

4.1   Pursuit of efficient coding in V1 by reducing higher order redundancy

As discussed in the previous chapter, the efficient coding principle suggests neural codes extract as much information as possible, at minimal resource cost. This can account for the properties of the input sampling and neural receptive fields in the retina, and also accommodate, via the degeneracy of the efficient code, the diversity of the RF properties in V1: tuning to orientation, color, ocularity, disparity, motion direction, scale, and the correlations between selectivities in different feature dimensions in individual cells. Thus, the principle appears to progress from the retina to V1.

If all signals are approximated as being Gaussian, and one ignores the fact that the number of neurons in V1 is about 100 times greater than the number of retinal ganglion cells (Barlow 1981), the V1 cortical code is no more efficient than the retinal code. This is true at least when space is viewed as one-dimensional. Specificallly, both codes transform receptor responses S to outputs $O = KS + $ noise via a receptive field transform K. The K for the retinal code has center-surround receptive fields, and the K for the V1 code has multiscale receptive fields. Both minimize the efficient coding objective (see Section 3.4)

$$E(K) = \sum_i \langle O_i^2 \rangle - \lambda I(O; S),$$

(4.1)

such that O contains as much information $I(O; S)$ as possible about signal S while consuming limited power $\sum_i \langle O_i^2 \rangle$. The difference between the retinal and V1 representations would be simply the two different degenerate choices of U to generate the transform $K = UgK_o$ by equation (3.145).

Now let us consider the 100-fold expansion from the retina to V1 in terms of the number of neurons. Let us say that n retinal ganglion cells transmit 10^7 bits/second of information, at 10 bits/second by each neuron. This information could be transmitted by $100 \times n$ V1 neurons at 0.1 bits/second each (Nadal and Parga 1993). This can happen if, for instance, V1 neurons are much less active than the ganglion cells, via a higher neural firing threshold. Then, one could say that the visual input information is represented much more sparsely in V1 than in the retina, in the sense that if one looks at the fraction of neurons active to represent a scene, this fraction should be smaller in V1. Such a sparser V1 representation, however, gains no coding efficiency for Gaussian signals, since it would transmit as much information $I(\mathbf{O}; \mathbf{S})$ given the same total output power cost $\sum_i \langle O_i^2 \rangle$ (if one ignores the cost of extra neurons and their associated intrinsic noise in the overcomplete representation).

Rather, at least in the noiseless limit, any improvement in coding efficiency has to come from reducing the higher order information redundancy, i.e., the redundancy due to the higher order input correlations between various O_i's. These higher order correlations are not accounted for by the Gaussian approximation (which we adopted in the last chapter to treats visual inputs and noise as Gaussian distributed), and their removal should enable the extraction of more information $I(\mathbf{O}; \mathbf{S})$ given a total output cost $\sum_i \langle O_i^2 \rangle$. Efficient coding in light of higher order statistics could break the degeneracy of the optimal efficient code based on Gaussian statistics.

Meanwhile, we should keep in mind that, even if V1 has a different representation from the retina in order to increase efficiency in light of the higher order statistics, it does not explain why this more efficient representation could not have been done in the retina and whether the overcomplete representation in V1 is necessary for the purpose. It could, for instance, be that performing encoding in the face of the higher order statistics is so complex that it has to be postponed to V1.

The degeneracy of the coding could also be broken if the cost function for the encoded neural responses $\mathbf{O} = (O_1, O_2, ...)$ is different from the total power $\sum_i \langle O_i^2 \rangle$ (see equation (3.85)). Is the cost (Laughlin 2001) in V1 different from that in the retina, such that it favors the multiscale representation? If the cost in each neuron i is a purely concave or purely convex function of $\langle O_i^2 \rangle$, then the multiscale representation is not favored.[13] We do not know enough to tell whether and why the cost in V1 should be different from that in the retina. We will henceforth focus on the question of whether V1 code is to deal with the higher order input statistics.

4.1.1 Higher order statistics contain much of the meaningful information about visual objects

Figure 4.1 demonstrates that higher order statistics underlie much of the visual perception of object forms. To see this, we analyze the relationship between the images Fig. 4.1 ABC. Let $S(x)$ describe the image in Fig. 4.1 A. Then, from the analysis in Section 3.6.1, $S(k)$, which is its Fourier component for wave number k, is the amplitude of a principal component of the ensemble of natural images. Averaged over the ensemble, the second order (pairwise) correlation is

$$\langle S(k)S^\dagger(k') \rangle \propto \delta(k - k') \quad \text{and}$$

$$R^S(x - x') \equiv \langle S(x)S(x') \rangle \propto \int dk \langle |S(k)|^2 \rangle e^{ik(x-x')}. \tag{4.2}$$

When an image $S(x)$ is whitened in the noiseless case, the whitened image is

$$O(x) \propto \int dx' \left[\int dk \left(\langle |S(k)|^2 \rangle \right)^{-1/2} e^{ik(x-x')} \right] S(x'), \tag{4.3}$$

shown in Fig. 4.1 C. According to equation (3.171), when averaged over the ensemble, different responses $O(x)$ and $O(x')$ (for $x \neq x'$) in the whitened image are decorrelated in the second order:

[13] If the cost is a purely convex function of $\langle O_i^2 \rangle$, the retinal code in which all the receptive fields have the same shape but differ in their center locations is favored. If the cost is a purely concave function of $\langle O_i^2 \rangle$, then the code in which each receptive field has the unique shape defined by the corresponding principal component is favored. However, the multiscale V1 code is neither of these two extremes.

A: Original image   B: A after randomizing   C: A after whitening
                          Fourier phases

 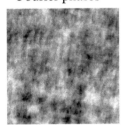 

Fig. 4.1: An original image in A becomes meaningless when the phases of its Fourier transform are replaced by random numbers, shown in B, as observed by Field (1989). Hence, A and B have the same first and second order statistics characterized by their common Fourier powers $S_k^2 \sim 1/k^2$, but the object information in the original image is not perceptible in B. Image C is obtained by whitening A; it thus has the power spectrum of white noise and has no second order correlation. However, C preserves the meaningful information about the original visual objects in the higher order statistics. Adapted with permission from Zhaoping, L., Theoretical understanding of the early visual processes by data compression and data selection, *Network: Computation in Neural Systems*, 17(4): 301–334, Fig. 7, copyright © 2006, Informa Healthcare.

$$\langle O(x)O(x')\rangle \propto \int dydy' \left[\int dk'dk \left(\langle |S(k)|^2\rangle \right)^{-1/2} \left(\langle |S(k')|^2\rangle \right)^{-1/2} \right.$$
$$\left. \cdot e^{ik(x-y)} e^{ik'(x'-y')} \right] \langle S(y)S(y')\rangle$$
$$\propto \int dydy' \left[\int dk'dkdk'' \left(\langle |S(k)|^2\rangle \right)^{-1/2} \left(\langle |S(k')|^2\rangle \right)^{-1/2} \right.$$
$$\left. \cdot e^{ik(x-y)} e^{ik'(x'-y')} \right] \langle |S(k'')|^2\rangle e^{ik''(y-y')}$$
$$= \int dk e^{ik(x-x')} \propto \delta(x-x'). \tag{4.4}$$

Assuming ergodicity (see equation (3.31)), within image Fig. 4.1 C, two responses $O(x)$ and $O(x')$ separated by a given (vector) displacement $d = x - x'$ should be decorrelated when averaged across the center location $((x+x')/2)$ of the pair in the whitened image.

However, we can recognize the peppers in Fig. 4.1 C, even though different responses O_i and O_j (for $i \neq j$) are decorrelated to second order. If different responses are uncorrelated to all orders, then the whitened image should look like white noise. Hence, the information about the peppers in the whitened image is present in the correlations between responses at higher orders. This could be the correlation between three responses $\langle O(x_i)O(x_j)O(x_k)\rangle$ for $x_i \neq x_j \neq x_k$ for three pixels, for instance associated with the possibility that they comprise part of a line or a smooth curve. In the original image in Fig. 4.1 A, the transition from black to white luminance at the left boundary of the long pepper are aligned along the edge of the pepper. Such an alignment is termed a "suspicious coincidence" by Barlow (1985); it gives rise to the higher order correlation in the original and whitened images Fig. 4.1 AC.

To examine whether the object information is present in the first and second order correlations, we created Fig. 4.1 B, which, as shown below, has first and second order correlations identical to those in Fig. 4.1 A. The Fourier spectrum $\mathcal{S}(k) = |\mathcal{S}(k)|e^{i\phi(k)}$ of the original image $S(x)$ in Fig. 4.1 A has magnitudes $|\mathcal{S}(k)|$ and phases $\phi(k)$. From this spectrum, we created another spectrum $\mathcal{S}'(k) = |\mathcal{S}(k)|e^{i\phi'(k)}$, which has the same magnitude spectrum but a random phase spectrum $\phi'(k)$ such that $\phi'(k_i)$ and $\phi'(k_j)$ are random numbers that are independent of each other for $k_i \neq k_j$. Figure 4.1 B shows the image $S'(x)$, which is the inverse Fourier transform of $S'(k)$. It looks like smoke, i.e., all the meaningful information about the objects in Fig. 4.1 A is perceptually lost. Since Fig. 4.1 A and Fig. 4.1 B have the same Fourier magnitude spectrum $|S(k)| = |S'(k)|$, we know from equation (4.2) that they have the same pairwise pixel correlation

$$\langle S'(x)S'(x')\rangle = \langle S(x)S(x')\rangle \propto \int \langle|S(k)|^2\rangle e^{ik(x-x')}. \tag{4.5}$$

They also have the same first order statistics $\langle S'(x)\rangle = \langle S(x)\rangle = 0$. Hence, image statistics up to the second order contain no object information that could support visible perception. The phase randomization apparently removed the higher order statistics, which are determined by the suspicious alignments that make up the object contours. This is because the image pixel value $S(x) \sim \int dk S(k)e^{ikx}$ at location x contains contribution $S(k)e^{ikx}$ from the $S(k)$ Fourier component. When the different Fourier components have some correlation between their phases, the contributions from various components can reinforce each other at object contour locations x, where image contrast tends to be higher. The whitening transform changes the amplitudes but not the phase of the Fourier components, and it thus maintains the phase alignments at the object contours. This suggests that the higher order statistics that contain the object information are transparent to the whitening transform.

4.1.2 Characterizing higher order statistics

Let us examine what the probability distribution $P(\mathbf{S})$ should be if it maximizes entropy $H(\mathbf{S}) = -\int d\mathbf{S} P(\mathbf{S}) \log_2 P(\mathbf{S})$, given the constraint of zero means $\langle S_i \rangle = 0$ and pairwise correlations $\langle S_i S_j \rangle = R_{ij}$ for the signals $\mathbf{S}$. This probability distribution $P(\mathbf{S})$ should maximize the value

$$E = H(\mathbf{S}) - \lambda_1 \left[\int d\mathbf{S} P(\mathbf{S}) - 1 \right] - \lambda_2 \sum_i \int d\mathbf{S} \left[P(\mathbf{S}) S_i \right]$$
$$- \sum_{ij} \lambda_{ij} \left[\int S_j S_j P(\mathbf{S}) d\mathbf{S} - R_{ij}^S \right], \tag{4.6}$$

in which λ_2 and λ_{ij} are respective Lagrange multipliers for the constraints to have the correct mean and second order correlations, and λ_1 is a Lagrange multiplier for the constraint that the distribution should be normalized. The required $P(\mathbf{S})$ is the solution to $\partial E/\partial P(\mathbf{S}) = 0$, which is

$$-\log_2 P(\mathbf{S}) - \log_2 e - \lambda_1 - \lambda_2 \sum_i S_i - \lambda_3 \sum_{S_i S_j} S_i S_j = 0. \tag{4.7}$$

Hence, this $P(\mathbf{S})$ is

$$P(\mathbf{S}) \propto \exp\left[-\frac{1}{\log_2 e} \left(\lambda_1 + \lambda_2 \sum_i S_i + \sum_{ij} \lambda_{ij} S_i S_j \right) \right]. \tag{4.8}$$

Setting λ_1 to normalize $P(\mathbf{S})$, λ_2 to make $\mathbf{S}$ have zero mean, and λ_{ij} to make $\langle S_i S_j \rangle = R_{ij}^S$ leads to the Gaussian distribution

$$P_{\text{Gaussian}}(\mathbf{S}) \equiv \frac{1}{2\pi\sqrt{\det \mathsf{R}^S}} \exp\left\{ -\sum_{ij} S_i S_j \left[(\mathsf{R}^S)^{-1} \right]_{ij} \right\}.$$

The nth order statistics for a random variable X is the nth order cumulant, defined as

$$C_n \equiv \frac{d^n}{dt^n}\left(\ln\langle e^{tX}\rangle\right)\Big|_{t=0}. \tag{4.9}$$

In the above expression, $|_{t=0}$ means that the mathematical expression to its immediate left is evaluated at $t = 0$. For instance, the first and second order cumulants are the mean and variance of X:

$$C_1 = \frac{d}{dt}\left(\ln\langle e^{tX}\rangle\right)\Big|_{t=0} = \frac{\langle X e^{tX}\rangle}{\langle e^{tX}\rangle}\Big|_{t=0} = \langle X\rangle,$$

$$C_2 = \frac{d^2}{dt^2}\left(\ln\langle e^{tX}\rangle\right)\Big|_{t=0} = -\frac{(\langle X e^{tX}\rangle)^2}{\langle e^{tX}\rangle^2}\Big|_{t=0} + \frac{\langle X^2 e^{tX}\rangle}{\langle e^{tX}\rangle}\Big|_{t=0}$$

$$= \langle X^2\rangle - (\langle X\rangle)^2 = \langle (X - \langle X\rangle)^2\rangle.$$

The fourth order cumulant is

$$C_4 = \left\langle (X - \langle X\rangle)^4\right\rangle - 3\left\langle (X - \langle X\rangle)^2\right\rangle^2 \xrightarrow{\langle X\rangle = 0} \langle X^4\rangle - 3\langle X^2\rangle^2. \tag{4.10}$$

For a Gaussian variable, all third and higher order statistics are zero. In particular, $C_4 = 0$, since $\left\langle (X - \langle X\rangle)^4\right\rangle = 3\left\langle (X - \langle X\rangle)^2\right\rangle^2$. Therefore, one often uses the deviation of the kurtosis

$$\text{kurtosis} \equiv \frac{\langle (X - \langle X\rangle)^4\rangle}{\left\langle (X - \langle X\rangle)^2\right\rangle^2}, \tag{4.11}$$

from a value kurtosis $= 3$ to assess how non-Gaussian a distribution is. If the kurtosis of a scalar variable X is larger than 3, the probability distribution $P(X)$ of X is likely more peaked at its mean, and it is likely to have a longer tail than a Gaussian distribution would have. Conversely, a kurtosis less than 3 suggests a distribution that is less peaked at the mean, with a thinner tail than a Gaussian distribution would have.

The nth order cumulant can be generalized to multiple variables in a vector

$$(X_1, X_2, ..., X_i, ...)$$

to give n^{th} order statistics. For example, the second order correlation between X_i and X_j is

$$C_{ij} \equiv \frac{d}{dt_i}\frac{d}{dt_j}\left(\ln\langle e^{\sum_a t_a X_a}\rangle\right)\Big|_{t_a=0,\,\text{all }a} = \langle X_i X_j\rangle - \langle X_i\rangle\langle X_j\rangle.$$

Hence, when the second order correlation $C_{ij} = 0$, we have $\langle X_i X_j\rangle = \langle X_i\rangle\langle X_j\rangle$.

Similarly, if S_1 and S_2 are two zero mean, uncorrelated, Gaussian variables, their fourth order correlation should be zero, giving $\langle S_1^2 S_2^2\rangle = \langle S_1^2\rangle\langle S_2^2\rangle$ (when $\langle S_1\rangle = \langle S_2\rangle = 0$). Therefore, a non-zero

$$\frac{\langle S_1^2 S_2^2\rangle}{\langle S_1^2\rangle\langle S_2^2\rangle} - 1 \tag{4.12}$$

indicates a non-zero fourth order correlation between them. Although higher order statistics should include cumulants of all higher (third or higher) orders, in practice, third and fourth order correlations are adequate indicators of higher order statistics in natural signals.

We can examine the higher order correlations in natural scenes by considering the whitened image $\mathbf{O}$ in Fig. 4.1 C. Using the ergodicity assumption (see equation (3.31)), the probability

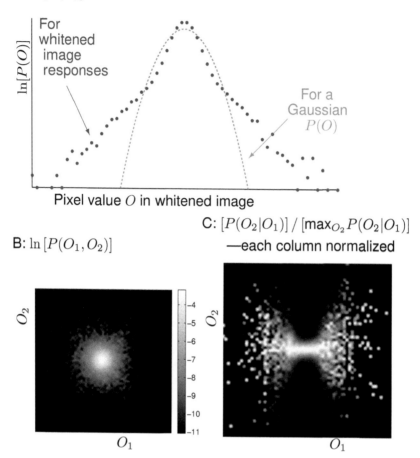

A: $\ln [P(O)]$ versus O

For whitened image responses

$\ln[P(O)]$

For a Gaussian $P(O)$

Pixel value O in whitened image

B: $\ln [P(O_1, O_2)]$

C: $[P(O_2|O_1)] / [\max_{O_2} P(O_2|O_1)]$
—each column normalized

O_2

O_1

O_2

O_1

Fig. 4.2: Higher order statistics in the photograph of Fig. 4.1 A. A: Probability distribution of the whitened image pixel values O_i (blue) and a Gaussian distribution (red) whose mean and variance match those of O_i. The whitened signal's kurtosis $\langle O_i^4 \rangle / \langle O_i^2 \rangle^2 = 12.7$. B: Grayscale plot of $\ln [P(O_1, O_2)]$, i.e., the log of the joint probability of responses O_1 and O_2 of two horizontally neighboring (whitened) pixel values. Here, $\ln [P(O_1, O_2)]$ instead of $P(O_1, O_2)$ is displayed, so that a larger range of probability values can be visualized. C: The grayscale plot of the conditional probability $P(O_2|O_1)$ of O_2 given O_1. Each column is normalized so that the brightest values are the same across columns. $\langle O_1^2 O_2^2 \rangle / (\langle O_1^2 \rangle \langle O_2^2 \rangle) = 5.1$. In both B and C, $O_1 = O_2 = 0$ is at the center of the image.

$P(O_i)$ of a pixel i is approximated by sampling the value O_i across pixels i. Fig. 4.2 A shows $\ln P(O_i)$, and $\ln P_{\text{Gaussian}}(O_i)$ for a Gaussian $P_{\text{Gaussian}}(O_i)$ with the same mean and variance as the pixel values O_i in the whitened image. The actual $P(O_i)$ is more peaked, with a fatter tail than the Gaussian distribution would have. This is manifested by a kurtosis value of 12.7, which is much larger than kurtosis = 3 for a Gaussian variable. Figure 4.2 B shows (the logarithm of) the joint probability distribution $P(O_1, O_2)$ of responses O_1 and O_2 from two horizontally neighboring whitened pixels. It is apparent that O_1 and O_2 are decorrelated to the second order. In particular, for each O_1 value, $O_2 = a$ and $O_2 = -a$ are apparently equally likely for any given a.

However, higher order correlations are revealed in the conditional probabilities $P(O_2|O_1)$ shown in Fig. 4.2 C. To better visualize the full range of $P(O_2|O_1)$ for each O_1, the columns are normalized so that the largest $P(O_2|O_1)$ for each value of O_1 is awarded the same brightness. This reveals that the variance $\langle O_2^2 \rangle$ is larger for larger O_1^2. Hence, $\langle O_2^2 \rangle$ and $\langle O_1^2 \rangle$ are correlated, giving $\langle O_1^2 O_2^2 \rangle / (\langle O_1^2 \rangle \langle O_2^2 \rangle) = 5.1$, which is much larger than the value of 1 that would arise if O_1 and O_2 were independent Gaussians. Such higher order correlations can also be revealed by taking outputs of the V1-like receptive fields that are tuned to orientation and scale (Schwartz and Simoncelli 2001, Buccigrossi and Simoncelli 1999).

4.1.3 Efforts to understand V1 neural properties from the perspective of reducing higher order redundancy

4.1.3.1 Higher order redundancy reduction by linear transforms

Much effort has been expended to try to understand V1's receptive fields, and in particular orientation selectivity, by assuming that they have the role of removing higher order redundancy. We refer to this work collectively as independent component analysis (ICA). The best-known examples are due to Olshausen and Field (1997) and by Bell and Sejnowski (1997), and we observed in Section 3.7 that these two examples are similar and in some situations equivalent.

ICA is typically applied to the case in which there is no input noise. The original input **S** is assumed to be the result of linearly mixing some independent sources $O_1, O_2, ...$ Let $\mathbf{O} = (O_1, O_2, ...)$. Explicitly, $\mathbf{S} = \mathbf{MO}$, based on a mixing matrix $\mathbf{M}$. If $\mathbf{M}$ is a square invertible matrix, then a transform $\mathsf{K} = \mathbf{M}^{-1}$ would make $\mathbf{O} = \mathsf{K}\mathbf{S}$ have independent components $(O_1, O_2, ...)$. The K obtained by ICA is identified as the desired V1 receptive fields. We examine next what these K should be.

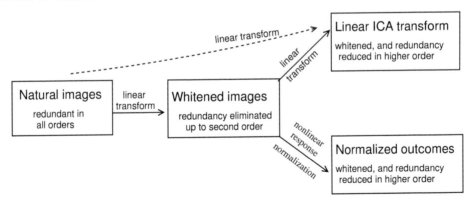

Fig. 4.3: Procedures of linear and nonlinear transforms, which are often used to reduce higher order redundancy in complete (rather than overcomplete) representations of visual inputs.

In a representation **O** with independent units O_i and O_j, the pairwise correlation vanishes. Without loss of generality, let us assume that each O_i has a zero mean and unit variance. Hence, **O** has an identity correlation matrix $\mathsf{R}^O = \mathbb{1}$. We thus deduce (from Chapter 3) that K must have the form $\mathsf{K} = \mathsf{U}\mathsf{K}_{\text{whitening}}$, in which $\mathsf{K}_{\text{whitening}}$ is the whitening transform in equation (3.171) that achieves the decorrelation $\langle O_i O_j \rangle = \delta_{ij}$ and U is a unitary matrix. The second order decorrelation is maintained by any U, since U simply transforms R^O to $\mathsf{U}\mathsf{R}^O\mathsf{U}^\dagger = \mathbb{1}$, because $\mathsf{U}\mathsf{U}^\dagger = \mathbb{1}$ by the definition of unitary matrices. Therefore, the special U aiming for independence between O_i and O_j should remove or reduce higher order correlations between O_i and O_j, e.g., to make $\langle O_i^2 O_j^2 \rangle \approx \langle O_i^2 \rangle \langle O_j^2 \rangle$.

The unitary transform U also leaves the entropy value $H(\mathbf{O})$ unchanged because $|\det U| = 1$ (see equation (3.24)). Meanwhile, since $H(\mathbf{O}) \leq \sum_i H(O_i)$ (see equation (3.28)), the U that performs higher order decorrelation, i.e., that makes $P(\mathbf{O}) = P(O_1)P(O_2)\cdots P(O_i)\cdots$, should be the one to minimize $\sum_i H(O_i)$. Minimizing $\sum_i H(O_i)$ given $H(\mathbf{O})$ is equivalent to minimizing the redundancy $\sum_i H(O_i)/H(\mathbf{O}) - 1$. Since there are no low order correlations between O_i and O_j, the redundancy to be minimized is the higher order redundancy.

Is it possible that the optimal K for ICA does not fully eliminate second order correlation in order better to reduce the dependence between components across all orders? We will see shortly that the higher order redundancy in natural scenes is only a tiny fraction of the lower order ones. Hence, to a good approximation, an optimal transform K for ICA will maintain the whitening so that $K = UK_{whitening}$. Therefore, finding the independent components in natural images by linear transforms is (approximately) equivalent to a procedure containing the following two steps: first follow the whitening procedure for efficient coding in Chapter 3 in the noiseless condition, and then, find the special U that minimizes $\sum_i H(O_i)$.

4.1.3.2 Higher order redundancy reduction by nonlinear transforms

The linear ICA procedure discussed above is only adequate if the visual inputs are indeed made by linear summations of independent components (at least approximately). However, some higher order correlations cannot be removed by linear transforms. This can be seen in the example in which $\mathbf{O} = (O_1, O_2)$ is two-dimensional and $P(\mathbf{O})$ only depends on $|\mathbf{O}|$: $P(\mathbf{O}) = f(|\mathbf{O}|) = f(\sqrt{O_1^2 + O_2^2})$. This example gives zero correlation in the first and second order:

$$\langle O_i \rangle = \int_{O_i \geq 0} d\mathbf{O} |O_i| P(\mathbf{O}) - \int_{O_i < 0} d\mathbf{O} |O_i| P(\mathbf{O}) = 0,$$

$$\langle O_1 O_2 \rangle = \int_{O_1 \geq 0} d\mathbf{O} |O_1| |O_2| P(\mathbf{O}) - \int_{O_1 < 0} d\mathbf{O} |O_1| |O_2| P(\mathbf{O}) = 0,$$

since $P(O_1, O_2)$ does not depend on the signs of O_1 and O_2. Hence $\mathbf{O}$ could be the outcome of whitening the input image $\mathbf{S}$. However, O_1 and O_2 are not independent unless $f(|\mathbf{O}|) \propto \exp(-\alpha(O_1^2 + O_2^2))$ which factorizes $P(\mathbf{O})$ as

$$P(\mathbf{O}) \propto \exp[-\alpha(O_1^2 + O_2^2)] = \exp(-\alpha O_1^2)\exp(-\alpha O_2^2) \propto P(O_1)P(O_2), \quad (4.13)$$

for a constant α. For example, if $f(|\mathbf{O}|) \approx (1 + O_1^2 + O_2^2)^{-2}$, it cannot be factorized, i.e., $P(\mathbf{O}) \neq P(O_1)P(O_2)$. Writing $\mathbf{O}$ as $(O_1, O_2) = |\mathbf{O}|(\cos(\theta), \sin(\theta))$, any two-dimensional rotation U (see equation (3.126) for such a rotation) simply gives

$$\mathbf{O}' = U\mathbf{O} = (O_1', O_2') = |\mathbf{O}'|(\cos(\theta'), \sin(\theta')), \quad |\mathbf{O}| = |\mathbf{O}'|.$$

Therefore, $P(\mathbf{O}') = f(|\mathbf{O}'|) = f\left(\sqrt{(O_1')^2 + (O_2')^2}\right) \neq P(O_1')P(O_2')$, without reducing the interdependence between the components. When $P(\mathbf{O}) = f(|\mathbf{O}|)$ while $P(\mathbf{O}) \neq P(O_1)P(O_2)$, then O_1^2 and O_2^2 are correlated. This correlation is observed in Fig. 4.2 C for natural images. It can be extended to multiple units $O_1, O_2, O_3, \ldots$ in whitened natural images $\mathbf{O}$, such that O_i^2 is correlated with local O_j^2 for multiple pixels $j \neq i$ near pixel i (Schwartz and Simoncelli 2001, Lyu and Simoncelli 2009). Hence, linear transforms cannot eliminate certain higher order correlations, even though for whitened images $\mathbf{O}$ of natural scenes, the probability density $P(\mathbf{O})$ is not fully determined by $|\mathbf{O}|$ alone (Lyu 2011).

These observations have motivated nonlinear transforms that reduce higher order redundancies. Denote the whitened image before the nonlinear transform as $\mathbf{L} = (L_1, L_2, ..., L_n)$, with pixels i. Let $\mathbf{L}$ be transformed nonlinearly to $\mathbf{O}$. If $P(\mathbf{L})$ is a function of $|\mathbf{L}|$, then a nonlinear transform from $|\mathbf{L}|$ to $|\mathbf{O}|$ such that $P(\mathbf{O}) \propto \exp(-\alpha|\mathbf{O}|^2) = \Pi_{i=1}^n P(O_i) \propto \exp(-\alpha \sum_i O_i^2)$ can factorize the joint probability distribution into individual marginal probabilities. This nonlinear transform is called Gaussianization and can be approximated by (Lyu and Simoncelli 2009)

$$O_i = \frac{L_i}{C + \sum_{j=1}^n L_j^2}, \quad \text{where } C \text{ is a suitable constant.} \tag{4.14}$$

The nonlinear transform resembles equation (2.43) and equation (2.84), the normalization model for V1's feature-unspecific contextual suppression in simple cells. The normalization model for the nonlinear transform is often expressed as (Schwartz and Simoncelli 2001)

$$O_i = \frac{L_i^2}{C + \sum_j L_j^2}. \tag{4.15}$$

The two transforms above have been observed to reduce the correlation between O_i^2 and O_j^2 for $i \neq j$ when the dimension n of the vector $\mathbf{L}$ is sufficiently large. However, this does not fully eliminate dependencies between the components O_i and O_j, unless the following assumptions hold. First, for second order decorrelation $\langle L_i L_{j\neq i} \rangle = 0$ to be achieved, $\mathbf{L} = (L_1, L_2, ..., L_n)$ must be a complete rather than an overcomplete representation of the original images $\mathbf{S}$. Second, the probability density $P(\mathbf{L})$ should depend on the magnitude $|\mathbf{L}|$ only. Third, the normalization transform above is the actual Gaussianization transform to factorize $P(\mathbf{O})$. The first condition is far from being true for the overcomplete V1 neural population, whereas the second and third conditions hold only approximately in some situations (Lyu 2011).

4.1.4 Higher order redundancy in natural images is only a very small fraction of the total redundancy

If appeal is to be made to higher order redundancy as a driver for the V1 RFs, it is important to estimate how significant this redundancy is over and above the redundancy that is removed by whitening up to the second order $\mathbf{O} = \mathsf{K}_{\text{whitening}}\mathbf{S}$. From the probability $P(O_i)$ of the whitened response O_i from a single pixel i across a single natural scene (using ergodicity), we obtain

$$\text{single pixel entropy} \quad H(O_i) = 3.4128 \text{ bits},$$

after discretizing the O_i values into 64 equal sized bins. The Gaussian distribution with the same mean and variance as O_i has entropy $H_{\text{Gaussian}}(O_i) = 3.6290$ bits, which is only a small fraction larger. Hence, considering entropy, the distribution of the single pixel response is well approximated by a Gaussian, despite its high kurtosis of 12.7. This is particularly the case since our $H(O_i)$ is an underestimate due to the small number (256×256 pixels) of samples in the whitened image used for estimation. Further, from two pixels O_1 and O_2 in the whitened image, we can estimate (using ergodicity)

$$\text{joint entropy of two horizontally neighboring pixels} \quad H(O_1, O_2) = 6.6404 \text{ bits}.$$

This value is an even more serious underestimate given the small sample size for the values of the joint responses. Meanwhile, if the responses had been independent, their joint entropy would be equal to $2H(O_i) = 6.8256$ bits; and, if they were two independent Gaussians with matched variance, their joint entropy would be $2H_{\text{Gaussian}}(O_i) = 7.2580$ bits. When considering the responses O_1 and O_2 only, the redundancy is as follows:

$$\text{Redundancy in whitened images} = 2H(O_i)/H(O_1, O_2) - 1 \lesssim 0.0279. \qquad (4.16)$$

Hence, if one transforms the whitened responses O_1, O_2, etc., to make a set of new responses O'_1, O'_2, etc., each a particular function of whitened responses (somewhat like combining the retinal outputs to form even more independent responses), the amount of redundancy that could be further reduced is only about this much. This is a a very small fraction of the total redundancy ≈ 0.49 (in equation (3.36)) between two image pixels in non-whitened images (also discretized to 64 gray levels). More detailed and elaborate estimates of the higher order redundancy, using more than two image pixels, have led to the similar conclusion that the contribution to the redundancy in natural scenes of higher order statistics is much smaller than that of the second order statistics (Li and Atick 1994b, Petrov and Zhaoping 2003, Bethge 2006).

4.2 Problems in understanding V1 solely based on efficient coding

From a more general viewpoint, there are two serious problems with the argument that V1 serves to improve coding efficiency. First, there is no apparent bit-rate bottleneck after the optic nerve, other than the attentional bottleneck (manifested in behavior). Rather than leading to data compression, which is the target of coding efficiency, the attentional bottleneck leads to data deletion. Second, efficient coding has great difficulty in explaining some major aspects of V1 processing.

Is the cortical code more efficient because it reduces or removes the residual redundancy due to higher order correlations in natural scenes? There is no quantitative demonstration that V1 significantly improves coding efficiency over the retina. Even if V1 coding is more efficient, then apart from issues to do with stereo coding, why isn't this code adopted by the retina? Even more puzzling is the fact that there is a 100-fold expansion from the number of retinal ganglion cells to the number of V1 neurons (Barlow 1981). This expansion gives a hugely overcomplete representation of visual inputs. For instance, to represent input orientation completely at a particular spatial location and scale, only two or three neurons tuned to two or three different orientations would be sufficient (Freeman and Adelson 1991). However, many more V1 cells tuned to many different orientations are actually used. It is thus highly unlikely that neighboring V1 neurons have decorrelated outputs, even considering the nonlinearity in the actual receptor-to-V1 transform. This contradicts the goal of efficient coding for high signal-to-noise ratios to reduce redundancy and reveal the independent entities of scenes. Nor does such an expansion improve signal recovery at low signal-to-noise ratios, since no retina-to-V1 transform could generate new information beyond that available at the retina.

It has been argued (Olshausen and Field 1997, Simoncelli and Olshausen 2001) that the expansion in the number of neurons can make the code even sparser, allowing neurons to be silent for most inputs except when they contain very specific features. Minimizing the energy consumed by neural signaling has also been proposed to drive sparser coding (Levy and Baxter 1996, Lennie 2003a), possibly favoring overcompleteness. However, these arguments remain hypothetical, since there is as yet no reliable quantitative measure of the data rate of V1 neurons to show whether or how the cost of having vastly more neurons is justified by the extent to which this representation sufficiently improves efficiency, saves on energy consumption, or exposes the underlying cognitive (putatively independent) components.

Additional difficulties for the efficient coding theory arise from observations (see Section 2.3.9) made since the 1970s that contextual stimuli outside a neuron's RF significantly modulate the neuron's response in a complex manner (Allman et al. 1985). There are extensive intracortical connections (Gilbert and Wiesel 1983, Rockland and Lund 1983) linking nearby cells with overlapping or non-overlapping classical receptive fields. These connections are plausible neural substrates for mediating the contextual influences. The contextual modulations are beyond the kind of suppressive activity normalization described in equations (4.14) and (4.15) because they can also be facilitative, depending on the strengths and configurations of the visual inputs. Ignoring the overcompleteness which counters redundancy reduction, there is as yet no satisfactory means to understand the contextual influence from a role of redundancy reduction. Such contextual influences are a nuisance to the classical view that individual neurons are local feature detectors. Until recently, they have not been the subject of a theoretical treatment.

4.3 Multiscale and overcomplete representation in V1 is useful for invariant object recognition from responses of selected neural subpopulations

We have seen that higher, but not lower, order redundancy is useful for object recognition, and yet higher order redundancy contributes only a small fraction to the total redundancy in the input information. For object recognition, the higher order redundancy should be kept; whereas for coding efficiency, the lower order redundancy should be removed. These observations motivated the proposal (Li and Atick 1994b) that V1's multiscale representation provides a translation- and scale-invariant representation that facilitates object recognition, rather than reducing higher order redundancy. This translation and scale invariant representation, expressed in equation (3.209), is useful to generate an underlying pattern of visual responses that is invariant when an object changes its position in the scene.

More explicitly, when objects change their position relative to the observer in a three-dimensional scene, their images on the retina move and scale. A translation and scale invariant representation has the property that the pattern of the responses evoked by a visual object in a subpopulation of neurons is replicated in another subpopulation of neurons when the object changes its position in the scene. For example, at one position, the object evokes responses O_1, O_2, etc., in neurons $i = 1$, $i = 2$, etc., and at another position, the object evokes responses O_{101}, O_{102}, etc., in neurons $i = 101$, $i = 102$, etc., such that $O_1 = O_{101}$, $O_2 = O_{102}$, etc. Such a representation is useful if the decoding process can selectively look at one subpopulation of neurons at a time (for example, one subpopulation involving neurons $i = 1, 2$, etc., and another involving neurons $i = 101, 102$, etc.) and recognize the same object at different positions by the invariant response pattern in different (selected) subpopulations.

If the V1 representation is complete rather than overcomplete, position and scale invariance as in the scheme above could only be realized for a limited set of position and scale changes. Recall from several equations starting from equation (3.213) that different subpopulations of neurons in a one-time complete representation are devoted to different spatial frequency bands whose center frequencies differ by a factor of $\alpha = 3$. This implies that scale invariance would hold only for scale changes by a power of 3, and an object scale change by another factor of (e.g.) 2 would evoke a very different response pattern regardless of which subpopulation of neurons was selected for examination. Similarly, position invariance would also be very limited for a complete (rather than overcomplete) representation. This is because the number of sampling positions in each scale, represented by the number of neurons devoted to this scale, would be much smaller than the number of sampling positions represented by the number of retinal ganglion cells, since the total number of V1 neurons for all scales should be no more than the number of retinal ganglion cells in a one-time complete V1 representation.

If V1 expresses an overcomplete representation, it can evade these limitations. The number of spatial sampling positions at each scale can be increased to allow approximate position invariance; many overlapping frequency bands can be used to achieve approximate scale invariance for many possible scale changes. The numbers of samples in position and scale need not be infinite for exact invariance for any position and scale change, since noise and retinal input resolution should set an upper limit to this number. The degree of overcompleteness in V1 may be partly determined by these considerations.

4.3.1 Information selection, amount, and meaning

As argued above, the translation- and scale-invariant multiscale representation in V1 should be useful for object recognition. However, this is under the condition that the information represented by subpopulations of neurons has been suitably selected. Information selection implies losing, rather than preserving, Shannon information, in order to devote the brain's resources to processing the most relevant input information. To an animal, a bit of information about the identity of a visual object typically has a very different relevance from another bit of information about the luminance of the light. Information theory can quantify the *amount* of information, and it is therefore useful for designing optimal codes for information *transmission*, a likely goal for the retina. However, it does not assess the *meaning* of information. The meaning should be associated with the relevance of the information to the animal, especially to its survival and well-being. Information selection plays an important role since only by selecting the relevant information can it serve the animal well.

Rather than being a nuisance for the classical view of input coding or feature detectors, intracortical interactions in V1 can be a useful means of implementing information selection. V1, the largest visual area in the brain, equipped with additional neural mechanisms unavailable to retina, ought to perform important cognitive tasks beyond information transmission. One of the most important and challenging visual tasks is segmentation; much of this involves information selection. To understand V1, we thus turn to the second data reduction strategy for early vision, namely to build a representation that facilitates bottom-up visual selection. As we proceed, we should keep in mind, however, that a representation that facilitates selection may compromise visual decoding or recognition after selection.

5 The V1 hypothesis—creating a bottom-up saliency map for preattentive selection and segmentation

In this chapter, we focus on bottom-up visual selection, the second stage in the three-stage process of vision: encoding, selection, and decoding. We describe the hypothesis that V1 creates a bottom-up saliency map to guide visual selection, and we show that this hypothesis can solve certain V1 puzzles which elude the efficient coding principle underpinning the encoding stage. This hypothesis is motivated, formulated in detail, and applied to account for existing behavioral data on visual selection. Non-trivial and surprising predictions that result from the hypothesis are presented, along with their experimental confirmation. A circuit model describing how V1 mechanisms might implement this hypothesis is outlined. Finally, V1's role in selection is discussed in relation to selection by brain areas beyond V1.

5.1 Visual selection and visual saliency

5.1.1 Visual selection—top-down and bottom-up selections

Recall from Section 1.2 that selection and decoding are two visual stages after the visual encoding stage in the three stages of vision. The resources for processing visual input are limited, leading to the attentional bottleneck. Selection enables vision to focus the processing resource, i.e., focus the decoding, on just a fraction of this input. Therefore, selection often makes vision unable to properly decode or perceive the visual input outside our attentional spotlight. Accordingly, the effect of visual selection is measured by the *cueing effect*, which is the improvement in performance and/or speed of visual tasks on the selected input (Posner 1980), and the presence of an cueing effect is often used to demonstrate the availability of visual selection.

Selection is most obviously manifest in the fact that we shift our gaze, or saccade, to the visual locations we select. This overt form of selection is referred to as orienting. Meanwhile, selection can also be done covertly. Responses of neurons in the extrastriate cortex to an attended stimulus are often enhanced relative to their responses to unattended stimuli. This enhancement can result in the responses of neurons being the same as if the unattended stimulus was absent (see Section 2.6).

Reflecting upon our own subjective visual experience, we are more aware of our own goal-directed or voluntary selections, such as when we attend to a book when reading and ignore visual space outside the book page. Goal-directed selection can also be based on prior knowledge or expectation, such as in directing gaze to one's bookshelf when looking for a book, by the knowledge about the location of the bookshelf and an expectation that the book is likely on the shelf. Hence, it is not surprising that most theories or research frameworks have emphasized this goal-directed selection (Treisman and Gelade 1980, Duncan and Humphreys 1989, Tsotsos 1990, Desimone and Duncan 1995), which is also called top-down selection.

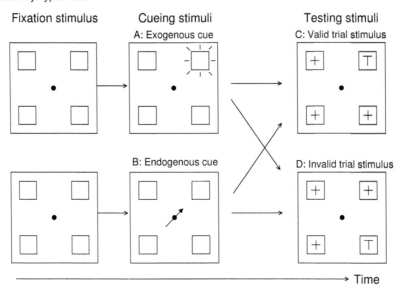

Fixation stimulus   Cueing stimuli   Testing stimuli

A: Exogenous cue   C: Valid trial stimulus

B: Endogenous cue   D: Invalid trial stimulus

→ Time

Fig. 5.1: A schematic of an example experimental design used to study top-down, or endogenous, guidance and bottom-up, or exogenous, guidance to attention. This schematic is
representative of many other similar studies. Each of the six large square boxes contains a
sketch of a stimulus on a fixed-location display. In a test trial, the fixation stimulus, a cueing
stimulus, and a testing stimulus are shown consecutively. Observers have to discriminate an
aspect of the brief test stimulus, e.g., the orientation of the letter "T" (whether its stem points
downwards, upwards, to the left, or to the right). The location of this letter can be at any one
of several (e.g., four in this figure) possible positions, unknown to the observer beforehand.
This location is cued by a brief exogenous or endogenous cue in the cueing stimulus, which
onsets at a time interval called stimulus onset asynchrony (SOA) before the test stimulus
onsets. An exogenous cue indicates this location typically by a flash at or near this location;
an endogenous cue does this by a symbol (e.g., an arrow pointing to the location) whose position (typically at the fixation point) is independent of the cued position. The actual location
of the letter "T" in the test stimulus is at the cued location in a valid trial, or otherwise in
an invalid trial. In typical experiments, observers have to keep their gaze fixed at the central
fixation point in each stimulus throughout a trial. The exogenous cue has been found to be
faster acting, more effective, and harder to ignore, and it can overwrite the endogenous cue
(Müller and Rabbitt 1989). Adapted with permission from Müller, H. J. and Rabbitt, P. M.,
*Reflexive and voluntary orienting of visual attention: time course of activation and resistance
to interruption*, *Journal of Experimental Psychology: Human Perception and Performance*,
15 (2): 315–330, Fig. 1, copyright © 1989, American Psychological Association.

We would be blind to unexpected things if selection was purely top down. There is
thus a vital role for an alternative form of visual selection driven directly by visual inputs,
or involuntary selection without the influence of task goals. These are called bottom-up
selection. In some situations, such as during an emergency, bottom-up selection should be
able to overwrite top-down selection, such that, e.g., we should direct our attention to a predator
pouncing at us even while reading. In this sense, bottom-up selection serves an ultimate top-
down goal of survival. However, this book follows the convention of referring to involuntary
selections as being bottom-up. Top-down selection is also referred to as endogenous or
reflective, since it is linked with internal goals or knowledge of the viewer. Bottom-up selection

is also referred to as exogenous, since it is driven by the external visual inputs, and is said to be reflexive.

A: A red bar attracts attention automatically

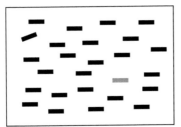

The task: to find a non-horizontal bar

B: A cue (red box) onset attracts attention additionally, even when its location is known a priori

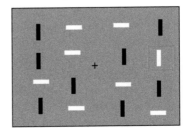

The task: to identify the bar in the red box

Fig. 5.2: Demonstration of the superior potency (A) and speed (B) of bottom-up over top-down selection. A: even if an observer's task is to find a non-horizontal bar in the image, the red non-target bar automatically distracts attention. B: The task for (human) observers is to keep gaze focused on the center cross, and to report whether the target bar in the red cue box (which is at the same, known, location in the image on each trial) is white vertical, black horizontal, or is like the non target bars (which are white horizontal and black vertical). Target and non-target bars are displayed simultaneously for a short time duration in each trial. Nakayama and Mackeben (1989) measured the shortest display duration necessary for the observers' report to be suitably accurate. This duration was longer when the red cue box remained on display throughout the experimental session to mark target location, compared to the case when the cue appeared about 50–150 ms ahead of the bars in each trial.

Bottom-up selection is often faster acting and more potent than top-down selection (Jonides 1981, Müller and Rabbitt 1989, Nakayama and Mackeben 1989), as one might expect from its role in emergencies. Figure 5.1 shows the kind of experimental setups used to investigate their respective characters. An exogenous cue, typically a brief flash, draws attention reflexively to the flashed location, such that discrimination of a test stimulus presented very soon after the cue is typically faster and more accurate at the flashed location than at another location. An endogenous cue, typically presented as symbols (e.g., an arrow to the northeast) to indicate a likely location of the upcoming test letter, is physically not at the same location as the test letter. It can also make discrimination better and faster at the cued rather than the uncued location. However, when instructed, observers can effectively ignore the endogenous cue, such that their performance is independent of whether the location is cued or uncued; but they are unable to ignore the exogenous cue (Jonides 1981). Furthermore, if the onsets of the cue and test stimuli differ by only a very brief stimulus onset asynchrony (SOA) such as 100–150 ms, the benefit of a valid cue, when the cued and tested locations agree, and the cost of an invalid cue, when the cued and tested locations disagree, are both larger with the exogenous than the endogenous cue. At longer SOAs, the difference between the performances at the cued and uncued locations decreases with SOA for exogenous cues but increases with SOA for endogenous cues, before this difference asymptotes at around SOA = 300–400 ms. Additionally, the performance benefit of a valid, endogenous, cue can be completely eliminated if, after the endogenous cue and about 100 ms before the test stimulus onset, an exogenous flash occurs at another location. This occurs even if this exogenous flash is completely uninformative about the location of the target letter (Müller and Rabbitt 1989).

Figure 5.2 A demonstrates that the red horizontal bar automatically attracts the attention of observers, even if they are intending to look for a non-horizontal bar. Such a distraction by a task-irrelevant salient color singleton is also hard to turn off voluntarily (Theeuwes 1992). Figure 5.2 B shows the stimulus patterns used by Nakayama and Mackeben (1989) to contrast top-down and bottom-up selection. Subjects were asked to discriminate a target bar in a briefly displayed, inhomogenous array of bars. They were unable to recognize the target bar reliably when the array was shown for only 33 ms, even though they knew the target location in the array long before the array appeared. This suggests that vision was too slow, or top-down selection was inadequate for this task, when the stimulus was displayed too briefly. However, task performance dramatically improved if a red box, the exogenous cue, surrounding the target location appeared about 50–150 ms before the array appeared. This bottom-up attraction by the red box, in addition to the top-down attraction due to the prior knowledge of the target location, caused a marked improvement in task performance. Without this bottom-up attraction, the array had to be shown for a much longer time in order to improve the task performance. This again suggests that, in this task, when bottom-up attention is available, vision is not too slow and the bottom-up attraction is faster acting.

Top-down and bottom-up attentional selection are often space-based, such that only visual input at or near a spatial location is selected. Selected locations are metaphorically referred to as being in the attentional spotlight. This also implies that the spotlight is spatially compact and has a finite size. Recall from Section 2.5 that spatial selection is closely linked with eye movements, such that the selected location typically coincides with the current gaze position or the destination of an impending saccade. Since our gaze can only be at one location at a time, it is not surprising that it is either impossible or very difficult to select two disjoint locations simultaneously (Cave and Bichot 1999), even for covert attention.

Selection can also be based on features of the input. Consider looking for a red cup. Selection can be based on a particular value of the feature of color, so that sensitivity to red-colored objects in the whole visual field is relatively enhanced. Obviously, feature-based selection is goal directed and is thus top down. Selection can also be object-based. For example, consider two visual locations in an image that are equally distant from the current gaze position, which is inside the image area of an object. If only one of the two locations is inside the image area of the same object, then sensitivity to inputs at this location is relatively higher. Since object-based selection relies on the perception of the object, knowledge about the object's shape or identity can influence the selection. Therefore, it is likely that object-based selection is not strictly bottom up.

To understand selection as a whole, we must clearly understand bottom-up selection, both because of its own potency and because of its competition and cooperation with top-down selection. This book focuses mostly on bottom-up (spatial) selection, mainly because more is known about its neural mechanisms. The book comparatively ignores feature-based or object-based attention.

5.1.1.1 Terminology: selection, attention, saliency, and priority

Voluntary or top-down visual selection is often colloquially referred to as "paying attention." When bottom-up selection overrides top-down selection, it is often said that attention has been distracted to a task-irrelevant input. In this colloquial sense, the word "attention" is understood as some sort of resource, which is applied to or spent on the selected input, enabling this input to be recognized or decoded. Hence, the term "preattentive" is understood as to refer to the processing stage before the resource of "attention" is applied, although exogenous selection process can be operative at the preattentive stage. Meanwhile, directing attention to somewhere or to something is also colloquially referred to as attending somewhere or something. In this sense, "attending" and "paying attention" refer to the act of selection, regardless of whether this is by top-down or bottom-up means. In this book, we will use the term "selection," "selecting," or "to select" to mean the act rather than the resource, to avoid the confusion.

In this book, we define the *saliency* of a visual location as the degree to which this location attracts selection by bottom-up mechanisms only. A location with a large saliency value is said to be salient. Following Egeth and Yantis (1997), the term *priority* is used to describe the degree to which this location attracts selection as a result of combining top-down and bottom-up mechanisms. A saliency map is a map of saliency values of the visual field, while a priority map is a map of priority values. Behaviorally, selection follows the priority map, such that attention or gaze is more likely to be directed first to locations of higher priorities. The temporal order in which spatial locations are selected should follow the order of their priorities, deterministically or stochastically, such that, when a scene is viewed, a location having a higher priority is more likely selected before a location having a lower priority. Often, a location is often more likely selected, i.e., having a higher priority, because it is closer to the currently attended location. This is likely caused by top-down rather than bottom-up factors.

5.1.1.2 Probing bottom-up saliency behaviorally even though selection is controlled by priority

According to the definitions above, a location's priority can often be assessed behaviorally by the reaction time (RT) associated with finding or identifying a target at its location. Alternatively, it can be assessed by measuring how well observers discriminate or identify a visual target at the location, given a fixed viewing duration. This latter measurement is called the accuracy, expressed as the probability that the task is performed correctly. Accuracy should increase with the amount of time the target spends in the attentional spotlight. Therefore, given a fixed viewing duration, greater accuracies should be coupled to shorter RTs for selecting the target location. Although saliency is only one of the contributing factors to priority, it can be studied in terms of the difference between RTs or accuracies for different tasks. For example, studies of visual search often assume that a shorter RT indicates a larger saliency at the location of the search target. Below we explain why and when this assumption is approximately valid.

Let RT_{task} be the behaviorally measured RT in a task, e.g., to find a certain target. Imagine an ideal world in which we could measure $RT_{selection}$, which is the RT to select the task relevant location, e.g., the location of the target. Let us define

$$RT_{other} \equiv RT_{task} - RT_{selection}; \quad RT_{other} \text{ is understood as the RT required} \atop \text{for all the non-selection processes to complete the task.} \tag{5.1}$$

For example, let visual task A be to look for a target bar tilted $20°$ anticlockwise from horizontal, among many non-target bars uniformly tilted $20°$ clockwise from horizontal. Experimentally, we measure RT_{task} as the time from the onset of the visual stimulus to the time when observers report the location of the target. This RT_{task} contains $RT_{selection}$ to put the target bar in the attentional spotlight, and RT_{other} to confirm that the bar in the spotlight is indeed the sought-after target and to execute the motor action to report the target's location. Note that when the absolute value of $RT_{selection}$ is difficult to measure, then so is RT_{other} since it is defined by $RT_{task} - RT_{selection}$. However, this should not affect our argument.

We can extract saliency from $RT_{saliency}$, defined as $RT_{saliency} \equiv RT_{selection}$ when the top-down contribution to selection is set to zero. However, eliminating all top-down contributions is impossible in typical behavioral experiments. In general, define

$$RT_{\text{top-down selection}} \equiv RT_{selection} - RT_{saliency}. \tag{5.2}$$

The term $RT_{\text{top-down selection}}$ could be negative or positive, depending on whether the top-down and bottom-up selection cooperate or compete. Putting the above together, we have

$$RT_{\text{task}} = RT_{\text{saliency}} + RT_{\text{top-down selection}} + RT_{\text{other}};$$

therefore, RT_{task} can be a proxy for RT_{saliency} when (5.3)

$$RT_{\text{top-down selection}} + RT_{\text{other}} = \text{constant for different tasks.}$$

One can design experiments to ensure that $RT_{\text{top-down selection}} + RT_{\text{other}}$ is a constant. For example, suppose that along with task A above, subjects execute task B, which requires finding the same target bar (tilted $20°$ anticlockwise from horizontal) but among non-target bars that are horizontal rather than being tilted $20°$ clockwise from horizontal. Provided that tasks A and B do not differ in other aspects (non-target numerosity, item sizes, etc.), it is reasonable to assume that the two tasks have approximately the same RT_{other}, since the target in the attentional spotlight is equally distinguishable from non-targets, and since the time it takes to report the target (after the target is recognized in the attentional spotlight) is unlikely to depend on the task. Since the target is the same, feature-based attention should also be set the same way, suggesting that $RT_{\text{top-down selection}}$ should also be the same. In total, if RT_{task} is shorter for tasks A than task B, then the target location can be considered to be more salient in task A, and vice versa.

Next, consider a slightly modified experiment in which the target feature is unknown ahead of time, e.g., when the task is simply to find a uniquely oriented bar in the image without specifying the orientation of this unique target bar. In this case, the top-down contribution will be more limited but will still be the same for the two tasks. In other experimental designs, the bottom-up saliency can be so strong that attention can be attracted to the target location automatically whether or not the target identity is known ahead of time, so that the top-down contribution to selection is negligible. Altogether, $RT_{\text{top-down selection}} + RT_{\text{other}}$ can be approximately independent of the task in many different situations. This can even be approximately true when the two tasks differ in both target and non-target identities.

Under the assumption that, among multiple tasks in a study, tasks with shorter RT_{task}s have more salient target locations, we can study how saliency is determined by input stimuli. This assumption is the basis for many behavioral and modeling studies into saliency, including those described in this book. One should nevertheless be wary of violations of the assumption, as sometimes (Zhaoping and Guyader 2007) a difference in RT_{task}s between tasks results from top-down rather than bottom-up effects.

Similar considerations and arguments apply when bottom-up saliency of a location is assessed by the accuracy of input discrimination at that location for sufficiently brief viewing durations.

Various features of experimental designs can minimize the impact of factors other than saliency in the measured RTs and accuracies. One is to minimize any a priori knowledge as to the possible positions and features of the visual inputs. Another is to make the visual input at a location task-irrelevant, to remove or minimize the top-down factors in selecting this location.

One can also make the input presentation very brief, such that there is insufficient time for more than one shift of attention, or one glimpse. Compared with this first shift, any second shift of visual attention is more influenced by the visual information gained during the initial glimpse, and such information can drive top-down influences. For example, take the case of asking an observer to find an apple in an image quickly but without revealing ahead of time what kind of image will be presented. When a picture of a kitchen scene is suddenly shown, the first gaze shift is likely to be governed by reflexes before the observer realizes what kind of scene is on show. However, the duration from the appearance of the image to the second gaze shift is often sufficient for the observer to realize the overall content or gist of the scene, and thence to exploit the top-down expectation that an apple is more likely to be on a kitchen counter than a kitchen floor. In turn, this can influence the second gaze shift. Conversely, a picture of a forest would direct the second gaze shift according to a different top-down factor.

Analogously, the effect of saliency is better manifested in shorter reaction times. A longer latency to respond after a brief display allows more top-down influences.

One should also design the task to minimize the contamination of the measured behavioral outcome by high level strategic factors. For example, in many conventional visual search tasks (Wolfe 1998), observers are asked to report the presence or absence of a target in an image. When observers do not find the target after an initial scan of the search display, they may decide to search further for the target, or report that the target is absent. Hence, the RT of the report is affected by the strategic decision of when to terminate the search, a strategic decision that may be task dependent. To avoid this, it helps to make the target present in all search trials and ask the observers to simply report an aspect of the target location, e.g., whether the target is in the left or right half of the visual display.

5.1.2 A brief overview of visual search and segmentation—behavioral studies of saliency

Studies of visual search are often used to examine how the saliency of a target location depends on the characteristics of the visual inputs, assuming that a shorter RT indicates a larger saliency at the location of the search target. In these experiments, human observers are asked to search for a target as quickly as possible, and their reaction time (RT) to report the target is recorded.

Visual search has been the target of extensive behavioral studies (Treisman and Gelade 1980, Duncan and Humphreys 1989, Julesz 1981, Wolfe, Cave and Franzel 1989, Wolfe 1998), and is introduced briefly in Fig. 5.3. It has been found, for instance, that if the target is characterized by a visual feature such as color or orientation that is sufficiently unique within a visual image or scene, the RT for finding it is often insensitive to the number of non-target items (distractors) in the scene. Visual search for which the target differs from all the non-targets in one unique feature is called *feature search*. Figure 5.3 E shows an example in which the target differs from the distractor not by a single feature but by a conjunction of two features: red and vertical. Each of these is present in the non-target items (which are blue-vertical or red-horizontal). Such a search is called a *conjunction search*, and it is usually more difficult than a feature search. RTs in conjunction searches usually grow quickly with the number of distractors. The total number of search items, target plus the distractors, is called the *set size* of the search. One can imagine that if a target were defined by a conjunction of more than two features, the search would be even more difficult.

Visual search in which the RT is almost independent of the set size is called *efficient*. By contrast, if RT increases with set size sufficiently quickly, the search is called *inefficient*. Efficient and inefficient searches have often been interpreted to depend on underlying neural processes that are, respectively, parallel and serial. Visual inputs in the scene are processed all at once for an efficient search and chunk by chunk for an inefficient search.

Efficiency in visual search has been used as an empirical definition of the notion of a basic feature dimension in visual input (Wolfe 1998). Accordingly, color, orientation, motion direction, stereoscopic depth, and sizes have been found to be among the basic feature dimensions, since a target that is sufficiently different from homogenous distractors in any one of these dimensions (e.g., Fig. 5.3 AD) can be found efficiently. Indeed, this has been used to define feature search in cases in which the target differs from the distractors by its feature value in a basic feature dimension. Feature searches are efficient if the non-targets have the same feature value in the basic feature dimension that distinguishes the target; they can be inefficient if their feature values in this dimension are sufficiently varied.

Often, a salient target in an efficient visual search is said to pop out of the scene *preattentively* or to require only preattentive mechanisms to be noticed by human observers. Here

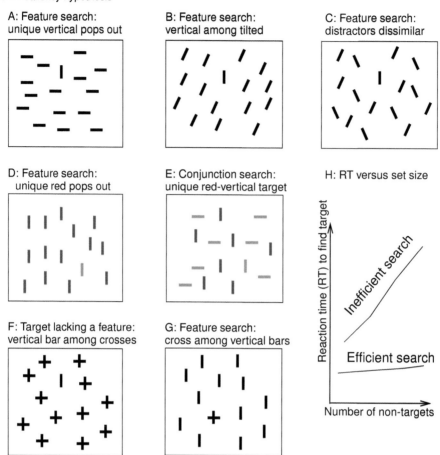

Fig. 5.3: A brief overview of visual search. A–G: Illustrative examples of visual search. The search target is a vertical bar in A–C and F, a red-vertical bar in D and E, and a cross in G. A–D and G are examples of feature search, when the target has a feature that is absent in the non-targets. E is an example of conjunction search when the target is defined by a unique conjunction of features, each of which is individually present in the non-targets. In F, the target is defined by the absence of a feature that is present in the non-targets. F and G together illustrate the asymmetry of visual search, when the ease of search changes by swapping identities of the targets and non-targets. H: Characteristics of efficient and inefficient searches in terms of how RTs depend on the number of non-target (distractor) items.

the term preattentive can be understood as without top-down attentional guidance. Hence, preattentive attentional guidance is by bottom-up attractions only. By contrast, an inefficient search requires something more than preattentive mechanisms. However, the meanings or definitions of these terms (e.g., pop-out, preattentive, etc.) may differ in the literature according to different research communities.

Efficiency in visual search can be affected by many factors, and there is a continuum rather than two discrete categories (efficient/parallel and inefficient/serial). Figure 5.3 AB demonstrate that searching for the vertical target is easier when the feature contrast (orientation contrast) between the target and distractors is larger; Fig. 5.3 BC demonstrate that search becomes more difficult when the distractors are not identical to each other (Duncan and

Humphreys 1989), even though the target's feature is unique in both examples. Figure 5.3 FG show a simple example of visual search asymmetry, in which the ease of search can change when the target and distractor swap identity. Figure 5.3 F is an example for which the target is more difficult to find when it is defined by the absence of a basic feature that is present in the distractors.

Further, efficiency, defined as the insensitivity of the search RT to the set size of the search, is insufficient to describe the ease of a search by itself. Some searches can require longish RTs even though these RTs are insensitive to the search set size.

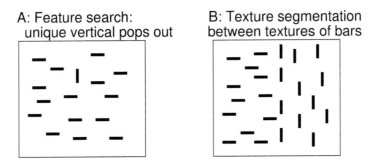

A: Feature search: unique vertical pops out

B: Texture segmentation between textures of bars

Fig. 5.4: Demonstration that visual search and segmentation are typically related. A: A vertical bar pops out among horizontal bars. B: A texture of vertical bars is readily segmented from a texture of horizontal bars.

Visual saliency is also manifest in texture segmentation behavior (Julesz 1981). Texture segmentation becomes easier when the border between two texture regions is more salient (Nothdurft 1991, Li 1999b, Li 2000b).

When a unique target pops out of non-targets in a visual search display, one texture region made of many of these target items is also typically easy to segment from another texture region made of the non-target items, as demonstrated in Fig. 5.4. This link between visual search and segmentation will be elaborated further in the chapter.

5.1.3 Saliency regardless of visual input features

One can compare the saliency induced by different input features such as color and orientation. For example, in Fig. 5.2 A, both the location of the red bar and the location of the non-horizontal bar are salient. If observers freely view this image in a task in which they do not have a preference for either feature, one can see which location attracts attention more strongly. Phenomenologically, it is as if there is a saliency map of the visual space, such that locations with higher saliency values in this map are more likely to attract attention, regardless of the visual input features that make those locations salient. In other words, once feature distinctions are converted into saliency values, or a raw visual image is turned into a saliency map, the original image feature values that caused these different saliencies are irrelevant as far as bottom-up attraction to attention is concerned. This is true even if the rules by which different features (or feature combinations) are converted into saliency values differ. In fact, this feature irrelevance is (implicitly) part of the definition of saliency, since the concept makes no reference to the feature dimensions or values that determine its values.

That saliency values are independent of input features may be a reason why traditional models (Koch and Ullman 1985, Wolfe et al. 1989, Itti and Koch 2000) compute saliency from visual inputs according to a framework which can be paraphrased as follows (Fig. 5.5 A).

A: The traditional framework for a bottom-up visual saliency map

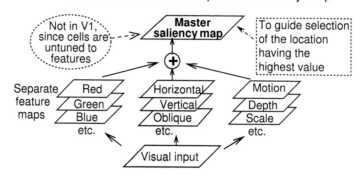

B: Application to feature-search (left) and conjunction-search input (right)

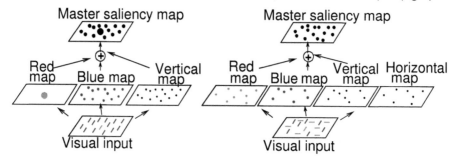

Fig. 5.5: A: Schematic of the framework for traditional models of visual saliency. This framework implies that a saliency map should be in a brain area (such as the lateral intraparietal area (LIP) (Gottlieb et al. 1998)) where cells are untuned to visual features. B: Application of this framework on feature-search (left) and conjunction-search (right) stimuli. Only the relevant feature maps are shown, and the activations in each feature map are higher when there are fewer items in that map. The master map has a hot spot at the location of the red bar in the image for the feature search to attract selection, but it has no hot spot for the conjunction search image. Adapted with permission from Zhaoping, L., Theoretical understanding of the early visual processes by data compression and data selection, *Network: Computation in Neural Systems*, 17(4): 301–334, Fig. 12, copyright © 2006, Informa Healthcare.

Visual inputs are analyzed by separate feature maps, e.g., red feature map, green feature map, vertical, horizontal, left tilt, and right tilt feature maps, etc., in several basic feature dimensions such as color, orientation, and motion direction. The activation of the unit representing an input at a particular location in its corresponding feature map decreases roughly with the number of the neighboring input items sharing the same feature value. Hence, in an image of a red bar among blue bars, as in the left example of Fig. 5.5 B, the red bar evokes a higher activation in the red map than that of each of the many blue bars in the blue feature map. Then, the saliency value for a location in the master map is the summation of the activations in all the separate feature maps associated with that location. The summation implies that the saliency values in the master map generalize across the actual input features. In the master saliency map for the left example of Fig. 5.5 B, the red bar evokes the highest activation at its location and attracts visual selection. By contrast, a unique red-vertical bar, among red-horizontal and blue-vertical bars, does not evoke a higher activation in any single feature map, red, blue, vertical, or horizontal, and thus not in the master map either.

The traditional framework provides a good phenomenological model of the saliency implied by behavior in feature and conjunctive searches. It has subsequently been made more explicit and implemented in computer algorithms (Itti and Koch 2000). It does not explicitly specify the neural mechanisms or the cortical area(s) underlying the feature and master maps. However, it implies that the master saliency map should be in a cortical area where neurons are not tuned to visual feature values, since combining the feature maps eliminates the feature selectivity in the master map. The LIP (lateral intraparietal area) or FEF (frontal eye field) could be candidates for this master map, since their neural activities are untuned to input features. This implication of feature irrelevance has had an obvious impact on the directions of experimental investigations—for many years, few experiments looked for the neural substrates of the saliency map in early visual areas, where neurons are feature selective.

Contrary to intuition, the fact that saliency is independent of particular input features does not mean that the cells reporting saliency *must* be untuned to input features. If saliencies are signaled by the activities or firing rates of neurons, then "signaling independent of input features" can simply mean that the neural firing rates associated with saliency are universal, independent of the neurons' feature preferences. For example, if one neuron prefers red color and another prefers vertical orientation, then the two neurons signal the same saliency value if they have the same firing rate, and the more active neuron signals a higher saliency than the less active neuron, *regardless* of their (different) feature preferences. This is just like the value of a pound sterling of English currency is independent of the race or gender of the currency holder.

In principle, according to this idea, V1 could contain the saliency map, since V1 neurons can use their activities to signal saliencies at the locations of their receptive fields, despite their respective feature preferences. This does not mean that the selectivities of the V1 neurons to features are useless for visual computation beyond saliency. For instance, these selectivities can be used for decoding visual inputs in a visual area downstream along the visual pathway. However, whether or not they are used as part of other computations should not be relevant for saliency signaling (Li 2002). Meanwhile, the saliency values represented in the firing rates of the saliency signaling neurons can be read out to execute visual selection, such as to execute a gaze shift to the most salient location. The neurons for saliency readout and execution, perhaps in the superior colliculus, can be untuned to features, but they are separate from the neurons computing and representing saliencies in the saliency map (we discuss this in more detail in Section 5.2.3).

The traditional framework, which uses separate feature maps to sum into a master saliency, also imposes an unnecessary and unjustified rule on the interaction between features for saliency. This is illustrated by the following example. Let there be two visual inputs containing bars, all of them are colored blue. One input contains a unique vertical (blue) bar among horizontal (blue) bars. The other input contains a unique rightward moving (blue) bar among leftward moving (blue) bars. Hence, there is an orientation singleton in the first input and a motion singleton in the other input. Let these two inputs be such that their respective master saliency maps by the traditional framework are identical to each other. In this saliency map, the singleton's location is the most salient location. Now let us change the color of the singleton in each input to red, with exactly the same red feature value for the two inputs. Hence, the original orientation and motion singletons are now, respectively, color-orientation and color-motion double-feature singletons. The traditional saliency framework predicts: (1) the saliency of each double-feature singleton is larger than that of the corresponding single-feature singleton; and, (2) the saliency increase from the single-feature to the corresponding double-feature singleton is the same in the two inputs, independent of the feature dimension which defines the original single-feature singleton. Prediction (1) arises from the feature summation rule; and prediction (2) arises from the separation of feature maps in making the master saliency map. Although the summation rule and the separation of feature maps seem natural, and sufficient, for achieving the property of saliency regardless of features, they are both unnecessary for this property. It will be shown in Section 5.5.3 that both predictions (1) and (2), which are consequences of the summation rule and the separation of the feature maps, are inconsistent with experimental data.

In contrast, when saliencies are signaled by the activities or firing rates of feature-tuned neurons, and when some neurons are tuned to more than one feature dimension (such as those in V1), there is no separation of feature maps in making the saliency map. This allows rich interactions between various features for saliency, when these neural activities are read out for their universal saliency values and for the visual locations (but not features) they represent. In this sense, V1 could hold the saliency map, rather than many coexisting feature maps in separate subpopulations of neurons; and the SC could be a saliency read-out area, rather than a master saliency map to combine various feature maps. This will be detailed in Section 5.2.

5.1.4 A quick review of what we should expect about saliencies and a saliency map

Before we proceed further, it is worth reviewing and drawing some conclusions from the definition and expected manifestations of saliency. To recap, the saliency of a visual location is defined as the degree in which this location attracts visual selection in a purely bottom-up manner, such that a location having a higher saliency should be more likely to attract bottom-up selection before rather than after another location having a lower saliency. Below we elaborate this definition somewhat, list a few immediate consequences, and discuss some implications.

1. The saliency of a location increases with the probability with which this location is selected bottom-up before the other locations in the scene. It is inversely related, either deterministically or stochastically, to the order in which this location is selected by bottom-up manner mechanisms in the scene.

2. Saliency should be context dependent, since saliency at a visual location is associated with, and is defined in the context of, the whole visual scene. Hence, the location of a red apple may be the most salient in one scene full of green leaves, but the same location and the same apple in another scene made of many other red apples is not salient. In particular, the RT for gaze to reach this red apple should be much shorter in the first scene, unless there is a strong top-down influence.

3. In typical behavioral settings, visual selection depends on both the saliencies of visual locations and top-down factors. Thus, empirical selection approaches selection by saliency only in the asymptotic limit at which top-down factors are eliminated. This asymptotic limit is an ideal, which is difficult to reach experimentally. This is because observers who are behaving consciously inevitably have internal top-down goals that influence selection, and because viewing a visual input typically triggers awareness and internal knowledge of the scene, which also affect selection in a top-down, knowledge-driven, manner. Nevertheless, we can still compare saliencies between visual locations (or visual inputs) when the top-down factors that influence selection are made equal. Furthermore, we can minimize top-down contributions to selection by minimizing expectation, knowledge, or awareness of the visual inputs, and we can also shorten input viewing duration to avoid visual knowledge being triggered or having time to exert its top-down effects on selection. In particular (as we will see later), saliency can be well manifested when it works against the top-down factors.

4. Even if the transformation rule from visual inputs to saliency values depends on the visual input features concerned, the rule to transform the responses of the saliency signaling neurons to saliency values, by definition, cannot depend on visual input features.

5. Any visual location within the visual field should have a saliency value. Therefore, a saliency map should cover the whole visual field. Consequently, one expects that brain area(s) computing and reporting the saliency map must be able to respond to the whole visual field.

6. Since attention shifts typically change the selected location from the center of the visual field to an eccentric location and since the center of the visual field is often associated with the current top-down goal of the observer, visual saliency should be mainly, or at least more strongly, operative at more eccentric locations away from the center of the visual field.

The above points will be further elaborated in this chapter.

5.2   The V1 saliency hypothesis

The V1 saliency hypothesis was originally proposed in the 1990s, and was elaborated over several following years (Li 1997, Li 1999a, Li 1999b, Li 2002, Zhaoping 2005b). It states that V1 creates a bottom-up saliency map of visual space such that, first, the saliency of a location is represented by the highest of the responses of the V1 neurons whose receptive fields cover that location; and second, the receptive field location of the most active V1 cell in response to a visual scene is the most salient location in the scene.

Usually, a small image feature evokes responses from many V1 cells which have overlapping classical receptive fields (CRFs) and may have different feature preferences. For example, a short, red vertical bar can excite a neuron preferring the color red, another neuron preferring vertical orientation, a third neuron preferring both the color red and vertical orientation, and a fourth neuron whose most preferred orientation is 5 degrees from vertical, and so on. The receptive fields of all these neurons include the location of the bar. According to the V1 saliency hypothesis, the saliency at this bar's location is represented by the response of the fastest firing neuron (i.e., the neuron with the highest response) with a receptive field covering this location in response to the image containing the bar, regardless of the feature preference of the neuron having this response. Locating the cell that is most responsive to a scene overall locates the most salient location. Saliency does not depend on extraneous processing, such as whether input features are decoded beforehand, simultaneously, or never, from the responses of the same cell population (potentially in a complex and feature-specific manner from the population responses (Dayan and Abbott 2001); decoding will be discussed in Chapter 6).

It is not economical to use cortical areas beyond V1 along the visual pathway to realize a saliency map, whether or not the cells in those areas are feature independent. That V1 cells have small CRFs implies that the spatial resolution of a V1-based saliency map can be better than a map based anywhere downstream. Furthermore, since V1 is at an early stage on the visual pathway, saliency can be signaled quickly. High spatial resolution and alacrity are both desirable for bottom-up visual selection.

It may come as a surprise to many experienced vision researchers, who are familiar with the traditional framework of saliency (Fig. 5.5) and its implications, that V1's activities could signal saliency. It has been known since the 1960s that V1 neurons are tuned to local visual features like orientation, color, motion direction, binocular disparity (Hubel and Wiesel 1968), and input scales (see Chapter 2). It was not obvious that V1 neurons could signal salience, which depends on global context—after all, a vertical bar is salient in the context of horizontal bars, but the same vertical bar is not salient among other vertical bars. Until recently, V1 had never been looked at as playing an essential role in computing saliency.

Figure 5.6 uses the metaphor of an auction to help explain this extended role of V1. An auction shop has the slogan "Attention auctioned here; no discrimination between your feature preferences; only spikes count;" three V1 neuron bidders are depicted, with one tuned to motion direction with one spike's worth of bidding "money," another tuned to red color with 3 spikes' worth, and the third one tuned to a tilted orientation with 2 spikes' worth. The auctioneer, although feature blind, can do his job perfectly provided that he can count the spike "money." Of course, a feature-blind auctioneer does not mean that the "attention" awarded to the highest bidder is feature blind—post-selectional decoding should recognize the features at the selected visual location. The superior colliculus (introduced in Section 2.5) could possibly play the role of the auctioneer—it receives monosynaptic inputs from V1; its neurons are poorly or not feature-tuned, but their receptive fields are retinotopically organized; and it directs eye movements, which can be seen as the ultimate manifestation of the attention that is awarded to the receptive field location of the highest bidder.

This metaphor also conveys an important aspect of the V1 saliency hypothesis: attention does not have a fixed price—it is just that the highest bidder wins. A given level of neural activity may signal the most salient location in one scene, when it is the highest among the responses of the population of V1 neurons, but the same activity level may signal only a mediocre saliency in another input scene, when it is only typical among the responses of the population. Hence, it is not sufficient to record the activity of a single V1 neuron to determine saliency; measurements across the neural population are required to determine whether one neuron signals the most salient location.

5.2.1 Detailed formulation of the V1 saliency hypothesis

A location with a larger scalar saliency value in the saliency map is more likely to be selected by bottom-up attentional mechanisms for further visual processing. Here, by selection, we always mean selection by bottom-up mechanisms only. According to the V1 saliency hypothesis, saliency values are represented by the firing rates of V1 neurons. Let $(x_1, x_2, ..., x_n)$ denote the centers of the RF locations of the V1 cells with responses $(O_1, O_2, ..., O_n)$. Given a location x, let $x_i \approx x$ mean that the receptive field of the neuron with response O_i covers location x. Let

$$\text{SMAP}(x) \equiv \max_{x_i \approx x} O_i, \quad \text{the highest response to } x;$$
$$\text{then SMAP}(x) \text{ as a function of } x \text{ is the saliency proto-map.} \tag{5.4}$$

The $\text{SMAP}(x)$ value in this saliency proto-map is the value that location x bids for selection in the sense of the attentional auction described in Fig. 5.6. Therefore, this proto-map completely determines, and can be translated into, the saliency map through the read-out of the auction. This makes the following hold.

1. If two scenes generate identical saliency proto-maps, then their saliency maps are identical.
2. Within a scene, if $\text{SMAP}(x) > \text{SMAP}(x')$, then the saliency at location x is higher than that at location x'.
3. If the saliency proto-map values for two scenes are identical to each other at all locations except location x, then this location is more salient in the scene with the larger proto-saliency value $\text{SMAP}(x)$.
4. The third point above is a special case of the following. Saliency at location x increases with the degree in which $\text{SMAP}(x)$ is relatively higher than $\text{SMAP}(x')$ at other locations

A: The theory of a bottom-up saliency map from V1

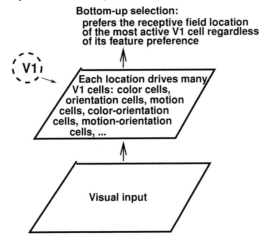

Bottom-up selection:
prefers the receptive field location
of the most active V1 cell regardless
of its feature preference

V1

Each location drives many
V1 cells: color cells,
orientation cells, motion
cells, color-orientation
cells, motion-orientation
cells, ...

Visual input

B: Its cartoon interpretation

**Attention auctioned here—
no discrimination between your
feature preferences, only spikes count!**

Hmm... I am
feature blind
anyway

Oh, no!
He only
cares about
money

Auctioneer

1 $pike   3 $pikes   2 $pikes

A
motion-
tuned
V1 cell

A
color-
tuned
V1 cell

An
orientation-
tuned
V1 cell

Fig. 5.6: A schematic summary and a cartoon interpretation of the V1 saliency hypothesis. In contrast with previous accounts, no separate feature maps, nor any summation of them, is needed in the V1 theory. V1 cells signal saliency despite their feature tuning. Adapted with permission from Zhaoping, L., Theoretical understanding of the early visual processes by data compression and data selection, *Network: Computation in Neural Systems*, 17(4): 301–334, Fig. 12, copyright © 2006, Informa Healthcare.

x'. Hence, for example, if $\bar{S}$ and σ_s are the mean and standard deviation of the proto-saliency $\text{SMAP}(x)$ across space, a location x tends to be more salient when $\text{SMAP}(x)/\bar{S}$ and $(\text{SMAP}(x) - \bar{S})/\sigma_s$ are larger.

Since the saliency proto-map completely defines the saliency map through the attentional auction, from here on, we will refer to the saliency proto-map $\text{SMAP}(.)$ as the saliency map and $\text{SMAP}(x)$ as the saliency value for location x, where it is not necessary to draw a distinction between the saliency map and the saliency proto-map. We will see that this notion of a saliency map leads to non-trivial, and experimentally testable, qualitative and quantitative predictions.

The most salient location in the scene is the location with the highest saliency value in the saliency map:

$$\text{the most salient location } \hat{x} = \text{argmax}_x \left[\text{SMAP}(x) \right]. \tag{5.5}$$

The most salient location can also be identified as the RF location of the most active V1 cell, i.e.,

$$\text{the most salient location } \hat{x} = x_{\hat{i}}, \quad \text{where } \hat{i} = \text{argmax}_i O_i. \tag{5.6}$$

One might worry that the most salient locations defined by the two equations above are not precisely the same. For example, let the response $O_1 > O_{i\neq 1}$ of the first neuron be the highest among all V1 neural responses, and let the receptive field of this neuron cover a circle of one degree in diameter centered at location x_1. From equation (5.6), $\hat{x} = x_1$. Meanwhile, from equation (5.4), the saliency value $\text{SMAP}(x)$ will be the same for all locations x within that one degree diameter circle centered at x_1. Then by equation (5.5), the most salient location includes all locations x within this circle, rather than just its center location x_1. This inconsistency however merely defines the spatial resolution of the saliency map. For the main functional role of saliency, which is to specify how attention should shift using saccades (Hoffman 1998), this resolution, as defined by the sizes of the V1 receptive fields, is adequate. In particular, it is no larger than the typical size of a saccadic error, which is the discrepancy between the target of a saccade and the actual gaze location brought by this saccade (Becker 1991). Note that the sizes of the V1 receptive fields scale with the eccentricity, and the saccadic error is about 10% of the eccentricity of the saccadic target (Becker 1991).

What we currently know does not determine whether "the V1 cells" that the V1 saliency hypothesis suggests to participate in the attentional auction include all cells in this area or just a subpopulation. Certainly, it is unlikely that these cells include the inhibitory interneurons in V1. However, "the V1 cells" should cover the whole visual field. For simplicity, in this book, we will not distinguish between a putative subpopulation and the other V1 cells.

5.2.2 Intracortical interactions in V1 as mechanisms to compute saliency

It has been suggested that intracortical interactions between V1 neurons are the neural mechanism by which saliency is computed. As seen in Section 2.3.9, the response of a V1 neuron to visual inputs in its classical receptive field (CRF) can be influenced by contextual inputs from outside this CRF. The intracortical neural connections, which link nearby neurons whose CRFs may or may not overlap, mediate these intracortical interactions between the neurons. These interactions have been suggested to underlie the contextual influences which underlie the context dependence of a neuron's response. This dependence is essential for computing saliency since, e.g., a vertical bar is salient in the context of horizontal bars but not other vertical bars.

The dominant contextual influence is iso-feature suppression, which is the suppression of the response to the input within a CRF when the context contains inputs with the same or similar features (Knierim and Van Essen 1992, Wachtler, Sejnowski and Albright 2003, Jones, Grieve, Wang and Sillito 2001). Iso-feature suppression is believed to be caused by the mutual antagonism between nearby V1 neurons tuned to similar features such as orientation and color. For the case of orientation, this suppression is known as iso-orientation suppression; see Section 2.3.9. Hence, e.g., a cell's response to its preferred orientation within its CRF is suppressed when the CRF is surrounded by stimuli sharing the same orientation.

A

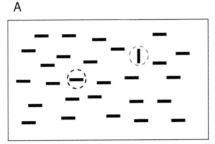

B

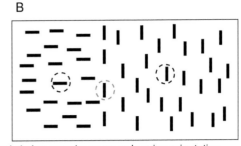

V1 neurons preferring and responding to bars in red circles experience no or less iso-orientation
suppression than neurons preferring and responding to bars in the black circles
(dashed circles mark classical receptive fields, not part of visual inputs)

Fig. 5.7: The responses to an orientation singleton or a bar at a texture border will be higher because iso-orientation is absent or weaker, respectively. The dashed circles mark the CRFs of the neurons responding to the bars enclosed. In A, the vertical bar is unique in having no iso-oriented neighbors. Hence, a neuron tuned to a vertical orientation and responding to this bar is free from the iso-orientation suppression which affects neurons that are tuned to a horizontal orientation and respond to the horizontal bars. In B, a bar at the texture border, e.g., the one within the red circle, has fewer iso-oriented neighbors than a bar that is far from the border (e.g., the two bars in the black circles). Hence, neurons responding to the border bars are less affected by iso-orientation suppression.

Figure 5.7 illustrates how iso-orientation suppression makes V1 responses to a salient orientation singleton or an orientation texture border higher than responses to the background bars. For the case of the visual input in Fig. 5.7 A, a cell preferring vertical orientations and responding to the vertical bar escapes any iso-orientation suppression, because there is no neighboring vertical bar to evoke activity in neighboring cells that are also tuned to vertical. By contrast, a neuron preferring horizontal orientations and responding to one of the background horizontal bars experiences suppression from other horizontally tuned neurons responding to the neighboring horizontal bars. Consequently, when the contrast of all input bars is the same, the V1 cell that is activated most strongly by this image is the one responding to the vertical bar. According to the V1 saliency hypothesis, its location is then the most salient in this image.

Similarly, the bars at the border of an orientation texture have fewer iso-oriented neighbors than those away from the texture border. Thus, neurons responding to a texture border bar in Fig. 5.7 B experience a weaker iso-orientation suppression than that experienced by neurons responding to the other texture bars.

Figure 5.7 A can be seen as a special case of Fig. 5.7 B in that an orientation singleton is a texture region with just one texture element. Hence, this singleton itself can be viewed as its own texture border, and, by the reasoning above for Fig. 5.7 B, is more salient. This is why, as demonstrated in Fig. 5.4, the facilities of visual search and visual segmentation are typically related, when the visual features involved correspond.

In just the same way, iso-color suppression between neighboring V1 neurons that prefer similar colors, and iso-motion-direction suppression between neighboring V1 neurons that prefer similar motion directions, should both make for relatively higher V1 responses to a singleton in color or motion direction. More generally, iso-feature suppression should make V1 responses relatively higher at locations of higher input feature contrast. Thus, even though the CRFs are small and the intracortical connections that mediate contextual influences have a finite range, this mechanism allows V1 to perform a *global* computation such that its neural responses reflect context beyond the range of the intracortical connections (Li 1997, Li 1999b, Li 2000b). By contrast, retinal neurons respond in a largely context independent manner, and so they could only adequately signal more specific and context independent forms of saliency, such as that caused by a bright image spot.

The neural mechanisms in V1 that mediate saliency have other properties. For instance, whether the saliency at the location of a red vertical bar is more likely signaled by a red-tuned cell or a vertically-tuned cell depends on the context. (For simplicity, we ignore neurons tuned both to the color red and to vertical orientation.) Both these cells respond to the red vertical bar; whichever one responds more vigorously should signal the saliency of this location (assuming this location has no other visual inputs). In the context of red horizontal bars in Fig. 5.8 A, it is the response of a vertical-tuned cell that determines its saliency; in the context of black vertical bars in Fig. 5.8 B, this is determined by a red-tuned cell. In either case, the most responsive neuron is determined by iso-feature suppression, and the saliency value depends only on the firing rate of the most responsive cell, regardless of whether it is color- or orientation-tuned.

A
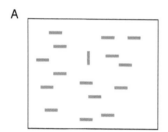

B
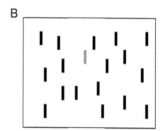

Fig. 5.8: Contextual dependence of the neurons signaling the saliency of the red vertical bar. This bar evokes responses in cells preferring red and in cells preferring vertical orientations (ignoring the cells tuned to the color red and vertical orientations simultaneously for simplicity of argument). In A, iso-orientation suppression makes the vertical-tuned cell the most responsive to the red vertical bar; in B, iso-color suppression makes the red-tuned cell the most responsive.

5.2.3 Reading out the saliency map

The saliency map SMAP(x) is read out in order to execute an attentional shift. In principle, the saliency map in V1 could be ignored, i.e., read-out is unnecessary unless there is a need to shift attention. It is important to distinguish a brain area that contains the saliency map SMAP(x) from the brain areas that read out the saliency map for the purpose of shifting attention. V1's saliency output may be read by (at least) the superior colliculus (SC) (Tehovnik, Slocum and Schiller 2003), which receives inputs from V1 and directs gaze (and thus attention). In this case, the SC is not considered to compute saliency, but is merely a read-out area which selects the most salient location to execute an attentional shift.

Operationally, selecting the most salient location $\hat{x}$ does not require SMAP(x) to be calculated by means of the maximum operation in equation (5.4) to find the highest response to each location x. Rather, it only needs a single maximum operation $\hat{i} = \text{argmax}_i O_i$ over all neurons i, regardless of their RF locations or preferred input features. This is algorithmically perhaps the simplest possible operation to read a saliency map, and it can thus be performed very quickly. Being quick is essential for bottom-up selection.

If the read-out of saliency is deterministic, the most salient location $\hat{x}$ should be the first one that bottom-up mechanisms select in the scene. If read-out is stochastic, this most salient location $\hat{x}$ is just most likely to be the first one selected.

Merely for the purpose of computing saliency, the maximum operation could be performed either in V1 or in the read-out area, or even both. The single maximum operation

$$\max_i O_i = \max_x(\text{SMAP}(x)) = \max_x(\max_{x_i \approx x} O_i) \qquad (5.7)$$

over all responses O_i is equivalent to cascading two maximum operations, the first one locally $\max_{x_i \approx x}(O_i)$ to get $\text{SMAP}(x)$ and then the second one globally $\max_x(\text{SMAP}(x))$. This is like selecting the national winner $\max_i O_i$ by having a two-round tournament: first, the local players near location x compete to get the local winner's score $\text{SMAP}(x) = \max_{x_i \approx x} O_i$; then, the local winners from different locations x compete globally to determine the overall winner $\max_x \text{SMAP}(x)$. If the local competition is performed in V1, and if the global competition is done in a read-out area such as the SC, then the explicit saliency map $\text{SMAP}(x)$ should be found in the activities of the neurons projecting to the SC. This would license just a numerically small number of projecting neural fibers from V1 to the SC, consistent with anatomical findings (Finlay, Schiller and Volman 1976). If the competition is done in a single-round tournament in the SC or if both rounds of a two-round tournament are performed in the SC, then the SC needs to extract the saliency map from the whole population of V1 responses. The V1 saliency hypothesis is agnostic as to where and how the maximum operations are performed; these questions can be investigated separately.

Since V1's responses $\mathbf{O} = (O_1, O_2, ..., O_n)$ most likely play additional roles beyond saliency, it is necessary that the maximum operation or competition that selects the most salient location does not prevent the original responses $\mathbf{O}$ from being sent to other brain areas such as V2. Therefore, multiple copies of the signals $\mathbf{O}$ should be sent out of V1 via separate routes: one to the saliency read-out area and the others to other brain areas for other visual computations. For saliency, the maximum operation is only needed en route (perhaps in the layer 5 of V1) to, or in, the saliency read-out area. This need not distort the $\mathbf{O}$ values projecting to other brain areas.

5.2.4 Statistical and operational definitions of saliency

A salient location, such as that of the orientation singleton or the texture border in Fig. 5.7, is typically a place where visual input deviates from its context to a statistically significant degree. Consider covering the unique vertical bar in Fig. 5.7 A. One would naturally expect the bar at the covered image location also to be horizontal. In other words, conditional on the contextual input, the probability that this location contains a vertical bar is very small. Similarly, given the horizontal bars in the left half of the image in Fig. 5.7 B, without the knowledge of the presence of vertical bars in the right half of the image, the probability that the orientations of the bars are vertical in the middle of the image is quite low. Within textures of uniformly oriented bars as in Fig. 5.7, input statistics are translationally invariant, i.e., are identical at all locations. Hence, saliency could be linked to the degree to which the translation symmetry of one of a class of input statistics is broken (Li 1997, Li 1998b, Li 1999b, Li 2000b). This notion is related to the one that saliency is associated with surprise or novelty (Itti and Baldi 2006, Lewis and Zhaoping 2005). Other related notions of saliency include: a salient location is where an "interest point" detector (for a particular geometric image feature like a corner) signals a hit or where local (pixel or feature) entropy (i.e., information content) is high (Kadir and Brady 2001).

Meanwhile, we saw in Section 5.2.2 that iso-feature suppression in V1 allows the neurons in this area to detect and highlight the salient locations where local statistics (such as the average of the values of basic features such as color, orientation, and motion direction over a neighborhood) change significantly. Ignoring any dependence on eccentricity for simplicity (or considering only a sufficiently small range of eccentricities), we assume that the properties of V1 CRFs and intracortical interactions are translation invariant. This implies that the input tunings and stimulus-bound responses to inputs within a neuron's CRF do not depend on the location of that CRF and that the interaction between two neurons depends only on the relative, but not the absolute, locations of their CRFs. In that case, the V1 responses should be translation invariant when the input is translation invariant (provided that there is no spontaneous symmetry breaking; discussed in Section 5.4.3). This input translation symmetry arises, for example, in an image comprising a regular texture of horizontal bars. It can also be generalized to cases such as the image of a slanted surface of a homogenous texture. However, when some statistics associated with these input features are not translation invariant, the responses of V1 neurons are expected to exhibit corresponding variabilities. Therefore, V1 mechanisms can often detect and highlight the locations where input symmetry breaks, and this will typically arise at the boundaries of objects. The V1-dependent process of highlighting salient object boundaries has also been termed as preattentive *segmentation without classification* (Li 1998b, Li 1999b), since the operation presumably occurs before object recognition or classification.

However, not all changes in visual input statistics make the corresponding input locations salient. Extensive studies have identified some of the kinds of input statistics whose change can make a location salient (Julesz 1981). For example, although there are exceptions (see Fig. 5.36 C), it has been observed that human observers can easily segment two textures which differ from each other according to their first and second order statistics, but not when the two textures differ in only higher order statistics. Hence, if one were to define saliency by the degree of change in some kind of input statistics, then this definition would need to include the individual sensitivity of saliency to changes in each kind of input statistics.

The basis of verifying whether a computational definition of saliency captures the reality should be our operational definition of saliency as the degree in which a visual location attracts attention by bottom-up mechanisms. This operational definition of saliency should also offer a basis to test any theory about saliency. In particular, the V1 saliency hypothesis should be tested against the behavioral data on saliency. For example, if a particular change in visual input statistics does not make the corresponding visual location behaviorally salient, then, if the hypothesis is correct, that location's saliency, as computed from V1 responses, should remain low, despite the change in the statistics of visual inputs.

As discussed in Section 5.1.1.2, saliency can be assessed by the reaction times and accuracies achieved in visual search and segmentation tasks. These tasks must be designed in such a way that the saliency of the visual location of interest is inversely related to the RT in the task (see Fig. 5.3), or monotonically related to the task performance accuracy using a brief visual display. To test the V1 saliency hypothesis, the predicted saliency from the V1 responses should be compared with that evident from the RTs and accuracies found in the behavioral experiments.

5.2.5 Overcomplete representation in V1 for the role of saliency

Chapter 4 discussed how the efficient coding principle could not readily explain the fact that V1 representation of visual input is highly overcomplete. This apparent over-representation greatly facilitates fast bottom-up selection by V1 outputs (Zhaoping 2006a). To see this more clearly, let us focus on orientation as a feature (ignoring other features such as color, motion, and scale). There is a large number of different cells in V1 tuned to many different orientations near the same location. This representation O helps to ensure that there is always a cell O_i at each location that *explicitly* signals the saliency value of this location (at least in the case that the saliency arises from the orientation feature).

For example, let there be 18 neurons whose receptive fields cover a location x; each neuron prefers a different orientation, $\theta_i = i \cdot 10^o$ for $i = 1, 2, ..., 18$, spanning the whole 180° range of orientation. The orientation tuning width of the neurons is sufficiently wide that presenting a bar of any orientation at x should excite some of the 18 neurons, with at least one being excited nearly optimally. This neuron will duly signal the saliency of the input. An input bar oriented at $\theta = 31^o$ would, for example, have its saliency signaled by the neuron tuned to 30°.

Imagine instead an alternative representation which has only three neurons covering this location, preferring orientations 0°, 60°, and 120° from vertical. To a 31° tilted bar, the most responsive neuron prefers 60°, the second most responsive one prefers 0°, but the actual response of neither comes close to that of their preferred 60° or 0° bars, respectively. The maximum neural response to a 60° bar would be higher than that to the 31° bar, and if saliency is defined by the maximum neural response, the 31° bar would appear less salient than the 60° bar. To rescue the calculation, the saliency of the 31° bar would have to be calculated as a function of responses from multiple underlying neurons (e.g., from the 0°- and 60°-preferring neurons, or from all three neurons), and this function would have to depend on the number of input bars at the same location x. The computational complexity of this calculation would increase dramatically when other feature dimensions and contextual inputs were also considered. It is completely unclear how such a saliency function could be computed and whether the computation would compromise the goal of fast saliency signaling along with adequate representation of the visual input.

It is likely that V1's overcomplete representation is also useful for other computational goals which could also be served by V1. Indeed, V1 also sends its outputs to higher visual areas for operations, such as recognition and learning, that go beyond selection. Within the scope of this chapter, I will not elaborate further upon our poor understanding of what constitutes the best V1 single representation for computing saliency as well as serving these other goals (although, as discussed, there can be different output channels for different goals).

5.3 A hallmark of the saliency map in V1—attention capture by an ocular singleton which is barely distinctive to perception

Everyday experience tells us that an item will only be salient if it is perceptually very distinct from its surroundings. This is the case, for example, for a red item among green ones or for a vertical bar among horizontal bars. It is thus surprising that the V1 saliency hypothesis predicts the following: an ocular singleton, which differs from surrounding items only by being shown to a different eye and is barely perceptually distinct, can be roughly as salient as a color or orientation singleton. For example, a horizontal bar shown to the left eye among surrounding horizontal bars shown to the right eye [14] is predicted to be highly salient.

This prediction arises because iso-feature suppression in V1, which is responsible for feature singleton pop-out, also applies for the feature that is the eye of origin of a visual input. This is underpinned by the many monocular neurons in this area, and indeed it is evident in

[14]Ocular singletons can be presented using stereo goggles. Some basic constraints need to be satisfied in such dichoptic presentations: vergence eye positions (which focus on specific locations in a three-dimensional scene) are anchored to elements of the display such as an image frame common to the two eyes; and the spacing of the elements should be such that neither binocular rivalry nor stereo matching occur for items in the two eyes (at least within a brief duration).

the observation that, when a V1 neuron responds to a monocular input, its response is more suppressed when the surrounding inputs are presented to the same, rather than to the other, eye (DeAngelis, Freeman and Ohzawa 1994).

However, an ocular singleton is not perceptually distinctive because few neurons downstream from V1 along the visual pathway are monocular. Various sources of evidence suggest that perception depends on the activity of extrastriate neurons; perception is thus blind to the eye of origin for monocular inputs. The lack of monocular cells in the extrastriate cortex is reflected in optical imaging of the ocular dominance columns. Figure 2.23 shows that these columns seen from the cortical surface stop abruptly at the border between V1 and V2, because V2 does not have enough monocular neurons. Therefore, the response of a binocular V2 neuron to a monocular input does not contain information regarding whether the input comes from the left or the right eye. This blindness at the neuron level is manifested behaviorally. For instance, in an experiment, observers were asked to report (and to guess, if necessary) whether there was a single item presented to the right eye among many background items presented to the left eye in a perceived image of these items (whose luminances are independently and randomly chosen). Their reports did not statistically differ from random guesses (Wolfe and Franzel 1988). Apparently, observers cannot distinguish an ocular singleton from the background items by its unique eye of origin. However, this does not mean that the singleton did not attract their attention. Indeed, this is an example for which the saliency of a visual input cannot be measured by the RT or accuracy of observers to find or identify this input.

In sum, the eye-of-origin signal is mainly, or exclusively, available in V1 among all the visual cortical areas. Furthermore, we know from Section 2.5 that upstream neurons are likely not involved in saliency. That is, in monkeys, the projection from the retina to the superior colliculus is normally not involved in visually guided eye movements; further, there is no projection from LGN to the superior colliculus. Therefore, the predicted high saliency of an ocular singleton (to guide attention or gaze shift) would be a clear fingerprint of the role of V1.

Despite the lack of reliable perception, one can probe the saliency of an ocular singleton by making it task irrelevant and testing if it interferes with visual search by distracting attention away from a true target. Figure 5.9 shows a visual input for observers who are asked to search for an orientation singleton among background bars. All bars are monocular, and one of the non-target bars is an ocular singleton. Observers perceive an image which is like the superposition of the image to the left eye and the image to the right eye. If gaze or attention is attracted in a bottom-up manner to the ocular singleton, it will interfere with the search, lengthening RTs. This was indeed observed. Take the case that the singletons are on opposite sides of the image, and $12°$ from the center of the display, which is where gaze initially pointed before the search begins. The first gaze shift during the search was directed to the task-irrelevant ocular singleton on 75% of the trials. This was the case even though the orientation singleton target was very salient, since it was tilted $50°$ away from 659 uniformly oriented non-target bars, and observers were told to search for it as quickly as possible (Zhaoping 2012). This is analogous to Fig. 5.2 A, in which the red non-target bar among black bars attracts attention automatically away from the target. However, unlike the ocular singleton, the red bar is highly perceptually distinctive.

We describe in more detail here the experiments showing that the ocular singleton could attract attention, even though observers could not perceive any visual difference between it and its neighboring bars. Three different dichoptic presentation conditions, monocular (M), dichoptic congruent (DC), and dichoptic incongruent (DI) are shown in Fig. 5.10. The superposition of the two monocular images is the same in these three conditions, and it resembles the perceived image, which has an orientation singleton bar in a background of uniformly oriented bars. The orientation singleton is the target of visual search. In the M

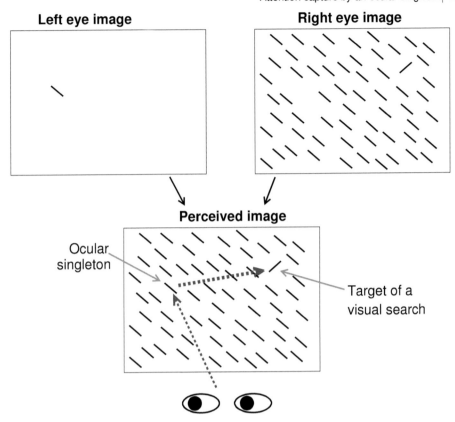

Fig. 5.9: An ocular singleton, though task-irrelevant and not perceptually distinct from background items, often attracts the first shift of gaze, before a subsequent shift to the target (the orientation singleton) of the visual search (Zhaoping 2012). The colored arrows are not part of the visual stimulus; they indicate the gaze shifts and point to the feature singletons.

condition, all bars are presented to the same single eye. In the DC condition, the target bar is an ocular singleton, since it is presented to a different eye than the other bars. In the DI condition, a non-target bar is an ocular singleton; it is presented to the opposite lateral side of the target from the center of the perceived image. In the search for the orientation singleton, the ocular feature is task irrelevant but could help or hinder the task in DC or DI conditions, respectively.

Figures 5.11 and 5.12 present experiments using such stimuli, together with their results (Zhaoping 2008). In Fig. 5.11, subjects had to report whether the orientation singleton, whose location the observers did not know ahead of each trial, was tilted clockwise or anticlockwise from horizontal. However, the images were presented so briefly that the task would be difficult unless attention was quickly guided to the location of the target. The degree of difficulty was measured by the error rate, which is the fraction of trials in which the observers performed the task erroneously. Figure 5.11 B shows that the error rate for this task was smaller in the DC trials than in the M and DI trials. This suggests that attention was guided to the target more effectively in the DC trials. One may see the DC or DI trials as ones in which an ocular singleton provides valid or invalid, respectively, guidance of attention to the target. In the M trials, there is no ocular singleton to guide attention.

Meanwhile, the second experiment in Fig. 5.11 revealed that the same observers were

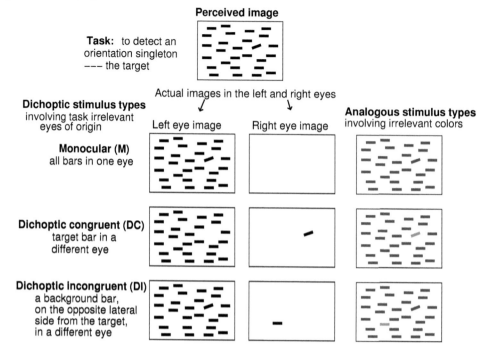

Fig. 5.10: Schematic of the stimulus used to test the automatic capture of attention by an eye-of-origin or ocular singleton, even though one can barely perceive any difference between inputs from different eyes. The ocular feature is irrelevant for the search for an orientation singleton, and the observers are not required to report it. They perceive an image with an orientation singleton target among background bars, but this perceived image could be made from three different dichoptic presentation conditions: monocular (M), dichoptic congruent (DC), and dichoptic incongruent (DI). The analogous case when color is the irrelevant feature disrupting the same task is shown on the right. If the ocular singleton is salient and attracts attention more strongly than the orientation singleton, it should help and hinder the task in the DC and DI conditions, respectively, by guiding attention to and away from the target.

not necessarily aware of these attention-guiding ocular singletons. When different bars in the display had randomly different luminances, observers could do no better than chance (error rate is 0.5) in reporting the presence or absence of the ocular singletons. They did better than chance when all the bars had the same luminance, because an ocular singleton can sometimes be identified by an illusory contrast different from the other bars. (Apparently, heterogenous luminances across the bars made this illusory contrast ineffective for identifying the ocular singleton by the observers.) Nevertheless, the ability of the task-irrelevant ocular singleton to guide attention in the first experiment did not depend on whether the luminance condition was such that, from the second experiment, we would expect subjects to have been able to identify the ocular singleton.

These findings suggest that RTs to locate the orientation singleton should be shorter in the DC trials and longer in the DI trials. This was indeed observed when observers were asked to report as quickly as possible whether the target was in the left or right half of the perceived image (which remained displayed until they made their choice); see Fig. 5.12. Let RT_M, RT_{DC}, and RT_{DI} denote the RTs for the monocular (M), dichoptic congruent (DC), and dichoptic incongruent (DI) stimulus conditions, respectively. The data show $RT_M > RT_{DC}$

A: Stimulus sequences (in a trial) of two tests

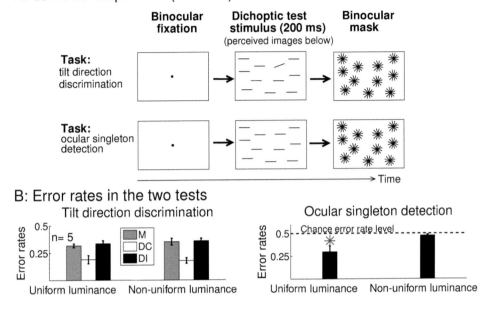

B: Error rates in the two tests

Fig. 5.11: Two experiments showing that an ocular singleton guides attention even when observers are unaware of its presence (Zhaoping 2008). A: Schematics of test trials in each of the two experiments. The dichoptic test stimulus, in which all bars are monocular, is binocularly masked after being displayed for only 200 ms. In one experiment (top), observers report whether an orientation singleton, tilted 20° from 659 horizontal bars, was tilted clockwise or anticlockwise from horizontal. As in Fig. 5.10, all test stimulus bars are monocular, and a given trial can be randomly monocular (M), dichoptic congruent (DC), or dichoptic incongruent (DI). In the second experiment (bottom), the test stimulus is the same as in the first experiment except that all bars are horizontal, and the ocular singleton has an equal chance of being present or absent (if present, it is randomly at one of the locations for the ocular singleton in the first experiment). Observers report whether the ocular singleton is present. In each trial in both experiments, either all stimulus bars have the same luminance or different bars have different random luminances. B: Error rates in the two experiments, averaged across five observers (who participated in both experiments), are shown separately for the two luminance conditions. In the right plot in B, an error rate significantly different from the chance level (0.5) is indicated by a "*".

and $RT_{DI} > RT_M$, in which $RT_M \approx 0.6$ seconds for typical observers. These relationships remained true whether or not the subjects were informed that different dichoptic stimulus types could be randomly interleaved in the trials. Even when they were informed that a non-target bar might distract them in some of the trials and were explicitly told to ignore it (experiment B of Fig. 5.12), RT_{DI} was still greater than their RT_M. This suggests that the bottom-up attraction of the irrelevant ocular singleton could not be easily suppressed by top-down control. The RT difference $RT_{DI} - RT_M$ was around 0.2–0.3 seconds on average, comparable to typical time intervals between two saccades in a visual search. This suggests that, in typical DI trials, attention or gaze was first attracted to the task-irrelevant ocular singleton before being directed to the target. This was later confirmed by tracking the gaze of subjects who were doing this task; see Fig. 5.9.

As mentioned in Section 5.1.2, orientation is one of the basic feature dimensions. That is,

Experimental design: an ocular singleton in orientation singleton search

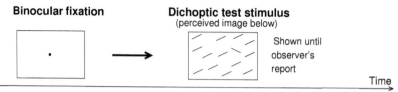

Binocular fixation

Dichoptic test stimulus
(perceived image below)

Shown until
observer's
report

Time

Task: report quickly whether the orientation singleton is in the left or right half of the perceived image.
   Three experiments: A, B, and C, each randomly interleaving trials of different dichoptic conditions.
   A: Did not include the DI trials.
   B: Observers informed of possible distractions away from the target.
   A & C: Observers uninformed of different dichoptic conditions.

Experimental results: reaction times and error rates in the search task

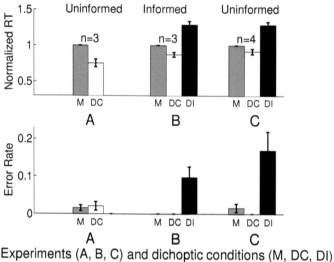

Experiments (A, B, C) and dichoptic conditions (M, DC, DI)

Fig. 5.12: An ocular singleton can speed up or slow down visual search for an orientation singleton. Each dichoptic search stimulus had 659 iso-oriented background bars and one orientation singleton bar, tilted $25°$ from horizontal in opposite directions. Subjects reported as soon as possible whether the target was in the left or right half of the perceived image. There were three experiments, A, B, and C, each of which randomly interleaved trials of various dichoptic conditions: monocular (M), dichoptic congruent (DC), and dichoptic incongruent (DI), as in Fig. 5.10. As indicated in the bar charts, experiment A contained M and DC trials, and experiments B and C each contained M, DC, and DI trials. Observers were not informed of the different dichoptic conditions except in experiment B, in which they were informed that some trials might contain a distracting non-target. RTs (normalized by RT_M, which is around 600 ms, of individual observers, so that $RT_M = 1$) and error rates are averaged across $n = 3$, 3, and 4 observers, respectively, for experiments A, B, and C.

an orientation singleton is sufficiently salient such that visual search for it is efficient, with RT being independent of the search set size. However, the experiment depicted in Fig. 5.12 showed that an ocular singleton can attract gaze more strongly than an orientation singleton tilted $50°$ from the background bars. Hence, the ocular singleton is more salient than the

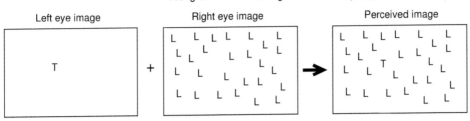

Fig. 5.13: Making the target "T" an ocular singleton renders efficient what is otherwise found to be an inefficient search for a "T" among "L"s (Zhaoping 2008).

orientation singleton, and so the eye-of-origin feature dimension must also be basic. Indeed, an inefficient search for a letter "T" among background letters "L" can be made efficient when "T" is an ocular singleton (Zhaoping 2008); see Fig. 5.13. This basic feature dimension of ocular origin was not recognized until the experimental findings described here, since this feature was not perceptually distinctive.

5.3.1   Food for thought: looking (acting) before or without seeing

At first, the prediction, that a visual item that is barely distinguishable from its neighbors can attract attention, like a red flower among green leaves, might seem surprising or even impossible. This reaction arises from our impression or belief, driven from experience, that seeing precedes looking, i.e., that we look at something after, or because, we have seen what it is. The confirmation of this counterintuitive prediction from the V1 saliency hypothesis invites us to ponder and revise our belief. Logically, one looks in order to see, and looking should be expected to precede seeing, at least for part of our visual behavior. This is analogous to the example in Fig. 1.4, when observers do the act of looking or shifting gaze before they know the identity of the visual input at the destination of their gaze shift. Looking should also be dissociable from seeing. Indeed, brain lesion patients who cannot recognize objects can still manipulate objects adequately (Goodale and Milner 1992). According to this analysis, it is likely that gaze is attracted to the location of the ocular singleton in the perceived image before the two monocular images have been combined to achieve the perception of the perceived image. Meanwhile, the perceived image, i.e., the image in observers' perception after combining the inputs from the two eyes, contains little or no information about the eye of origin that could influence gaze.

5.4   Testing and understanding the V1 saliency map in a V1 model

This section presents a model of V1. This model is intended to serve two main purposes: first, as a substitute for the real V1 to test the relationship between V1 activities and behavioral saliencies. The experiments on ocular singletons offer convincing support for the V1 saliency hypothesis. However, we should also examine whether the link between V1 responses and saliency also applies in general cases, including those for which the saliency effects are subtle. The literature (Wolfe 1998) contains a wide range of behavioral data on saliency in terms of reaction times or task difficulties in visual search and segmentation. However, physiological data based on stimuli used in the behavioral experiments are few and far between. Furthermore, according to the V1 saliency hypothesis, predicting the saliency of a location requires us to compare the V1 responses of neurons with that location as their classical RFs to the responses of neurons favoring other locations. This would require simultaneous recordings of many V1 units responding to many locations, a very daunting task with current technology. Examining the responses of all neurons in a simulation of the model provides a simpler, though obviously inferior, alternative to recording in the actual V1. Figure 5.14 shows an outline of the V1 model and its function.

Second, examining the model neural circuit can help us understand how intracortical interactions in V1 lead to the computation of context-dependent saliency from local contrast inputs. Qualitative arguments such as those in Section 5.2.2 suffice for us to envisage how iso-feature suppression could explain the relative enhancement of V1 responses to very salient feature singletons and texture borders. However, they are insufficient for knowing whether or how V1 interactions could also account for subtle saliency effects and indeed whether the neural circuit dynamics are well-behaved. Obviously, the actual V1 neural dynamics *are* well behaved. However, testing whether a model of V1 mechanisms identified by us as responsible for saliency computation has well behaved dynamics enables us to test whether our understanding is correct. Simulating the intracortical interactions, showing how they produce iso-feature suppression as well as the less dominant interactions, can allow us to verify our intuitions and help to identify how various intracortical mechanisms shape visual saliency.

The material in this section is mostly adapted and extended from papers in the literature (Li 1998a, Li 1999a, Li 1999b, Li 2000b, Li 2001, Li 2002, Zhaoping 2003). These were all published before the confirmation of the ocular singleton effect and of other predictions of the V1 saliency hypothesis (described later this chapter). Therefore, this model actually served the purpose of assessing whether V1 mechanisms could feasibly subserve the computation of saliency. That is, it tested the V1 saliency hypothesis using behavioral data already known before the model was constructed, or using self-evident behavioral phenomena. In this section, we will show simulated V1 responses to representative visual inputs whose behavioral saliency profiles are well known. Furthermore, we will show examples in which model responses highlight locations where input statistics break translation symmetry.

5.4.1 The V1 model: its neural elements, connections, and desired behavior

V1 neurons can be tuned to orientation, color, scale, motion direction, eye of origin, disparity, and combinations of these. This is in addition to the selectivity to input spatial locations (and spatial phase/form) by their receptive fields. As an initial attempt to study whether it is feasible for V1 mechanisms to compute saliency, the model here focuses on only two features, spatial location and orientation. Thus, each model neuron is characterized by its preferred orientation and spatial location; all model receptive fields have the same size; and the centers of the receptive fields sit on a regular spatial grid. Hence, the model ignores other visual cues such as color, motion, and depth. Emulating and understanding the dependence of saliencies on the spatial configurations of oriented bars is arguably more difficult than emulating and understanding the dependence on luminance and color features. Once the feasibility of V1 mechanisms for saliency can be established in this simplified V1 model, the model can then be extended to include the other feature dimensions and neural selectivities (see Section 5.8.3).

Since the model focuses on the role of intracortical interactions in computing saliency, the model mainly includes orientation selective neurons in layers 2–3 of V1. These are coupled by intracortical connections, which are sometimes also called horizontal or lateral connections. The model ignores the mechanism by which the neural receptive fields are generated. Inputs to the model are images seen through the model classical receptive fields (CRFs) of V1 complex cells, which are modeled as edge or bar detectors (we use "edge" and "bar" interchangeably). (To avoid confusion, here the term "edge" refers only to local luminance contrast. Meanwhile,

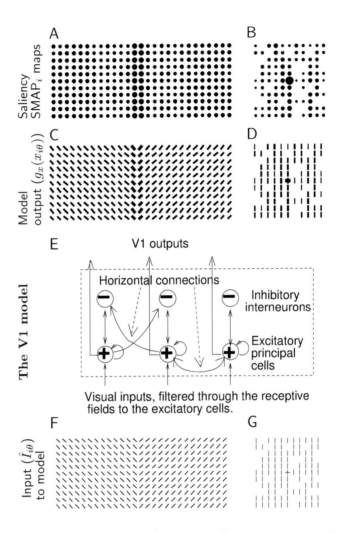

Fig. 5.14: The V1 model and its operation. The model (E) focuses on the part of V1 responsible for contextual influences: excitatory pyramidal (principal) cells in layers 2–3 of V1, interneurons, and intracortical (horizontal) connections. A pyramidal cell can excite another pyramidal cell monosynaptically, and/or inhibit it disynaptically via the inhibitory interneurons. The model also includes general and local normalization of activities. F and G are two example input images. Their evoked model responses, C and D, are those of the pyramidal cells preferring the corresponding positions and orientations. As in many figures in the rest of this chapter, the input contrast (strength) or output responses are visualized by (in proportion to) the thicknesses of the bars in the input or output images. A principal cell receives direct visual input only from the input bar within its CRF. Its response depends both on the contrast of the bar and the stimuli in the context, the latter via the intracortical connections. Each input/output/saliency image that is shown is only a small part of a larger, extended input/output/saliency image. At the top (A, B) are saliency maps, in which each location i is for a hypercolumn. The size of the disk at location i visualizes the highest response SMAP_i among the pyramidal cells responding to this location. A location is highly salient if this disk is much larger (assessed by a z score) than the other disks in the map. The notations $\hat{I}_{i\theta}$ and $g_x(x_{i\theta})$ will be defined shortly. Adapted with permission from Zhaoping, L., Theoretical understanding of the early visual processes by data compression and data selection, *Network: Computation in Neural Systems*, 17(4): 301–334, Fig. 8, copyright © 2006, Informa Healthcare.

a boundary of a region is termed "boundary" or "border", which, especially in textures, may or may not correspond to any actual luminance edges in the image.) Intracortical connections (Rockland and Lund 1983, Gilbert and Wiesel 1983) mediate interactions between neurons such that patterns of direct inputs to the neurons via their CRFs are transformed into patterns of contextually modulated responses (firing rates) from these neurons.

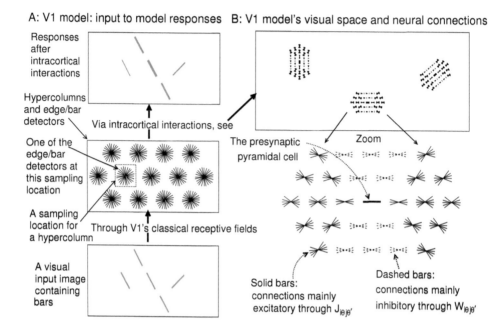

Fig. 5.15: Schematic of the V1 model. A: An example visual input contains five bars of equal contrast (marked by red color, to distinguish them from black bars visualizing neurons and the neural connection pattern); the (black) rectangle (not part of input image) frames the input image. In the middle is the V1 model, which contains many classical edge or bar detectors; each detector is visualized by a black bar and is modeled by a pair of mutually connected neurons: an excitatory pyramidal cell and an inhibitory interneuron (see Fig. 5.16). A single hypercolumn occupies a spatial sampling location and comprises many detectors preferring various orientations that span $180°$. Without intracortical interactions, five edge/bar detectors (shown in red in the middle frame) are equally excited by the five equal contrast input bars through their respective CRFs; no other detector is as substantially activated directly. Through intracortical interactions, the eventual responses from the five detectors are unequal, visualized by different thicknesses of the (red) bars in the top frame. B: A schematic of the lateral connections in the model. The rectangle frames the visual space. Three groups of neural connections (translated and rotated versions of each other) radiating from three presynaptic cells are shown. In the zoomed view of one group, the central horizontal bar marks the presynaptic pyramidal cell preferring horizontal orientations. The thin bars mark the locations and preferred orientations of the postsynaptic pyramidal cells: the solid ones are for cells mainly excited by the presynaptic cell through monosynaptic $J_{i\theta,j\theta'}$ connections; dashed ones are for cells mainly disynaptically inhibited by the presynaptic cell, via connections $W_{i\theta,j\theta'}$; see text.

Figures 5.15 and 5.16 show the elements of the model and the way they interact. Following original literature, we denote a spatial sampling location by i rather than x, which will instead denote membrane potentials of pyramidal cells. At each spatial sampling location i, there is a model V1 hypercolumn composed of cells whose CRFs are centered at i. Each of these cells is tuned to one of $K = 12$ different orientations θ spanning $180°$. Based on experimental data (White 1989, Douglas and Martin 1990), each edge or bar detector at location i and preferring orientation θ is modeled by one pair of interconnected model neurons: one excitatory pyramidal cell and one inhibitory interneuron; detailed in Fig. 5.16. Hence, altogether, each hypercolumn consists of 24 model neurons. Each model pyramidal cell or interneuron is a simple rate-based neuron (see Section 2.1.2). It could model abstractly, say, 1000 spiking pyramidal cells or 200 spiking interneurons with similar CRF tuning (i.e., similar i and θ) in the real cortex. Therefore, a 1:1 ratio between the numbers of pyramidal cells and interneurons in the model does not imply such a ratio in the cortex. We often refer to the cells tuned to θ at location i as simply the edge or bar element $i\theta$. The image that is shown is represented as inputs $I_{i\theta}$ across various $i\theta$. Each $I_{i\theta}$ models the visual input image seen through the CRF of a complex (pyramidal) cell preferring location i and orientation θ.

Although readers can follow the rest of this section without any equations, the following equations summarize the neural interactions in the model (see Section 2.1.2 on neuron models):

$$\dot{x}_{i\theta} = -\alpha_x x_{i\theta} - g_y(y_{i,\theta}) - \sum_{\Delta\theta \neq 0} \psi(\Delta\theta) g_y(y_{i,\theta+\Delta\theta})$$

$$+ J_o g_x(x_{i\theta}) + \sum_{j \neq i, \theta'} J_{i\theta,j\theta'} g_x(x_{j\theta'}) + I_{i\theta} + I_o + I_{\text{noise}}, \qquad (5.8)$$

$$\dot{y}_{i\theta} = -\alpha_y y_{i\theta} + g_x(x_{i\theta}) + \sum_{j \neq i, \theta'} W_{i\theta,j\theta'} g_x(x_{j\theta'}) + I_c + I_{\text{noise}}. \qquad (5.9)$$

In the above equations, $x_{i\theta}$ and $y_{i\theta}$ model the membrane potentials of the pyramidal cell and the interneuron, respectively, for edge or bar element $i\theta$; $g_x(x)$ and $g_y(y)$ are sigmoid-like functions modeling cells' firing rates or responses given membrane potentials x and y for the pyramidals and interneurons; $-\alpha_x x_{i\theta}$ and $-\alpha_y y_{i\theta}$ model the decay to resting potentials with time constants $1/\alpha_x$ and $1/\alpha_y$; $\psi(\Delta\theta)$ models the spread of inhibition within a hypercolumn; $J_o g_x(x_{i\theta})$ models self-excitation; $J_{i\theta,j\theta'}$ and $W_{i\theta,j\theta'}$ are neural projections from pyramidal cell $j\theta'$ to excitatory and inhibitory postsynaptic cell $i\theta$; I_c and I_o are background inputs modeling the general and local normalization of activities; and I_{noise} is input noise which is independent between different neurons. The pyramidal cell outputs $g_x(x_{i\theta})$ (or temporal averages over these) represent the V1 responses. Equations (5.8) and (5.9) specify how the pyramidal activities $g_x(x_{i\theta})$, which are initialized by external inputs $I_{i\theta}$, are modified by the contextual influences via the neural connections. This model can be reproduced using the complete details in the appendix of this chapter (see Section 5.9).

Note that notations in this chapter often have different semantics from those in other chapters. For example, K means the number of preferred orientations in a hypercolumn, and should not be confused with the kernels or filters in the previous chapters. This book tries to balance between self-consistency within its own notation, and consistency with the notation used in the original literature.

The pyramidal responses or output activities $g_x(x_{i\theta})$, which are sent to higher visual areas as well as subcortical areas such as the superior colliculus, will be used to quantify the saliencies of their associated locations and edge elements. The inhibitory cells are treated as interneurons. The input $I_{i\theta}$ to pyramidal cell $i\theta$ is obtained by filtering the input image through the CRF associated with $i\theta$. Hence, when the input image contains a bar of contrast $\hat{I}_{i\gamma}$ at location i and oriented at angle γ, this bar contributes to $I_{i\theta}$ by the amount

$$\hat{I}_{i\gamma}\phi(\theta - \gamma), \quad \text{where } \phi(\theta - \gamma) \text{ is the orientation tuning curve of the neurons.} \tag{5.10}$$
$$\text{(See Section 5.9 for the actual } \phi(x) \text{ used.)}$$

To visualize the strength of the input (contrast) and the model responses, the widths of the bars plotted in each figure are made to be larger for stronger input strength $I_{i\theta}$, or greater pyramidal responses $g_x(x_{i\theta})$ (or their temporal averages).

In the absence of intracortical interactions between different edge elements $i\theta$, the reciprocal connections between each pyramidal cell and its partner inhibitory interneuron would mainly provide a form of gain control for the direct input $I_{i\theta}$ (and make the response transiently oscillatory). The response $g_x(x_{i\theta})$ from the pyramidal cell $i\theta$ would only be a function of this direct input, in a context independent manner. With intracortical interactions, the influence of one pyramidal cell on the response of its neighboring pyramidal cell is excitatory via monosynaptic connections and inhibitory via disynaptic connections through the interneurons. Consequently, a pyramidal cell's response depends on inputs outside its CRF, and the pattern of pyramidal responses $\{g_x(x_{i\theta})\}$ is typically not just a scaled version of the input pattern $\{I_{i\theta}\}$ (see Fig. 5.15 A).

Figure 5.15 B shows the structure of the lateral connections in the model (Li 1999b). Connection $J_{i\theta,j\theta'}$ from pyramidal cell $j\theta'$ to pyramidal cell $i\theta$ mediates monosynaptic excitation. It is present if these two segments are tuned to similar orientations $\theta \approx \theta'$ and the centers i and j of their CRFs are displaced from each other roughly along their preferred orientations θ and θ'. Connection $W_{i\theta,j\theta'}$ from pyramidal cell $j\theta'$ to the inhibitory interneuron $i\theta$ mediates disynaptic inhibition from pyramidal cell $j\theta'$ to pyramidal cell $i\theta$. It tends to be present when the preferred orientations of the two cells are similar $\theta \approx \theta'$, but the centers i and j of their CRFs are displaced from each other along a direction roughly orthogonal to their preferred orientations. This V1 model has a translation invariant structure, such that all neurons of the same type have the same properties, and the neural connections $J_{i\theta,j\theta'}$ (or $W_{i\theta,j\theta'}$) have the same structure from all the presynaptic neurons $j\theta'$ except for translation and rotation to suit the position and orientation of the presynaptic receptive field $j\theta'$ (Bressloff, Cowan, Golubitsky, Thomas and Wiener 2002). The structure of the connections from a single pyramidal cell resembles a bow-tie.

Figure 5.16 illustrates the intracortical connections and their functions in further detail. The input image in Fig. 5.16 contains just horizontal bars. Hence, neurons preferring non-horizontal orientations are not strongly excited directly and are omitted from the figure. Here, the monosynaptic connections J link neighboring horizontal bars displaced from each other roughly horizontally, and the disynaptic connections W link those bars displaced from each other more or less vertically in the visual input image plane. The full lateral connection structure from a cell preferring a horizontal bar to cells preferring other bars (including bars that are not horizontal) is shown in Fig. 5.15 B.

In the input image, the five horizontal bars have the same input contrast, giving equal strength input $I_{i\theta}$ to the five corresponding pyramidal cells. Nevertheless, the output responses from these five pyramidals are different from each other, illustrated in the top plate of Fig. 5.16 by the different widths of the bars. The three horizontally aligned bars evoke higher output responses because the corresponding neurons facilitate each other's activities via the monosynaptic connections $J_{i\theta,j\theta'}$. The other two horizontal bars evoke lower output responses because the corresponding neurons receive no monosynaptic lateral excitation but receive disynaptic lateral inhibition from (neurons responding to) the neighboring horizontal bars displaced vertically from, and not co-aligned with, them. (To avoid excessive words, we sometimes use the term "bars" to refer to "neurons receiving direct inputs from the bars" when the meaning is clear from context). The three horizontally aligned bars, especially the middle one, also receive disynaptic inhibitions from the two vertically displaced bars.

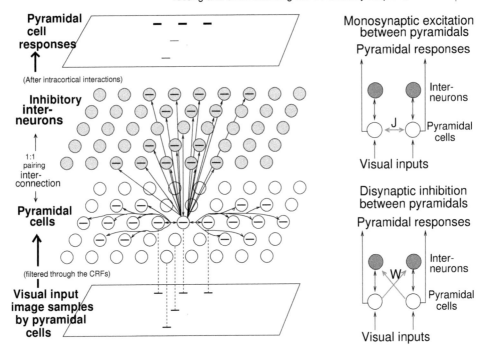

Fig. 5.16: Model elements. To avoid excessive clutter, only cells tuned to horizontal orientations are shown; and only connections to and from the central pyramidal cell are drawn. A horizontal bar, marking the preferred orientation, is drawn on the central pyramidal cell and the postsynaptic cells to which it is linked via lateral connections. In the input image plane, the central pyramidal neuron sends axons to other pyramidal cells displaced from it locally in a roughly horizontal direction, and to the interneurons which are also displaced locally, but in a roughly vertical direction. These axons are, respectively, for the monosynaptic excitation and disynaptic inhibition between the pyramidal cells (illustrated in the plots on the right). Five horizontal bars of equal contrast are shown in the input image in the bottom plane; each excites a pyramidal cell with the corresponding CRF (the correspondences are indicated by the dashed lines). The three aligned bars evoke higher responses, while two bars displaced vertically from them evoke lower responses (shown in the top plate). These differential responses are caused by facilitation between the three aligned bars via the monosynaptic connections J and the suppression between the vertically displaced bars by the disynaptic inhibition mediated by W. Adapted with permission from Li, Z., Pre-attentive segmentation in the primary visual cortex, *Spatial Vision*, 13(1): 25–50, Fig. 2C, copyright © 2000, Koninklijke Brill NV.

When the input image is a homogenous texture of horizontal bars, each bar receives monosynaptic lateral excitation from its (roughly) left and right neighbors but disynaptic lateral inhibition from its (roughly) top and bottom neighbors. The intracortical connections in the model are designed so that the sum of the disynaptic inhibition overwhelms the sum of the monosynaptic excitation in an iso-orientation texture. Hence, the net contextual influence on any bar in an iso-oriented and homogenous texture will be suppressive—this is iso-orientation suppression. Therefore, it is possible for the same neural circuit to exhibit iso-orientation suppression when the input image is a uniform texture, and to exhibit colinear facilitation, or contour enhancement, when the input image is an isolated contour made of multiple co-aligned bar segments. This is what has been observed in physiological experiments (Knierim and Van Essen 1992, Kapadia et al. 1995); see Section 2.3.9. Note that an iso-orientation texture can be seen as an array of parallel contours or lines.

Figure 5.17 illustrates how a smooth contour in a noisy background or along a texture border should evoke higher V1 responses than the same contour lying within a texture. The contextual influence depends on both the orientation and the spatial configuration of the context due to the following reasons: first, lateral connections tend to link bars having similar orientations; and second, the interaction between these similarly oriented bars tend to be monosynaptic and excitatory when they are co-aligned but disynaptic and inhibitory when they are not co-aligned. Each of the three vertical bars in dashed circles in Fig. 5.17 is part of a vertical contour. However, the contours are either along a texture border, within an iso-oriented texture, or embedded in a random background. Each enjoys the monosynaptic excitation from its co-aligned neighbors. However, iso-orientation suppression in V1 implies that this monosynaptic excitation is overwhelmed by the disynaptic inhibition when the contour is in the center of an iso-oriented texture. Meanwhile, when the contour is along a texture border such that it has fewer parallel contours as neighbors, this disynaptic inhibition should be reduced. The disynaptic inhibition should be minimal when there is no parallel contour neighbor, such as when the contour is isolated or embedded in a random background, as in Fig. 5.17 B. These intuitions will be confirmed by model simulations later in this chapter.

In most, or all figures of this chapter, we only show a small segment of the actual visual inputs and model responses, and the actual spatial extent of the input and response patterns should be understood to extend spatially well beyond the boundaries of the plotted regions. The model has a periodic or wrap-around boundary condition to simulate an infinitely large visual space; this is a conventional idealization of reality.

5.4.2 Calibration of the V1 model to biological reality

We intend to use the V1 model as a substitute for the real V1 to test whether saliency computations can feasibly be carried out by V1 mechanisms. Thus, we need to ensure that the relevant behaviors of the model resemble those of real V1 as much as possible. This is just like calibrating an experimental instrument in order to be able to trust subsequent measurements taken with this instrument. This does not mean that the model should include parts to model neural spikes and ionic channels on the neural membrane. (Later on in this chapter, in Section 5.8, it will be argued that equations (5.8) and (5.9) give a minimal model for V1's saliency computation.) However, when we use the visual inputs for which the firing rate responses from the real V1 are known, the model neuron's firing rate response, which will be used to predict saliency, should qualitatively resemble the real V1 responses.

More specifically, we examine representative visual input cases (see Fig. 2.24), in which contextual influences in real V1 have been studied. We simulate V1 model responses to these inputs, and compare the average firing rates of model and real V1 units to assess the qualitative resemblance (see Fig. 5.18 and Fig. 5.19). Figure 5.18 A–D model the contextual suppression that was seen physiologically by Knierim and Van Essen (1992). Figure 5.18 E–H model the contextual facilitation that Kapadia et al. (1995) recorded. To make model responses and the real V1 responses agree with each other, a neural circuit containing separate excitatory and inhibitory neurons is employed for the model V1, and a bow-tie pattern of the neural connections has been designed (see Fig. 5.15 B).

The model neurons' responses, in particular their dependence on the input context, varies with the strength or contrast of the input bar on which the contextual influence is being examined. As in physiological data, stronger and weaker input contrast are associated with stronger suppression and facilitation, respectively (see Section 5.4.6).

A: A vertical bar inside a texture or on a vertical texture border

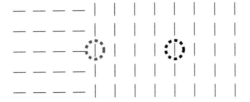

B: A bar in a smooth contour

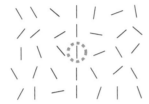

Fig. 5.17: Colinear facilitation and iso-orientation suppression arising from excitation and inhibition between V1 neurons. All three vertical bars, in blue, black, and red dashed circles, respectively (the circles are for illustration; they are not present in the visual input), receive strong monosynaptic excitation, because each is co-aligned with its top and bottom neighboring vertical bars. Meanwhile, these three vertical bars receive different degrees of disynaptic inhibition. Inhibition increases with the number of neighboring bars parallel to, but not co-aligned with, each of them. The bar in the red circle is minimally affected by disynaptic inhibition; the bar in the black circle is maximally affected, and inhibition can overwhelm the monosynaptic excitation.

5.4.2.1 Some conventions in displaying the model behavior

Figure 5.18 also illustrates some conventions used to display the model behavior in many figures of this chapter. To display model input and responses, only a limited spatial range of the locations i of model units $i\theta$'s is shown for illustration. This limited region should be understood as being only a part of an infinitely large image, and the plotted image content should extrapolate beyond the plotted region. (Otherwise, translation invariance of inputs breaks at the outer boundary of the plotted images, and this break should also manifest in substantial non-homogeneities in the response levels.)

In addition, unless otherwise stated explicitly, the model is always simulated in a two-dimensional visual space in a wrap-around or periodic boundary condition. In particular, let location $i = (i_x, i_y)$ of the model neural units $i\theta$ have the horizontal and vertical components i_x and i_y, respectively, in a Manhattan grid, such that $i_x = 1, 2, ..., N_x$ and $i_y = 1, 2, ..., N_y$; then location $i = (i_x = 1, i_y)$ is the horizontal neighbor of location $i' = (i_x = N_x, i_y)$, and location $i = (i_x, i_y = 1)$ is the vertical neighbor of location $i' = (i_x, i_y = N_y)$. Analogous conditions apply if the visual inputs are sampled in a hexagonal grid. Furthermore, N_x and N_y are much larger than the maximum length $|i - j|$ of the lateral connections $J_{i\theta, j\theta'}$ and $W_{i\theta, j\theta'}$.

Furthermore, to avoid clutter in plots to visualize model responses, we only show bars whose output responses $g_x(x_{i\theta})$ exceed a threshold. For example, due to the finite width of orientation tuning curves (see equation (5.10)), a bar $\hat{I}_{i\beta}$ at location i in the input image actually provides direct inputs $I_{i\theta}$ to multiple model neurons with similar, but not identical, preferred orientations θ. When input $\hat{I}_{i\beta}$ and contextual facilitations are sufficiently strong,

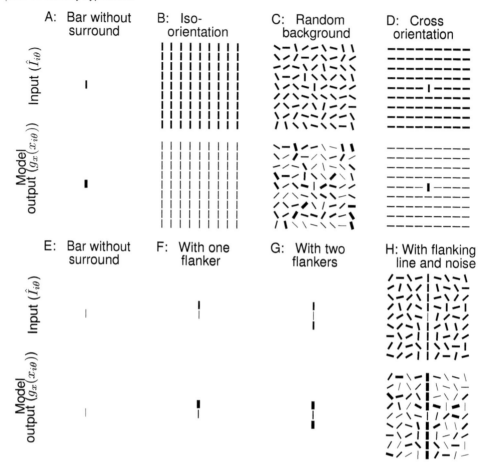

Fig. 5.18: The V1 model qualitatively reproduces representative observations of contextual influences in V1. Each model input pattern has a central vertical (target) bar with or without contextual stimuli. All visible bars are presented at the same high contrast ($\hat{I}_{i\theta} = 3.5$) except for the target bar in E, F, G, H where $\hat{I}_{i\theta} = 1.05$ is near threshold. Input and output strengths are visualized by the widths of the bars, using the same scale in all plots. Isolated high and low contrast bars are presented in A and E. B, C, and D simulate various forms of contextual suppression of the response to the high contrast target. F, G, and H simulate various forms of contextual facilitation of the response to the low contrast target. Note that the response to the near threshold target bar in H is stronger than that to the high contrast target bar in B. Output responses weaker than a threshold are not plotted to avoid clutter. Adapted with permission from Li, Z., Pre-attentive segmentation in the primary visual cortex, *Spatial Vision* 13(1): 25–50, Fig. 3A–3H, copyright © 2000, Koninklijke Brill NV.

more than one model neuron at location i can be activated (making $g_x(x_{i\theta}) > 0$). However, the responses of the less activated bars at this location are often below the threshold we use for visualization, and so these bars do not appear in the plots. Similarly, model input plots are typically plotted according to the values of $\hat{I}_{i\theta}$ (i.e., the actual input image) rather than $I_{i\theta}$ (the direct inputs to individual model neurons).

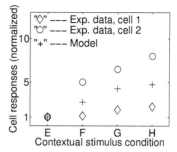

Fig. 5.19: Comparison between the output of the model in Fig. 5.18 and physiological observations. The labels A, B, C, D, E, F, G, and H on the horizontal axes mark the various contextual configurations in the subplots of Fig. 5.18. Responses are normalized relative to the response to the isolated bar. In the left plot, data points "o" and "◇" are taken from Knierim, J.J. and Van Essen, D.C. Neuronal responses to static texture patterns in area V1 of the alert macaque monkey, *Journal of Neurophysiology*, 67(4):961–980, figures 4b and 11, 1992. In the right plot, data points "o" and "◇" are taken from the two cell examples in figures 12B and 12C of Kapadia, M.K., Ito, M., Gilbert, C.D., and Westheimer, G. Improvement in visual sensitivity by changes in local context: parallel studies in human observers and in V1 of alert monkeys, *Neuron*, 15(4):843–56, 1995. Adapted with permission from Li, Z., Pre-attentive segmentation in the primary visual cortex, *Spatial Vision*, 13(1): 25–50, Fig. 3I–3J, copyright © 2000, Koninklijke Brill NV.

5.4.3 Computational requirements on the dynamic behavior of the model

The V1 model should be applied to visual inputs which have not been used in physiological experiments. Hence, in addition to calibrating the model to the existing physiological observations, the model should also be designed such that it is well behaved in a manner expected for a visual system that computes saliency appropriately. This imposes the following requirements on the model; some of them also help to ensure that the model is properly calibrated to existing physiological data.

The first requirement is that when the input is not translation invariant, and if the location where the input changes is conspicuous, the model should give relatively higher responses to this location. Figure 5.17 A presents an example for which the orientations of the bars change at the texture border. Elevated responses to the texture border bars highlight the conspicuous input locations, consistent with their higher saliency. As we have argued, this can be achieved by mutual suppression between neurons responding to neighboring iso-oriented bars. Border bars have fewer iso-oriented neighbors and so experience less suppression and have relatively higher responses. Hence, iso-orientation suppression should be sufficiently strong to make the degree of highlights sufficient; indeed as strong as that observed physiologically.

The second requirement is that, when the model is exposed to an homogenous texture, the population response should also be homogenous. In particular, this means that if inputs $I_{i\theta}$ to the model are independent of the spatial location i, then the outputs $g_x(x_{i\theta})$ should also be (neglecting the response noise, which should be such that they do not cause qualitative differences). If this requirement was not satisfied by real V1, then we would hallucinate inhomogenous patterns even when the input image did not contain them, or we would hallucinate

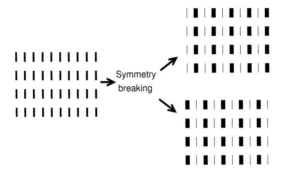

Fig. 5.20: Spontaneous symmetry breaking. Given sufficient mutual suppression between vertical arrays of bars, the output in response to the homogenous texture input (on the left) can evolve to one of the two inhomogenous response patterns on the right. Which pattern will emerge depends on how the initial activities deviate from the homogeneity—an initial deviation (caused by noise) toward one of the final patterns will be amplified to increase the chance of the emergence of the corresponding final pattern. The real V1 avoids such symmetry breaking (in normal conditions); it should therefore also be avoided in the model.

salient locations when there are none. This requirement has to be satisfied in order to obtain the model behavior demonstrated in Fig. 5.18 B.

It may seem that this requirement should be satisfied automatically, since translation invariant (i.e., homogenous) inputs might seem obviously to give rise to translation invariant outputs when the intracortical connections $J_{i\theta,j\theta'}$ and $W_{i\theta,j\theta'}$ are all translation invariant. However, translation invariant dynamical systems are subject to spontaneous symmetry breaking, and so they could generate non-homogenous responses even when fed with homogenous inputs. For instance, a thin stick standing vertically has a strong tendency to fall sideways to one side or another. The symmetric equilibrium position of upright standing is dynamically unstable—a small perturbation of the stick to one side will be amplified further by the dynamics under gravity.

In the V1 model, just as for the case of the stick, a homogenous response to a homogenous texture input pattern (such as the regular texture of vertical bars in Fig. 5.20) is also an equilibrium point in a dynamic system. Again, as for the stick, this equilibrium point can be unstable if the neural dynamics are incorrectly modeled. In particular, iso-orientation suppression between neighboring vertical bars in Fig. 5.20 makes neighboring vertical arrays of bars suppress each other. Consider the case that one array has slightly higher response than the other because of noise included in the dynamics. Then this array will suppress neighboring arrays more; those arrays could then suppress the first array less, making the first array's responses higher still. Thus, the perturbation could be amplified by a form of positive feedback in the dynamics. If this positive feedback is too strong, spontaneous pattern formation occurs, as schematized in Fig. 5.20. The mutual suppression between the arrays is caused by iso-orientation suppression. Intuitively, reducing the strength of this suppression should help reduce the instability.

However, reducing the strength of the iso-orientation suppression will compromise the first requirement to highlight conspicuous input locations where input changes. Hence, there is a conflict between the need to have strong iso-orientation suppression to highlight conspicuous input locations, e.g., at a texture border or a feature singleton, and the need to have a weak iso-orientation suppression in order to prevent spontaneous symmetry breaking to homogenous inputs. Mathematical analysis of the dynamic system of neural circuits, explained in detail in Section 5.8, shows that resolving this conflict imposes the following requirement on the model's neural circuit: mutual suppression between principal neurons should be mediated disynaptically by inhibitory interneurons, as in the real V1. This circuit requirement precludes implementing iso-orientation suppression by direct inhibition between the principal units, as is often the case in artificial neural networks or computer vision algorithms (such as the Markov random field model).

Thirdly, the strength of mutual excitation between neurons should be limited, in order to prevent ubiquitous non-zero responses of pyramidal neurons to zero direct input given contextual inputs. In particular, the colinear facilitation implied by Fig. 5.18 FGH should not be so strong as to activate a neuron whose most preferred stimulus bar is absent in the input but is an extrapolation of a straight line present in the input image. Otherwise, the visual system would hallucinate the eternal growth of short, unchanging input lines.

If V1 does not create a saliency map in the form proposed by the V1 saliency hypothesis, then the above requirements for a well-behaved model for saliency computation is not expected to be consistent with the requirement that the model being calibrated to sufficiently resemble the real V1 (as in Fig. 5.18 and Fig. 5.19). Nevertheless, a single set of model parameters (presented in the appendix to this chapter; see Section 5.9) has been found that satisfies both sets of requirements, reinforcing the plausibility of the hypothesis that V1 creates a bottom-up saliency map. The design and analysis of the recurrent neural circuit are mathematically somewhat challenging. Hence, I separate the mathematical details into a separate section (Section 5.8) for readers interested in the nonlinear neural dynamics (Li 1999b, Li 2001, Li and Dayan 1999). However, the challenging mathematics is far less formidable than simultaneous *in vivo* recordings from hundreds of primate V1 neurons using visual search stimuli and the current technology in physiological experiments.

5.4.4 Applying the V1 model to visual search and visual segmentation

The model parameters include the neural connections $J_{i\theta,j\theta'}$ and $W_{i\theta,j\theta'}$, the activation functions $g_x(.)$ and $g_y(,)$, the neurons' decay constants, the way the model activities are normalized, the local interactions within a hypercolumn, and characteristics of the input noise. Following the design and calibration, all these parameters were fixed (to the values presented in the appendix to this chapter; see Section 5.9), and the model's response to a variety of input stimuli (including stimuli not used for calibration) can be tested.

In particular, we examine representative visual inputs for which the saliency properties, e.g., which locations are salient and how saliency depends on input characteristics, are known from visual experience or behavioral experiments. We compare these saliency properties with those predicted from the responses of the V1 model.[15] These representative inputs and saliency properties are:

1. Images containing orientation singletons, or borders between iso-orientation textures;
2. images contrasting saliencies of visual search targets in feature and conjunction searches;
3. images demonstrating visual search asymmetry;
4. images demonstrating how saliencies depend on input feature contrasts, spatial densities of input items, or regularities of texture elements;
5. images demonstrating the conspicuousness of a hole in the visual input pattern, or of a missing input;

---

[15] In principle, one could design the model such that the model's predicted saliency behavior agrees with those observed in visual behavior or experience. If so, this agreement should be included as one of the computational requirements for the model in Section 5.4.3. In practice, the model was designed, and its parameters were fixed, without first ensuring this agreement.

6. images containing more complex textures whose boundaries are conspicuous to varying degrees.

Because the model parameters are fixed, the differences in model responses arise solely from the differences in the input stimuli $\hat{I}_{i\theta}$ (and, sometimes, the difference between Manhattan and hexagonal input grids, which we use to sample the input more proficiently).

To illustrate the function of the intracortical interactions, many model simulations use input patterns in which all visible bars $i\theta$ have the same underlying input contrast $\hat{I}_{i\theta}$, such that differential responses to different visible bars can only arise systematically from the intracortical interactions. For each bar element $i\theta$, the initial model response $g_x(x_{i\theta})$ is dictated only by the external inputs $I_{i\theta}$ to this bar. However, due to intracortical interactions, the response $g_x(x_{i\theta})$ is significantly affected by inputs $I_{j\theta'}$ to other bar elements $j\theta'$ within about one membrane time constant $1/\alpha_x$ after the initial neural response. (The current model implementation has the parameter $\alpha_y = \alpha_x$.) This agrees with physiological observations (Knierim and Van Essen 1992, Kapadia et al. 1995, Gallant, Van Essen and Nothdurft 1995), if this time constant is assumed to be of the order of 10 milliseconds (ms).

5.4.4.1 Model behavior, and additional conventions in its presentation, in an example: two neighboring textures

Figure 5.21 shows an example of the temporal evolution of the model responses. The activities of units in each texture column initially rise quickly to an initial response peak and then decrease. The initial responses at time $t = 0.7$ (in units of the membrane time constant, and excluding latency from retinal input to LGN output) after stimulus onset are roughly the same across the columns, since they are mainly determined by the direct, rather than the contextual, input to the receptive fields. By time $t = 0.9$, responses to the horizontal bars near the vertical texture border are relatively weaker than responses elsewhere, because these bars enjoy less monosynaptic colinear facilitation. Neural responses reach their initial peak at around $t = 1.2$. Then, iso-orientation suppression starts to manifest itself. This suppression lags the colinear facilitation since it is mediated disynaptically. The suppression is most obvious away from the texture border, where each bar has more iso-orientation neighbors. The black curve plots the mean responses after input onset as a function of the column, averaging over many cycles of the oscillating neural responses. This curve is another version of the plot in Fig. 5.22 C.

In the rest of this chapter, we generally omit the temporal details of the responses, and so we report just the temporal averages of the neural activities $g_x(x_{i\theta})$ after the model has evolved for several time constants after the onset of the visual input $I_{i\theta}$. (For simplicity, we often use "outputs," "responses," or "$g_x(x_{i\theta})$" to mean the temporal averages of the model pyramidal responses $g_x(x_{i\theta})$.) Further, inputs $I_{i\theta}$ are typically presented to the model at time 0 and persist, unless stated otherwise.

This focus on static inputs and temporally averaged model responses is motivated by the following considerations: (1) most of the behavioral data on saliency are from experiments using static visual images presented for a much longer time duration than the time constant of neurons; (2) even though the initial presentation of the image will lead to a strong impulse of saliency at locations of all image items, the behavioral effects that are typically measured depend on the differences between saliencies at locations of different input items, and these should be most pronounced *after* this impulse has subsided. (Of course, the model can also be applied to temporally varying inputs or asynchronously presented inputs. For example, if all except one item in an homogenous array are presented simultaneously, the temporally unique item, if presented with a sufficiently long delay, should make its location very salient by the saliency impulse associated with its onset.)

For each model simulation, the input contrasts, which are represented by $\hat{I}_{i\theta}$, are adjusted

A: Input image ($\hat{I}_{i\theta}$) to model (texture column numbers at bottom)

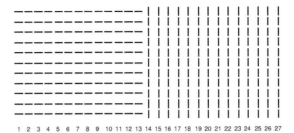

1 2 3 4 5 6 7 8 9 10 11 12 13 14 15 16 17 18 19 20 21 22 23 24 25 26 27

B: Responses $g_x(x_{i\theta})$ versus texture columns above at various time since visual input onset, or temporal average responses (black).

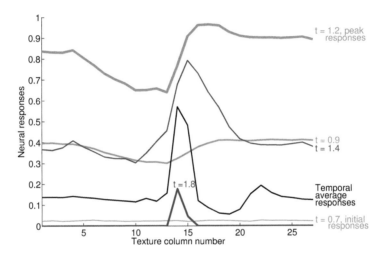

Fig. 5.21: The temporal evolution of the model responses to an input pattern. A: The input pattern contains two regions (excluding the texture column numbers indicated at the bottom); each visible bar has the same input strength $\hat{I}_{i\theta} = 2.0$. The input pattern is presented at time $t = 0$ and remains presented thereafter. Only 11 rows by 27 columns of the input bars are plotted out of a larger image (of 22 rows by 60 columns, using wrap-around boundary conditions). B: The response traces are the average of $g_x(x_{i\theta})$ across positions i within the same texture column, where θ is the orientation of the input bars in the column. Red and blue curves plot responses at various time points t (indicated to the right of each curve) during the rising and decaying phases, respectively, of the initial phase of the responses. Time t is in the units of membrane time constant, and it excludes the latency from retina input to LGN output. The black curve plots the responses averaged over a duration from $t = 0$ to $t = 12$. The initial responses (at $t = 0.7$) are not context dependent; but contextual influences are apparent within half a time constant after the initial response.

to mimic the corresponding conditions in physiological and psychophysical experiments. In the model, the input dynamic range is $\hat{I}_{i\theta} = (1.0, 4.0)$, which will allow an isolated bar to drive an excitatory neuron from threshold activation to saturation. Hence, low contrast input bars, which are typically used to demonstrate colinear facilitation in physiological experiments, are represented by $\hat{I}_{i\theta} = 1.05$ to 1.2. Intermediate or high contrast inputs (e.g., $\hat{I}_{i\theta} = 2 - 4$)

A: Input image ($\hat{I}_{i\theta}$) to model

B: Model output ($g_x(x_{i\theta})$)

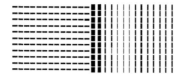

C: Bar plot of the neural response levels versus texture columns above

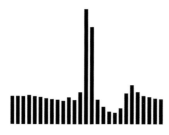

D: Thresholded version of the model output in B

Fig. 5.22: Texture segmentation. A is the same as the input pattern in Fig. 5.21 A. B: Model output responses to A, i.e., temporal averages of $g_x(x_{i\theta})$ for the bars. C: The average model response in each column in B (considering only the most responsive neuron at each texture element location) is represented by the height of the bar for each column location. This plot shows the same information as the black curve in Fig. 5.21 B. D: The result of applying a threshold of half of the maximum response among all bars to the responses $g_x(x_{i\theta})$ in B. Adapted with permission from Li, Z., Visual segmentation by contextual influences via intracortical interactions in primary visual cortex, *Network: Computation in Neural Systems*, 10(2): 187–212, Fig. 3, copyright © 1999, Informa Healthcare.

are used for all the visible bars in other input images, including those illustrating texture segmentation and feature pop-out. Meanwhile, the neural output $g_x(x_{i\theta})$ ranges from 0 to 1.

Figure 5.22 B further illustrates the model response to the same input as in Fig. 5.21 by showing the average responses $g_x(x_{i\theta})$ for a substantial patch of the input texture. Figure 5.22 C plots the (temporal averaged) responses $g_x(x_{i\theta})$ to the bars averaged in each column in Fig. 5.22 B. It shows that the most salient bars are indeed near the region boundary. Figure

5.22 D confirms that the boundary can be identified by thresholding the output responses using a threshold parameter, $thresh = 0.5$, set to be a proportion of the maximum response to the image, to eliminate weak outputs that would otherwise clutter the figure. Thresholding is not performed by V1 but is only used for visualization.

According to the V1 saliency hypothesis, the visual locations surviving the thresholding are more likely to be selected first by bottom-up mechanisms. If the diameter of the attentional spotlight is smaller than the length of the texture border, then only a part of the border can be selected first. We might reasonably assume that the reaction time for an observer to complete the overall task of segmenting two neighboring textures decreases with the time it takes until any part of the border is first selected. Thus, we can use this reaction time to probe the saliency of a texture border, without addressing how the full task of segmentation is completed after the selection of only a small part of the texture border.

Here we also briefly point out something beyond the scope of investigating saliency. In Fig. 5.22 B, the response highlights are not distributed symmetrically around the texture border. This could make the viewers perceive the location of the texture border as being biased slightly to the right of the border. This has indeed been observed psychophysically (Popple 2003), although there may be additional causes for such biases beyond V1, such as the perception of figure and ground. This is a demonstration that the V1 saliency mechanisms make V1 responses distort the visual input image. The ultimate percept is likely the outcome of additional processing based on the V1 responses.

5.4.4.2 Assessing saliency from model V1 responses: illustrated by the effect of the orientation contrast at a texture border

According to equation (5.4), the saliency value at each location is the highest (pyramidal) response to inputs at that location. A location in the V1 model is denoted by i, and various neurons $i\theta$ give responses $g_x(x_{i\theta})$. Therefore, saliency value at location i is

$$\text{SMAP}_i \equiv \max_\theta \left[g_x(x_{i\theta}) \right]. \tag{5.11}$$

As discussed in Section 5.2.1, these pseudo-saliency values at various locations need to be compared with each other in order to determine the most salient location in an image. The actual saliency value of a location should reflect this comparison. For this purpose, let

$$\bar{S} \equiv \text{the average of SMAP}_i \text{ over } i \quad \text{and}$$
$$\sigma_s \equiv \text{the standard deviation of SMAP}_i \text{ over } i \tag{5.12}$$

be the mean and standard deviation of the SMAP_i values at all locations i, or alternatively, at all locations i with non-zero neural responses. The salience of a location i can then be assessed by

$$r_i \equiv \frac{\text{SMAP}_i}{\bar{S}} \quad \text{and} \quad z_i \equiv \frac{\text{SMAP}_i - \bar{S}}{\sigma_s}. \tag{5.13}$$

In our plots of the model responses, quantities r can be visualized by the thickness of the plotted output bars. Meanwhile, z models the psychological z score.

The quantities $\bar{S}$ and σ_s in equation (5.12) could alternatively be defined as

$$\bar{S} \equiv \text{average of } g_x(x_{i\theta}) \text{ over } (i, \theta) \quad \text{and}$$
$$\sigma_s \equiv \text{standard deviation of } g_x(x_{i\theta}) \text{ over } (i, \theta). \tag{5.14}$$

This alternative is conceptually and algorithmically simpler, since it omits the intermediate step of obtaining $\text{SMAP}_i = \max_\theta \left[g_x(x_{i\theta}) \right]$, in which neurons are grouped according to their receptive field location i. Using the alternative should only make quantitative rather than

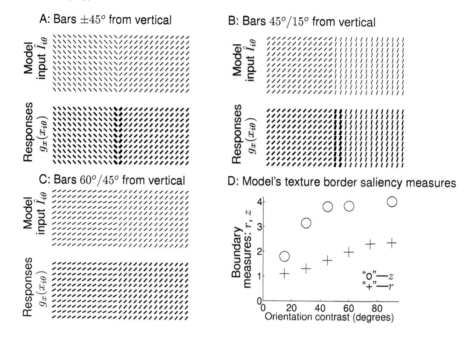

Fig. 5.23: A, B, C: Additional examples of model behavior at orientation texture borders. Each example contains two neighboring textures, in which texture bars have orientation θ_1 and θ_2, respectively, meeting in the middle at a vertical border. In A, B, and C, the saliency measures for the borders are $(r, z) = (1.4, 3.4)$, $(r, z) = (1.7, 3.7)$, and $(r, z) = (1.03, 0.78)$. D: Texture border saliency measures r, z (indicated by "+" and "o", respectively) from the model as a function of the orientation contrast at the border. Each data point is the averaged measure from borders of all possible pairs of θ_1 and θ_2 for a given $|\theta_1 - \theta_2|$. The most salient column in B is in fact the second left column in the texture region on the right. In C, the texture border is barely detectable without close scrutiny. Although the texture border bars are among the most salient ones, their evoked responses are only slightly ($\sim 10\%$) higher than those of the other bars (this is imperceptible in the line widths shown in the output). Adapted with permission from Li, Z., Visual segmentation by contextual influences via intracortical interactions in primary visual cortex, *Network: Computation in Neural Systems*, 10(2): 187–212, Fig. 4, copyright © 1999, Informa Healthcare.

qualitative difference to r and z. In this book, the r and z values are obtained by using $\bar{S}$ and σ_s in equation (5.12), with the locations i used to obtain the mean $\bar{S}$ and σ_s only including the locations which have non-zero responses $g_x(x_{i\theta})$ for at least one θ.

To assess the saliency of a texture border, we replace the SMAP$_i$ in equation (5.13) by the average SMAP$_i$ in the most salient grid column parallel to, and near, the texture boundary. A salient texture border should give large values for (r, z). For instance, in Fig. 5.22, $(r, z) = (3.7, 4.0)$ at the texture border.

V1 does not (and does not need to) calculate r and z. These two values just help us characterize the saliencies of visual locations in order to compare them with our visual experience or behavior, e.g., to see whether locations with high r and z values indeed correspond to the locations that are more conspicuous. In particular, locations with smaller z scores are expected to take longer to select, due to the competition between multiple locations for selection. A z score larger than three makes a location quite salient and indeed, likely to be the most salient in the scene. An example is the texture border in Fig. 5.22. Meanwhile a location with $z \sim 1$ is not so salient, even if it has the largest z score in the scene.

Consider applying these tools to the examples of orientation textures shown in Fig. 5.23. One can see for oneself how conspicuous is each texture border. Texture borders with orientation contrasts of 90° (Fig. 5.23 A) or 30° (Fig. 5.23 B) are quite conspicuous, i.e., salient. However, a border with an orientation contrast of only 15° (Fig. 5.23 C) is rather difficult to notice without scrutiny. These observations agree with the model's z scores. The z score for this 15° contrast border is indeed only $z = 0.78$. Other 15° contrast borders will lead to higher z scores—for instance, if one texture comprises vertical bars, and the other texture comprises bars that are 15° clockwise from vertical.

Psychologically, the just-noticeable orientation contrast for a texture border to be detected quickly is indeed about 15°. In this model, a border with a 15° orientation contrast has an average $z \approx 1.8$ (averaged over all possible orientations θ_1 and θ_2 for the bars in the two textures, given $|\theta_1 - \theta_2| = 15^o$); see Fig. 5.23 D. This is expected for a border with only a moderate saliency psychophysically.

The dependence of the border saliency on the orientation contrast is mainly caused by the decrease in the suppression between two neighboring bars as the orientation difference between them increases. This suppression is strongest between parallel, but not co-aligned, bars, it remains substantial when the two bars are similarly but not identically oriented, and it is much reduced when the two bars are orthogonal to each other. This is reflected in the bow-tie connection pattern between the V1 neurons shown in Fig. 5.15 B, and it is manifest in the contextual influences that are observed physiologically; see Fig. 5.18 ABD.

Henceforth, the model saliency of a visual location i is assessed by the z score only. In particular, the z score for the location of a target of a visual search will be assessed this way, to link with psychophysical data on visual search tasks (Li 1997, Li 1999b, Li 1999a, Li 2002).

5.4.4.3 Feature search and conjunction search by the V1 model

Figure 5.24 demonstrates the model's behavior for a feature search and a conjunction search. The same target "⟋" is presented in two different contexts in Fig. 5.24 A and Fig. 5.24 B. Against a texture of " ⟍", it is highly salient because its horizontal bar is unique. Against a texture of " ⟍" and "⟍", it is much less salient because only the conjunction of "—" and "⟋" distinguishes it. This is consistent with psychophysical findings (Wolfe et al. 1989, Treisman and Gelade 1980). In the V1 model, the unique horizontal target bar in Fig. 5.24 A leads to the response of a V1 neuron that is not subject to iso-orientation suppression. All the other input bars are suppressed in this way. Thus, the horizontal bar evokes the highest response among all V1 neurons and makes the target location salient. Meanwhile, in Fig. 5.24 B, the V1 responses to both bars in the target suffer from iso-orientation suppression, just like all the other bars in the image. Hence, neither of the bars in the target evokes a response that is significantly greater than typical responses to the other bars, and so the location of the target is not salient.

Therefore, V1 mechanisms can be the neural substrate underlying the psychological "rule" that feature searches are typically easy and conjunction searches are difficult (Treisman and Gelade 1980).

Two kinds of feature tunings

Our observations suggest the following: consider a visual characteristic such as orientation or color that psychophysical rules deem to be a "feature" dimension, supporting such phenomena as easy or efficient search. Then, we can expect two neural properties to be tuned to feature values in this feature dimension. The first is that (the responses of) some V1 neurons should be

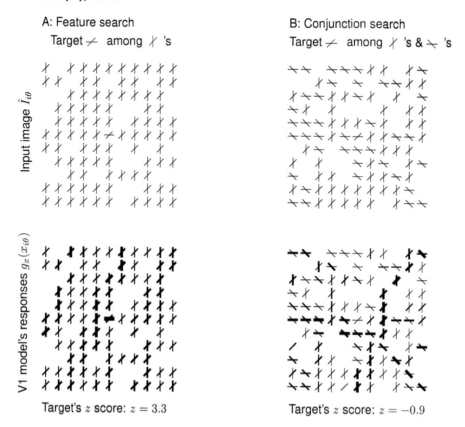

Fig. 5.24: The behavior of the model in feature (A) and conjunction (B) searches. Stimulus (top), model responses (bottom), and the z scores for the targets (displayed in the center of each pattern for convenience) are shown for the two examples. The target in both A and B is made of a horizontal bar and a $45°$ (tilted clockwise from vertical) oblique bar intersecting each other. A: The target is unique in having a horizontal bar, making it a case of orientation feature search, and leading to a high z score, $z = 3.3$. B: Each target feature, i.e., the horizontal or the oblique bar, is present in the distractors; these differ from the target only in the conjunctions of the two orientations. This leads to a low z score for the target, $z = -0.9$.

tuned to this dimension—this is of course a classical concept. Orientation tuning is an example; it enables some V1 neurons to signal the saliencies caused by their preferred features.

The other neural property that should be tuned is the intracortical connection pattern between V1 neurons, such that the strength of the intracortical connection between neurons roughly decays with the difference between the preferred features of the two neurons. In other words, the intracortical connections are tuned to the preferred features of the linked V1 cells. The most critical aspect of this tuning should be iso-feature suppression. Tuned intracortical suppression makes a feature singleton salient. There can also be a feature tuning width analogous to that in the neural response tuning to features; this tuning width should be compatible to the minimum feature difference necessary between a feature singleton and the background feature values to make the singleton sufficiently salient. For example, an orientation singleton can be viewed as having a sufficiently unique orientation for salient pop-out if the intracortical connections between neurons most activated by the singleton and background features are absent or insignificant.

The V1 model explains the results of the feature and conjunction search tasks in Fig. 5.24 without any explicit representation of the conjunctions between features. According to the argument above, a lack of explicit representation of the conjunction, i.e., a lack of tuning to the conjunction feature, prevents the conjunction feature from behaving like a basic feature in terms of saliency. Therefore, a target whose uniqueness is only defined by a feature conjunction cannot be salient. On the other hand, the target in Fig. 5.24 A is salient not because the whole object item "⤢" is recognized or signaled by a single neuron; instead, it is salient because one of its component features, namely the horizontal bar, is a unique basic feature and is sufficiently salient to attract attention strongly by itself. As far as saliency is concerned, the oblique bar in the target is not visible to the saliency system, which only looks at the highest response at each location.

In Fig. 5.24, the background items are not spatially uniform or regular, and so responses to the background bars are not uniformly low. The response to each bar is determined by its particular contextual surround. An accidental alignment of a given bar with its local context facilitates (or at least reduces the suppression of) the final response. On the other hand, if the bar has more iso-orientation neighbors with which it is not aligned, then the response will be more greatly suppressed. Despite the heterogeneity in the population responses, the response to the target horizontal bar in Fig. 5.24 A is still substantially higher than most of the background responses, making the ultimate z score high.

5.4.4.4 A trivial example of visual search asymmetry through the presence or the absence of a feature in the target

Given the observation that a feature search is easier than other searches, it is straightforward to understand the simple example of visual search asymmetry in Fig. 5.25. Search asymmetry is the phenomenon that the ease of a visual search can change when the target and distractors are swapped—for instance, searching for a cross among vertical bars is easier than vice versa. The target cross is easier to find in Fig. 5.25 A because it can be found by feature search. The horizontal bar in the cross is the unique feature, and it evokes the highest V1 response since it is the only one which lacks iso-orientation neighbors. Meanwhile, in Fig. 5.25 B, the target fails to possess any unique feature lacking in the non-targets; hence, it cannot be found by feature search. The target vertical bar and the vertical bars in the background crosses are almost equally suppressed; thus the target's z score is too low for it to pop out.

As in Fig. 5.24 A, the target cross in Fig. 5.25 A is easier to find not because the whole cross is recognized; instead, it is because one of its components, namely the horizontal bar, evokes the highest overall V1 response. Its other component, the vertical bar, does not contribute to the z score for the target.

Note that the search asymmetry between the cross and the vertical bar cannot be predicted from the idea that the ease of finding a target depends on how different it is from the distractors, since this difference does not change when target and distractors swap identity. A long-standing psychological rule (Treisman and Gelade 1980) is that a target having an unique (basic) feature which is lacking in the non-targets (as in Fig. 5.25 A) is easier to find than a target defined by lacking a (basic) feature which is present in the non-targets (as in Fig. 5.25 B). We suggest V1 saliency mechanisms provide the neural substrate of this rule.

5.4.4.5 The ease of a visual search decreases with increasing background variability

The formula for a search target's z score, $z = (\text{SMAP}_i - \bar{S})/\sigma_s$, suggests that increasing σ_s, by increasing the heterogeneity of the responses to non-targets, should decrease the target's z score when the target is at least minimally salient, i.e., when its highest evoked response SMAP_i is above the average response $\bar{S}$ to the scene. This is demonstrated in Fig. 5.26. A

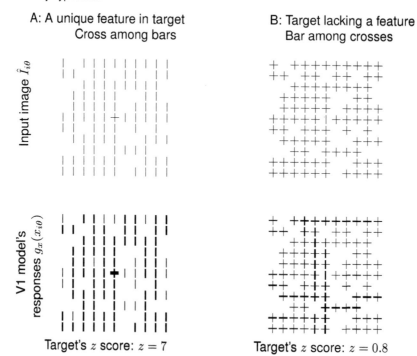

Fig. 5.25: A simple example (Li 1999b, Li 1999a) of search asymmetry in the V1 model. Searching for a cross among vertical bars (A) is easier than searching for a vertical bar among crosses (B). This figure is shown using the same format as that in Fig. 5.24. These examples also demonstrate that a target is easier (or more difficult) to find when it is defined by having (or lacking) a feature (e.g., the horizontal bar) that is absent (or present) in the distractors. The horizontal bar in the target in A is the only one in the image to evoke a V1 response that is not suppressed by iso-orientation suppression; the target vertical bar in B, however, suffers the same iso-orientation suppression experienced by other vertical bars.

target's saliency according to the model decreases when the non-targets are more variable, either because the non-targets are irregularly positioned in space, as in Fig. 5.26 A, or because the non-target feature values are heterogeneous, as in Fig. 5.26 B. Psychological observations have previously led to the rule that a target is more difficult to find when the background variabilities increase in these ways (Duncan and Humphreys 1989); and it has been suggested that random background variability acts as noise and limits the performance of visual search (Rubenstein and Sagi 1990).

Contextual influences can arrange for two identical visual items to evoke different V1 responses when in different contexts. This effect underlies the heterogeneous responses to non-targets in Fig. 5.26 A. Meanwhile, heterogeneous non-targets placed in a regular grid, as in Fig. 5.26 B, also evoke heterogeneous responses, since the contextual influences depend on the feature similarity between neighboring input items.

The model responses in Fig. 5.26 AB are more heterogeneous than those in Fig. 5.26 C. Therefore σ_s is larger in Fig. 5.26 AB. For example, if the maximum response SMAP$_i$ to the target at location i is 10% above the average response $\bar{S}$, it will still stand out, making the target very salient if no other item in the scene evokes a response more than 5% above $\bar{S}$. However, if the background responses vary between 50% to 150% of the average $\bar{S}$, the target

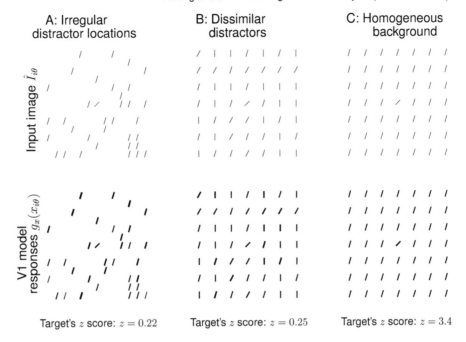

A: Irregular
distractor locations

B: Dissimilar
distractors

C: Homogeneous
background

Input image $\hat{I}_{i\theta}$

V1 model
responses $g_x(x_{i\theta})$

Target's z score: $z = 0.22$ Target's z score: $z = 0.25$ Target's z score: $z = 3.4$

Fig. 5.26: The effect in the model of background variability on the saliency of a target (Li 2002). A, B, and C show visual search images and model responses. The target bar, tilted $45°$ clockwise from vertical (and shown in the center of each image for convenience), is among distractors, which are irregularly placed identical bars tilted $15°$ clockwise from vertical (A), or regularly placed bars randomly drawn from a selection of those tilted $0°$, $15°$, or $30°$ clockwise from vertical (B), or regularly placed identical bars tilted $15°$ clockwise from vertical (C). The z scores for the targets are listed immediately below each example.

would not be salient, since a response of only 10% above the average would be comparatively mediocre.

Of course, if the $\text{SMAP}_i < \bar{S}$, the target is not at all salient anyway, regardless of the variability in the background responses.

5.4.4.6 Saliency by feature contrast decreases with a decreasing density of input items

Contextual influences are mediated by intracortical connections, which are known to extend over only a finite range. These influences are thus reduced when the visual input density decreases, since this reduces the number of contextual neighbors within the reach of each visual input item via the intracortical connections. In turn, this reduces many saliency effects. For instance, it is apparent in the images in Fig. 5.27 that it is more difficult to segment two neighboring textures when the texture density is lower. This has also been observed in more rigorous behavioral experiments (Nothdurft 1985). The V1 model shows the same behavior (the right column of Fig. 5.27). The ease of the segmentation is reflected in the highest z score among the texture columns near the texture border. This z score is $z = 4.0$ in the densest example in Fig. 5.27 A, and it is $z = 0.57$ in the sparsest example in Fig. 5.27 D, which is quite difficult to segment without scrutiny.

To be concrete, iso-orientation suppression is weaker in sparser textures. Therefore, the dependence of V1-evoked responses on contextual inputs is weaker in sparser textures, and

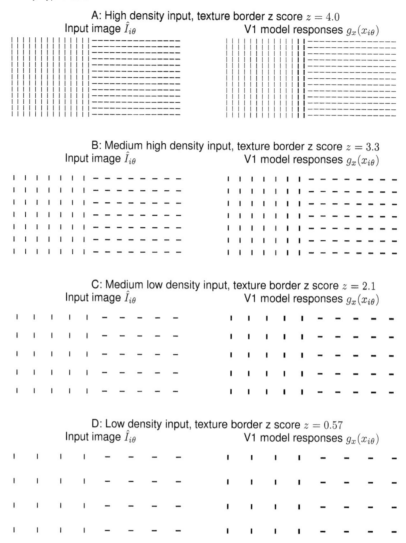

Fig. 5.27: Texture segmentation is more difficult in sparser textures. This is evident from examining the input images, and from the z scores of the texture columns at the borders that are obtained from the V1 model's responses (shown in the right column). All texture bars have input value $\hat{I}_{i\theta} = 2.0$. The average responses $g_x(x_{i\theta})$ to all texture bars are 0.15 (A), 0.38(B), 0.56(C), and 0.54(D).

so the response to a texture bar will be less sensitive to the proximity of this bar to a texture border. More specifically, the highlight at a texture border is caused by the difference between the contextual suppression of the border bars and that of the non-border bars (as explained in Fig. 5.17 A). When the distance between any two texture bars is longer than the longest intracortical connection, there should be zero iso-orientation suppression and so no saliency highlight at the texture border. For each background texture bar, the strength of iso-orientation suppression is largely determined by the number of iso-orientation neighbors that are within reach of the intracortical connections responsible for the suppression. Denser textures provide more iso-orientation neighbors to make this suppression stronger, making the texture border more salient. Indeed, in Fig. 5.27, the average response to all the texture bars is lowest in the densest texture and higher in sparser textures. This argument also applies to the saliency of a feature singleton in a homogenous background texture. Indeed, such a singleton is easier to find in denser textures (Nothdurft 2000).

5.4.4.7 How does a hole in a texture attract attention?

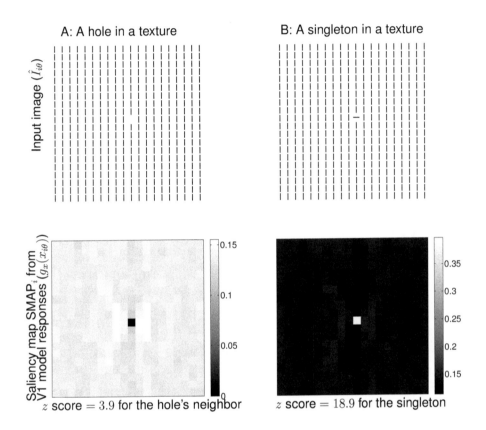

Fig. 5.28: Comparison between the conspicuousness of a hole (A) and a singleton (B). This figure uses the same format as in previous figures, except that the model responses are visualized by grayscale images, in which the gray level at each pixel i represents the maximum response magnitude $SMAP_i$ according to the scale bar on the right of the plot. (Gray scales rather than widths of the bars are used to visualize model responses, since otherwise the small but significant differences in the responses in A would be difficult to manifest as differences in the widths of the bars.) The two grayscale plots have different scale bars, although the average $SMAP_i$ values across i are similar around $SMAP_i \sim 0.136$. Much of the fluctuations in the responses further away from the hole or the singleton are caused by the input noise. In A, attention can be guided to the hole by first being attracted to its most salient neighbor.

It is apparent from Fig. 5.28 A that a hole in a texture is also conspicuous when the background is homogenous. Since a hole, or a missing bar in a texture, does not evoke any V1 response, how can its location attract attention? This can be understood from the observation that the hole still destroys the homogeneity of the texture. In particular, the bars near the hole are subject to weaker iso-orientation suppression because they have one fewer iso-orientation neighbor

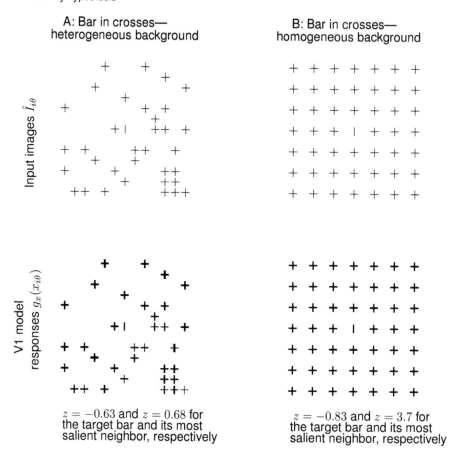

A: Bar in crosses—
heterogeneous background

B: Bar in crosses—
homogeneous background

Input images $\hat{I}_{i\theta}$

V1 model
responses $g_x(x_{i\theta})$

$z = -0.63$ and $z = 0.68$ for
the target bar and its most
salient neighbor, respectively

$z = -0.83$ and $z = 3.7$ for
the target bar and its most
salient neighbor, respectively

Fig. 5.29: Two additional examples of a target bar in distractor crosses (Li 2002), which are analogous to a hole in a texture as in Fig. 5.28 A. The distractor crosses are more regularly placed in B than A. Although the z score of the target vertical bar is higher in A than B, the most salient neighbor of the target bar has a higher z score in B than A. This underpins the observation that the target is more conspicuous in B than A, guided by the salient neighbor.

due to the hole. Although the suppression is reduced by only a small fraction, this fraction can generate a sizable z score when the background responses are sufficiently homogenous. In the example of Fig. 5.28 A, the mean and standard deviation of the responses over all the texture bars are 0.136 and 0.005 respectively. Meanwhile the response to the most salient neighbor of the hole is 0.155, giving this neighbor a z score of $z = (0.155 - 0.136)/0.005 = 3.9$. This salient neighbor attracts attention; although this attraction is weaker than that of an orientation singleton in the same background texture (Fig. 5.28 B). If the size of the attentional window is sufficiently large (as is suggested by experimental data (Motter and Belky 1998)), the hole can be contained within this window centered on the salient neighbor. Consequently, it may appear to awareness that our attention is attracted by the hole.

From the above interpretation, one prediction is that, in a visual search for a hole, gaze might land on a neighbor of the hole before making a corrective saccade to land on the target. Another prediction is that the conspicuousness of the hole can be manipulated by manipulating the input strength of its neighbors. In particular, the hole would be less conspicuous if its neighbors have slightly weaker input strength than those of the background texture elements.

This prediction has been supported by some preliminary observations (Zhaoping 2004, Zhou and Zhaoping 2010).

If the background texture is not so homogenous, as in the case of Fig. 5.64 B in which the non-homogeneity is caused by multiple holes randomly distributed in the texture, then the z score would be lower and the hole would be less conspicuous. In such cases, the missing input at the hole may be viewed as having been filled-in because it escapes attention. Note that this form of filling-in is not caused by a response to the hole, as would happen if there was a texture element at the location of the hole. This will be discussed more when analyzing Fig. 5.64.

Looking for a hole in a texture can be viewed as a special case of searching for a target lacking a feature that is present in the non-targets. Therefore it is natural that searching for a hole is more difficult than searching for a singleton target defined by the presence of a feature. This is seen in Fig. 5.28: the singleton target in the same texture generates a much higher z score. In the example of a target bar among crosses in Fig. 5.25 B, the target bar's z score $z = 0.8$ is in fact lower than the z score $z = 1.4$ of its left neighbor, although this more salient neighbor is not as salient as the horizontal bar in the target cross in Fig. 5.25 A. In general, the neighbors of a target lacking a feature present in the non-targets are not necessarily more salient than the target, because the actual responses depend on the contextual configurations of the visual input.

Figure 5.29 shows two additional examples of a bar among background crosses. In both examples, the z scores of the target location are negative, indicating that the responses to the target location are below the average responses (maximized at each location) at the locations of other visual items. Comparing Fig. 5.29 A and Fig. 5.29 B, the target has a higher z score in the former but appears to attract attention more strongly in the latter. This is because the most salient neighbor of the target has a higher z score $z = 3.7$ in the latter. The responses to the horizontal bars above and below the target vertical bar in Fig. 5.29 B are slightly higher than most of the other responses, because the missing horizontal bar in the target reduces the iso-orientation suppression on these neighboring horizontal bars by a small but significant fraction.

So far, all the examples of behavior of the model can be more or less intuitively and qualitatively understood from iso-feature suppression, which is the dominant intracortical interaction in V1. This intuition has been used to understand feature versus conjunction searches, search asymmetry between cross and bar, and the saliency effects by texture density, input heterogeneity, a hole, and the orientation contrast between textures. The model simulations merely confirm our intuitive understanding. However, it is desirable to test whether V1 mechanisms can also explain more complex and subtler saliency effects that cannot be intuitively or qualitatively understood from only the effects of iso-orientation suppression. Therefore, we next apply the V1 model to some complex examples, and we will see that these subtler saliency effects are often the net outcome from multiple balancing factors.

5.4.4.8 Segmenting two identical abutting textures from each other

Figure 5.30 A shows that the V1 model responses can even highlight a texture border between two identical textures. Perceptually, the texture border in Fig. 5.30 B seems more salient than that in Fig. 5.30 A, as if there were an illusory vertical border cutting between the two textures. However, the V1 model provides a z score that is somewhat larger for the texture border in Fig. 5.30 A. The reason for this may be that the perception of the illusory contour, rather than saliency, is more likely to arise in V2 rather than V1, as suggested by experimental data (von der Heydt et al. 1984, Ramsden, Hung and Roe 2001). The perception of the illusory contour could be mistaken as the saliency effect.

In each of these examples, all texture bars have about the same number of iso-orientation

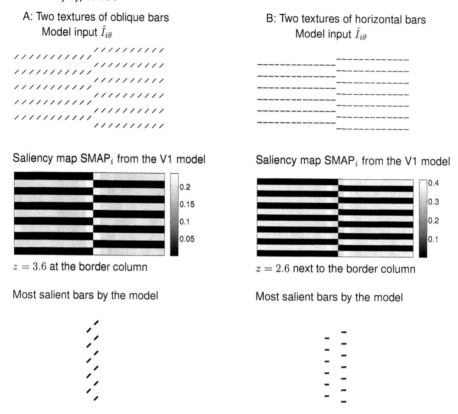

Fig. 5.30: Segmenting two identical textures by detecting the salient border where input statistics change. In both A and B, the two neighboring textures are identical but are displaced from each other vertically. The top two rows of the figure use the format in Fig. 5.28, with the grayscale at each pixel i in the middle row representing the SMAP_i value. The bottom row visualizes the most salient bars. All visible bars have $\hat{I}_{i\theta} = 2$ and $\hat{I}_{i\theta} = 3.5$ in A and B, respectively. In A, the most responsive locations are at the texture border; the bars there have $\text{SMAP}_i = 0.23$ against a background $\bar{S} = 0.203$. In B, the most responsive locations are one column away from the border, with $\text{SMAP}_i = 0.4$, against a background $\bar{S} = 0.377$.

neighbors regardless of their positions relative to the texture border. It is no longer obvious whether the border bars should be less subject to iso-orientation suppression. Nevertheless, the spatial configuration of the context of each texture bar depends on whether this bar is close to the texture border. This configuration is an aspect of the input statistics and determines the contextual influence. Although the net influence from all the iso-orientation neighbors is typically suppressive, some iso-orientation neighbors can give rise to colinear facilitation when they are co-aligned with the central bar. Apparently, the configurations of the contextual surrounds are such that the net suppression is weaker for a texture bar at or near the texture border.

In both of the examples in Fig. 5.30, the subtle changes in the spatial configuration of the surround are such that the model V1 responses to the locations near the borders are relatively higher. This is consistent with the experience of conspicuous borders.

5.4.4.9 More subtle examples of visual search asymmetry

Some example visual search asymmetries, shown in Fig. 5.31, are much more subtle than that in Fig. 5.25. In each example, the ease of the visual search changes slightly upon swapping the target and the distractor. For example, in Fig. 5.31 E, it is slightly easier to find an ellipse among circles than a circle among ellipses. Readers can examine them to see which target-distractor condition seems easier for finding the target.

The asymmetry between bars and crosses in Fig. 5.25 involves a clear case of the absence versus presence of a basic feature, namely orientation, in the target. Both the neurons and intracortical connections in V1 are tuned to this feature dimension. Hence, via V1 mechanisms, this orientation feature drives a strong asymmetry in an obvious manner. By contrast, there is not a clear V1 feature that distinguishes a circle and an ellipse. If the sizes of the circle and ellipse are comparable to those of the CRFs of the V1 neurons which are not tuned to orientation (see Fig. 3.32), then the circle and the ellipse should evoke similar response levels, if anything perhaps slightly favoring the circle (i.e., opposite to the direction of the asymmetry). Most individual V1 neurons only respond to the line or curve segments in the circles and ellipses, according to their own oriented receptive fields. The V1 model treats the circle as eight line segments, oriented in four different orientations, in a particular spatial arrangement; and the ellipse as ten line segments in five different orientations. None of the ten line segments in the ellipse is oriented sufficiently differently from all the line segments in the circle, and vice versa. So the asymmetry between circle and ellipse cannot be realized in the model in terms of an obvious differential presence of a feature in one versus the other target.

The V1 model indeed generates the asymmetry. The largest z score for the bars in the target ellipse among circles is larger than that for the bars in the target circle among ellipses. The asymmetry arises as the net result of many sources of contextual influences, including iso-orientation suppression, colinear facilitation, and general surround suppression, which is independent of the orientations concerned. None of these contextual influences obviously weighs for or against the direction of the asymmetry. This is similar to the two examples in Fig. 5.30, where the relatively higher saliencies at the texture borders arise not from an obvious change in iso-orientation suppression but from a net result of subtle changes in both contextual suppression and contextual facilitation.

The V1 model was applied to all the search images in Fig. 5.31 and their random variations (such as the random changes in the spatial arrangements of the visual items). As in all the examples of the model application, the model parameters had already been fixed beforehand by the requirements from model calibration and dynamic behavior (described in Sections 5.4.2 and 5.4.3). The z score of the target is calculated as the maximum z score among the line segments which make up the target. In all the five examples of visual search asymmetry, the directions of the asymmetry predicted by the V1 model agree with those observed behaviorally (Treisman and Gormican 1988, Li 1999a).

Note that if V1 responses are not responsible for these subtle examples of asymmetry, then a prediction from the V1 saliency hypothesis on the direction of the asymmetry would only match the behavioral direction by chance. Whether the predicted directions match the behavioral ones in all the five examples provides a stringent test of the V1 saliency hypothesis.

Conventional psychological theories (Treisman and Gormican 1988, Wolfe 2001) presume that each target-distractor pair that exhibits search asymmetry implies the presence and absence of a preattentive basic feature in the easier and the more difficult, respectively, search of the pair. For example, since the ellipse is easier to find among circles than vice-versa, one should conclude that the ellipse has an "ovoid" feature that is absent in a circle (i.e., the ellipse is seen as a departure from the circle). This, of course, leads to feature proliferation. That the

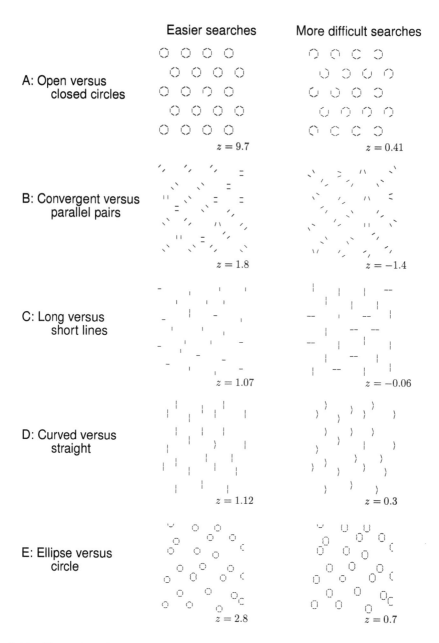

Easier searches   More difficult searches

A: Open versus
   closed circles
z = 9.7   z = 0.41

B: Convergent versus
   parallel pairs
z = 1.8   z = −1.4

C: Long versus
   short lines
z = 1.07   z = −0.06

D: Curved versus
   straight
z = 1.12   z = 0.3

E: Ellipse versus
   circle
z = 2.8   z = 0.7

Fig. 5.31: Five pairs of images for the subtle examples of visual search asymmetry. They resemble those studied by Treisman and Gormican (1988). The V1 model can account for the directions of all these asymmetries. Stimulus patterns ($\hat{I}_{i\theta}$) are shown with the targets' z scores (as the largest z score for the bar segments which comprise the target) from the model marked underneath. Adapted with permission from Zhaoping, L., Theoretical understanding of the early visual processes by data compression and data selection, *Network: Computation in Neural Systems*, 17(4): 301–334, Fig. 10, copyright © 2006, Informa Healthcare.

Box 5.1: **Some examples of visual search asymmetries are due to higher level mechanisms**

Another example of search asymmetry is shown in Fig. 5.32: a target letter "N" is more difficult to find among mirror images of "N"s than the reverse (Frith 1974). The letter "N" and its mirror image differ only in the direction of the oblique bar in their shape, and there are no known mechanisms in V1 to break this mirror-reflection symmetry. To explain this asymmetry, conventional psychological theories suggest that a more familiar letter "N" lacks a novelty feature which is present in its mirror image. It seems difficult to envision that V1 mechanisms might account for any such feature based on object familiarity or novelty.

However, later observations (Zhaoping and Frith 2011) indicate that there is little asymmetry between the reaction times for gaze to reach the respective targets, the letter "N" and its mirror image, during the visual searches. This suggests that the search

Find a target "N" amomg its mirror images

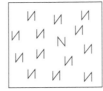

Find a mirror image of "N" among "N"s

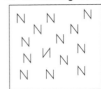

Fig. 5.32: Object shape confusion, not saliency, makes the search on the left more difficult (Zhaoping and Frith 2011).

asymmetry does not result from the initial visual selection by bottom-up saliency of the targets. (Note that either target is very salient, having an uniquely oriented oblique bar in the image.) Instead, the asymmetry would originate from confusing the target as a non-target, since all items in the search image have the same viewpoint-invariant shape. This confusion is of the kind we saw in Fig. 1.4, occurring at the shape recognition stage *after* a visual input location is selected. Apparently, this confusion is more effective when the target is "N" in its more familiar, rather than the unfamiliar, view; the familiarity makes the shape recognition faster, allowing an earlier onset of confusion during the task execution (Zhaoping and Frith 2011).

The asymmetry between the "N" and its mirror image as the target-distractor pair is an example in which the reaction time $RT_{task} = RT_{saliency} + RT_{top\text{-}down \; selection} + RT_{other}$ (see equation (5.3)) to report the search target is not indicative of the relative degree of bottom-up saliency of the targets in the two searches. This is because this RT's non-saliency component, $RT_{top\text{-}down \; selection} + RT_{other}$, is not a constant between the two different searches. Among all the known examples of visual search asymmetry, it has yet to be worked out which examples are mainly caused by bottom-up saliency processes to test the V1 saliency hypothesis more extensively.

V1 model can successfully predict all these asymmetries suggests that it is unnecessary to introduce a feature for each such target-distractor, since an asymmetry can also be caused by the complexity of V1 circuit dynamics in response to the spatial configurations of primitive bar/edge segments in visual inputs.

The V1 model also suggests that it is not necessary to have custom neural detectors for a circle, ellipse, cross (see Fig. 5.25), curvedness, parallelness, closure, or perhaps even a face, in order to exhibit saliency effects associated with these input shapes. V1 detectors for primitive bars and edges, and the associated intracortical interactions, enable V1 responses to exhibit response properties which can be specific to spatial configurations of bar/edge

Input images $\hat{I}_{i\theta}$ to the V1 model

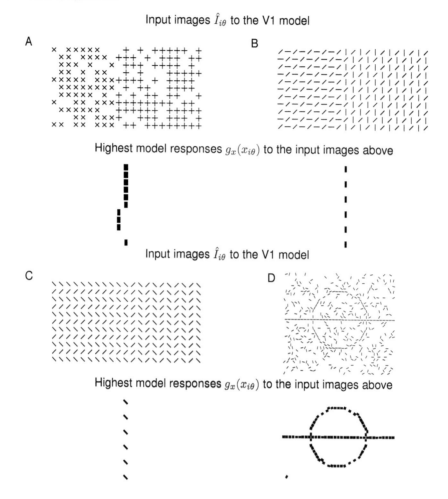

Fig. 5.33: Four examples of V1 model's response to highlight the input locations where input statistics deviate from the statistics of the context.

segments. In principle, these configurations could include many meaningful object shapes such as those of crosses and ellipses.

However, since the V1 model is a poor imitation of the real V1, the z scores of the search targets predicted by the V1 model in the stimuli in Fig. 5.31 can be quantitatively quite different from what is suggested by the behavioral data. A better test of whether V1 mechanisms can account for the asymmetries is to examine the response of the real V1 to the stimuli concerned while preventing top-down interference.

5.4.4.10 Complex examples where V1 responses highlight input locations where input statistics deviate from that of the context

We have argued that places where visual input statistics deviate significantly from those of the context are often predominant examples of salient locations. These locations are often at boundaries of objects, such as the border between two textures. Figures 5.23 and 5.27 show that the V1 model works well to highlight these input deviations at the borders between simple textures, each made of iso-oriented bars. Figure 5.30 shows that this also works in two examples when the borders are between identical textures. Figure 5.33 shows that it also works in more complex examples.

The V1 model can highlight borders between two textures that are stochastic (Fig. 5.33 A), that involve checkerboard patterns of elements (Fig. 5.33 BC), or that have identical individual elements, but different second order correlations between texture elements (Fig. 5.33 C). Like the real V1 (Li, Piëch and Gilbert 2006), the V1 model can also highlight a contour in a noisy background (Fig. 5.33 D). The V1 saliency hypothesis and the behavior of the V1 model in Fig. 5.33 ABC suggest that the real V1 should also detect such complex input deviations from surrounding statistics. This suggestion is consistent with subsequent observations by functional magnetic resonance imaging of the cortex (Joo, Boynton and Murray 2012) in response to complex arrays of Gabor patterns.

5.4.5 Other effects of the saliency mechanisms—figure-ground segmentation and the medial axis effect

The foreground (figure) of a visual scene typically attracts attention more strongly than the background (often called just the "ground"). When both are textures made from iso-oriented bars, the figure has been observed to evoke higher V1 responses than the ground (Lamme 1995, Lee, Mumford, Romero and Lamme 1998), a phenomenon known as the *figure-ground effect*. Equally, V1 responses to a figure grating can sometimes be higher when it is presented against a background grating having a different (e.g., orthogonal) orientation, versus when a blank background is used (Sillito et al. 1995); this is termed *cross-orientation enhancement*. Finally, V1 responses to the central or medial axis of a figure texture can sometimes be higher than its responses to other regions of the figure that are not borders (Lee et al. 1998). This is called the *medial axis effect*.

Medial axes can be useful for characterizing deformable object shapes, e.g., to represent a human body as a stick figure. Hence, the figure-ground and medial axis effects appear to provide tantalizing hints that V1 could play a role in figure-ground segmentation and higher-order object representation—operations that go beyond highlighting salient border regions (which is the *border effect*). In this section, we show that these effects can be explained (Zhaoping 2003) as side effects of V1 saliency mechanisms that stress image locations where input translation invariance breaks down. This analysis explains why such effects are weaker than the border effect, and, furthermore, predicts that these side effects occur only for particular sizes of figures.

For example, when the figure is sufficiently small, as in Fig. 5.34 B, the responses its bars evoke should all be higher than those of the larger background texture, since each bar is part of the border. This is analogous to the pop-out of an orientation singleton. This "figure-ground" effect is observed electrophysiologically when the RF of the recorded neuron is in the figure region. However, it was predicted (Li 2000a) that this effect should disappear when the figure is large enough such that the RF of the recorded neuron is no longer a part of the texture border. This prediction was subsequently confirmed physiologically (Rossi, Desimone and Ungerleider 2001), and is illustrated in the V1 model simulations shown in Fig. 5.34 C–E.

In fact, the higher responses to the texture border enhance the iso-orientation suppression suffered by the figure regions immediately next to the border; this region is thus termed the *border suppression region*. We may refer to the suppression of the border suppression region by the salient border as the *border's neighbor effect*. Hence, when the figure size is such that the center of the figure is also within the border suppression region flanked by two or more border sides, as in Fig. 5.34 C, the response to the center of the figure should even be weaker than the typical responses to the background texture.

However, when the figure is large enough, as in Fig. 5.34 E, the distance between the center

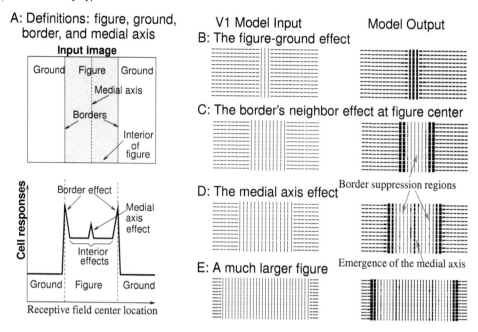

Fig. 5.34: Appropriate sizes of the figures evoke figure-ground effect, border's neighbor effect, and medial axis effect as side effects of the *border effect*, which is the relatively higher response to texture borders caused by V1 saliency mechanisms. A defines the terms. B, C, D, and E show V1 model responses to the figure texture for various figure sizes. The figure-ground effect, defined as higher responses to the figure, emerges in B when the figure size is small enough, making the whole figure its own border. In C–E, responses to the border suppression regions—the figure texture region next to the figure border—are low due to the *border's neighbor effect*, the stronger iso-orientation suppression from the salient figure border. D manifests the medial axis effect, since the axis escapes the suppression from the borders by virtue of (1) being sufficiently far from both borders and (2) being subject to weaker iso-orientation suppression from the two flanking border suppression regions that have themselves been suppressed by the salient borders. E shows the V1 model responses when the figure size is much larger. Adapted with permission from L. Zhaoping, V1 mechanisms and some figure-ground and border effects, *Journal of Physiology-Paris*, 97(4–6): 503–515, figure 1 and figure 4, copyright © 2003, Elsevier.

of the figure and either border is much longer than the typical length of the intracortical V1 connections responsible for iso-orientation suppression. In this case, the response to the figure center becomes indistinguishable from typical responses to the background texture. Feedback from higher visual areas could subsequently enhance the responses to the figure center, as suggested physiologically (Lamme, Rodriguez-Rodriguez and Spekreijse 1999, Scholte, Jolij, Fahrenfort and Lamme 2008), and could partly explain behavioral aspects of the figure-ground effect. However, modulation of V1 responses by the immediate context, which is responsible for the border effect, remains intact after V2 inactivation (Hupé et al. 2001) and is present whether the animal is awake or under anaesthesia (Knierim and Van Essen 1992, Nothdurft et al. 1999).

Figure 5.34 D shows how the medial axis effect can arise as a further consequence of the border effect. The response to the medial axis will be enhanced when the figure is just the right size such that the following two conditions are satisfied: first, the figure center is out of

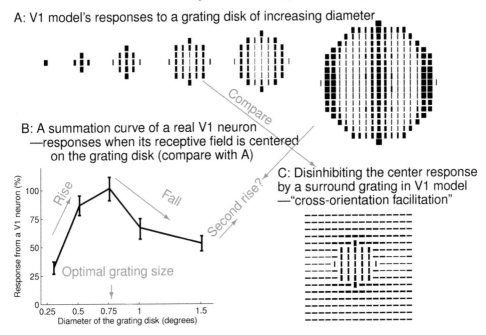

A: V1 model's responses to a grating disk of increasing diameter

B: A summation curve of a real V1 neuron
—responses when its receptive field is centered
on the grating disk (compare with A)

Compare

Rise

Fall

Second rise?

Optimal grating size

Response from a V1 neuron (%)

100

75

50

25

0

0.25   0.5   0.75   1   1.5
Diameter of the grating disk (degrees)

C: Disinhibiting the center response
by a surround grating in V1 model
—"cross-orientation facilitation"

Fig. 5.35: V1 responses to a disk grating, and cross-orientation facilitation (Zhaoping 2003). A: V1 model's responses to coarse-sampled disk gratings. As the disk size increases, the center of the grating changes from being part of the border (with a high response), to being part of the border suppression region (with a suppressed response), to being included in the emerging medial axis (with an enhanced response). B: Responses of a real V1 neuron to a disk grating as a function of the diameter of the disk (this curve is called a summation curve). The model's response to the largest disk grating in A predicts a second rise in this summation curve associated with the medial axis effect. C: When the central disk grating is larger than the optimal size, such that the center of the disk is in the border suppression region (the fourth grating in A), a surrounding grating can suppress the responses to the border of the central grating by general (orientation-unspecific) surround suppression, thereby releasing the central response from the border suppression. This may explain some physiological observations (Sillito et al. 1995) of cross-orientation facilitation. Data in B from Jones, H. E., Grieve, K. L., Wang, W., and Sillito, A. M. Surround suppression in primate V1, *Journal of Neurophysiology*, 86(4): 2011–28, 2001.

reach of iso-orientation suppression caused by both lateral borders of the figure; and second, the figure center is within reach of the relatively weaker suppression occasioned by the border suppression regions associated with these borders. The resulting suppression of the media axis could then be weaker than that suffered by typical background texture bars. Therefore, the medial axis effect should only be observed for certain figure sizes, as is indeed the case physiologically (Lee et al. 1998).

The dependence of the neural response on the size of the figure is also manifest in the way that a V1 neuron's response to a disk grating varies with the disk's diameter (Zhaoping 2003); see Fig. 5.35. When the orientation of the grating is that preferred by the neuron whose RF is centered on the disk, the neural response increases with the diameter for small diameters when the disk is smaller than the classical receptive field of the neuron, and then the response decreases with the diameter when the center of the disk moves out of the disk border and into the border suppression zone. The overall relationship between response level and disk diameter is called a *summation curve*.

When the disk grating is even larger, its center moves out of the border suppression zones, and the response it evokes should therefore rise again. This is the medial axis effect. It should lead to a second rise in the summation curve; see Fig. 5.35 B. This prediction is supported by recent physiological data.[16]

Consider the case that the disk grating is somewhat larger than the optimal size (where the summation curve peaks), such that its center is in the border suppression region associated with the disk border. Then, adding a surrounding grating of a different (e.g., orthogonal) orientation should suppress the responses to the border of the grating disk, via general, orientation-unspecific, surround suppression. In turn, this should lessen the iso-orientation suppression by the disk grating's border onto the disk center. In other words, the surround grating disinhibits the response to the center of the figure grating; see Fig. 5.35 AC. This could explain physiological observations of cross-orientation enhancement (Sillito et al. 1995), which is indeed often observed when the figure grating is somewhat larger than the optimal size where the summation curve peaks.

5.4.6   Input contrast dependence of the contextual influences

Contextual influences are dependent on the contrast or strength of visual inputs, with suppression decreasing as the input contrast decreases. As will be explained later in equation (5.94), this is because the inhibitory interneurons, which mediate the suppression, are less sensitive to inputs from the excitatory cells at lower input contrast. Since salience depends on iso-orientation suppression, this implies that orientation singletons and texture borders are less salient at low input contrast. This is why most simulations of the V1 model use medium or high input contrast, as indeed is also true of most behavioral experiments into saliency.

Since figure-ground and medial axis effects arise from border effects, they are also weaker at lower input contrast. Consequently, the radius of the grating where a neuron's summation curve (Fig. 5.35 B) peaks tends to be larger when the input contrast is weaker. This is true in the model and is also observed physiologically.

5.4.7   Reflections from the V1 model

In total, building and applying the V1 model has led to the following conclusions.

1. It is possible to build a V1 model which can simultaneously satisfy two requirements: (1) reproducing the contextual influences that are observed physiologically; and (2) being able to amplify selective deviations from homogeneity in the input, without hallucinating heterogeneous responses to homogenous visual inputs. Therefore, V1 mechanisms are plausible neural substrates for saliency.

2. Under the V1 saliency hypothesis, the responses of the V1 model to representative visual inputs produce saliency maps that are consistent with subjective visual experience and previous behavioral observations.

3. The V1 model confirms the intuition that iso-orientation suppression is the dominant mechanism underlying various saliency effects. Such effects include the qualitative distinction between feature and conjunction searches, the greater saliency of locations where feature contrast is greater, and the dependence of saliency on visual input density and heterogeneity. The V1 model also demonstrates that other intracortical interactions,

---

[16] Private communication from Kenneth D. Miller (2013), who collaborated with Dan Rubin and Stephen Van Hooser on an investigation which revealed these data.

including colinear facilitation and general, feature-unspecific, contextual suppression, also play essential roles in shaping saliency. This is especially the case for visual inputs that are more similar to typical visual inputs and so are more complex than those used in feature searches.

4. The V1 model can signal saliency at locations of complex shapes such as ellipses and crosses, even though there is no V1 cell tuned to such shapes. This reaffirms our understanding that selection of a visual location can occur before recognition of inputs or objects at this location. One may even ask whether V1 mechanisms can also contribute to attentional attraction of, e.g., a face, which, like a cross, is a particular spatial configuration of image elements (like bars and patches of luminance and color) that activate V1 neurons.

5. Saliency mechanisms have side effects, and these can be understood.

Recall from Fig. 1.1 that an important role for a model is to be an intermediary between a theoretical hypothesis and experimental data. This role can be fulfilled, for example, by demonstrating the theory in particular instances or by fitting data to a particular manifestation of the theory. In our current example, the theory is the V1 saliency hypothesis, the data are observations of bottom-up visual selection, and the V1 model played a role of verifying the theory by testing the ability of a restricted set of V1 mechanisms to account for some behavioral observations. The restrictions include: (1) that the model contains only neurons tuned to spatial locations and orientations; (2) that the model ignores many physiological details; and, (3) that all the model neurons have the same receptive field size. These restrictions imply that the model can only be tested against a restricted set of saliency data. For example, the model is not expected to account well for the saliency of a scale singleton because it omits the multiscale property of V1. Nevertheless, we can ask whether the V1 model can be successful when applied to an appropriately restricted set of data.

The success of the V1 model with the restricted data suggests that one can extrapolate and generalize beyond the current model. For example, the model can be extended to include model V1 neurons tuned to feature dimensions other than orientation, such as color, motion direction, scale, disparity, and ocularity (tuning to ocularity can be defined as a relative sensitivity to inputs from the two eyes). One can expect, and verify, that iso-feature suppression should work in the same way for these feature dimensions as it does for orientation. Indeed, one such extension has been carried out for the case of color (Li 2002, Zhaoping and Snowden 2006). Similarly, although the model has mostly been applied to synthetic images (with a few exceptions (Li 1998a, Li 1999b, Li 2000b)), one can expect, and test, that the theory also applies to more realistic visual inputs such as those from natural scenes.

V1 neurons that have large and unoriented receptive fields (see Fig. 3.32) can also be included in the model. Extending iso-feature suppression to the feature of the round shapes of these receptive fields (of a given scale), one would expect mutual suppression between all nearby neurons of this class. These neurons are likely to play an important role in saliency for round shapes or patches, perhaps contributing to the attentional attraction of a face (of a similar size to the receptive fields).

While a model can be used to build confidence in a theory, the theory should be able to stand despite a model's inaccuracies or fail despite a model's fit to many details. Furthermore, a theory should be ultimately tested against experiment data rather than just against model simulations (Fig. 1.1). The test in Section 5.3 of the predicted high saliency of an ocular singleton is an example of a direct test of the theory without the aid of the V1 model. We next turn to more such tests.

5.5 Additional psychophysical tests of the V1 saliency hypothesis

This section presents additional non-trivial predictions and their behavioral tests. Each of these predictions exploits either a distinctive characteristic of V1 physiology, to test the specifically V1 nature of the hypothesis, or a qualitative difference between the V1 saliency hypothesis and conventional ideas about saliency. The V1 model is not necessary as an intermediary between the hypothesis and the link between physiology and saliency behavior. This is because the hypothesis is so explicit, and because the knowledge about V1's physiology is extensive, that it is easy to predict from physiology to behavior via the medium of the hypothesis. It is also typically easier to test behavioral predictions using psychophysical experiments than to test physiological predictions by electrophysiological experiments.

One prediction is based on the feature-blind, "auction" nature of selection by saliency that is depicted in Fig. 5.6. It states that texture segmentation should be more severely impaired than traditional theories would imply if a task-irrelevant texture is superposed. This prediction arises because the saliency value at a location can be hijacked by the irrelevant features whose evoked V1 responses are higher than those of the task-relevant features. This prediction cannot be derived from the traditional saliency frameworks that are depicted in Fig. 5.5, so it allows us to test them against the V1 saliency hypothesis.

The second prediction arises from colinear facilitation, which is a characteristic of V1 physiology. Via the saliency hypothesis, this implies how the ease of texture segmentation can be influenced by the degree of spatial alignment between the texture bars.

The third prediction arises from the observation that whereas some V1 neurons are tuned simultaneously to color and orientation (see Section 3.6.6.3), and some V1 neurons are tuned simultaneously to orientation and motion direction, very few V1 neurons are tuned simultaneously to color and motion direction (Horwitz and Albright 2005); see Section 3.6.9. Based on this, according to the V1 saliency hypothesis, we can predict whether the RTs for finding a feature singleton that is unique in two feature dimensions should be shorter than the statistically appropriate combinations of the RTs for finding feature singletons that are unique in just one of the two feature dimensions.

These three predictions are qualitative, in that they anticipate that the RT in one situation (RT_1) should be shorter than another RT in a different situation (RT_2). They do not predict a precise value for the difference $RT_2 - RT_1$. The fourth prediction is quantitative, based on the assumption that there are no V1 neurons (or an insignificant number of V1 neurons) tuned simultaneously to the three feature dimensions: color, orientation, and motion direction. It derives a precise relationship among the RTs for a single observer for finding feature singletons that differ from a uniform background in one, two, or three of these dimensions, and it uses this relationship to predict the whole distribution of one of these RTs from the distributions of the other RTs. This is a quantitative prediction that is derived without any free parameters. Therefore, the V1 saliency hypothesis could be easily falsified if it is incorrect, since there is no freedom to fit data to the prediction. We will show an experimental confirmation of this prediction.

5.5.1 The feature-blind "auction"—maximum rather than summation over features

According to the V1 saliency hypothesis, the saliency of a location is signaled by the highest response to this location, regardless of the feature preference of the neurons concerned. For instance, the cross among bars in Fig. 5.25 A is salient due to the response of the neuron tuned to the horizontal bar, with the weaker response of a different neuron tuned to the vertical bar being ignored. Therefore, the "less salient features" at any location are invisible to bottom-up saliency or selection, even though they are visible to attention attracted to the location due to the response to another feature at the same location. This leads to the following prediction: a visual search or a segmentation task can be severely interfered with by task-irrelevant stimuli that evoke higher V1 responses than the task-relevant stimulus does at the same location. This is because the task-irrelevant stimulus, which makes the task-relevant stimulus invisible to saliency, will determine the saliency values of the stimulus and thereby control bottom-up selection. This attention control by task-irrelevant stimuli makes the task performance inefficient.

Figure 5.36 shows texture patterns that illustrate and test the prediction. Pattern A has two iso-orientation textures, activating two populations of neurons, one tuned to left tilts and one to right tilts. Pattern B is a checkerboard, evoking responses from another two groups of neurons tuned to horizontal and vertical orientations.

With iso-orientation suppression, neurons responding to the texture border bars in pattern A are more active than those responding to the background bars, since each border bar has fewer iso-orientation neighbors to exert contextual iso-orientation suppression (as explained in Fig. 5.7). For ease of explanation, let us say that the responses from the most active neurons to a border bar and a background bar are 10 and 5 spikes/second respectively. This response pattern renders the border location more salient, making texture segmentation easy. Each bar in pattern B has as many iso-orientation neighbors as a texture border bar in pattern A, and so it also evokes a response of (roughly) 10 spikes/second.

The composite pattern C, which is made by superposing patterns A and B, activates all neurons responding to patterns A and B. For simplicity (and without changing the conclusions), we ignore interactions between neurons tuned to different orientations. Therefore, the neurons tuned to oblique orientations respond roughly to the same degree as they do to A alone (we call these relevant responses); and the neurons tuned to horizontal or vertical orientation respond roughly to the same degree as they do to B alone (irrelevant responses). This implies that all texture element locations evoke the same maximum response of 10 spikes/second, which is the largest of the relevant and irrelevant responses to each location.

According to the feature-blind auction framework of the V1 hypothesis, it is this maximum response to a location x, $\text{SMAP}(x) = \max_{x_i \approx x} O_i$ (from equation (5.4)) that determines the saliency $\text{SMAP}(x)$ at that location, where x_i is the center of the receptive field (which covers location x) of neuron i giving the response O_i. Thus, by the V1 hypothesis, all locations are equally salient (or non-salient), without a saliency highlight at the texture border. Therefore texture segmentation is predicted to be much more difficult in C than A, as indeed is apparent in Fig. 5.36. Any saliency signal associated with the task-relevant, oblique, bars is swamped by the uniform responses to the task-irrelevant horizontal and vertical bars.

If saliency was instead determined by the summation rule $\text{SMAP}(x) \propto \text{sum}_{x_i \approx x} O_i$ (this is a modification of equation (5.4)), responses to the various orientations at each texture element location in pattern C could sum to preserve the border highlight as $20 = 10 + 10$ (spikes/second) against a background of $15 = 10 + 5$ (spikes/second). This predicts that texture segmentation should be easy (Zhaoping and May 2007). This summation rule is the basis of traditional saliency models (Itti and Koch 2001, Wolfe et al. 1989) (depicted in Fig. 5.5). By the maximum rule, it may seem a waste not to include the contributions of "less salient features" to obtain a "more informative" saliency measure of locations, as in the summation rule. However, reaction times for locating the texture border[17] confirmed the prediction of the maximum rather than the summation rule; see Fig. 5.36 D.

---

[17]In the experiment (Zhaoping and May 2007), each stimulus display consisted of 22 rows × 30 columns of items (of single or double bars) on a regular grid with unit distance 1.6° of visual angle. Observers were instructed to press

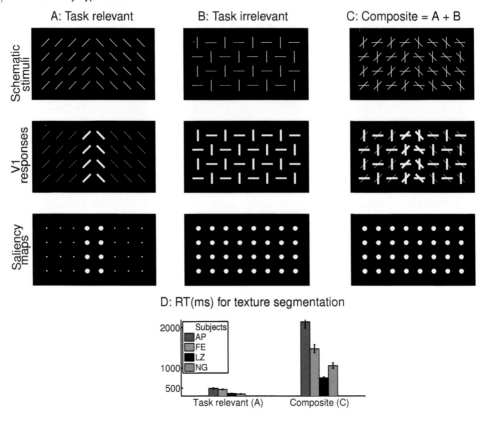

Fig. 5.36: Psychophysical confirmation of the maximum rule used by the V1 saliency hypothesis, instead of the summation rule used by traditional models of saliency. A, B, C: Schematics of texture stimuli (extending continuously in all directions beyond the portions shown), each followed by schematic illustrations of V1's responses and a saliency map, formulated as in Fig. 5.14. Each dot in the saliency map scales with the maximum V1 response to the corresponding location, rather than the sum of all V1 responses there. Every bar in B, or every texture border bar in A, experiences less iso-orientation suppression. The composite stimulus C, made by superposing A and B, is predicted to be difficult to segment, since the task-irrelevant features from B interfere with the task-relevant features from A, giving no saliency highlights to the texture border. D: Reaction times of four observers (subjects) for the texture segmentation task using stimuli similar to A and C. Adapted from Zhaoping, L. and May, K. A., Psychophysical tests of the hypothesis of a bottom-up saliency map in primary visual cortex, *PLoS Computational Biology*, 3(4):e62, Fig. 1, copyright © 2007, Zhaoping, L. and May, K. A.

The two halves of Fig. 5.36 C have very different second order statistics of visual inputs.[18] This is an example for which the breakdown in the translation symmetry of the input statistics, even though it involves low (i.e., second) order statistics, does not lead to high saliency.

---

a left or right button as soon as possible to indicate whether the texture border was in the left or right half of the display.

[18] These two halves can be easily distinguished by a standard texture segregation model (Bergen and Landy 1991), which works by examining whether two textures have identical visual inputs in matching orientation channels.

5.5.1.1 Further discussion and exploration of interference from task-irrelevant features

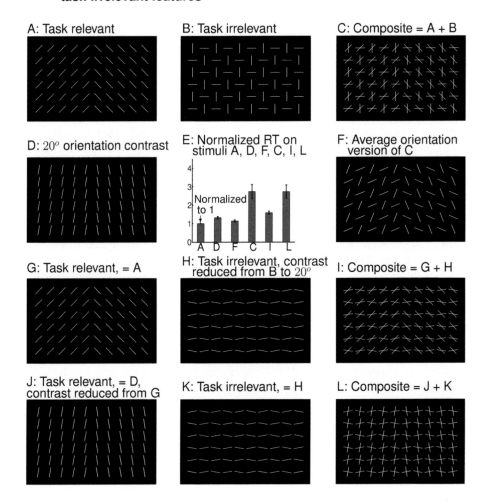

A: Task relevant

B: Task irrelevant

C: Composite = A + B

D: 20° orientation contrast

E: Normalized RT on stimuli A, D, F, C, I, L

F: Average orientation version of C

G: Task relevant, = A

H: Task irrelevant, contrast reduced from B to 20°

I: Composite = G + H

J: Task relevant, = D, contrast reduced from G

K: Task irrelevant, = H

L: Composite = J + K

Fig. 5.37: Further illustrations of the interference wrought by task-irrelevant features. A, B, and C are the schematic stimuli from Fig. 5.36. D is a version of A, with bars being $10°$ from vertical, thus reducing the orientation contrast at the texture border to $20°$. F is derived from C by replacing each texture element of two intersecting bars by one bar whose orientation is the average of the two intersecting bars. G, H, and I are derived from A, B, and C by reducing the orientation contrast (to $20°$) in the interfering bars; each is $10°$ from horizontal. J, K, and L are derived from G, H, and I by reducing the task-relevant contrast to $20°$. E plots the average of the normalized reaction times for three subjects, on stimuli A, D, F, C, I, and L (which were randomly interleaved within a session). Each normalized RT is obtained by dividing the actual RT by that of the same subject for stimulus A. Error bars denote standard error of the mean. Adapted from Zhaoping, L. and May, K. A., Psychophysical tests of the hypothesis of a bottom-up saliency map in primary visual cortex, *PLoS Computational Biology*, 3(4):e62, Fig. 2, copyright © 2007, Zhaoping, L. and May, K. A.

One might wonder whether the composite texture elements in Fig. 5.36 C (each of which comprises two intersecting bars) are acting (for saliency) as single bars having the average orientation at each location; see Fig. 5.37 F. This would make the relevant orientation feature noisy and impair performance. The control experiment reported in Fig. 5.37 E demonstrates that this would not have caused so large an impairment. The RT for this "orientation-averaged" stimulus (Fig. 5.37 F) is at least 37% shorter than that for the composite stimulus (Fig. 5.36 C).

Box 5.2: Alternative accounts for the interference by task-irrelevant features

One may seek alternative explanations for the observed interference by task-irrelevant features that is predicted by the V1 saliency hypothesis. For instance, in Fig. 5.36 C, one may assign a new feature type, let us call it feature "X", to "two bars crossing each other at 45°." Then, each texture element is this "X" at a particular orientation, and each texture region in Fig. 5.36 C is a checkerboard of two orientations of "X". So the segmentation could be more difficult in Fig. 5.36 C, in the same way that it is more difficult to segment the texture of "ABABAB" from "CDCDCD" in a stimulus pattern "ABABABABABCDCDCDCDCD" than it is to segment "AAA" from "CCC" in "AAAAAACCCCCC." This approach of creating new feature types to explain hitherto unexplained data could of course be extended to accommodate other cases, such as double-feature conjunctions (e.g., color-orientation conjunction), triple, quadruple, and other multiple feature conjunctions, or even complex stimuli like faces. It is not clear how long this list of new feature types would have to be. By contrast, the V1 saliency hypothesis is a parsimonious account, since it explains all these data without invoking additional free parameters or mechanisms. It was also used in Section 5.4 to explain visual searches for, e.g., a cross among bars or an ellipse among circles without any detectors for crosses or circles/ellipses. Our aim should be to explain the most data with the fewest necessary assumptions or parameters. Additionally, the V1 saliency hypothesis is based on substantial physiological findings. When additional data reveal the limitation of V1 for bottom-up saliency, the search for additional mechanisms for bottom-up saliency can be guided by following conclusions suggested by visual pathways and cortical circuits in the brain (Shipp 2004).

From the analysis above, one can see that the V1 saliency hypothesis also predicts a decrease in the interference if the irrelevant feature contrast is reduced. This is evident from comparing Fig. 5.37 GHI with Fig. 5.37 ABC, and it is confirmed by the RT data (Fig. 5.37 E). The neighboring irrelevant bars in Fig. 5.37 I have more similar orientations, inducing stronger iso-feature suppression. Consequently, their evoked responses are decreased, say, from 10 to 7 spikes/second. (Colinear facilitation is also greater for this stimulus; however, iso-orientation suppression dominates colinear facilitation physiologically, so the net effect is a decreased response to each task-irrelevant bar.) Consequently, the relevant responses to the texture border, i.e., the border highlights, are no longer submerged by the irrelevant responses. The irrelevant responses interfere less with the relevant responses, although the fact that the former (at 7 spikes/second) still dominate the latter (5 spikes/second) in the background implies that there would still be some interference (with the border highlight being reduced from 5 to 3 spikes/second).

Analogously, interference can be increased by decreasing the task-relevant orientation contrast at the texture border. This is demonstrated by comparing Fig. 5.37 JKL and Fig. 5.37 GHI, and it is also confirmed in experimental data (Fig. 5.37 E). Reducing the relevant feature contrast makes the relevant responses to the texture border weaker, say from 10 to 7 spikes/second. Consequently, these relevant responses are even more vulnerable to being submerged by the irrelevant responses. Therefore, interference is stronger in Fig. 5.37 L than in Fig. 5.37 I.

In sum, the existence and strength of the interference depend on the relative levels of responses to task-relevant and task-irrelevant features, with these responses depending on the corresponding feature contrasts and direct input strengths. When the relevant responses dictate saliency everywhere, and when their values are barely affected by the presence or absence of the irrelevant stimuli, there should be little interference. Conversely, when the irrelevant responses dictate saliency everywhere, interference with the visual selection required for the task is strongest. When the relevant responses dictate the saliency value at the location of the texture border but not in the texture background, the degree of interference is intermediate. In both Fig. 5.37 C and Fig. 5.37 L, the irrelevant responses (approximately) dictate the saliency everywhere, so the texture borders are predicted to be equally non-salient. This is confirmed in the data (Fig. 5.37 E). However, the RT performance of subjects for Fig. 5.37 CL varies widely, perhaps because the bottom-up saliency is so weak for these two stimuli that subject-specific top-down factors contribute significantly to the RTs.

Additional data (Zhaoping and May 2007) confirmed analogous predictions from the V1 theory, such as predictions of interference by irrelevant color with orientation-based tasks.

5.5.1.2 Contrasting the V1 saliency hypothesis and traditional frameworks for saliency

As mentioned, in traditional models, the saliency values in the master map come from $\text{SMAP}(x) \propto \text{sum}_{x_i \approx x} O_i$, i.e., summing the activations in various feature maps, each based on one visual feature such as a particular color or orientation; see Fig. 5.5.

The V1 saliency theory differs from traditional theories, partly because one of its motivations was to understand V1. It also aims for fast computation, and thus it calculates saliency without requiring separate feature maps or decoding of input features. Indeed, many V1 neurons are tuned to more than one feature dimension (Livingstone and Hubel 1984, Lennie 1998) (e.g., to orientation and motion direction), making it impossible that separate groups of V1 cells represent separate feature dimensions or separate feature maps.

In contrast, the traditional theories were motivated by explaining behavioral data. They do not specify the cortical location of the feature maps or the master saliency map, or aim for algorithmic simplicity. For example, although the summation rule seems natural for the feature-blind saliency, it is in practice more complex to implement. The value of $\text{SMAP}(x) \propto \text{sum}_{x_i \approx x} O_i$ is sensitive to exactly which activations O_i should be included in the sum, considering that the receptive fields of different neurons have different sizes, shapes, center locations, and sharpness of their boundaries. If the boundary of the receptive field of neuron i is vague, and if this vague boundary barely covers location x, should the neural activation O_i be included in the summation for $\text{SMAP}(x)$? Should the summation rule be implemented as a weighted summation of neural activations, and if so, what weights should be used? The summation step has to be carried out in order to find the most salient location to direct attention to. In comparison, the most salient location by maximum rule can be easily found by finding the neuron with the highest neural response.

From the perspective of the feature-blind auction process, feature maps, and thus a master map, are unnecessary. The observations in Fig. 5.36 thus motivate the framework of visual selection without separate feature maps.

5.5.2 The fingerprints of colinear facilitation in V1

Two nearby V1 neurons can facilitate each other's responses if their preferred bars or edges are aligned with each other such that these bars or edges could be parts of a single smooth contour (Nelson, and Frost 1985, Kapadia et al. 1995). Although such colinear facilitation is much weaker than the iso-feature suppression which is mainly responsible for singleton pop-out in bottom-up saliency, it also has consequences for saliency behavior.

Figure 5.38 shows the first such consequence. Figures 5.38 A and 5.38 B both have two orientation textures with a $90°$ contrast between them. The texture borders pop out automatically. However, in Fig. 5.38 B, the vertical texture border bars in addition enjoy full colinear facilitation, since each has more colinear neighbors than the other texture border bars in either Fig. 5.38 A or Fig. 5.38 B. The vertical texture border bars are thus more salient than other border bars. We call a texture border made of bars that are parallel to the border a colinear border. In general, for a given orientation contrast at a border, a colinear border is more salient than other borders (Li 1999b, Li 2000b). This is also seen in the output of the V1 model—compare Fig. 5.22 and Fig. 5.23 A.

Hence, one can predict that it takes longer to detect the border in Fig. 5.38 A than in Fig. 5.38 B. This prediction was indeed confirmed (Fig. 5.38 E, in the same experiment reported in Fig. 5.36 D). A related observation (Wolfson and Landy 1995) is that it is easier to discriminate the curvature of a colinear than a non-colinear texture border.

Since both texture borders in Fig. 5.38 A and Fig. 5.38 B are so salient that they require very short RTs, and since RTs cannot be shorter than a certain minimum for each subject, even a large difference between the saliencies of these borders will only be manifest as a small difference in the RTs to detect them. However, the saliency difference can be unmasked by the interference caused by task-irrelevant bars. This is shown in Fig. 5.38 CD, involving the superposition of a checkerboard pattern of task-irrelevant bars tilted $45°$ away from the task-relevant bars. This manipulation is the same as that to induce interference in Fig. 5.36. Again, for convenience, let us refer to relevant bars as leading to relevant responses from relevant neurons, and similarly for the irrelevant bars. As argued in Fig. 5.36, the irrelevant responses in the background texture region are higher than the relevant responses, and so they dictate the saliency of the background in both Fig. 5.38 C and Fig. 5.38 D. Meanwhile, the RT for detecting the texture border in Fig. 5.38 D is much shorter than that for Fig. 5.38 C, since the interference is much weaker in Fig. 5.38 D.

For concreteness, let us say, as we did when analyzing Fig. 5.36, that the relevant responses in Fig. 5.38 C are 10 spikes/second at the border and 5 spikes/second in the background, and that they are 15 spikes/second and 5 spikes/second, respectively, in Fig. 5.38 D. Meanwhile, the irrelevant responses are roughly 10 spikes/second at all locations in both Fig. 5.38 CD (as in Fig. 5.36). At the colinear vertical border bars in Fig. 5.38 D, the relevant responses (15 spikes/second) are much higher than the irrelevant responses (10 spikes/second), and so are less vulnerable to being submerged. However, because the irrelevant responses dictate and raise the background saliency, the irrelevant texture still causes interference by reducing the ratio between the maximum responses to the border and background, from a ratio of $15/5 = 3$ to $15/10 = 1.5$. This interference is much weaker than that in Fig. 5.38 C, whose border-to-background response ratio is reduced from 10/5 to 10/10.

Figure 5.39 demonstrates another, subtler, fingerprint of colinear facilitation. The task-relevant stimulus component is as in Fig. 5.38 A. The task-irrelevant stimulus consists of horizontal bars in Fig. 5.39 A and vertical bars in Fig. 5.39 B. Away from the border, both relevant and irrelevant bars lack orientation contrast. Thus, they have comparable iso-orientation suppressions and comparable final responses there. However, some readers might notice that the border in Fig. 5.39 A is slightly easier to notice than in Fig. 5.39 B. This can be understood by considering three types of intracortical interactions: iso-orientation suppression between the relevant responses, a general contextual suppression between the relevant and irrelevant responses regardless of the orientation preferences of the neurons, and colinear facilitation between irrelevant responses. The effects of the first two interactions are the same in Fig. 5.39 A and Fig. 5.39 B, but the effect of the third differs between the two stimuli.

Consider iso-orientation suppression from the relevant responses to the texture border to the relevant responses to the border suppression region next to the border (see Fig. 5.34 BC for an illustration of the border suppression region). Because the relevant responses to the

A: Two textures of oblique bars   B: A vertical texture and a horizontal one

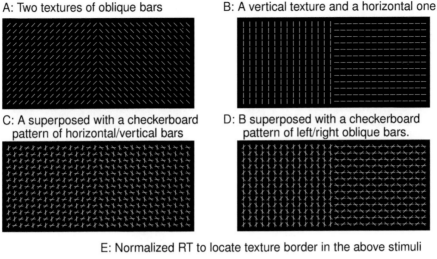

C: A superposed with a checkerboard
pattern of horizontal/vertical bars

D: B superposed with a checkerboard
pattern of left/right oblique bars.

E: Normalized RT to locate texture border in the above stimuli

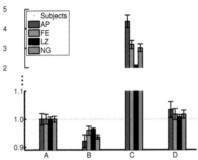

Fig. 5.38: Fingerprint of colinear facilitation in V1: a texture border with texture bars parallel to the border (called a colinear border) is more salient (Zhaoping and May 2007). A and B: Stimulus patterns for texture segmentation; each contains two neighboring orientation textures with 90° orientation contrast at the texture border. The texture border in B appears more salient. The interference by task-irrelevant bars in C (as schematized in Fig. 5.36) is analogous to that in D. Nevertheless, the interference is much less effective in D since the more salient, task-relevant, colinear border bars are less vulnerable to interference. E: Normalized RT of subjects to localize the texture borders, given by the ratio of actual RT to each subject's (trial averaged) RT for stimulus condition A (493, 465, 363, 351 ms for AP, FE, LZ, and NG, respectively).

border bars are the strongest among the relevant responses, the iso-orientation suppression that they exert is also strongest, making the relevant responses in the border suppression region weaker than those further away from the border. In turn, these weaker relevant responses in the border suppression region generate less general suppression on the local irrelevant neurons, making the local irrelevant responses slightly higher than the other irrelevant responses. Hence, in the border suppression region, the relevant responses are slightly weaker, and the irrelevant responses slightly stronger, than their respective values in the homogenous region further away from the border. In this region, the irrelevant responses therefore dictate the local saliencies; furthermore, because these saliencies are slightly higher than those in the background, they induce interference for the task by reducing the relative saliency of the texture border. Figures 5.39 A and 5.39 B differ in the direction of the colinear facilitation

A: Texture segmentation with
translation invariant horizontal bars

B: Same as A, but with
translation invariant vertical bars

 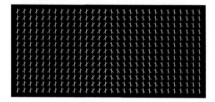

C: Normalized RT for A and B

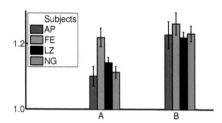

Fig. 5.39: Differential interference by irrelevant bars due to colinear facilitation (Zhaoping and May 2007). Stimuli A and B are made by superposing task-irrelevant horizontal (A) or vertical (B) bars on top of the relevant stimulus from Fig. 5.38 A. C: Normalized reaction times to locate the texture border in A and B (using the same subjects and presented the same way as in Fig. 5.38). The RT for B is significantly longer than that for A ($p < 0.01$) in three out of four subjects. By matched sample t-test across subjects, the RT for B is significantly longer than that in A ($p < 0.01$). For each subject, RTs for both A and B are significantly longer ($p < 0.0005$) than that for Fig. 5.38 A.

among the irrelevant bars. This direction is perpendicular to the border in Fig. 5.39 A and parallel with it in Fig. 5.39 B. Mutual facilitation between neurons tends to equalize their response levels, thereby smoothing away the response variations in the direction of colinear facilitation. Consequently, the local peaks in the irrelevant responses in the border suppression region should be somewhat smoothed away in Fig. 5.39 A but not in Fig. 5.39 B. This predicts stronger interference in Fig. 5.39 B than in Fig. 5.39 A, as indeed is confirmed by the segmentation RTs; see Fig. 5.39 C.

5.5.3 The fingerprint of V1's conjunctive cells

Figure 5.40 shows that a bar that is unique in color, orientation, or in both color and orientation can pop out of (at least statistically) identical backgrounds made of bars with uniform orientation and color. We call the first two cases single-feature singletons and single-feature pop-outs, and the third case a double-feature singleton in a double-feature pop-out. If it takes a subject a reaction time of $RT_C = 500$ ms to find the color singleton, and another reaction time of $RT_O = 600$ ms to find the orientation singleton, one may wonder whether the reaction time RT_{CO} for finding the double-feature singleton should be 500 ms or less.

Let us consider an extremely ideal case, when $RT_C = 500$ ms and $RT_O = 600$ ms always, without any stochasticity or trial to trial fluctuations. Then, if

$$RT_{CO} = \min(RT_C, RT_O) = 500 \text{ ms},$$

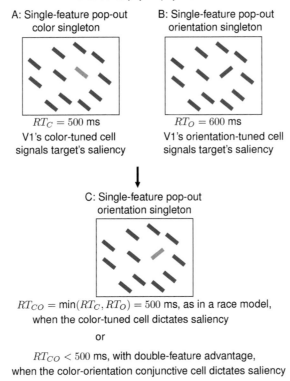

A: Single-feature pop-out
color singleton

B: Single-feature pop-out
orientation singleton

$RT_C = 500$ ms

V1's color-tuned cell
signals target's saliency

$RT_O = 600$ ms

V1's orientation-tuned cell
signals target's saliency

C: Single-feature pop-out
orientation singleton

$RT_{CO} = \min(RT_C, RT_O) = 500$ ms, as in a race model,
when the color-tuned cell dictates saliency

or

$RT_{CO} < 500$ ms, with double-feature advantage,
when the color-orientation conjunctive cell dictates saliency

Fig. 5.40: Schematic of single- and double-feature pop-out in color and/or orientation. A: The saliency of the color (C) singleton is dictated by the response of the cell tuned to a red color, which is the only cell free from iso-feature suppression for this input stimulus. B: Similarly, the saliency of the orientation (O) singleton is dictated by the response of the cell tuned to its orientation. C: The color-orientation (CO) double-feature singleton highly activates all three cell types: color-tuned, orientation-tuned, and conjunctive color-orientation–tuned cells; the most activated among them should dictate the singleton's saliency. Consider the simplest case when A, B, and C do not differ in the neural responses to the background bars; furthermore, let the color (only)-tuned cell respond identically to the singletons in A and C, and let the orientation (only)-tuned cell respond identically to the singletons in B and C. When the RTs to find the singletons are, for example, $RT_C = 500$ ms for the color singleton and $RT_O = 600$ ms for the orientation singleton, whether RT_{CO} for the color-orientation singleton is less than or equal to $\min(RT_C, RT_O) = 500$ ms depends on whether the conjunctive cell is the most active cell responding to the CO singleton and is more active than its activation in A and B.

we call RT_{CO} an outcome of a *race model*, as if RT_{CO} is the result of a race between two racers with times RT_C and RT_O, respectively. If $RT_{CO} < \min(RT_C, RT_O)$, we say that there is a double-feature advantage. The idealization to treat RTs as deterministic, rather than stochastic, will be removed later when we work with real behavioral RTs. Meanwhile, for the ease of explanation, we use this idealization without changing the conclusions. We will explain below that the V1 saliency hypothesis predicts a double-feature advantage when V1 has cells tuned conjunctively (or simultaneously) to features in both feature dimensions—in this example, color and orientation.

V1 has conjunctive neurons tuned to color (C) *and* orientation (O), or to orientation *and* motion direction (M). However, experiments have observed few V1 neurons tuned to color *and* motion direction (Horwitz and Albright 2005). Therefore, the V1 saliency hypothesis predicts that a double-feature advantage should exist for a color-orientation (CO) double feature and a motion-orientation (MO) double feature, but this double-feature advantage should be absent for a color-motion (CM) double feature. It is known that V2, receiving inputs from V1, has neurons selective to all three types of feature conjunctions: CO, MO, and CM (Gegenfurtner, Kiper and Fenstemaker 1996). Thus, if V2, or visual areas that are further downstream, are responsible for bottom-up saliency, then one would predict double-feature advantage for all three types of double-feature singletons. Therefore, we refer to the prediction of a double-feature advantage for CO and MO singletons but not for CM singleton as a V1 fingerprint.

Below, we provide a rigorous argument for the prediction, starting with the example of CO singleton. For intuition, though, consider the activity of neurons whose relevant tuning is solely to C or O, or conjunctively to CO. Due to iso-feature suppression, a C (only)-tuned neuron should, by definition, respond identically to the CO singleton and a C singleton, but it should be less activated by an O singleton. Similarly, an O (only)-tuned neuron should respond identically to the CO singleton and an O singleton, but it should be less activated by a C singleton. Finally, the response from a CO-tuned neuron to a CO singleton should be no less than its response to a C or O single-feature singleton. Thus, among all neurons, whether they are tuned to C, O, or CO, the highest response to the CO singleton should be no less than the highest response to the C singleton or the O singleton. Provided that, for different singletons, the statistical properties, e.g., the average and standard deviation, of the V1 neural responses to the background bars are sufficiently similar, the V1 saliency hypothesis predicts that the CO singleton will be no less salient than the C and O singletons. Since a singleton's saliency should relate inversely to the RT for finding it, $RT_{CO} \leq \min(RT_C, RT_O)$ follows.

For ease of notation, in this section we eschew the usual notation O_i for the output or response of a V1 neuron indexed by i. Instead, let α denote an input bar, and let C_α, O_α, or CO_α, respectively, denote the highest response to this bar from a population of neurons tuned solely to C, or O, or conjunctively to CO (and these neurons have their RFs cover the location of this input bar). The value α can be $\alpha = C, O$, or CO for a C, O, or CO singleton or $\alpha = B$ for a background non-singleton bar. Hence (C_C, O_C, CO_C) is the triplet of responses to a color singleton, (C_O, O_O, CO_O) to an orientation singleton, $(C_{CO}, O_{CO}, CO_{CO})$ to a CO double-feature singleton, and (C_B, O_B, CO_B) to one of the many bars in the background. The maximum rule states that the saliency of the bar indexed by $\alpha = C, O, CO$, or B is

$$\text{SMAP}_\alpha \equiv \max(C_\alpha, O_\alpha, CO_\alpha). \tag{5.15}$$

Note that, among the neurons responding to the bar α, the number of neurons tuned to C may not be the same as the number of neurons tuned to O (or CO). However, this does not matter in our formulation since C_α, O_α, or CO_α marks the highest response to the bar from a subpopulation of neurons having a particular tuning property regardless of the number of neurons in this subpopulation.

For a neuron tuned only to color or orientation, its response should be independent of any feature contrast in other feature dimensions. Hence

$$C_{CO} = C_C, \quad O_{CO} = O_O, \tag{5.16}$$
$$C_O = C_B, \quad O_C = O_B. \tag{5.17}$$

(Note that, although, e.g., $C_O = C_B$, this C neuron is still tuned to color.) Furthermore, iso-color and iso-orientation suppression, and the strong saliency of the singletons, imply

$$C_C > C_B \quad \text{and} \quad O_O > O_B. \tag{5.18}$$

Generalizing iso-feature suppression to the conjunctive cells, we expect

$$CO_{CO} \geq CO_O, \quad CO_{CO} \geq CO_C, \tag{5.19}$$
$$CO_O \geq CO_B, \quad CO_C \geq CO_B. \tag{5.20}$$

Since the singletons $\alpha = C, O,$ or CO pop out, we have

$$\text{SMAP}_\alpha \gg \text{SMAP}_B \quad \text{for } \alpha = C, O, \text{ or } CO. \tag{5.21}$$

Since $O_C = O_B$ (by equation (5.17)), then

$$\text{SMAP}_C = \max(C_C, O_C, CO_C) = \max(C_C, O_B, CO_C).$$

This, combined with $\text{SMAP}_C \gg \text{SMAP}_B$ and $\text{SMAP}_B \geq O_B$, leads to

$$\text{SMAP}_C = \max(C_C, CO_C), \quad \text{and analogously,} \quad \text{SMAP}_O = \max(O_O, CO_O). \tag{5.22}$$

Then we can derive

$$\text{SMAP}_{CO} = \max(C_{CO}, O_{CO}, CO_{CO}) \tag{5.23}$$
$$= \max(C_C, O_O, CO_{CO}) \qquad \{\text{by equation (5.16)}\} \tag{5.24}$$
$$= \max(C_C, O_O, \max(CO_{CO}, CO_C, CO_O)) \qquad \{\text{by equation (5.19)}\} \tag{5.25}$$
$$= \max(\max(C_C, CO_C), \max(O_O, CO_O), CO_{CO}) \tag{5.26}$$
$$= \max(\text{SMAP}_C, \text{SMAP}_O, CO_{CO}) \qquad \{\text{by equation (5.22)}\} \tag{5.27}$$
$$\geq \max(\text{SMAP}_C, \text{SMAP}_O). \tag{5.28}$$

In the above, each $\{...\}$ is not part of the equation, but it contains text pointing out the equation used to arrive at the mathematical expression to its left. Equations (5.27) and (5.28) mean that the double-feature singleton CO can be more salient than both the single-feature singletons C and O if there are conjunctive cells whose response CO_{CO} has a non-zero chance of being larger than both SMAP_C and SMAP_O to dictate the saliency of the CO singleton (this is achieved when CO_{CO} is larger than O_O, C_C, CO_C, and CO_O). When there is no conjunctive cell CO, we can simply make $CO_\alpha = 0$ in the above equations, eliminating its ability to dictate the saliency value. Then, inequality (5.28) becomes an equality:

$$\text{SMAP}_{CO} = \max(\text{SMAP}_C, \text{SMAP}_O) \quad \text{when there is no conjunctive CO neuron.} \tag{5.29}$$

The saliency SMAP_α is taken as determining the RT_α for detecting the singleton α via a monotonic function $f(.)$:

$$RT_\alpha = f(\text{SMAP}_\alpha), \quad \text{such that } f(x_1) > f(x_2) \text{ when } x_1 < x_2, \tag{5.30}$$

by the definition of saliency. Equations (5.28–5.30) then lead to

$$RT_{CO} = \min[RT_C, RT_O],$$
$$\text{the race model, when there is no conjunctive CO cell.} \tag{5.31}$$

$$RT_{CO} = \min[RT_C, RT_O, f(CO_{CO})]$$
$$\leq \min[RT_C, RT_O], \tag{5.32}$$
$$\text{double-feature advantage, with conjunctive CO cells.}$$

Hence, without conjunctive CO cells, RT_{CO} to detect a CO double-feature singleton can be predicted by a race model between two racers SMAP_C and SMAP_O with their respective racing times as RT_C and RT_O for detecting the corresponding single-feature singletons. With conjunctive cells, there may be a double-feature advantage. The RT_{CO} can be shorter than predicted by the race model between the two racers SMAP$_C$ and SMAP$_O$, since there is now a third racer, CO_{CO}, with its RT as $f(CO_{CO})$; see equation (5.32). Note that, when we say the race model for the RT of a double-feature singleton, we mean a race between *only two* racers whose racing times are the RTs for the two corresponding single-feature singletons, *without* any additional racers.

Now let us remove the deterministic idealization and treat the RTs and the V1 neural responses as stochastic, as they are in reality. The neural responses in single trials can be seen as being drawn from a probability distribution function (pdf). Thus, SMAP$_C$, SMAP$_O$, and CO_{CO} are really all random variables drawn from their respective pdfs, making SMAP$_{CO}$ another random variable. Accordingly, the RTs are also random variables by their respective pdfs. In particular, when the race model holds, i.e., when there is no CO conjunctive cell, Monte Carlo simulation methods based on equation (5.31) can be used to predict RTs for the double-feature singleton as follows. Let us denote RT_{CO}(race) as the RT_{CO} by the race model. We randomly sample one RT each from the distribution of RT_C and that of RT_O, respectively, and call these samples RT_C(sample) and RT_O(sample). This gives a simulated sample, RT_{CO}(race sample), of RT_{CO}(race) according to the race model as

$$RT_{CO}(\text{race sample}) \equiv \min[RT_C(\text{sample}), RT_O(\text{sample})] \qquad (5.33)$$

by equation (5.31). Using a sufficient number of such samples, one can generate a distribution of RT_{CO}(race). We can then test whether human RTs to detect a CO singleton is statistically shorter than RT_{CO}(race) predicted by the race model.

The response CO_{CO} of the CO neuron to the CO singleton is also stochastic, and its corresponding (would-be) RT $f(CO_{CO})$ also follows a pdf. Averaged over trials, according to equations (5.27) and (5.32), as long as this additional racer CO_{CO} has a non-zero chance of being larger than both SMAP$_C$ and SMAP$_O$, the trial-averaged RT_{CO} should be shorter than the one predicted by the race model. Note that this double-feature advantage can happen even when the average response of the CO neurons are no larger than those of the C and O neurons.

We also note that, even when there is no CO cell, the race-model predicted RT_{CO}(race) can be on average (over the trials) shorter than both the average RT_C and the average RT_O (unless the distributions of RT_C and RT_O do not overlap), since RT_{CO}(race) is always the shorter one of the two single-feature RT samples.

The derivation and analysis above can be analogously applied to the double-feature singletons MO and CM, involving the motion-direction feature. Hence, the fingerprints of V1's conjunctive cells are predicted to be as follows: compared to the RT predicted by the race model from the RTs for the corresponding single-feature singletons, RTs for the CO and MO double-feature singletons should be shorter, but the RT for the CM double-feature singleton should be the same as predicted.

This fingerprint was tested (Koene and Zhaoping 2007) in a visual search task for a singleton bar (among 659 background bars) regardless of the features of the singleton, using stimuli as schematized in Fig. 5.40. Each bar is about $1 \times 0.2°$ in visual angle, takes one of the two possible isoluminant colors (green and purple) against a black background, is tilted from vertical in either direction by a constant amount, and moves left or right at a constant speed. All background bars are identical to each other in color, tilt, and motion direction, and the singleton pops out by virtue of its unique color, tilt, or motion direction, or any combination of these features. The singleton has an eccentricity $12.8°$ from the initial fixation point at the center of the display in the beginning of each search trial. Subjects have to press a button as soon as possible to indicate whether the singleton is in the left or right half of the display,

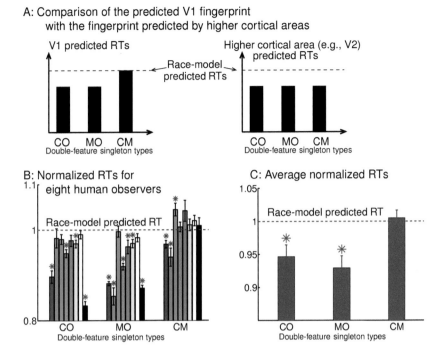

Fig. 5.41: Testing the fingerprint of V1 conjunctive cells in bottom-up saliency. A: The predicted V1 fingerprint, depicted in the left plot, compared with the prediction by V2/higher cortical areas (right plot). The dashed lines indicate the value of the predicted RT for the double-feature singletons by the race model. If bottom-up saliency in these tasks were computed by higher cortical areas, double-feature advantage, by an RT shorter than predicted from the race model, should occur in all double-feature singletons CO, MO, and CM. By contrast, V1 predicts a double-feature advantage for CO and MO singletons but not for the CM singleton. This is because V1 has conjunctive CO and MO cells but no CM cells, but V2/higher areas have all the three cell types. B,C: Experimental findings by Koene and Zhaoping (2007). The plotted bars show normalized mean RTs across trials for each subject (in B) or the average of these means across subjects (in C). The normalization factor comes from the predictions of the race model. Error bars indicate the standard errors of the means. In C, by matched sample two-tailed t-tests, the observed RT_{CO} and RT_{MO} for the double-feature singletons CO and MO are significantly shorter than those predicted by the race model, whereas the observed RT_{CM} for the double-feature singleton CM is not significantly different from the race-model prediction. In B and C, a "*" above a data bar indicates a significant difference between the RT data and that predicted from a race model.

regardless of the singleton conditions, which are randomly interleaved and unpredictable. Trials with incorrect button presses or with RTs shorter than 0.2 seconds or longer than three standard deviations above the average RTs were excluded from data analysis.

The experiment by Koene and Zhaoping (2007) was designed such that there was a symmetry between the two possible feature values in each feature dimension, i.e., between the isoluminant green and purple colors, between left-tilt and right-tilt orientations, and between the leftward and rightward movements. Hence, for our derivation, the highest responses from the V1 neurons to a bar is assumed to be regardless of whether the bar takes one or the other of the two possible feature values, e.g., C_C (the highest response of V1 neurons tuned to color only to a color singleton bar) is regardless of whether a color singleton is the unique green bar among purple bars or the other way around (even though the highest response to a green singleton is from a cell tuned to green and that to a purple singleton is from a cell tuned to purple). This feature symmetry allows us to pool together the RT data for symmetry-related singletons, e.g., a green singleton and a purple singleton, in the data analysis.

Figure 5.41 BC plot the observed RTs for the double-feature singletons, normalized by the RTs predicted by the race model. For example, for each observer, a distribution of RT_{CO}(race) can be predicted from the histograms of the behavioral RT_C and RT_O data from this observer, using the Monte Carlo method above. The normalized RT of this observer for the CO singleton is his/her behavioral RT to detect the CO singleton divided by the average RT_{CO}(race) predicted by the race model for the same observer. Therefore, a double-feature advantage is manifest in a normalized RT smaller than unity, and a race-model predicted RT gives a normalized RT equal to unity. The results confirm the predicted V1 fingerprint. A double-feature advantage for the CO singleton has been previously observed (Krummenacher, Müller and Heller 2001). Similarly, a lack of double-feature advantage has also been observed when both features are in the orientation dimension (Zhaoping and May 2007), consistent with the V1 saliency hypothesis, since there is no V1 cell conjunctively tuned to two different orientations.

Note that traditional models of saliency would predict that a double-feature singleton should, if anything, be more salient than the single-feature singletons. In particular, recall from Section 5.1.3 that the traditional models should predict that, in the experiment by Koene and Zhaoping, a CO singleton should be more salient than an O singleton in the same way, and by the same amount, as a CM singleton is more salient than an M singleton. These predictions arise from the separation between feature maps and from the summation rule. The observations shown in Fig. 5.41 refute these predictions.

5.5.4 A zero-parameter quantitative prediction and its experimental test

Equation (5.31) shows that if there were no V1 neuron tuned simultaneously to both C and O, then one could quantitatively predict RT_{CO} from RT_C and RT_O. Hence, for example, an $RT_C = 500$ ms and an $RT_O = 600$ ms could together predict $RT_{CO} = \min(RT_C, RT_O) = 500$ ms from the race model, without any free parameters (see Fig. 5.40). As both RT_O and RT_C follow their respective probability distributions, the probability distribution of RT_{CO} could also be quantitatively derived without any parameters, by drawing random samples of RT_{CO} as $RT_{CO} = \min(RT_O, RT_C)$ from sample pairs (RT_O, RT_C).

However, because V1 has neurons tuned conjunctively to C and O, the measured probability distribution of RT_{CO} is different from this race-model prediction derived just from the distributions of RT_C and RT_O. This is shown in Fig. 5.42 B.

We mentioned in the previous section that few CM V1 neurons have been found that are tuned conjunctively to color (C) and motion direction (M). Hence, when RT_{CM} is the RT to find a double-feature singleton unique in both color (C) and motion direction (M) and RT_M is the RT to find a singleton unique in motion direction, the distribution of RT_{CM} is the same as that of $\min(RT_C, RT_M)$. This is consistent with the observation that, across observers, the average RT_{CM} is indeed the same as that predicted by the race model (by drawing random samples of RT_{CM} as $RT_{CM} = \min(RT_C, RT_M)$) from sample pairs (RT_C, RT_M). However, a closer observation of Fig. 5.41 B suggests that $RT_{CM} < \min(RT_C, RT_M)$ (averaged across trials) for two out of the eight observers. Indeed, findings by different researchers differ as to the existence of CM cells in V1 (Horwitz and Albright 2005, Michael

1978)—perhaps they exist in some observers but are just less numerous than CO and MO double-feature conjunctive cells.

Even if there are some CM neurons in V1, there has yet to be a report of triple feature conjunctive cells, CMO, which are simultaneously tuned to color (C), motion direction (M), and orientation (O) features. (Note that the CMO cells should be a subset of the CM cells, and hence they cannot be more numerous than the CM cells.) Indeed, this dearth is as expected from the input signal-to-noise considerations discussed in Section 3.6.9 that preclude substantial numbers of V1 neurons from being tuned to multiple feature dimensions. Hence, we can use exactly the same idea as the one that led to $RT_{CO} = \min[RT_O, RT_C]$ in equation (5.31) to derive the following parameter-free equation (Zhaoping and Zhe 2012b) from the V1 saliency hypothesis:

$$\min(RT_C, RT_M, RT_O, RT_{CMO}) = \min(RT_{CM}, RT_{CO}, RT_{MO}). \tag{5.34}$$

In this expression, CO, MO, and CM are the conjunctions of two features indicated by the respective letters (C, M, and O), and CMO is the triple feature conjunction of C, M, and O, and RT_α is the RT to find a single-, double-, or triple-feature singleton denoted by α, which can take values $\alpha = C, M, O, CM, MO, CO,$ or CMO.

To derive equation (5.34), we proceed as in the last section. Let $C, M, O, CM, CO,$ and MO denote V1 neurons tuned to a single or double (conjunctive) feature(s) indicated by the respective letters, and let $C_\alpha, M_\alpha, O_\alpha, CM_\alpha, CO_\alpha,$ and MO_α be the responses of these neurons to singleton α or a background item $\alpha = B$. Then, as in equation (5.15), the saliency at the location of item α is

$$\text{SMAP}_\alpha = \max(C_\alpha, M_\alpha, O_\alpha, CM_\alpha, CO_\alpha, MO_\alpha). \tag{5.35}$$

Just as in equations (5.16–5.20), the neurons should respond more vigorously to a singleton whose feature uniqueness matches more of its preferred tuning, and its response should be indifferent to feature contrast in a dimension to which it is not tuned. Hence, statistically,

$$
\begin{aligned}
C_C = C_{CO} = C_{CM} = C_{CMO} &> C_B = C_O = C_M = C_{MO}, \\
O_O = O_{CO} = O_{MO} = O_{CMO} &> O_B = O_C = O_M = O_{CM}, \\
M_M = M_{CM} = M_{MO} = M_{CMO} &> M_B = M_C = M_O = M_{CO}, \\
CM_{CM} = CM_{CMO}, \quad CM_M = CM_{MO}, \quad CM_C &= CM_{CO}, \quad CM_B = CM_O, \\
CO_{CO} = CO_{CMO}, \quad CO_C = CO_{CM}, \quad CO_O &= CO_{MO}, \quad CO_B = CO_M, \\
MO_{MO} = MO_{CMO}, \quad MO_M = MO_{CM}, \quad MO_O &= MO_{CO}, \quad MO_B = MO_C.
\end{aligned}
\tag{5.36}
$$

Furthermore, since the singletons are very salient, we have, for $\alpha = C, M, O, CM, CO, MO$ and CMO,

$$\text{SMAP}_\alpha > \text{SMAP}_B = \max(C_B, M_B, O_B, CM_B, CO_B, MO_B). \tag{5.37}$$

From these equations, one can derive

$$
\begin{aligned}
\max(\text{SMAP}_C, \text{SMAP}_M, \text{SMAP}_O, \text{SMAP}_{CMO}) = \\
\max(\text{SMAP}_{CM}, \text{SMAP}_{CO}, \text{SMAP}_{MO}),
\end{aligned}
\tag{5.38}
$$

which is the saliency equivalent of the RT equation (5.34) due to the monotonically inverse relationship between SMAP_α and RT_α. The above equation can be verified by substituting equation (5.35) for each SMAP_α in equation (5.38), using properties in equation (5.36) and noting that

$$\max[\max(a, b, \ldots), \max(a', b' \ldots), \ldots] = \max[a, b, \ldots, a', b', \ldots] \tag{5.39}$$

for various quantities $a, b, a',$ and b', etc.

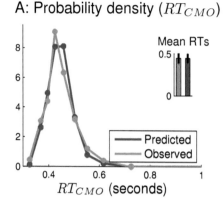

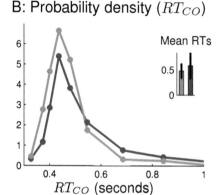

Fig. 5.42: Testing a quantitative prediction from the V1 saliency hypothesis using data collected by Koene and Zhaoping in the same experiment as in Fig. 5.41. Only six out of the eight observers in the experiment had the complete set of data on all feature singletons. A: The distribution of RT_{CMO} for one of the six observers is predicted from the other RTs of the same observer according to equation (5.34). The predicted and observed quantities are plotted in blue and red, respectively. In comparison, B shows the disconfirmation (using data from the same observer) of the incorrect prediction, $RT_{CO} = \min(RT_C, RT_O)$ (i.e., predicting RT_{CO} from RT_C and RT_O by a race model), which does not arise from the hypothesis, because of the presence of CO neurons in V1. In A, but not in B, the predicted distribution is not significantly different from the observed distribution.

Let $RT_1 \equiv \min(RT_C, RT_M, RT_O, RT_{CMO})$ and $RT_2 \equiv \min(RT_{CM}, RT_{CO}, RT_{MO})$. Then equation (5.34) states that $RT_1 = RT_2$. Because neural responses are stochastic, the actual equality is between the probability distributions of RT_1 and RT_2, respectively. Given the observed distributions of $RT_C, RT_O, RT_M, RT_{CM}, RT_{CO}$, and RT_{MO}, one can derive the distribution of RT_{CMO} by finding the one which minimizes the difference (quantified by some appropriate measure) between the probability distributions of RT_1 and RT_2, respectively.

Figure 5.42 A shows that the predicted and observed distributions of RT_{CMO} agree with each other quantitatively, up to the noise in estimating these distributions that comes from the finite number of search trials. Statistical analysis confirms that the difference between the predicted and observed distributions is not significant for any of the six observers. By contrast, a test of the incorrect prediction $RT_{CO} = \min(RT_C, RT_O)$ of RT_{CO} from a race model that assumes no CO neuron fails, since there is a significant difference between the predicted and the observed distributions of RT_{CO} for most of the six observers. Figure 5.42 shows an example. Two other (examples of) incorrect predictions (which cannot be predicted by the V1 theory without additional requirements on V1 physiology) of RT_{CMO} by the race equations $\min(RT_C, RT_{MO}, RT_{CMO}) = \min(RT_{CM}, RT_{CO}, RT_M, RT_O)$ and $RT_{CMO} = \min(RT_C, RT_M, RT_O)$, respectively, also fail to agree with data for at least some observers.

The agreement between the quantitative prediction and experimental data further supports the idea that V1 is the substrate for saliency computation. This is because higher cortical areas (such as V2) downstream along the visual pathway do have neurons tuned to the triple or more conjunctions of simple features,[19] as expected from the general observation that neural selectivities become more complex in higher visual cortical areas. Since our prediction

[19]Private communication from Stewart Shipp, 2011.

requires an absence of these triple conjunctive cells, it is unlikely that the higher cortical areas, instead of V1, used the maximum rule (from equation (5.4)) to compute saliency for the feature singletons in our search stimuli. Otherwise, the predicted RT_{CMO} would be statistically longer than the observed RT_{CMO}, just like the race-model predicted RT_{CO} is longer than the RT_{CO} in reality.

5.5.5 Reflections—from behavior back to physiology via the V1 saliency hypothesis

Recall from Chapter 1 that one of the aims of a theory is to link physiological and behavioral observations. So far, we have used the V1 saliency hypothesis to predict behavior in visual search and segmentation tasks from the physiological properties of V1. From a surprisingly "impossible" prediction of a salient ocular singleton to a quantitative prediction derived without any free parameters, experimental confirmations of these predictions build confidence in the theory.

These successes encourage us to reverse the direction of prediction by applying this theory to predict V1 neural properties from behavioral data. For example, a significant difference between the distribution of the behavioral RT_{CO} and that of the race model $RT_{CO} = \min(RT_C, RT_O)$, seen in Fig. 5.42 B, predicts a non-trivial contribution to the saliency of the CO singleton from CO neurons in V1; see equations (5.27) and (5.32). This predicted contribution is the portion that is beyond that by the same CO cells to the saliency of the single-feature (C and O) singletons, and it can be quantitatively assessed from the behavioral RT data (Zhaoping and Zhe 2012a). We predict that the V1 CO neurons should respond to their preferred conjunction of C and O features more vigorously when this conjunction is a double-feature rather than a single-feature singleton. Consequently, the contextual suppression on a CO cell responding to this conjunction is predicted to be weaker when the contextual inputs differ from this conjunction in both, rather than just one of, the C and O dimensions. Analogous predictions hold for the contextual influences on the MO neurons in V1.

5.6 The roles of V1 and other cortical areas in visual selection

According to Fig. 2.3 and Fig. 2.29, V1 is just one of the visual areas that send signals to the superior colliculus (SC) to control gaze. The SC also receives inputs from the retina, extrastriate areas, lateral intraparietal cortex (LIP), and the frontal eye field (FEF). Figure 5.43 shows a simplified schematic of the brain areas involved in gaze control. If we identify gaze control with the control of selection (ignoring covert selection), then it is likely that other brain areas must also contribute to selection, i.e., the guidance of attention.

It is instructive to imagine that the decision as to where, or to what, to direct attention is by a committee. Various brain areas, including V1, are the committee members which send their contributions to the decision; and the SC transforms the decision by the committee to motor responses. The impact of each brain area on the decision is determined by various factors, including the strength and timeliness of its contribution. Hence, some decisions are dominated by top-down effects, while others by bottom-up ones.

It is generally believed that the frontal and parietal brain areas are involved in many top-down aspects of attentional selection (Desimone and Duncan 1995, Corbetta and Shulman 2002). By contrast, observations in Section 5.3 and Section 5.5.4 suggest that cortical areas beyond V1 play little role in the bottom-up control of selection mediated by the saliency of the very salient singletons in the eye of origin, color, orientation, and motion direction. However, the V1 saliency hypothesis does not preclude additional influences from other cortical areas to bottom-up selection for more complex stimuli. It is therefore important to understand the extent of their contribution.

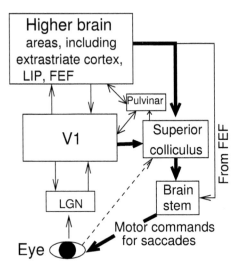

Fig. 5.43: Brain areas governing gaze control (Schiller 1998). V1 is only one of a number of areas contributing to the control of gaze. In monkeys and cats, the retina plays a very limited role in controlling saccades driven by visual inputs (Schiller et al. 1974, Schiller 1998). The role of higher visual areas can be assessed by investigating the influence on gaze control of aspects of visual perception that are not processed by V1.

First, various observations suggest that the retina plays little role in visually guided saccades in normal (non-lesioned) monkeys and cats, even though it can play a role in stabilizing retinal images during viewer or scene motion via their projection to the accessory optic system (Schiller 1998). In monkeys, a very small and relatively poorly understood fraction of retinal ganglion cells, called W cells, projects to the superficial layers of the SC (Schiller 1998). The axons of these neurons conduct spikes more slowly than parvo- and magnocellular ganglion cells (Schiller and Malpeli 1977). When V1 in monkeys or cats is removed or cooled, neurons in the intermediate and deep layers of the SC, and in particular, those eye-movement cells which activate to evoke saccades, can no longer be driven by visual stimuli. This is the case even though the animals can still make saccades in the dark (but not in response to visual stimulation) and even though these cells still fire before non-visually guided saccades (Schiller et al. 1974, Schiller 1998), which can be controlled by FEF, which can bypass the SC to control saccades. Also, monkeys suffering V1 lesions do not have proper visually guided saccades for up to two months after the lesion (Isa and Yoshida 2009).

The LGN lacks direct input to the SC. Thus, apart from the retina, only V1 and brain areas downstream along the visual pathway can be responsible for visually guided saccades or selection for normal primates. One key difference between V1 and downstream areas is latency—the latter typically have longer latencies than V1 in response to visual input, and so their contributions to bottom-up selection, or to top-down selection contingent on visual input, are likely to lag behind that of V1 (Bullier and Nowak 1995, Schmolesky, Wang, Hanes, Thompson, Leutgeb, Schall and Leventhal 1998). One can imagine situations in which V1's contribution is so strong and fast that the non-V1 contributions could be too slow to have an impact. The non-V1 contribution could also be ignored if it is too weak, or if it and V1's contributions are redundant. Conversely, it could be substantial when V1's contribution is too weak to reach a quick decision.

To investigate the respective contributions by V1 and other brain areas in selection, we focus on those selections which are contingent on external visual inputs, assuming that V1 plays little role in the other selections. Note that top-down and task-driven factors can influence input contingent selections. This is because, when selection has a sufficiently long latency after visual input, the gist of the scene obtained by observers during the latency could exert influence associated with the knowledge of the scene or ongoing tasks.

5.6.1 Using visual depth feature to probe contributions of extrastriate cortex to attentional control

To explore contributions to bottom-up selection beyond V1, it helps to identify visual processes that are carried out in higher visual areas but not in V1, and to investigate how these visual processes guide selection. A good candidate is stereo vision, which analyzes surfaces and their depth orders to achieve the perception of three-dimensional (3D) surfaces. Even though V1 cells are tuned to binocular disparities, 3D perception requires stereo processes to suppress false matches, which occur between visual inputs to the two eyes arising from two different object features in the scene (see Fig. 6.7 C). It is known that these stereo matching processes aimed at surface perception are centered outside V1, notably in V2 (Cumming and Parker 2000, Bakin et al. 2000, von der Heydt et al. 1984, von der Heydt et al. 2000, Qiu and von der Heydt 2005, Janssen, Vogels, Liu and Orban 2003). Hence, attentional guidance by depth or 3D cues should reflect contributions coming from beyond V1.

It has been shown (Nakayama and Silverman 1986, He and Nakayama 1995) that searching for a target defined by a unique conjunction of depth and another feature is much easier than typical conjunction searches that lack the depth feature (e.g., the color-orientation conjunction in Fig. 5.3 E). This suggests that 3D cues can help direct attention to task-relevant locations. We can measure and compare selection with and without 3D cues while the 2D cues are held constant. The speed-up of attentional guidance by the 3D cues is identified as a contribution from beyond V1.

One such study (Zhaoping, Guyader and Lewis 2009) is an extension to the experiment shown in Fig. 5.36, which was used to test the maximum rule in V1 saliency computations. In that experiment, the segmentation of a task-relevant texture was subject to interference from a superposed task-irrelevant texture surface. Denote the task-relevant image (texture A in Fig. 5.36) by I_{rel}, the task-irrelevant image (texture B in Fig. 5.36) by I_{ir}, and the composite image (texture C in Fig. 5.36) as $I_{com} = I_{rel} + I_{ir}$. The interference from I_{ir} can be reduced when I_{ir}'s position is slightly shifted horizontally from I_{rel} by a displacement x. Let us denote this shifted version of I_{ir} as $I_{ir}(x)$, and the resulting composite image as $I_{com}(x) = I_{rel} + I_{ir}(x)$; see Fig. 5.44. The RT for segmenting $I_{com}(x)$ is less than that for segmenting the original composite I_{com}. (One can also simulate the V1 model in Section 5.4 and confirm that the V1 saliency value at the texture border is higher in the 2D offset images $I_{com}(\pm x)$ than in the original I_{com}.) A horizontal shift $-x$ of I_{ir} in the opposite direction, giving a composite image $I_{com}(-x)$, would reduce the RT just as effectively. These RT reductions are not caused by any 3D cues, since exactly identical textures $I_{com}(\pm x)$ are presented to the two eyes.

If $I_{com}(x)$ and $I_{com}(-x)$ are viewed dichoptically by the two eyes, the percept is 3D: the two texture surfaces I_{rel} and I_{ir} separate in depth (see Fig. 5.44). Whether I_{rel} appears in front of or behind I_{ir} depends on whether the right or left eye sees $I_{com}(x)$. If the separation in depth makes segmentation faster than for the 2D percept arising from binocular (sometimes called bioptic) viewing of $I_{com}(x)$, the post-V1 3D effect should be credited.

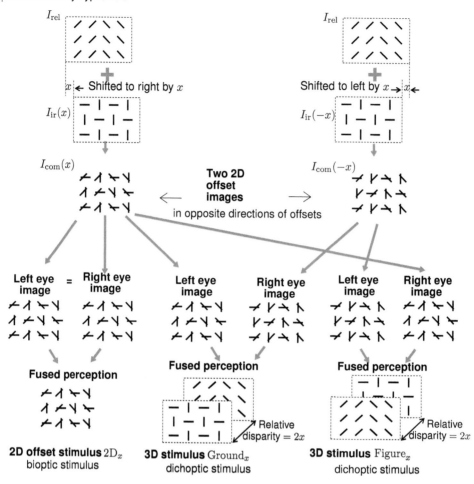

Fig. 5.44: Construction of 2D and 3D stimuli used to assess the contribution to selection of 3D processes in brain areas beyond V1. The texture images I_{rel} are as textures A in Fig. 5.36, and texture images $I_{ir}(\pm x)$ are spatially shifted (horizontally by $\pm x$) versions of texture B in Fig. 5.36. Superposing I_{rel} and $I_{ir}(\pm x)$ makes $I_{com}(\pm x)$, and $I_{com}(x = 0)$ is as texture C in Fig. 5.36. The bottom row shows the 2D offset stimulus $2D_x$, created by presenting the 2D offset image $I_{com}(x)$ (or $I_{com}(-x)$) identically to both eyes, and the 3D stimuli $Ground_x$ and $Figure_x$, created by presenting $I_{com}(x)$ to one eye and $I_{com}(-x)$ to the other. The relative disparity between I_{rel} and I_{ir} in the 3D stimuli is $2x$. Adapted with permission from Zhaoping, L., Guyader, N., and Lewis, A., Relative contributions of 2D and 3D cues in a texture segmentation task, implications for the roles of striate and extrastriate cortex in attentional selection, *Journal of Vision*, 9(11), article 20, doi: 10.1167/9.11.20, Fig. 2, copyright © 2009, ARVO.

Denote the 2D bioptic stimulus when the 2D offset image $I_{com}(x)$ or $I_{com}(-x)$ is presented identically to both eyes as $2D_x$, and the 3D dichoptic stimuli when $I_{com}(x)$ and $I_{com}(-x)$ are presented to different eyes as $Figure_x$ and $Ground_x$, when I_{rel} is in the foreground or background, respectively. These stimuli share the same 2D cues, notably the same 2D positional offset between the task-relevant and task-irrelevant textures. However, the 3D stimulus has an additional 3D cue, the depth separation between the two textures, to which

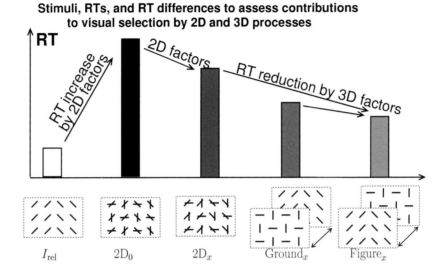

Fig. 5.45: The contributions of 2D and 3D processes to selection are manifest in the differences between RTs for texture segmentation using five different types of texture stimuli: I_{rel}, $2D_0$, $2D_x$, Figure$_x$, and Ground$_x$. Each of the last four stimuli (created as in Fig. 5.44) contains two texture surfaces, I_{rel}, which is task relevant, and I_{ir}, which is task irrelevant. These two surfaces are placed at the same depth, as in $2D_0$ and $2D_x$, or at different depths, as in Figure$_x$ and Ground$_x$, in which I_{rel} is in the foreground and background respectively. The 2D offset stimulus $2D_x$ has a spatial offset $\pm x$ between textures I_{rel} and I_{ir}; this offset is zero in $2D_0$. The contribution of 3D processes to selection should be manifested in the RT difference $RT(2D_x) - RT(\text{Figure}_x)$, and it is perhaps also manifested in the RT difference $RT(\text{Ground}_x) - RT(\text{Figure}_x)$ regardless of the eye dominance. Adapted with permission from Zhaoping, L., Guyader, N., and Lewis, A., Relative contributions of 2D and 3D cues in a texture segmentation task, implications for the roles of striate and extrastriate cortex in attentional selection, *Journal of Vision*, 9(11), article 20, doi: 10.1167/9.11.20, Fig. 3, copyright © 2009, ARVO.

3D perception is sensitive. If $RT(2D_x)$, $RT(\text{Figure}_x)$, and $RT(\text{Ground}_x)$ are the RTs for the segmentation task for the corresponding stimuli (see Fig. 5.45), then any 3D contribution to selection is likely to manifest as the following two differences between the RTs,

$$\text{the first RT difference} \equiv RT(2D_x) - RT(\text{Figure}_x) \quad \text{and}$$
$$\text{the second RT difference} \equiv RT(\text{Ground}_x) - RT(\text{Figure}_x),$$

being positive. The second RT difference may be positive if the task-relevant surface in the foreground helps steer attention.

The result was that these differences were only significantly positive, indicating a contribution from 3D processes, for RTs that were at least 1 second long; see Fig. 5.46. This RT is the time it takes for observers to press one of the two buttons to report whether the texture border in I_{rel} component is in the left or right half of the visual display. Assuming that it takes subjects around 300–400 ms after making the decision actually to press the button, this one second RT implies a roughly 600–700 ms RT for the task decision. The data in Fig. 5.46 thus suggest that if the saliency signal from V1 is sufficiently fast and adequate for the task, then a decision can be made quickly without waiting for contributions from higher brain areas. This

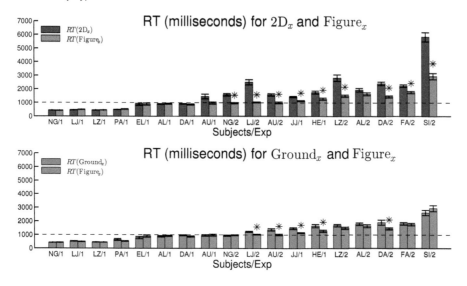

Fig. 5.46: Contributions from 3D visual processes to selection in a segmentation task are absent unless observers take at least 1000 ms to register their decision. The task is depicted in Fig. 5.45. RTs are plotted for various subjects in experiments 1 and 2. These two experiments differ in that the orientation contrast at the texture border in I_{rel} is 90° in experiment 1 (with I_{rel} as in Fig. 5.45), and 14° in experiment 2 (I_{rel}, not shown, has roughly oblique texture bars). Experiment 2 is designed to reduce the saliency of the texture border so that a longer RT is required. The horizontal axes label the subjects and the experiments in which these RTs were obtained, e.g., Subject/Exp = "LZ/2" means subject LZ in experiment 2. A "*" on top of the red data bar (for $RT(\mathrm{Figure}_x)$) indicates a significantly larger (two-tailed t-test) $RT(2D)$ or $RT(\mathrm{Ground}_x)$ than $RT(\mathrm{Figure}_x)$. Data from Zhaoping, L., Guyader, N., and Lewis, A., Relative contributions of 2D and 3D cues in a texture segmentation task, implications for the roles of striate and extrastriate cortex in attentional selection, *Journal of Vision*, 9(11), article 20, 2009, doi: 10.1167/9.11.20.

situation should apply for cases in which subjects can respond manually within 1 second for the task, regardless of whether the stimuli are 2D or 3D.

However, if the visual input is such that the V1 saliency signal is inadequate for a quick and confident task decision, contributions from higher brain areas can be important. This situation should apply to 3D stimuli like Figure_x and Ground_x, when their monocular 2D component images do not give rise to adequate V1 saliency signals for the task within a short time. Then, depth perception by higher brain areas aids attentional guidance. If the additional contribution from higher visual areas is absent, or it and the contribution from V1 are redundant, and the V1 contribution is weak, then the task RT can be long. This should apply to the situation in which observers take longer than 1 second to respond to the stimuli $2D_x$ or $2D_0$ without 3D cues. In other words, the findings suggest that, at least for depth processing, extrastriate areas do not contribute to input contingent selection immediately upon visual input, but they do contribute around several hundred milliseconds after the input onset. Meanwhile, V1 dominates in controlling selection within this initial time window after visual input onset or after a sudden unpredicted change to the visual scene.

It has been observed (Einhäuser, Spain and Perona 2008) that human saccades on static photographs are better predicted by visual objects (i.e., recognizable objects which are likely meaningful to the viewers) than by saliency. However, the first few saccades are very similar to those made by observers who have visual object agnosia (Mannan, Kennard and Husain 2009), suggesting that the early saccades are primarily controlled by bottom-up saliency rather than object processes occurring outside V1. These findings are consistent with our observations using the depth feature.

In everyday life, we may divide our visual experience into separate episodes, each being defined by our visual exposure to a constant environment, such as a room or an outdoor field. A typical visual episode may last many seconds, minutes, or hours. Many of the eye movements in an episode, such as directing gaze to a kitchen counter to look for a kettle, are controlled by our knowledge of the environment. This knowledge can be obtained very quickly, within perhaps a second or so after we enter a new environment (and aided by our stored knowledge of what typical kitchens, or other environments, look like). Viewed from the perspective of temporal durations of selection control, V1's contribution to attentional guidance is confined to just the first second, and will only be a tiny fraction of the total contributions from all brain areas in typical daily situations. However, by exerting initial control, V1's role in selection is special.

5.6.2 Salient but indistinguishable inputs activate early visual cortical areas but not the parietal and frontal areas

Neural activities correlated or associated with bottom-up saliency (which we call saliency-like signals) have been observed in LIP (Gottlieb et al. 1998, Bislay and Goldberg 2011, Arcizet, Mirpour and Bisley 2011) and FEF (Thompson and Bichot 2005). Areas such as the pulvinar and V4 have also been observed to be linked with attentional guidance or maintenance (Schiller and Lee 1991, Robinson and Petersen 1992, Mazer and Gallant 2003). These areas, and particularly a fronto-parietal network, are often considered to organize the guidance of attention (Corbetta and Shulman 2002) or the maintenance of information associated with attention, especially in task-dependent manners. Thus, it is important to know whether their computational role is confined to controlling top-down and task-dependent attention, and whether V1 is the origin of the saliency-like signals which are received in order to be combined with top-down signals in the computation. Importantly, saliency-like signals have always been evoked in experiments using salient inputs which are highly distinctive perceptually. Hence, the relevant activities in these regions might come from recognizing or perceiving the input, rather than saliency itself.

Attentional capture by an eye-of-origin singleton, shown in Fig. 5.9, indicates that some inputs can be quite salient without evoking any perceptual awareness of how they are distinct from background items. If there is a difference between the neural response to such a salient input and the response to a perceptually identical, yet non-salient, background input, this difference should represent a relatively pure signal for saliency, with minimal contamination by non-saliency factors. Identifying brain areas that exhibit such a pure saliency signal can help us to identify brain areas involved in computing saliency. Assessing this was the intent of an experiment (Zhang, Zhaoping, Zhou and Fang 2011) in which functional magnetic resonance imaging (fMRI) and event-related electroencephalography potentials (ERPs) were used to probe neural responses to salient inputs that are not perceptually distinctive.

In the experiment, we used an input texture containing a regular array of iso-oriented bars except for a foreground region of 2×2 bars tilted differently from the background bars; see Fig. 5.47. This whole texture was shown for only 50 ms and was quickly replaced by a high-contrast mask texture (which itself lasted for just 100 ms). Meanwhile, the observers directed their gaze to an ever-present fixation point. The foreground texture could only be at one of the two possible locations, which were $7.2°$ to the lower left or lower right of the fixation point. However, since the presentation was so brief and the mask so powerful, observers could not

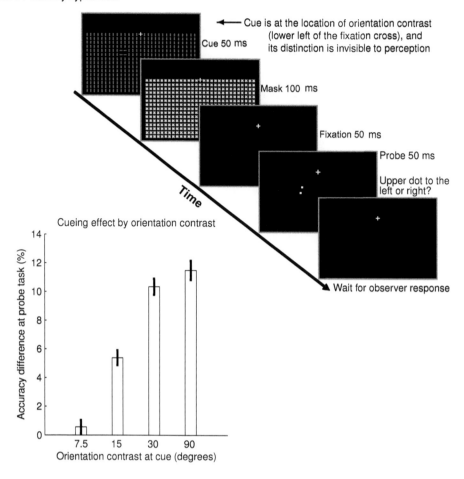

Cue is at the location of orientation contrast (lower left of the fixation cross), and its distinction is invisible to perception

Cue 50 ms

Mask 100 ms

Fixation 50 ms

Probe 50 ms

Upper dot to the left or right?

Wait for observer response

Cueing effect by orientation contrast

Fig. 5.47: A salient cue, which improves discrimination at the cued location but whose distinction from background inputs is invisible to perception. Upper: stimuli for the cue and probe. The cue, at the orientation contrast (which may be $0°$, i.e., no cue, or $7.5°$, $15°$, $30°$, or $90°$), and the probe, the two dots, appeared at the same (in this figure) or different locations when the probe was cued or uncued, respectively. The cross in each image was the fixation point. Observers reported whether the upper dot was to the left or right of the lower dot in the probe. Lower: the cueing effect was the increase in the percentage of the correct reports by observers in the cued relative to the uncued condition. Adapted from *Neuron*, 73 (1), Xilin Zhang, Li Zhaoping, Tiangang Zhou, and Fang Fang, Neural Activities in V1 Create a Bottom-Up Saliency Map, pp. 183–92, figures 1 and 2, copyright © (2012), with permission from Elsevier.

tell whether the foreground region was at one or the other location, even if forced to guess. Nevertheless, given sufficient orientation contrast, the foreground was salient in that it could serve as an effective exogenous cue to influence the discrimination of an input probe stimulus that was shown for 50 ms, starting 50 ms after the mask.

The probe consisted of two small dots; observers had to report whether the upper dot was to the left or right of the lower dot. This task was difficult, such that the accuracy of the task performance, i.e., the percentage of the reports that were correct, was typically only about 70%. However, performance on cued trials, when the probe was shown at the same location

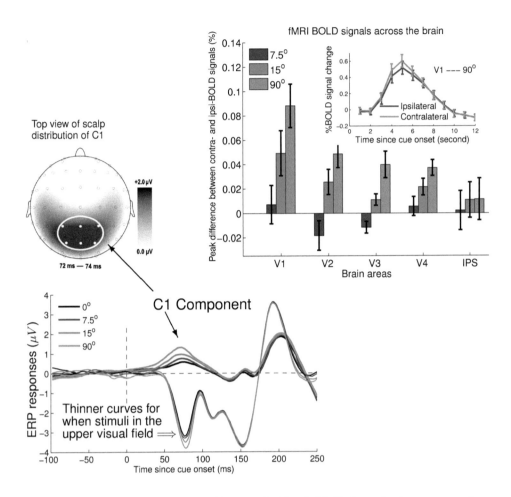

Fig. 5.48: Brain responses to the salient cue in Fig. 5.47. Brain activations averaged over observers were seen in the ERPs (lower left). The thinner curves were ERP responses (initially negative) to visual stimuli shown in the upper visual field for comparison. The C1 component in ERP, emerging around 55 ms and peaking around 73 ms from the cue onset, was mainly caused by V1 activities in the occipital region at the back of the scalp. Its polarity depended on whether the evoking stimulus was in the upper or lower visual field. The upper right plot shows cue-evoked brain activations probed by fMRI in V1–V4 and the intraparietal sulcus (IPS). They (plotted as colored bars) were significant in V1–V4 but not in the IPS. Each bar marks the difference between the peak BOLD (blood-oxygenation-level-dependent) responses to the cue at cortical locations (in a particular retinotopic brain area) contralateral and ipsilateral to the cue. The inset shows the time courses of the BOLD responses in the region of the retinotopic V1 for the cue location (contralateral) and in the corresponding region in the opposite hemisphere (ipsilateral), when the orientation contrast at the cue was 90°. All data in this figure are for when the visual stimuli were in the lower visual field, except for the thinner curves in the ERP responses (lower left). Data from Zhang, X., Zhaoping, L., Zhou, T., and Fang, F., Neural activities in V1 create a bottom-up saliency map, *Neuron*, 73(1): 183–192, 2012.

as the cue, was up to more than 10% better than on uncued trials, when the probe was shown at the other location where the cue could have been. This increase in the accuracy is called the cueing effect, and it makes manifest the saliency of the cue, despite the fact that the cue's distinction was invisible to the observers.

The cueing effect was significant when the orientation contrast between the bars at the cue and those in the background was more than $7.5°$. The basic cueing effect, and its increase with orientation contrast, was also observed using other probe tasks, e.g., discriminating the motion direction of moving dots or the orientation of a Gabor patch.

We used fMRI to compare the blood-oxygenation-level-dependent (BOLD) signals in regions of the brain processing the visual stimuli for the two possible locations of the salient but non-distinctive cue. A difference between these signals at these two locations is seen as a saliency signal. Areas V1, V2, V3, and V4 exhibited significant differences, but the intraparietal sulcus (IPS, which is thought to contain the human homologue of LIP) did not.

Briefly, increased neural processing in any brain region is thought to increase its demand for blood and thus the local BOLD signal (which has a slow time constant of a few seconds). The difference between the BOLD signals with and without any sensory input may be seen as the BOLD response to this sensory input. In a retinotopic cortical area contralateral to the cue location, one can identify the region of interest (ROI) as the cortical surface patch that responds to visual inputs at the cue location. Then, one can quantify a form of saliency signal by comparing BOLD signals in the ROI to those in the equivalent cortical surface patch in the ipsilateral hemisphere. We define this to be the maximum difference between the BOLD responses in these two surface patches over the time course of the fMRI responses for a trial. This saliency signal is averaged over all trials for a given orientation contrast at the cue location, with the uniform orientation of the background texture bars randomly chosen for each trial. These signals are plotted in the upper right panel in Fig. 5.48 for different retinotopic brain regions (V1, V2, V3, V4, and the IPS), and different orientation contrasts of the cue. In FEF and LGN, where retinotopy and/or the spatial resolution of the fMRI are too poor, the lack of a saliency signal was determined by a lack of a difference between the BOLD responses to cues of different saliencies.

We measured ERPs in a separate experiment. These showed that the earliest scalp potential response evoked by the invisible cue was a C1 component. This component emerged around 55 ms after the onset of the cue, reaching its peak response around 73 ms after the onset, and crucially, had an amplitude that increased significantly with the orientation contrast at the cue. Its polarity depended on whether the cue (with the whole accompanying texture) was in the upper or lower visual field (see the lower left panel in Fig. 5.48). This C1 component is believed to be generated mainly by neural activities in V1 in the occipital lobe (at the back of the brain), because it has a short latency and because of the location-dependence of its polarity.[20]

The fMRI activations in V1–V4 evoked by saliency also increased significantly with the orientation contrast of the cue, but to a degree that decreased from V1 to V4. This suggests that the saliency activation in V1 is unlikely to be caused by that in V4. Further evidence for this is that lesioning V4 in a monkey impairs its visual selection of non-salient objects but does not impair the selection of salient objects (Schiller and Lee 1991); see Fig. 2.28 B.

Furthermore, for a given orientation contrast, the fMRI activation evoked by saliency also decreased from V1 to V4. This finding contrasts with that of another study (Melloni, Van Leeuwen, Alink and Müller 2012), in which the stimulus contained four oriented grating

---

[20]The lower and upper visual fields are mapped to the retinotopic V1 at the upper and lower banks, respectively, of the calcarine fissure. They activate neurons with geometrically opposite orientations in their spatial layout (Jeffreys and Axford 1972).

patches and observers had to find a perceptually distinct orientation-singleton patch (which was oriented orthogonally to the non-target patches). In that study, the difference between the fMRI responses to the target and non-target patches *increased* from V1 to V4. The contrast between the findings suggests that recognition, and perhaps also the task-dependency, of the orientation contrast are the causes for higher saliency signals in higher rather than lower visual areas. Furthermore, in that study, fMRI responses in IPS were stronger when there was a need to suppress the distraction of a salient color singleton which was not a target; while fMRI responses in FEF were stronger to enhance the less salient targets. These observations are consistent with task-oriented functions in the IPS and FEF regions.

5.7 V1's role beyond saliency—selection versus decoding, periphery versus central vision

Crowding: the letter "T" is harder to recognize in the right image while fixating on "+".

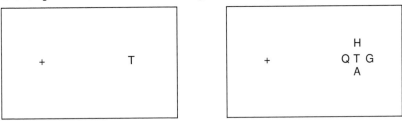

Fig. 5.49: Crowding is an impairment of the ability to recognize a stimulus in visual periphery that is caused by surrounding stimuli. In both the left and right images, the letter "T" does not overlap with any other image elements. However, fixating on the "+," you will find the "T" in the right image much harder to recognize.

Our overall framework sees vision in terms of encoding, selection, and decoding. In Chapters 3 and 4, we discussed extensively the possible role of V1 in encoding. Here, we have discussed how V1 influences selection via its output to the superior colliculus. V1 also projects to higher cortical areas, receives feedbacks from them, and sends feedback to the LGN (Fig. 5.43), so it is important to consider how it contributes to post-selectional decoding.

Visual spatial selection, and in particular saliency, shifts attention from the current focus to a new location. Since the current focus generally lies within the fovea, saliency mechanisms should primarily operate outside it. Take iso-orientation suppression, which is the essential contextual influence underpinning how orientation contrast leads to saliency. Behavioral studies suggest that this is mostly absent in the fovea (Petrov, Carandini and McKee 2005). Following this argument, iso-feature suppression for other feature dimensions should also be absent in the fovea. This absence of contextual influence at the fovea is beneficial for decoding stimuli shown there, since contextual influences would distort the relevant V1 responses. Just such distortion makes it hard to decode peripheral inputs that are surrounded by context. This might be part of the cause for crowding, which is an impairment of the ability to recognize or discriminate peripheral inputs that is caused by having contextual inputs nearby (Levi 2008); for a demonstration, see Fig. 5.49. Crowding makes it very difficult to read text more than a few characters down the line from the current fixation.

Hence, V1 saliency mechanisms operate in the periphery to help select the next focus of attention and bring it to the fovea by a saccade. Meanwhile, the lack of saliency mechanisms in the fovea allows the visual representation there to be faithful to the input to facilitate post-selectional visual decoding; see Fig. 5.50.

A: Differentiating the visual field for V1's roles

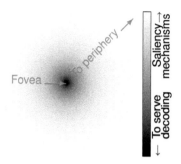

B: V1 serves decoding and bottom-up selection via its interactions with other brain areas

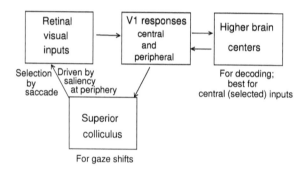

Fig. 5.50: The distinct roles of V1 in peripheral and central vision. A: The visual field is shown in light and dark shades to represent preferential involvement of V1 in saliency and decoding, respectively. B: V1's contribution to bottom-up selection is mediated through the superior colliculus; its contribution to decoding involves higher visual areas.

The next target of visual selection is typically influenced by the stimulus that is currently being decoded. Consequently, along with bottom-up saliency, selection is affected by other factors including the knowledge of decoded visual objects and the ongoing task. Since decoding is better for foveal inputs, the relative impact of saliency in selection increases with increasing eccentricity. In the example depicted in Fig. 5.9, both the ocular and orientation singletons are highly salient because of the contrast between the singleton features and the background features; however, the ongoing task of finding an orientation singleton makes top-down selection favor the latter. As the singletons become more peripheral, recognizing them (i.e., decoding) becomes more difficult, diminishing the task-dependent advantage of the orientation-singleton target in the competition for selection. Consequently, the non-target ocular singleton should become an increasingly effective distractor, damaging performance at the outset of the search. This was indeed observed (Zhaoping 2012)—75% of the first saccade during search were directed to the lateral side of the ocular singleton when both singletons had an eccentricity of 12 degrees from the initial gaze position (at the center of the displays). By contrast, only 50% were inappropriately directed when the eccentricity was 7.3 degrees.[21]

As shown in Fig. 5.50 B and argued in Chapter 6, feedback from higher visual areas to early visual areas such as V1 is expected to help with decoding, i.e., visual recognition. Given that decoding is favored in the central visual field, one might expect that feedback, particularly from ventral visual areas associated with "what" vision, is more extensive at and near the fovea to help recognizing object features. This suggestion can be tested empirically. It is consistent with recent observations[22] that feedbacks to central and peripheral regions of primate V1 are predominantly from the cortical areas in the ventral and dorsal streams, respectively. It is also supported by observations in a recent behavioral study (Zhaoping 2013a). The stimuli in this behavioral study were adapted from the dichoptic stimuli in the experiment shown in Fig. 3.15 designed to test efficient stereo coding. The percepts induced by such stimuli are ambiguous, such that two likely percepts correspond respectively to the sum (S_+) and difference (S_-) of the visual inputs to the two eyes. Top-down expectations would favor the sum over the difference, in view of the normal correlations between binocular inputs in daily visual experience, and is known to influence perception (see Chapter 6). In the fovea, there was indeed a bias to perceive the sum rather than the difference signal, consistent with this top-down influence; the bias was weak or absent in the periphery (at about 10° eccentricity), even though the stimulus was adequately enlarged at periphery to compensate for the drop in visual acuity.

5.7.1 Implications for the functional roles of visual cortical areas based on their representations of the visual field

We have argued above that, at least to a first approximation, visual selection by saccades brings the selected visual inputs to the central visual field to be decoded. Consequently, the neural representation of the central and peripheral fields in all visual cortical areas should depend on whether the area is mainly concerned with selection, decoding, or both (Zhaoping 2011). If an area is concerned with bottom-up selection, then the whole visual field should be represented in its neurons, since salient locations can arise unexpectedly anywhere in the visual field. However, areas that are mainly concerned with decoding or top-down or task-dependent selection (which is typically dependent on the current attended object or on knowledge and memory rather than external sensory stimulus) should have neurons that respond mainly or solely to central visual inputs.

In particular, if a cortical area contains a bottom-up saliency map, then it should represent the whole visual space, including the full ranges of eccentricity and polar angles. Conversely, cortical areas further downstream along the visual pathway are likely to devote their resources to the attended, i.e., near foveal, regions, since they are more likely involved in post-selectional processing.

These arguments allow the experimental literature to be used to discriminate among brain regions as to which are likely to contain a saliency map. V1 and V2 respond to the whole visual space (Gattass, Sousa and Rosa 1987, Rosa, Sousa and Gattass 1988) up to at least 80° eccentricity. Nevertheless, a recent fMRI study showed that V2 (and V3) devotes more cortical area than V1 to the central 0.75 degree of the visual field (Schira, Tyler, Breakspear and Spehar 2009). However V3 and V4 represent only the central 35–40 degrees (Gattass, Sousa and Gross 1988). Toward the culmination of the ventral visual pathway, which is

---

[21] The stimulus pattern was also different for the two different eccentricity cases, but the densities of the background bars (when the spatial dimension is measured in the unit of bar length) were comparable (Zhaoping 2012).

[22] Private communication from Henry Kennedy (2013).

devoted to processing object features or "what" processing, neurons in the IT cortex (area TE) have very large receptive fields, typically covering the central gaze region and extending to both the left and right half of the visual field. However, they are devoted to inputs within 40° eccentricity (Boussaoud, Desimone and Ungerleider 1991). Along the dorsal visual pathway, which is more concerned with "where" visual processing, the visual field representation may be expected to reflect visual selection better. However, area MT has no neural receptive field beyond 60° (Fiorani, Gattass, Rosa and Sousa 1989). Experiments differ as to the maximum eccentricities of the RFs of neurons in area LIP, with one (involving a central fixation task) finding few neurons with receptive fields centered beyond 30° eccentricity (Ben Hamed, Duhamel, Bremmer and Graf 2001), and another (Blatt, Andersen and Stoner 1990), using anesthetized monkeys, finding few neurons with receptive fields beyond 50° (albeit with those having receptive-field radii of up to 20°). V6 (the parieto-occipital area PO), which has many non-visual neurons and neurons influenced by eye positions (Galletti, Battaglini and Fattori 1995), is substantially devoted to peripheral vision. Its neurons can respond to visual inputs up to 80° in eccentricity (Galletti, Fattori, Gamberini and Kutz 1999). The FEF receives inputs from both the ventral and dorsal streams; however, the spatial coverage of neurons in FEF is poorly studied. Few receptive fields beyond 35° eccentricity have so far been mapped; the receptive fields concerned have been seen as being open-ended in the periphery since their true extent is unclear (Mohler, Goldberg and Wurtz 1973).

These observations are collectively consistent with the idea that V1 creates a bottom-up saliency map to guide attention. They also imply that some higher visual cortical areas such as V4 and IT along the ventral pathway are less likely to be involved in bottom-up selection than in decoding. Since some cortical areas in the dorsal visual stream cover a large extent of the peripheral visual field, their role in bottom-up visual selection cannot be excluded, although it is also likely that the peripheral coverage serve the purpose of visually guided action (such as grasping and reaching). V2's coverage of the whole visual field suggests that it may also be involved in bottom-up selection. This motivates future investigations. In sum, applying the perspective of inferring functional roles from visual field representations (Zhaoping 2011), V1 is the most likely candidate to compute a bottom-up saliency map.

The functional role of V1 should have direct implications on the role of V2 and other downstream cortical areas along the visual pathway. If V1 creates a saliency map to guide attention in a bottom-up manner, then the downstream areas might be better understood in terms of computations in light of the exogenous selection, and these computations include endogenous selection and post-selectional decoding (Zhaoping 2013b). This is consistent with observations (see Section 2.6) that top-down attention associated with an ongoing task can typically modulate the neural responses more substantially in the extrastriate cortices than in V1.

5.7.2 Saliency, visual segmentation, and visual recognition

Selection by (bottom-up) saliency may be the initial step in visual segmentation, which is the problem of separating out from the rest of the scene those image locations that are associated with visual objects that need to be recognized; see Fig. 5.51 A. Most non-trivial visual functions involve object recognition and localization to enable motor responses and memory storage; segmentation is a fundamental issue for recognition because it needs to be carried out before and during this operation. Computer vision approaches have been tried to solve the problem of image segmentation for decades without a satisfactory solution in the general input situation. The crux of the problem is the following dilemma: to segment the image area containing an object, it helps to recognize it first; while to recognize the object requires having

A: Image of an apple
and a house

B: Two texture regions to be segmented

C: Segmentation by classification

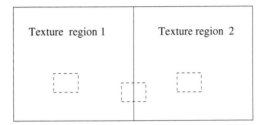

Fig. 5.51: Demonstration of the segmentation-classification dilemma. A: To recognize the apple, it helps to segment the image area associated with it; however, to segment this image area, it helps to recognize the apple. B: Segmenting the two texture regions from each other is hard, since the two regions do not obviously differ by mean luminance or another simple measure. Characterizing local image areas by measures such as smoothness, regularity, orientation, spectrum of spatial frequencies, etc., could help to distinguish different texture regions. C: To segment the image into a priori unknown regions, each local image area, denoted by dashed boxes, needs to be characterized by some such measures.

first segmented the image area that contains it. (Here, we exclude recognizing the gist of a scene without recognizing individual objects.)

Thus, the many computer vision algorithms that have been developed for image segmentation can all be viewed as performing "segmentation by recognition" or "segmentation by classification." Consider segmenting the two texture regions in Fig. 5.51 B; this is not trivial since the two texture regions do not differ in some obvious measure (such as the mean luminance or color), and it is not even known a priori whether the image contains one or two or more regions. Conventional algorithms start by taking any image area, e.g., one of the dashed boxes in Fig. 5.51 C, and trying to characterize it by some image-feature measures. These measures might quantify the mean pixel luminance, regularity, smoothness, dominant spatial frequency, dominant orientations, characteristics of the histogram of the image pixel values, or other aspects of the local image area. Each measure is called a "feature" of the image area, and the image area can be described by a feature vector whose components are the various feature measurements. When two image areas differ sufficiently by their feature vectors, they are presumed to belong to different surface regions. Hence, such algorithms perform "segmentation by classification," i.e., they segment by classifying the feature vectors.

However, these algorithms operate under the assumption that each image area chosen

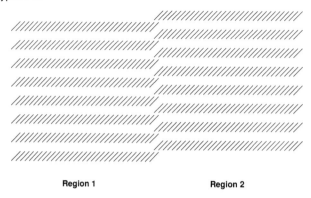

Region 1   Region 2

Fig. 5.52: An example demonstrating that biological vision can operate without performing segmentation by classification. We can readily see two regions in this image, even though these regions share all the same feature values. Thus, feature classification is neither sufficient nor necessary to segment the two regions. There is also no vertical contrast edge at the vertical region border, so algorithms using edge-based approaches for segmentation would also fail. Reproduced with permission from Li, Z., Visual segmentation by contextual influences via intracortical interactions in primary visual cortex, *Network: Computation in Neural Systems*, 10(2): 187–212, Fig. 1, copyright © 1999, Informa Healthcare.

to be classified happens to fall into a single surface region to be segmented. This is not guaranteed since we do not know a priori where the region boundaries are. If a chosen area, e.g., the central image area bounded by the central dashed box in Fig. 5.51 C, falls on the border between two regions, it would be hard to characterize its features. The chance of such an event can be reduced by making the individually inspected image areas smaller. This inevitably makes the feature vector values less precise, since many feature values, such as the value of the dominant spatial frequency, require the image area to be large enough for them to be quantified with sufficient precision. This problem stems ultimately from the dilemma that segmentation requires classification and classification requires segmentation.

The "classification" of the image patches in the above example is not the same as recognizing an object, although it can provide clues to the underlying object (e.g., inferring a leopard by its skin) or at least a surface of the object. Nevertheless, the fundamental interdependence between recognition and segmentation remains.

Figure 5.52 demonstrates that biological vision can operate without employing segmentation-by-classification, since classifying the two identical texture regions flanking the texture border is neither necessary nor sufficient for the segmentation. One may argue that special image processing operators could be constructed to detect the border between these two textures. However, such image processing operators would almost certainly be bespoke for this particular image example. Different examples analogous to this one would require different tailored operators to achieve segmentation. It is not desirable to build a big bag of many tricks to tackle this problem, since one can build many special examples that require special purpose operators and so make the bag infeasibly large. Apparently, human vision can carry out segmentation without classification (Li 1999b). This is analogous to making a saccade to a visual location before recognizing the object at that location (see in Fig. 1.4).

Selection by saliency can underpin segmentation without classification. If the border between the two texture regions in Fig. 5.52 is salient, it attracts selection. Locating the border between two objects might be the first step to segmenting them. This first step can be coarse but can nevertheless provide an initial condition in what could be an iterative process, alternating between segmentation and recognition. In other words, the initial segmentation, by the selection of the border due to its high saliency, can lead to preliminary recognition which can refine segmentation. In turn, this can refine the recognition, and so on. This iterative process is likely to involve both V1 and other cortical areas. Understanding the underlying process is a challenge.

5.8   Nonlinear V1 neural dynamics for saliency and preattentive segmentation

The credibility of the hypothesis that V1 creates a saliency map is significantly bolstered by the demonstration in Section 5.4 that a model using plausible V1 mechanisms could realize the computation concerned. In this section, we show how this model was designed through the analysis of neural circuit dynamics. Readers not interested in these details can skip this section.

The computation of saliency transforms one representation of visual inputs based largely on image contrast to another representation based instead on saliencies. We identify V1 with this transformation, suggesting that its input, the visual stimulus filtered through the classical receptive fields of the V1 neurons, is transformed to an output represented by the activities from the V1 output cells, such that the output can be read out for saliency through the maximum rule in equation (5.4); and the mechanisms it employs are the intracortical interactions mediated by its nonlinear recurrent neural circuit.

There are two characteristics of this V1 saliency transform. First, we focus on cases in which top-down feedback from higher visual areas does not change during the course of the saliency transform but merely sets a background or operating point for V1. In such cases, V1's computation is autonomous, consistent with its being bottom-up or preattentive. Of course, more extensive computations can doubtlessly be performed when V1 interacts dynamically with other visual areas.

Second, the saliency of a location should depend on the global context. Making the output of a V1 neuron depend non-locally on its inputs would be hard to achieve in a purely feed-forward network with retinotopically organized connections and local receptive fields. Rather, the recurrent dynamics enable computations to occur at a *global* scale despite the local neural connectivity.

Nonlinear dynamics involving many recurrently connected neurons is typically difficult to understand and control. As we have seen from Fig. 5.15 B, V1 pyramidal neurons are generally engaged in mutual excitation or mutual inhibition (via interneurons). Since mutual excitation or mutual inhibition involves a positive feedback loop, a recurrent neural network with both interactions is typically unstable against random fluctuations unless the interactions are very weak. The difficulty of understanding such nonlinear recurrent networks in order to properly control and design them is apparent in many previous works (Grossberg and Mingolla 1985, Zucker, Dobbins and Iverson 1989, Yen and Finkel 1998). Nevertheless, harnessing this dynamics is essential to realize the saliency computation.

In this section, we summarize analysis from various research papers (Li 1997, Li 1998a, Li 1999b, Li and Dayan 1999, Li 2001) which addressed the following central issues: (1) computational considerations regarding how a saliency model should behave; (2) a minimal model of the recurrent dynamics for computing saliency, i.e., to achieve (1); (3) the specific constraints on the recurrent neural connections; and (4) how recurrent dynamics give rise to phenomena such as region segmentation, figure-ground segregation, contour enhancement, and filling-in. In addressing these issues, we perform a stability analysis of nonlinear dynamics to examine the conditions governing neural oscillations, illusory contours, and (the absence of) visual hallucinations. By contrast, single neural properties such as orientation tuning that are less relevant to computations at a global scale will not be our focus. Some of the analytical techniques, e.g., the analysis of the cortical microcircuit and the stability analysis of the translation-invariant networks, can be applied to study other cortical areas that share similar neural elements and neural connection structures with V1's canonical microcircuit (Shepherd 1990).

5.8.1 A minimal model of the primary visual cortex for saliency computation

A minimal model is the one which has just barely enough components to execute the necessary computations without anything extra. This criterion is inevitably subjective, since there is no fixed recipe for a minimalist design. However, as a candidate, I present a model that performs all the desired computations but for which simplified versions fail. Since the minimal model depends on the desired computation to be carried out by the model, I will also articulate this saliency computation as to what this model should do and what this model should not do.

Throughout the section, we try to keep our analysis of the characteristics of the recurrent dynamics general. However, to illustrate particular analytical results, approximations, and simplification techniques, I often use a model of V1 whose specifics and numerical parameters are presented in the appendix to this chapter, so that the readers can try out the simulations.

We use notation such as $\{x_{i\theta}\}$ to denote a vector containing components $x_{i\theta}$ for all $i\theta$. Hence, $\{I_{i\theta}\}$ is the input, and $\{g_x(x_{i\theta})\}$ is the response. The V1 model should transform $\{I_{i\theta}\}$ to $\{g_x(x_{i\theta})\}$, with higher responses $g_x(x_{i\theta})$ to input bars $i\theta$ which have higher perceptual saliency. This is achieved through recurrent interactions between neurons. What kind of recurrent model is needed?

5.8.1.1 A less-than-minimal recurrent model of V1

A very simple recurrent model of the cortex can be described by this equation

$$\dot{x}_{i\theta} = -x_{i\theta} + \sum_{j\theta'} T_{i\theta,j\theta'} g_x(x_{j\theta'}) + I_{i\theta} + I_o, \tag{5.40}$$

where $-x_{i\theta}$ models the decay (with a time constant of unity) in membrane potential, and I_o is the background input. The recurrent connection $T_{i\theta,j\theta'}$ links cells $i\theta$ and $j\theta'$. Visual input $I_{i\theta}$ (taken as being static for illustration) initializes the activity levels $g_x(x_{i\theta})$ and also persists after onset. The activities are then modified by the network interaction, making $g_x(x_{i\theta})$ dependent on input $I_{j\theta'}$ for $(j\theta') \neq (i\theta)$. The connections are translation invariant in that $T_{i\theta,j\theta'}$ depends only on the vector $i - j$ and on the angles of this vector (in 2D space) to the orientations θ and θ'. Reflection symmetry (e.g., when a horizontal bar facilitates another horizontal bar to its right, so should the latter facilitate the former, with the same facilitation strength) gives the constraint $T_{i\theta,j\theta'} = T_{j\theta',i\theta}$.

Many previous models of the primary visual cortex (e.g., Grossberg and Mingolla (1985), Zucker, Dobbins, and Iverson (1989), and Braun, Niebur, Schuster, Koch (1994)) can be seen as more complex versions of the one described above. The added complexities include stronger nonlinearities, global normalization (e.g., by adding a global normalizing input to the background I_o), and shunting inhibition. However, they are all characterized by reciprocal or symmetric interactions between model units, i.e., $T_{i\theta,j\theta'} = T_{j\theta',i\theta}$. It is well known (Hopfield 1984, Cohen and Grossberg 1983) that in such a symmetric recurrent network, the dynamic trajectory $x_{i\theta}(t)$ (given a static input pattern $\{I_{i\theta}\}$) will converge in time t to a fixed point. This fixed point is a local minimum (attractor) in an energy landscape

$$E(\{x_{i\theta}\}) = -\frac{1}{2}\sum_{i\theta,j\theta'} T_{i\theta,j\theta'} g_x(x_{i\theta}) g_x(x_{j\theta'}) - \sum_{i\theta} I_{i\theta} g_x(x_{i\theta}) + \sum_{i\theta} \int_0^{g_x(x_{i\theta})} g_x^{-1}(x)dx,$$

$$(5.41)$$

where $g_x^{-1}(x)$ means the inverse function of $g_x(x)$. Empirically, convergence to attractors typically occurs even when the complexities in the previous models mentioned above are included.

The fixed point $\bar{x}_{i\theta}$ of the motion trajectory, or the minimum energy state $E(\{x_{i\theta}\})$ where $\partial E/\partial g_x(x_{i\theta}) = 0$ for all $i\theta$, is given by (when $I_o = 0$)

$$\bar{x}_{i\theta} = I_{i\theta} + \sum_{j\theta'} T_{i\theta,j\theta'} g_x(\bar{x}_{j\theta'}). \tag{5.42}$$

Without recurrent interactions ($T = 0$), this fixed point $\bar{x}_{i\theta} = I_{i\theta}$ is a faithful copy of the input $I_{i\theta}$. Weak but non-zero T makes pattern $\{\bar{x}_{i\theta}\}$ a slightly modified version of the input pattern $\{I_{i\theta}\}$. However, sufficiently strong interactions T can make $\bar{x}_{i\theta}$ dramatically unfaithful to the input. This happens when T is so strong that one of the eigenvalues λ^T of the matrix T with elements $\mathbb{T}_{i\theta,j\theta'} \equiv T_{i\theta,j\theta'} g_x'(\bar{x}_{j\theta'})$ satisfies $Re(\lambda^T) > 1$ (here, g_x' is the slope of $g_x(.)$ and $Re(.)$ means the real part of a complex number). For instance, when the input $I_{i\theta}$ is translation invariant such that $I_{i\theta} = I_{j\theta}$ for all $i \neq j$, there is a translation-invariant fixed point $\bar{x}_{i\theta} = \bar{x}_{j\theta}$ for all $i \neq j$. Strong interactions T could destabilize this fixed point, such that it is no longer a local minimum of the energy landscape $E(\{x_{i\theta}\})$. Consequently, the recurrent dynamics pulls $\{x_{i\theta}\}$ into an attractor in the direction of an eigenvector of T that is not translation invariant, i.e., $x_{i\theta} \neq x_{j\theta}$ for $i \neq j$ at the attractor.

Computationally, a certain unfaithfulness to the input, i.e., making $g_x(x_{i\theta})$ not to be a function of $I_{i\theta}$ alone, is actually desirable. This is exactly what is required for unequal responses $g_x(x_{i\theta})$ to be given to input bars of equal contrast $I_{i\theta}$ but different saliencies (e.g., a vertical bar among horizontal bars when all bars have the same input contrast). However, this unfaithfulness should be driven by the nature of the input pattern $\{I_{i\theta}\}$ and in particular, driven by how the input pattern deviates from homogeneity (e.g., smooth contours or figures against a background). If, instead, spontaneous or non-input-driven network behavior—*spontaneous pattern formation or symmetry breaking*—occurs, then visual hallucinations (in this case we mean saliency outcomes which are drastically different from the saliency values of the input) would result (Ermentrout and Cowan 1979). Such hallucinations, whose patterns are not meaningfully determined by external inputs, should be avoided.

For example, given homogenous input $I_{i\theta} = I_{j\theta}$, if $\{x_{i\theta}\}$ is an attractor, then so is a translated state $\{x_{i\theta}'\}$ such that $x_{i\theta}' = x_{i+a,\theta}$ for any translation a. This is because $\{x_{i\theta}\}$ and $\{x_{i\theta}'\}$ have the same energy value E. The two possible patterns after symmetry breaking on the right part of Fig. 5.20 are instances of this, being translations of each other. (When the translation a is one-dimensional, such a continuum of attractors has been called a "line attractor" (Zhang 1996). For two- or more dimensional patterns, the continuum is a "surface attractor.") That the absolute positions of the hallucinated patterns are random, can even shift, and are not determined by the sensory input $\{I_{i\theta}\}$ implies a degree of unfaithfulness that is undesirable for saliency.

To illustrate, consider the case that $T_{i\theta,j\theta'}$ is non-zero only when $\theta = \theta'$, i.e., the connections only link cells that prefer the same orientation. (This is a limit of the observations (Gilbert and Wiesel 1983, Rockland and Lund 1983) that the lateral interactions tend to link cells preferring similar orientations.) The network then contains multiple, independent, subnetworks, one for each θ. Take the $\theta = 90°$ (vertical orientation) subnet, and for convenience, drop the subindex θ. We have

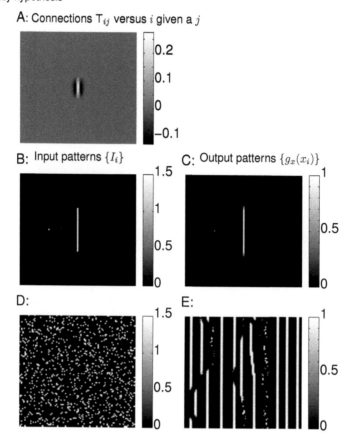

Fig. 5.53: A reduced model consisting of symmetrically coupled cells tuned to vertical orientation ($\theta = 90°$), as in equation (5.43). Five grayscale images are shown; each has a scale bar on the right. The network has 100×100 cells arranged in a two-dimensional (2D) array, with wrap-around boundary conditions. Each cell models a cortical neuron tuned to vertical orientation, arranged retinotopically. The function $g_x(x)$ gives $g_x(x) = 0$ when $x < 1$, $g_x(x) = x-1$ when $1 \leq x < 2$, and $g_x(x) = 1$ when $x > 2$. A: The connection pattern T between the center cell j and the other cells i. This pattern is local and translation invariant, with excitation or inhibition between i and j which are, respectively, roughly vertically or horizontally displaced from each other. B: An input pattern $\{I_i\}$, consisting of an input line and a noise spot. C: Output response $\{g_x(x_i)\}$ to the input in B. The line induces a response that is $\sim 100\%$ higher than the response to the noise spot. D: A sample noisy input pattern $\{I_i\}$. E: Output response $\{g_x(x_i)\}$ to the input in D, showing hallucinated vertical streaks. Adapted with permission from Li, Z., Computational design and nonlinear dynamics of a recurrent network model of the primary visual cortex, *Neural Computation*, 13(8): 1749–1780, Fig. 1, copyright © 2001, MIT Press.

$$\dot{x}_i = -x_i + \sum_j T_{ij} g_x(x_j) + I_i, \tag{5.43}$$

in which T_{ij} is still symmetric, $T_{ij} = T_{ji}$, and translation invariant. As an example, let T be a simple, center-surround pattern of connections in a Manhattan grid (for which grid location $i = (m_i, n_i)$ has horizontal and vertical coordinates m_i and n_i, respectively). Let

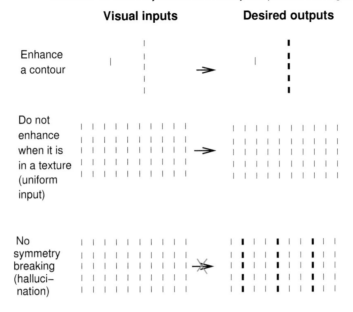

Fig. 5.54: Desired input-output mapping for saliency computation in three special input cases. Adapted with permission from Li, Z. and Dayan, P., Computational differences between asymmetrical and symmetrical networks, *Network: Computation in Neural Systems*, 10(1): 59–77, Fig. 1, copyright © 1999, Informa Healthcare.

$$T_{ij} \propto \begin{cases} T & \text{if } i = j, \\ -T & \text{if } (m_j, n_j) = (m_i \pm 1, n_i) \text{ or } (m_i, n_i \pm 1), \\ 0 & \text{otherwise.} \end{cases} \tag{5.44}$$

If T is sufficiently strong, then even with an homogenous input $I_i = I_j$ for all i, j, the network can settle into an "antiferromagnetic" state in which neighboring units x_i exhibit one of the two different activities $x_{m_i,n_i} = x_{m_i+1,n_i+1} \neq x_{m_i+1,n_i} = x_{m_i,n_i+1}$, arranged in a checkerboard pattern. This pattern $\{x_i\}$ is just a spatial array of the replicas of the center-surround interaction pattern T. Note that the patterns after the spontaneous symmetry breaking in Fig. 5.20 are simply one-dimensional checkerboard patterns.

Intracortical interaction that are more faithful to V1 (Kapadia et al. 1995, Polat and Sagi 1993, Field, Hayes and Hess 1993) have T_{ij} depend on the orientation of $i - j$. However, this T_{ij} will still be translation invariant, i.e., independent of the absolute value of i and depending only on the magnitude and orientation of $i - j$. In the subnet of vertical bars, such V1-like interactions specify that two nearby bars i and j excite each other when they are co-aligned and inhibit each other otherwise. More directly, $T_{ij} > 0$ between i and j which are close and roughly vertically displaced from each other, and $T_{ij} < 0$ between i and j which are close and more horizontally displaced. Figure 5.53 shows the behavior of such a subnet. Although the network enhances an input (vertical) line relative to an isolated (short) bar, it also hallucinates other vertical lines when exposed to noisy inputs.

Instead, the recurrent network should have the properties illustrated in Fig. 5.54. First, its response to a smooth contour should be higher than to a bar segment that is either isolated or is an element in a homogenous texture. Second, it should not respond inhomogenously to a homogenous texture. In other words, the network should selectively amplify certain inputs against some other inputs. The ability of the network to achieve this property can be measured by the gain (or sensitivity) to a contour relative to a homogenous texture. We call this the *selective amplification ratio* (Li and Dayan 1999):

$$\text{selective amplification ratio} = \frac{\text{gain to contour input}}{\text{gain to texture input}}. \tag{5.45}$$

A higher selective amplification ratio makes it easier to distinguish salient input (such as a contour) from the less salient inputs (such as homogenous textures). For instance, if the level of noise in the neural responses is comparable to the mean response to the homogenous texture, a selective amplification ratio comfortably larger than two is desirable to make the response to a contour stand out relative to the responses to a background texture. Physiological data (Nelson et al. 1985, Knierim and Van Essen 1992, Kapadia et al. 1995) shown in Fig. 5.19 suggest that the selective amplification ratio is up to at least four to five.

The competition between internal interactions T and the external inputs $\{I_i\}$ to shape $\{x_i\}$ makes it impossible to achieve a high selective amplification ratio. For analysis, consider the following simple pattern of interaction in the vertical bar subnet:

$$\begin{cases} T_{ij} > 0, \text{ when } i \text{ and } j \text{ are nearby and in the same vertical column,} \\ T_{ij} < 0, \text{ when } m_j = m_i \pm 1, \text{ i.e., } i \text{ and } j \text{ are in the neighboring columns,} \\ T_{ij} = 0, \text{ otherwise.} \end{cases}$$

Furthermore, denote the total excitatory connection to a neuron from its contour as

$$T_0 \equiv \sum_{j, m_j = m_i} T_{ij} > 0, \tag{5.46}$$

and denote the total suppressive connection to a neuron from neighboring contours as

$$T' \equiv - \sum_{j, m_j = m_i \pm 1} T_{ij}. \tag{5.47}$$

In addition, for simplicity, take a piece-wise linear function for $g_x(x)$:

$$g_x(x) = \begin{cases} x - x_{\text{th}} & \text{if } x_{\text{th}} \leq x \leq x_{\text{sat}}, \text{ where } x_{\text{th}} \text{ is the threshold,} \\ & \text{and } x_{\text{sat}} > x_{\text{th}} \text{ is the point of saturation,} \\ x_{\text{sat}} - x_{\text{th}} & \text{if } x > x_{\text{sat}}, \\ 0 & \text{otherwise.} \end{cases} \tag{5.48}$$

A vertical contour input has $I_i = I > x_{\text{th}}$ for i with $m_i = 1$, and $I_i = 0$ otherwise. Call the neurons i with $m_i = 1$ on the vertical line the "line units." We can ignore all other neurons since they will be at most suppressed by the line unit, and so none can be activated beyond threshold. By symmetry, at the fixed point, all the line units i have the same state $\bar{x}_i = \bar{x}$, where

$$\bar{x} = I + \sum_{j, m_j = m_i} T_{ij} g_x(\bar{x}) = I + T_0 g_x(\bar{x}), \tag{5.49}$$

$$\rightarrow \quad g_x(\bar{x}) = \frac{I - x_{\text{th}}}{1 - T_0}, \tag{5.50}$$

when $I > x_{\text{th}}$ is not too large. Thus, a large $T_0 < 1$ helps to give the following high gain

$$\frac{\delta g_x(\bar{x})}{\delta I} = \frac{1}{1 - T_0} \quad \text{to an isolated input contour, and } 1 > T_0 \text{ is required for stability.}$$

$$\tag{5.51}$$

By contrast, for a homogenous texture with $I_i = I > x_{th}$ for all units i, the fixed point $\bar{x}_i = \bar{x}$ of the response is

$$\bar{x} = I + \left(\sum_{j, |m_j - m_i| \leq 1} T_{ij} \right) g_x(\bar{x}) = I + (T_0 - T')g_x(\bar{x}) \tag{5.52}$$

$$\rightarrow \quad g_x(\bar{x}) = \frac{I - x_{th}}{1 + (T' - T_0)}. \tag{5.53}$$

This means that

$$\text{the gain} \quad \frac{\delta g_x(\bar{x})}{\delta I} = \frac{1}{1 + (T' - T_0)} \quad \text{to a homogenous input texture} \tag{5.54}$$

can be made small when the net suppression $T' - T_0$ is made large. Note that $T' - T_0$ quantifies the net iso-orientation suppression in a homogenous texture. Then,

$$\text{the selective amplification ratio} = \frac{\text{gain to contour input}}{\text{gain to texture input}} = \frac{1 + (T' - T_0)}{1 - T_0} \tag{5.55}$$

increases with increasing net iso-orientation suppression $T' - T_0$ and increasing contour facilitation T_0.

However, in a homogenous texture, a large net suppression destabilizes the homogenous fixed point $x_i = \bar{x}$ to fluctuations. Consider the fluctuation $x_i' = x_i - \bar{x}$, and assume that this lies within the linear range of $g_x(.)$, i.e., $x_{th} < x_i' + \bar{x} < x_{sat}$. Then

$$\dot{x}_i' = -x_i' + \sum_j T_{ij} \left[g_x(\bar{x} + x_j') - g_x(\bar{x}) \right]$$

$$= -x_i' + \sum_j T_{ij} x_j'. \tag{5.56}$$

This linear equation has an inhomogenous eigenmode, $x_i' = x' \cdot (-1)^{m_i}$ for all i, which is a spatially oscillating pattern (with amplitude x') of activities like those in the symmetry-breaking solution of Fig. 5.20. To see this, substitute $x_i' = x' \cdot (-1)^{m_i}$ into the above equation to obtain the equation of motion for the amplitude x':

$$\dot{x}' = -x' + \left(\sum_{j, m_j = m_i} T_{ij} - \sum_{j, m_j = m_i \pm 1} T_{ij} \right) x' \tag{5.57}$$

$$= -x' + (T' + T_0) x' \tag{5.58}$$

$$= (T' + T_0 - 1) x'. \tag{5.59}$$

This has a solution in which x' evolves with time t as

$$x'(t) \propto \exp[(T' + T_0 - 1) t]. \tag{5.60}$$

When $T' + T_0 > 1$, the amplitude x' of the inhomogenous eigenmode grows exponentially with time t. Given an initial deviation $\{x_i'(0)\}$ at $t = 0$ from the homogenous equilibrium state $x_i = \bar{x}$, the projection of this deviation pattern on this inhomogenous eigenmode grows exponentially, driving the activity pattern $x_i = \bar{x} + x_i'$ inhomogenous very quickly after the onset of visual input. The growth will eventually saturate because of the nonlinearity in $g_x(x)$. Hence, given homogenous input, the network spontaneously breaks symmetry from the homogenous fixed point and hallucinates a saliency wave whose period is two columns. As shown in Fig. 5.20, there are two such waves—one has $x'_i = x' \cdot (-1)^{m_i}$ and the other has $x'_i = x' \cdot (-1)^{m_i+1}$—and they are a $180°$ phase apart. Both of these inhomogenous states are also equilibrium points of the network even for homogenous input; however, they are stable to fluctuations, and they arise when $T' + T_0 > 1$ to make the homogenous fixed point unstable. Whether the network state will approach one or the other stable fixed point depends on the direction of the initial fluctuation pattern.

In sum, contour enhancement makes the network prone to "see" ghost contours whose orientations and widths match the interaction structure in T. Avoiding such hallucinations requires $T' + T_0 < 1$. This, together with $1 > T_0$ required by equation (5.51), implies that the selective amplification ratio is limited to the following:

$$\text{selective amplification ratio} = \frac{1 + (T' - T_0)}{1 - T_0} < 2. \tag{5.61}$$

Consequently, enhancement of contours relative to the background is insufficient (Li and Dayan 1999). Similar numerical limits on the selective amplification ratio apply to the cases of a general $g_x(x)$.[23] Although symmetric recurrent networks are useful for associative memory (Hopfield 1984), which requires significant input errors or omissions to be corrected or filled-in, they imply too much distortion for early visual tasks that require the output to be more faithful to the input.

5.8.1.2 A minimal recurrent model with hidden units

The strong tendency to hallucinate input in the symmetrically connected model of equation (5.40) is largely dictated by the symmetry of the neural connections. Hence, this tendency cannot be removed by introducing more complex cellular and network mechanisms without removing the symmetry of the neural connections. (These cellular and network mechanisms include, for instance, ion channels, spiking rather than firing rate neurons, multiplicative inhibition, global activity normalization, and input gating (Grossberg and Mingolla 1985, Zucker et al. 1989, Braun, Niebur, Schuster and Koch 1994), which are used by many neural network models.) Furthermore, attractor dynamics are untenable in the face of the well-established fact of Dale's law, namely that real neurons are overwhelmingly either exclusively excitatory or exclusively inhibitory. It is obviously impossible to have symmetric connections between excitatory and inhibitory neurons.

Mathematical analysis (Li and Dayan 1999) showed that asymmetric recurrent EI networks with separate excitatory (E) and inhibitory (I) cells can perform computations that are inaccessible to symmetric networks. In particular, EI networks support much larger selective amplification ratios without degenerating into hallucination. To illustrate this, start again with the simplification of a separated subnet (equations (5.43)) and piece-wise linear $g_x(x)$, as in equation (5.48). Then, replace neural units and connections (as in the example in Fig. 5.55):

> neural unit $x_i \rightarrow$ an EI pair [excitatory x_i, inhibitory y_i with time constant τ_y],
>
> connection $\mathsf{T}_{ij} \rightarrow \mathsf{J}_{ij}$ from x_j to x_i, and W_{ij} from x_j to y_i,

such that the circuit's equation of motion becomes

---

[23] In such cases, let $\bar{x}_{\text{contour}}$ and $\bar{x}_{\text{texture}}$ be the fixed points for the contour and texture inputs, respectively. The response gains for a contour and a homogenous texture are $\delta g_x(\bar{x})/\delta I = g'_x(\bar{x}_{\text{contour}})/[1 - T_0 g'_x(\bar{x}_{\text{contour}})]$ and $\delta g_x(\bar{x})/\delta I = g'_x(\bar{x}_{\text{texture}})/[1 + (T' - T_0)g'_x(\bar{x}_{\text{texture}})]$, respectively, and the requirement for avoiding hallucinations becomes $(T' + T_0)g'_x(\bar{x}_{\text{texture}}) < 1$. Consequently, the selective amplification ratio is limited by an upper bound $2\frac{g'_x(\bar{x}_{\text{contour}})}{g'_x(\bar{x}_{\text{texture}})}\frac{[1 - T_0 g'_x(\bar{x}_{\text{texture}})]}{[1 - T_0 g'_x(\bar{x}_{\text{contour}})]}$. If $g'_x(\bar{x}_{\text{contour}}) = g'_x(\bar{x}_{\text{texture}})$, this upper bound is again 2.

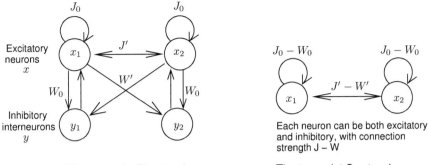

The two-point EI network   The two-point S network

Fig. 5.55: Two-point EI (as in equations (5.66) and (5.67)) and S networks. There are austere models to elucidate the essential computation in a recurrent V1 sub-network involving only neurons tuned to a single orientation. The two networks are exact counterparts when the interneurons y_1 and y_2 are linear, with $g_y(y) = y$. The fixed points of the dynamics in one network are also the fixed points in the other, but the stabilities of the fixed points, and thus the computational power, differ in the two networks.

$$\dot{x}_i = -x_i - g_y(y_i) + \sum_j J_{ij}g_x(x_j) + I_i \tag{5.62}$$

$$\tau_y \dot{y}_i = -y_i + \sum_j W_{ij}g_x(x_j) \tag{5.63}$$

In this circuit, x_i is the excitatory unit and conveys the output of network, and y_i is the inhibitory interneuron (with output $g_y(y_i)$), which acts as an auxiliary or hidden unit of the network.

The fixed point $\{\bar{x}_i, \bar{y}_i\}$ of this EI network satisfies $\dot{x}_i = \dot{y}_i = 0$. The network can be designed such that these fixed points are identical (ignoring the y dimension) to the fixed points $\{\bar{x}_i\}$ of the original symmetric network in equation (5.43). This EI network is then a formal counterpart of the symmetric network (which we call the S network). This is particularly simple in the case when $g_y(y) = y$ is linear. Then, as the time constant τ_y of the interneurons approaches zero, such that $y_i = \sum_j W_{ij}g_x(x_j)$, equation (5.62) becomes

$$\dot{x}_i = -x_i + \sum_j (J_{ij} - W_{ij})g_x(x_j) + I_i. \tag{5.64}$$

Hence, an EI network with very fast interneurons is equivalent to the S network when

$$J_{ij} - W_{ij} = T_{ij}, \quad \tau_y = 0, \quad \text{and} \quad g_y(y) = y. \tag{5.65}$$

If $\tau_y > 0$, these two networks are counterparts of each other, with the same fixed points but different dynamics for the motion trajectories. From now on, for simplicity, we always take $\tau_y = 1$, and use this simple model to compare EI and S networks.

Consider an EI network that is the counterpart of the particular S sub-network that only involves vertical bars (described in equations (5.43–5.48)), and with translation-invariant J and W, where $J_0 \equiv \sum_{j,m_j=m_i} J_{ij}$, $J' \equiv \sum_{j,m_j=m_i\pm1} J_{ij}$, and similarly for W_0 and W'. We have $T_0 = J_0 - W_0$ and $T' = W' - J'$. Since the EI network has the same fixed points as the S network, the selective amplification ratio, which is evaluated at fixed points, is identical for the two networks. A high value for this ratio cannot be realized in the S network because of the tendency for hallucination, resulting from the instability of the homogenous fixed point. However, in the EI network, under homogenous input, all the three fixed points, one homogenous and two inhomogenous, are unstable. As a result, the primary mode of instability of the homogenous solution in the EI network (to the homogenous input) is a spatially homogenous, temporal oscillation, because the network state cannot approach the non-homogenous, unstable, fixed points.

These conclusions can be understood by considering a highly simplified problem. In this simplification, $I_i \equiv I_1$ in all the odd columns have the same strength, and $I_i \equiv I_2$ in all the even columns also have the same strength. To quantify the selective amplification ratio, we need to consider responses to two input cases: a homogenous input pattern and a single contour. The former can be straightforwardly modeled by setting $I_1 = I_2$. For the latter, since the connections T_{ij} do not span more than a single column, we can set $I_1 > I_2 = 0$ or $0 = I_1 < I_2$, and thus consider non-interacting contours on either the odd or even columns.

Taking advantage of this simplification and assuming (as is arranged by the dynamics of the network) that all excitatory and inhibitory units in each column have the same activities, the input and state variables can be described by two-dimensional vectors $(I_1, I_2)^T$, $(x_1, x_2)^T$ and $(y_1, y_2)^T$ for the various quantities associated with the odd and even columns. We call this simplified system the two-point system (see Fig. 5.55) which captures the essence of our problem. The equations of motion of the two-point system are

$$\dot{x}_a = -x_a - y_a + J_0 g_x(x_a) + J' g_x(x_{a'}) + I_a, \tag{5.66}$$
$$\dot{y}_a = -y_a + W_0 g_x(x_a) + W' g_x(x_{a'}), \tag{5.67}$$

where $a = 1, 2$ and $a' \neq a$. Thus, the EI network has been reduced to two pairs of EI units, one for the odd columns and the other for the even columns of units. The 2×2 connection matrices for this reduced EI network are

$$J = \begin{pmatrix} J_0 & J' \\ J' & J_0 \end{pmatrix} \quad \text{and} \quad W = \begin{pmatrix} W_0 & W' \\ W' & W_0 \end{pmatrix}. \tag{5.68}$$

The S network that is the counterpart of this EI network has just two neurons (rather than four), and the connection matrix $T = J - W$.

From Fig. 5.54, we require relatively higher responses to the one-point input, $(I_1, I_2) \propto (1, 0)$, which corresponds to a contour input, and lower responses to the (uniform) two-point input, $(I_1, I_2) \propto (1, 1)$, which corresponds to a homogenous texture. The symmetry of the responses must be preserved for the two-point input $(I_1, I_2) \propto (1, 1)$.

In the two-point S system, the input response function to the one-point and two-point inputs are the same as those in equations (5.50) and (5.53) respectively, with a selective amplification ratio as in equation (5.55).

We can linearize the EI network to examine the approximate evolution of the deviations

$$(x'_a, y'_a) \equiv (x_a, y_a) - (\bar{x}_a, \bar{y}_a)$$

from the homogenous fixed point in response to homogenous input $(I_1, I_2) \propto (1, 1)$. The deviation (x'_a, y'_a) follows the equations (for $a \neq a'$)

$$\dot{x}'_a = -x'_a - y'_a + J_0 x'_a + J' x'_{a'}, \tag{5.69}$$
$$\dot{y}'_a = -y'_a + W_0 x'_a + W' x'_{a'}. \tag{5.70}$$

In comparison, in the two-point S network, the deviations x'_a follow

$$\dot{x}'_a = -x'_a + (J_0 - W_0)x'_a + (J' - W')x'_{a'}. \tag{5.71}$$

Note that matrices J, W, and T commute with each other, with common eigenvectors

$$V^{(+)} \equiv \frac{1}{\sqrt{2}} \begin{pmatrix} 1 \\ 1 \end{pmatrix} \quad \text{and} \quad V^{(-)} \equiv \frac{1}{\sqrt{2}} \begin{pmatrix} 1 \\ -1 \end{pmatrix}, \tag{5.72}$$

which are called plus and minus modes, respectively, (note that we used analogous modes for the stereo summation and opponency signals in the stereo encoding in Section 3.5) or spatial synchronous and anti-phase modes, respectively. The corresponding eigenvalues of J, W, and T are $\lambda_{\pm}^J = J_0 \pm J'$, $\lambda_{\pm}^W = W_0 \pm W'$, and $\lambda_{\pm}^T = \lambda_{\pm}^J - \lambda_{\pm}^W$, respectively. Then states (x_1', x_2') and (y_1', y_2') can be represented by their projections $x_{\pm}$ and $y_{\pm}$ onto these eigenmodes

$$\begin{pmatrix} x_1' \\ x_2' \end{pmatrix} = x_+ V^{(+)} + x_- V^{(-)}, \quad \begin{pmatrix} y_1' \\ y_2' \end{pmatrix} = y_+ V^{(+)} + y_- V^{(-)}. \tag{5.73}$$

Equations (5.69) and (5.70) can then be transformed to

$$\dot{x}_{\pm}' = -x_{\pm}' - y_{\pm}' + \lambda_{\pm}^J x_{\pm}', \tag{5.74}$$
$$\dot{y}_{\pm}' = -y_{\pm}' + \lambda_{\pm}^W x_{\pm}'. \tag{5.75}$$

Eliminating $y_{\pm}'$ from these equations, the EI dynamics follow

$$\ddot{x}_{\pm}' + (2 - \lambda_{\pm}^J)\dot{x}_{\pm}' + (\lambda_{\pm}^W - \lambda_{\pm}^J + 1)x_{\pm}' = 0. \tag{5.76}$$

Similarly, the S network dynamics is

$$\dot{x}_{\pm}' = -x_{\pm}' + (\lambda_{\pm}^J - \lambda_{\pm}^W)x_{\pm}'. \tag{5.77}$$

The solutions to the linear equations are

$$x_{\pm}'(t) \propto \exp(\gamma_{\pm}^{EI} t) \quad \text{for the EI network}, \tag{5.78}$$
$$x_{\pm}'(t) \propto \exp(\gamma_{\pm}^{S} t) \quad \text{for the S network}, \tag{5.79}$$

where $\gamma_{\pm}^{EI}$ and $\gamma_{\pm}^{S}$ are the eigenvalues of the linear system in equation (5.76) for the EI network and equation (5.77) for the S network, respectively,

$$\gamma_{\pm}^{EI} = -1 + \frac{1}{2}\lambda_{\pm}^J \pm \left(\frac{1}{4}(\lambda_{\pm}^J)^2 - \lambda_{\pm}^W \right)^{1/2}, \quad \text{and} \tag{5.80}$$
$$\gamma_{\pm}^{S} = -1 - \lambda_{\pm}^W + \lambda_{\pm}^J. \tag{5.81}$$

The fixed point for the EI or S network is unstable if $Re(\gamma_{\pm}^{EI})$ or $Re(\gamma_{\pm}^{S})$ is positive, as then fluctuations will grow. Note that, since $\lambda_-^J = J_0 - J'$, $\lambda_-^W = W_0 - W'$, $T_0 = J_0 - W_0$, and $T' = W' - J'$, it follows that

$$\gamma_-^S = -1 - \lambda_-^W + \lambda_-^J = -1 + W' - J' + (J_0 - W_0) = -1 + T_0 + T'.$$

The above equation and equation (5.60) are showing the same thing: when $T_0 + T' > 1$, the S network is unstable against fluctuations in the x_- mode, which is opposing fluctuations in the odd and even columns (or in x_1 and x_2).

Although the fixed points for the two networks are the same, their eigenvalues $\gamma_{\pm}^S$ and $\gamma_{\pm}^{EI}$ are different, and so their stabilities can also differ. Since $\lambda_{\pm}^J$ and $\lambda_{\pm}^W$ are real, $\gamma_{\pm}^S$ is always

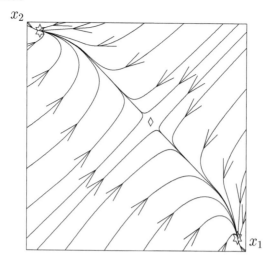

Fig. 5.56: The motion trajectory of the two-point S network under input $I \propto (1,1)$. The symmetric fixed point (marked by $\diamond$) becomes unstable when the two asymmetric fixed points (marked by ☆'s) appear. The symmetric fixed point is a saddle point, and the asymmetric ones are energy minima, of an energy landscape; the former repels and the latter attract nearby trajectories. The counterpart EI network has the same three fixed points but different dynamics (and no energy landscape). When the asymmetric fixed points in the EI networks are also unstable (and thus unapproachable), the network state can oscillate along the diagonal line $x_1 = x_2$ around the symmetric fixed point into the y dimensions without breaking symmetry. Adapted with permission from Li, Z. and Dayan, P., Computational differences between asymmetrical and symmetrical networks, *Network: Computation in Neural Systems*, 10(1): 59–77, Fig. 4, copyright © 1999, Informa Healthcare.

real. However, $\gamma_{\pm}^{EI}$ can be a complex number, leading to oscillatory behavior if its imaginary part $Im\left(\gamma_{\pm}^{EI}\right)$ is non-zero. One can derive that, for $k = +$ or $k = -$,

when $\gamma_k^S > 0$, then $Re\left(\gamma_k^{EI}\right) > 0$,

   i.e., the EI net is no less stable than the S net; (5.82)

when $Im\left(\gamma_k^{EI}\right) \neq 0$, then $\gamma_k^S < 0$,

   i.e., the S net is stable when the EI net is oscillatory (stable or not). (5.83)

These conclusions hold for any fixed point. Equation (5.82) can be proven by noting that $\gamma_k^S = -1 - \lambda_k^W + \lambda_k^J > 0$ gives $\lambda_k^W < -1 + \lambda_k^J$; hence $\left[\frac{1}{4}(\lambda_k^J)^2 - \lambda_k^W\right]^{1/2} > \left|-1 + \frac{1}{2}\lambda_k^J\right|$, and thus $Re\left(\gamma_k^{EI}\right) > 0$ (for one of the roots). Equation (5.83) can be proven by noting that

$$\frac{1}{4}\left(\lambda_k^J\right)^2 < \lambda_k^W \text{ leads to } \gamma_k^S < -1 - \lambda_k^W + 2\sqrt{\lambda_k^W} = -\left(1 - \sqrt{\lambda_k^W}\right)^2 \leq 0.$$

Now we can understand how the EI network maintains spatial symmetry under homogenous input $(I_1, I_2) \propto (1,1)$, even when $T' + T_0$ is large enough for the S network to break symmetry. Figure 5.56 shows the energy landscape and motion trajectory for the two-point S network under the homogenous input, with symmetry breaking. As analyzed above, the symmetry breaking is accompanied by three fixed points: one symmetric $\bar{x}_1 = \bar{x}_2$ and two asymmetric $\bar{x}_1 \neq \bar{x}_2$. The symmetric one is a saddle point, stable against fluctuations of x_+ (i.e., synchronous fluctuations in x_1 and x_2) from it but unstable against fluctuations of x_-

(i.e., opposing fluctuations in x_1 and x_2). The x_- fluctuation grows, with its initial value determining to which of the two asymmetric fixed points (x_1, x_2) will converge.

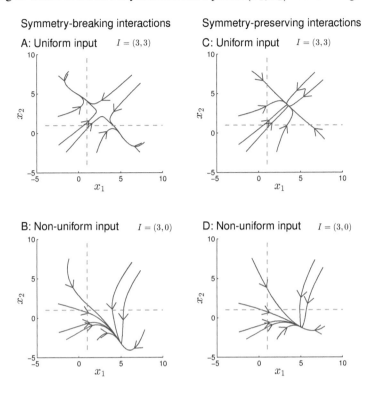

Fig. 5.57: Motion trajectories of the two-point S network. The interactions in A and B are symmetry breaking, with $T_0 = 0.5$ and $T' = 0.8$, so that the responses to uniform inputs converge to asymmetric fixed points (A). C, D: Lowering the inter-unit suppression to $T' = 0.3$ allows the network to preserve symmetry; however, the selective amplification ratio is now quite small. The function $g_y(y) = y$, and $g_x(x)$ is a threshold linear function with $x_{th} = 1$ and no saturation. The red dashed lines mark the threshold (x_{th}). Adapted with permission from Li, Z. and Dayan, P., Computational differences between asymmetrical and symmetrical networks, *Network: Computation in Neural Systems*, 10(1): 59–77, Fig. 2, copyright © 1999, Informa Healthcare.

The same three fixed points in the EI network can be all unstable. In particular, synchronous fluctuations x_+ from the symmetric fixed point $\bar{x}_1 = \bar{x}_2$ can be made unstable and oscillatory by

$$-1 + (J_0 + J')/2 > 0 \quad \text{and} \quad W_0 + W' > (J_0 + J')^2/4, \tag{5.84}$$

and the asymmetric fixed point can be made unstable by

$$-1 + J_0 > 0. \tag{5.85}$$

Note that, at the asymmetric fixed point, the non-active neural pair contributes nothing to the dynamics, and so the network becomes a one-point system, albeit a two-neuron, one-point system.

Given this, no fluctuation from the symmetric fixed point can converge—the only other fixed points (the asymmetric ones) are themselves unstable. Consequently, the fluctuations around the symmetric fixed point tend to be symmetric along a trajectory $x_1(t) = x_2(t)$ and $y_1(t) = y_2(t)$, and oscillate in the (x, y) phase space. Small fluctuations in the x_- direction are also unstable; however, in the nonlinear system, they are strongly squashed below the threshold x_{th} and above saturation at x_{sat}. Overall, this oscillation preserves the symmetry in the (x_1, x_2) space. This allows a very large selective amplification ratio without inducing any hallucination.

Figure 5.57 shows the trajectory of motion in the (x_1, x_2) space for two S networks. One network strongly amplifies an asymmetric input $I_1 \neq I_2$, but it spontaneously breaks symmetry by responding non-homogenously, $x_1 \neq x_2$, to homogenous input $I_1 = I_2$ (i.e., it strongly amplifies noise). Another network does not spontaneously break symmetry when $I_1 = I_2$, but it cannot amplify asymmetric input to nearly such a degree. Recall that the symmetric and asymmetric input represent the homogenous texture and isolated contours, respectively, of visual inputs in the expanded subnetwork for vertical bars. Hence, the S network cannot realize a sufficiently large selective amplification ratio, and so it is inadequate as a model of V1.

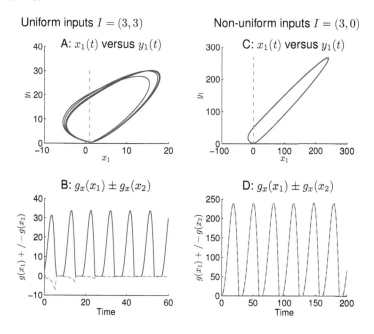

Fig. 5.58: Oscillatory trajectories of a two-point EI network with a high selective amplification ratio. The connections are $J_0 = 2.1$, $J' = 0.4$, $W_0 = 1.13$, and $W' = 0.9$. In B and D, the plot of $g_x(x_1) + g_x(x_2)$ is in blue, and $g_x(x_1) - g_x(x_2)$, in red. In the symmetric (uniform) input case, $g_x(x_1) - g_x(x_2)$ quickly decays in time (B). With asymmetric (non-uniform) input (C,D), the red and blue curves lie on top of each other (D). Here, $g_x(x)$, $g_y(y)$, and $x_{th} = 1$ are the same as in Fig. 5.57. The red dashed lines in A and C mark the thresholds (x_{th}). Adapted with permission from Li, Z. and Dayan, P., Computational differences between asymmetrical and symmetrical networks, *Network: Computation in Neural Systems*, 10(1): 59–77, Fig. 3, copyright © 1999, Informa Healthcare.

Figure 5.58 shows the evolution of a two-point EI network. The responses to both the symmetric and asymmetric inputs are oscillatory, but there is no spontaneous symmetry breaking to homogenous inputs, even though the selective amplification ratio is high. Hence, the EI network is the minimal network architecture for our V1 computation.

We next expand the toy model subset for one particular orientation θ into a full network including more orientations θ and interactions between orientations. In this case, the dynamical equations are

$$\dot{x}_{i\theta} = -x_{i\theta} - g_y(y_{i,\theta}) + J_o g_x(x_{i\theta}) - \sum_{\Delta\theta\neq0} \psi(\Delta\theta)g_y(y_{i,\theta+\Delta\theta})$$

$$+ \sum_{j\neq i,\theta'} J_{i\theta,j\theta'}g_x(x_{j\theta'}) + I_{i\theta} + I_o, \tag{5.86}$$

$$\dot{y}_{i\theta} = -\alpha_y y_{i\theta} + g_x(x_{i\theta}) + \sum_{j\neq i,\theta'} W_{i\theta,j\theta'}g_x(x_{j\theta'}) + I_c, \tag{5.87}$$

which are the same as equations (5.8–5.9) (except for the lack of I_{noise}). The neural connections $T_{i\theta,j\theta'}$ in the original S network are now replaced by various components including J, W, J_o, and ψ.

Although this analysis suggests that, unlike S networks, an EI network might be able to model V1, it is necessary for the connections J and W to be set appropriately in order to realize the necessary computations. The next section provides an analytical understanding of the nonlinear dynamics concerned and provides specific constraints on the J and W.

5.8.2 Dynamic analysis of the V1 model and constraints on the neural connections

The model state is characterized by $\{x_{i\theta}, y_{i\theta}\}$, or simply $\{x_{i\theta}\}$, omitting the auxiliary units $\{y_{i\theta}\}$. The interaction between excitatory and inhibitory cells makes $\{x_{i\theta}(t)\}$ intrinsically oscillatory in time (Li and Hopfield 1989), although whether the oscillations are damped or sustained depends on the external input patterns and neural connections. Thus, given an input $\{I_{i\theta}\}$, the model often does not convergence to a fixed point where $\dot{x}_{i\theta} = \dot{y}_{i\theta} = 0$. However, if $\{x_{i\theta}(t)\}$ oscillates periodically around a fixed point, then after the transient following the onset of $\{I_{i\theta}\}$, the temporal average of $\{x_{i\theta}(t)\}$ can characterize the model output and approximate the encircled fixed point. We henceforth use the notation $\{\bar{x}_{i\theta}\}$ to denote either the fixed point, if it is stable, or the temporal average, and denote the computation as $I \to g_x(\bar{x}_{i\theta})$.

5.8.2.1 A single pair of neurons

An isolated single pair $i\theta$ follows equations

$$\dot{x} = -x - g_y(y) + J_o g_x(x) + I, \tag{5.88}$$

$$\dot{y} = -y + g_x(x) + I_c, \tag{5.89}$$

(omitting the redundant index $i\theta$) where we set $\alpha_y = 1$ for simplicity and $I = I_{i\theta} + I_o$. The gain in the input-output transform $(I, I_c) \to g_x(\bar{x})$ at a fixed point $(\bar{x}, \bar{y})$ is

$$\frac{\delta g_x(\bar{x})}{\delta I} = \frac{g'_x(\bar{x})}{1 + g'_x(\bar{x})g'_y(\bar{y}) - J_o g'_x(\bar{x})}, \qquad \frac{\delta g_x(\bar{x})}{\delta I_c} = -g'_y(\bar{y})\frac{\delta g_x(\bar{x})}{\delta I}, \tag{5.90}$$

where $g'_x(\bar{x})$ and $g'_y(\bar{y})$, respectively, are the derivatives of the functions $g_x(.)$ and $g_y(.)$ at the fixed point $\bar{x}$ and $\bar{y}$.

Figure 5.59 illustrates an example when both $g_x(x)$ and $g_y(y)$ are piece-wise linear functions. In this case, the input-output transform $I \to g_x(\bar{x})$ is also piece-wise linear; see

A: Excitatory cell output $g_x(x)$ versus x

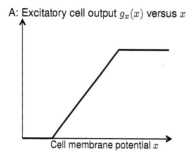

Cell membrane potential x

B: Interneuron output $g_y(y)$ versus y

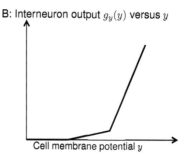

Cell membrane potential y

C: Response $g_x(\bar{x})$ versus I depends on I_c

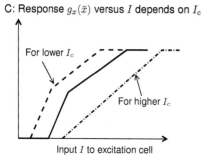

For lower I_c

For higher I_c

Input I to excitation cell

D: Minimum $\Delta I/\Delta I_c$ for contextual facilitation

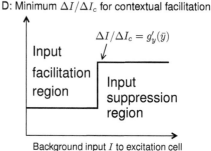

$\Delta I/\Delta I_c = g'_y(\bar{y})$

Input
facilitation
region

Input
suppression
region

Background input I to excitation cell

Fig. 5.59: A,B: Examples of $g_x(x)$ and $g_y(y)$ functions. C: Input-output function $I \to g_x(\bar{x})$ for an isolated neural pair without inter-pair neural interactions, under different levels of I_c. D: The overall effect of any additional external or contextual inputs $(\Delta I, \Delta I_c)$ on a neural pair is excitatory or inhibitory depending on whether $\Delta I/\Delta I_c > g'_y(\bar{y})$; this depends on background input I. Adapted with permission from Li, Z., Computational design and nonlinear dynamics of a recurrent network model of the primary visual cortex, *Neural Computation*, 13(8): 1749–1780, Fig. 3, copyright © 2001, MIT Press.

Fig. 5.59 C. The threshold, input gain control, and saturation in $I \to g_x(\bar{x})$ are apparent. The slope $\frac{\delta g_x(\bar{x})}{\delta I}$ is non-negative; otherwise, $I = 0$ gives non-zero output $x \neq 0$. The slope increases with $g'_x(\bar{x})$, decreases with $g'_y(\bar{y})$, and depends on I_c. Shifting (I, I_c) to $(I + \Delta I, I_c + \Delta I_c)$ changes $g_x(\bar{x})$ by

$$\Delta g_x(\bar{x}) \approx (\delta g_x(\bar{x})/\delta I)\left(\Delta I - g'_y(\bar{y})\Delta I_c\right), \tag{5.91}$$

which is positive or negative depending on whether $\Delta I/\Delta I_c > g'_y(\bar{y})$. Hence, a more elaborate model could allow that the interneurons also be partially activated by the external visual input, as is suggested by physiology (White 1989). It would be necessary that $\Delta I/\Delta I_c > g'_y(\bar{y})$.

5.8.2.2 Two interacting pairs of neurons with non-overlapping receptive fields

Consider two vectors (x_1, y_1) and (x_2, y_2) denoting the states of two interacting EI pairs whose connections are $J_{12} = J_{21} = J'$ and $W_{12} = W_{21} = W'$. Then

$$\dot{x}_a = -x_a - g_y(y_a) + J_o g_x(x_a) + J'g_x(x_b) + I_a + I_o,$$
$$\dot{y}_a = -y_a + g_x(x_a) + W'g_x(x_b) + I_c$$

where $a, b = 1, 2$ and $a \neq b$. Hence, to the pair (x_1, y_1), the effect of the pair (x_2, y_2) is the same as adding $\Delta I = J'g_x(x_2)$ to the I_1 and adding $\Delta I_c = W'g_x(x_2)$ to I_c. This gives, according to equation (5.91),

$$\Delta g_x(\bar{x}_1) \approx (\delta g_x(\bar{x}_1)/\delta I) \left(\Delta I - g'_y(\bar{y}_1) \, \Delta I_c \right) \tag{5.92}$$

$$= (\delta g_x(\bar{x}_1)/\delta I) \left(J' - g'_y(\bar{y}_1)W' \right) g_x(x_2). \tag{5.93}$$

Hence,

the net effective connection from x_2 to x_1 is $\quad J' - g'_y(\bar{y}_1)W'. \tag{5.94}$

This connection strength depends on how active the interneuron is, $\bar{y}_1$, and thus it also depends on the input activation of the bar associated with (x_1, y_1). Therefore, since $g'_y(\bar{y}_1)$ tends to increase with the direct input I_1 (and with I_c), the influence on x_1 from the contextual input I_2 becomes more suppressive as the direct input I_1 to the bar becomes stronger. This explains some of the contrast dependence of the contextual influences that is observed physiologically (Sengpiel, Baddeley, Freeman, Harrad and Blakemore 1998). In the simplest case, that $I \equiv I_1 = I_2$, the two bars (associated with these two EI pairs) suppress each other more strongly as input contrast increases, but they can facilitate each other's response when the input contrast is sufficiently weak and J' sufficiently strong. Figure 5.59 D shows an example of how the contextual inputs can switch from being facilitatory to being suppressive as I increases (Stemmler, Usher and Niebur 1995, Somers, Todorov, Siapas, Toth, Kim and Sur 1998).

This very simple model of contextual influence, with only two EI pairs, can be applied to account for various perceptual phenomena involving only single test and contextual bars. For example, a contextual bar can alter the detection threshold (Polat and Sagi 1993, Kapadia et al. 1995) or perceived orientation (Gilbert and Wiesel 1990, Li 1999b) of a test bar.

5.8.2.3 A one-dimensional array of identical bars

Figure 5.60 ABC shows sample input stimuli comprising infinitely long horizontal arrays of evenly spaced, identical bars. These can be approximated as

$$I_{i\theta} = \begin{cases} I_{\text{array}} & \text{for } i = (m_i, n_i = 0) \text{ on the horizontal axis and } \theta = \theta_1, \\ 0 & \text{otherwise.} \end{cases} \tag{5.95}$$

The approximation is to set $I_{i\theta} = 0$ for $\theta \neq \theta_1$; this is reasonable when the input contrast is weak and the neurons have small orientation tuning widths. When bars $i\theta$ outside the array are silent (i.e., $g_x(x_{i\theta}) = 0$) due to insufficient excitation, we omit them and treat the one-dimensional system that only contains the activated neurons. Omitting index θ and using i to index locations, we get

$$\dot{x}_i = -x_i - g_y(y_i) + J_o g_x(x_i) + \sum_{j \neq i} J_{ij} g_x(x_j) + I_{\text{array}} + I_o, \tag{5.96}$$

$$\dot{y}_i = -y_i + g_x(x_i) + \sum_{j \neq i} W_{ij} g_x(x_j) + I_c. \tag{5.97}$$

Translation symmetry implies that all units have the same equilibrium point $(\bar{x}_i, \bar{y}_i) = (\bar{x}, \bar{y})$, and

$$\dot{\bar{x}} = 0 = -\bar{x} - g_y(\bar{y}) + \left(J_o + \sum_{i \neq j} J_{ij} \right) g_x(\bar{x}) + I_{\text{array}} + I_o, \tag{5.98}$$

$$\dot{\bar{y}} = 0 = -\bar{y} + \left(1 + \sum_{i \neq j} W_{ij} \right) g_x(\bar{x}) + I_c. \tag{5.99}$$

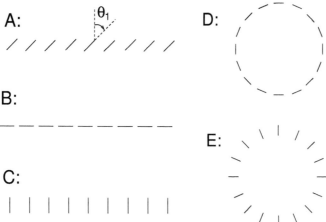

Fig. 5.60: Examples of one-dimensional input stimuli. A: Horizontal array of identical bars oriented at angle θ_1. B: A special case of A when $\theta_1 = \pi/2$ and, in C, when $\theta_1 = 0$. D: An array of bars arranged as being tangential to a circle; the pattern in B is a special case of this circle when the radius is infinitely large. E: Same as D except that the bars are perpendicular to the circle's circumference; the pattern in C is a special case of E when the radius is infinitely large. Reproduced with permission from Li, Z., Computational design and nonlinear dynamics of a recurrent network model of the primary visual cortex, *Neural Computation*, 13(8): 1749–1780, Fig. 4, copyright © 2001, MIT Press.

This array is then equivalent to a single EI neural pair (cf. equations (5.88) and (5.89)), making the substitution $J_o \rightarrow J_o + \sum_j J_{ij}$ and $g'(\bar{y}) \rightarrow g'_y(\bar{y}) \left(1 + \sum_j W_{ij}\right)$. The response to bars in the array is thus higher than to an isolated bar if the net extra excitatory connection

$$\mathcal{E} \equiv \sum_j J_{ij} \tag{5.100}$$

is stronger than the net extra inhibitory (effective) connection

$$\mathcal{I} \equiv g'_y(\bar{y}) \sum_j W_{ij}. \tag{5.101}$$

The input-output relationship $I \rightarrow g_x(\bar{x})$ is qualitatively the same as that for a single bar, but with a quantitative change in the gain

$$\frac{\delta g_x(\bar{x})}{\delta I} = \frac{g'_x(\bar{x})}{1 + g'_x(\bar{x})\left[g'_y(\bar{y}) - (\mathcal{E} - \mathcal{I})\right] - J_o g'_x(\bar{x})}. \tag{5.102}$$

When $\mathcal{E} - \mathcal{I} = 0$, the gain reverts back to that of a single bar. The connections $\mathcal{E}$ and $\mathcal{I}$ depend on the angle θ_1 between the bars and the array; see Fig. 5.60 A. Consider connections as in the association field in Fig. 5.15 B. When the bars are parallel to the array, making a straight line (Fig. 5.60 B), $\mathcal{E} > \mathcal{I}$. The condition for enhancing the responses to a contour is

$$\text{contour facilitation} \quad F_{\text{contour}} \equiv (\mathcal{E} - \mathcal{I})g_x(\bar{x}) > 0. \tag{5.103}$$

When the bars are orthogonal to the array (Fig. 5.60 C), $\mathcal{E} < \mathcal{I}$, and the responses are suppressed. This analysis extends to other one-dimensional, translation-invariant arrays like

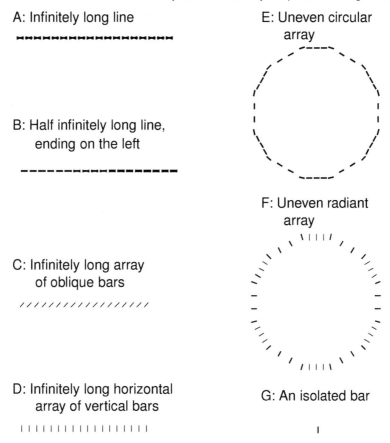

Fig. 5.61: The response $g_x(x_{i\theta})$ of the V1 model (visualized by thickness) to one-dimensional arrays of bars. The input $\hat{I}_{i\theta} = 1.5$ is of low/intermediate contrast for all visible bars. Compared with the isolated bar in G, contextual facilitation causes higher outputs in A, B, E; contextual suppression causes lower outputs in C, D, F. The uneven spacings between the bars (E, F) or at an end of a line (at the left end of B) cause deviations from the translation invariance of responses. Note that the responses taper off near the end of the line in B, and the responses are noticeably weaker to bars that are more densely packed in F. In A and B, cells preferring orientations that are nearly (but not exactly) horizontal are also excited above threshold. This goes beyond the approximate treatment in the text. Adapted with permission from Li, Z., Computational design and nonlinear dynamics of a recurrent network model of the primary visual cortex, *Neural Computation*, 13(8): 1749–1780, Fig. 5, copyright © 2001, MIT Press.

those in Fig. 5.60 DE. The straight line in Fig. 5.60 B is in fact the limit of a circle in Fig. 5.60 D when the radius goes to infinity. Similarly, the pattern in Fig. 5.60 C is a special case of the one in Fig. 5.60 E.

How good the approximations in equations (5.95–5.99) are depends on the input. This is illustrated in Fig. 5.61. In Fig. 5.61 A, cells whose RFs are centered on the line, oriented close to, but not exactly, horizontal, are also excited above threshold. This is not consistent with our approximation $g_x(x_{i\theta}) = 0$ for non-horizontal θ. (This should not cause perceptual problems, though, given population coding.) The cells for these non-horizontal bars can be activated by direct visual input $I_{i\theta}$ for $\theta \neq \theta_1$ ($\theta \approx \theta_1$), due to the finite width of orientation tuning *and* by the colinear facilitation from other bars in or along the line. In Fig. 5.61 B, the approximation of translation invariance $\bar{x}_i = \bar{x}_j$ for all bars is compromised by the fact that the array comes to an end. The bars at or near the left end of the line are less enhanced since they receive less or no contextual facilitation from their left. Uneven spacing between bars in Fig. 5.61 EF also compromises translation invariance. In Fig. 5.61 F, the more densely spaced bars are more strongly suppressed by their neighbors.

5.8.2.4 Two-dimensional textures and texture boundaries

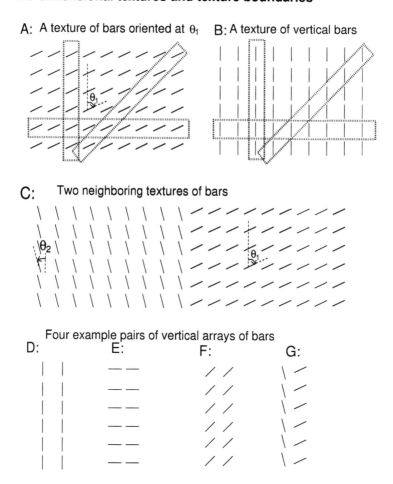

Fig. 5.62: Examples of two-dimensional textures and their interactions. A: Texture made of bars oriented at θ_1 and sitting on a Manhattan grid. This can be seen as a horizontal array of vertical arrays of bars, or indeed as a vertical or oblique array of arrays of bars, as in the dotted boxes. B: A special case of A when $\theta_1 = 0$. C: Two nearby textures with a boundary. D, E, F: Examples of nearby, identical, vertical arrays. G: Two nearby but different vertical arrays. When each vertical array is seen as an entity, one can calculate the effective connections J' and W' between them (see the definitions in the text). Adapted with permission from Li, Z., Computational design and nonlinear dynamics of a recurrent network model of the primary visual cortex, *Neural Computation*, 13(8): 1749–1780, Fig. 6, copyright © 2001, MIT Press.

The analysis of infinitely long one-dimensional arrays can be extended to an infinitely large two-dimensional texture of uniform inputs $I_{i\theta_1} = I_{\text{texture}}$ in the case that $i = (m_i, n_i)$ sits on a regularly spaced grid (Fig. 5.62 A). The sums $\mathcal{E} = \sum_j J_{ij}$ and $\mathcal{I} = g'_y(\bar{y}) \sum_j W_{ij}$ are now taken over all j in that grid.

Physiologically, the response to a bar is reduced when the bar is part of a texture (Knierim and Van Essen 1992). This arises when $\mathcal{E} < \mathcal{I}$. Consider, for example, the case when $i = (m_i, n_i)$ form a Manhattan grid with integer values of m_i and n_i (Fig. 5.62). The texture can be seen as a horizontal array of vertical arrays of bars, e.g., a horizontal array of vertical contours in Fig. 5.62 B. The effective connections between two vertical arrays (Fig. 5.62 DEF) spaced apart by a are:

$$J'_a \equiv \sum_{j,m_j=m_i+a} J_{ij}, \qquad W'_a \equiv \sum_{j,m_j=m_i+a} W_{ij}. \qquad (5.104)$$

Then $\mathcal{E} = \sum_a J'_a$ and $\mathcal{I} = g'_y(\bar{y}) \sum_a W'_a$. The effective connection within a single vertical array is J'_0 and W'_0. One should design J and W such that contour enhancement and texture suppression can occur using the same neural circuit (V1). That is, when the vertical array is a long straight line ($\theta_1 = 0$), contour enhancement (i.e., $J'_0 > g'_y(\bar{y})W'_0$) occurs when the line is isolated, but overall suppression (i.e., $\mathcal{E} = \sum_a J'_a < \mathcal{I} = g'_y(\bar{y}) \sum_a W'_a$) occurs when that line is embedded within a texture of lines (Fig. 5.62 B). This can be satisfied when there is sufficient excitation within a line and sufficient inhibition between the lines.

Computationally, contextual suppression within a texture region endows its boundaries with relatively higher responses, thereby making them salient. The contextual suppression of a bar within a texture is

$$C^{\theta_1}_{\text{whole-texture}} \equiv \sum_a \left[g'_y(\bar{y}_{\theta_1}) W'^{\theta_1}_a - J'^{\theta_1}_a \right] g_x(\bar{x}_{\theta_1}) = (\mathcal{I} - \mathcal{E}) g_x(\bar{x}_{\theta_1}) > 0, \qquad (5.105)$$

where $\bar{x}_{\theta_1}$ denotes the (translation-invariant) fixed point for all texture bars in an infinitely large texture, and $J'^{\theta_1}_a$ and $W'^{\theta_1}_a$ are as the quantities in equation (5.104) with the addition of the superscript θ_1 to indicate the orientation of the texture bars, i.e., the connections J_{ij} and W_{ij} refer to $J_{i\theta_1 j\theta_1}$ and $W_{i\theta_1 j\theta_1}$, respectively, in the full V1 model. Consider the bars on the vertical axis $i = (m_i = 0, n_i)$. Removing the texture bars to their left, $j = (m_j < 0, n_j)$, removes the contextual suppression from them and thereby highlights the boundary bars $i = (m_i = 0, n_i)$. Then the activity $(\bar{x}_{i\theta_1}, \bar{y}_{i\theta_1})$ depends on m_i, the distance of the bars from the texture boundary. As $m_i \to \infty$, the responses $(\bar{x}_{i\theta_1}, \bar{y}_{i\theta_1})$ approach those $(\bar{x}_{\theta_1}, \bar{y}_{\theta_1})$ to the bars within an infinitely large texture. The contextual suppression of the boundary bars ($m_i = 0$) is

$$C^{\theta_1}_{\text{half-texture}} \equiv \sum_{m_j \geq 0} \left[g'_y(\bar{y}_{i\theta_1}) W'^{\theta_1}_{m_j} - J'^{\theta_1}_{m_j} \right] g_x(\bar{x}_{j\theta_1}) \qquad (5.106)$$

$$\approx \sum_{a \geq 0} \left[g'_y(\bar{y}_{\theta_1}) W'^{\theta_1}_a - J'^{\theta_1}_a \right] g_x(\bar{x}_{\theta_1}) < C^{\theta_1}_{\text{whole-texture}}, \qquad (5.107)$$

making the approximation $(\bar{x}_{j\theta_1}, \bar{y}_{j\theta_1}) \approx (\bar{x}_{\theta_1}, \bar{y}_{\theta_1})$ for all $m_j \geq 0$.

The boundary highlight persists when there is a neighboring texture whose bars have a different orientation θ_2, with bar positions $i = (m_i < 0, n_i)$ (Fig. 5.62 C). To analyze this, define connections between arrays in different textures (Fig. 5.62 G) as

$$J'^{\theta_1\theta_2}_a \equiv \sum_{j,m_j=m_i+a} J_{i\theta_1 j\theta_2}, \qquad W'^{\theta_1\theta_2}_a \equiv \sum_{j,m_j=m_i+a} W_{i\theta_1 j\theta_2}. \qquad (5.108)$$

When $\theta_1 = \theta_2$, then $J_a^{\prime\theta_1\theta_2} = J_a^{\prime\theta_1}$ and $W_a^{\prime\theta_1\theta_2} = W_a^{\prime\theta_1}$. The contextual suppression from the neighboring texture (θ_2) on the texture boundary ($m_i = 0$) is

$$C_{\text{neighbor-half-texture}}^{\theta_1,\theta_2} \equiv \sum_{m_j < 0} \left[g_y'(\bar{y}_{i\theta_1}) W_{m_j}^{\prime\theta_1\theta_2} - J_{m_j}^{\prime\theta_1\theta_2} \right] g_x(\bar{x}_{j\theta_2}).$$

With connections as for the association field, $J_{i\theta_1,j\theta_2}$ and $W_{i\theta_1,j\theta_2}$, $J_a^{\prime\theta_1\theta_2}$ and $W_a^{\prime\theta_1\theta_2}$ tend to link similarly oriented bars $\theta_1 \sim \theta_2$. Consequently, $C_{\text{neighbor-half-texture}}^{\theta_1,\theta_2}$ is minimum or zero when $\theta_1 \perp \theta_2$ and increases with decreasing $|\theta_1 - \theta_2|$. Hence, the boundary highlight is expected to increase with the orientation contrast $|\theta_1 - \theta_2|$. The net contextual suppression on the border, contributed by both textures, is

$$C_{\text{two half-textures}}^{\theta_1,\theta_2} \equiv C_{\text{half-texture}}^{\theta_1} + C_{\text{neighbor-half-texture}}^{\theta_1,\theta_2}.$$

Hence, the border enhancement, or the reduction of contextual suppression at the border relative to regions further inside the texture is

$$\delta C \equiv C_{\text{whole-texture}}^{\theta_1} - C_{\text{two half-textures}}^{\theta_1,\theta_2} \tag{5.109}$$

$$\approx C_{\text{neighbor-half-texture}}^{\theta_1,\theta_2=\theta_1} - C_{\text{neighbor-half-texture}}^{\theta_1,\theta_2} \tag{5.110}$$

$$\approx \sum_{a<0} \left[g_y'(\bar{y}_{\theta_1}) W_a^{\prime\theta_1} - J_a^{\prime\theta_1} \right] g_x(\bar{x}_{\theta_1}) \tag{5.111}$$

$$- \sum_{a<0} \left[g_y'(\bar{y}_{\theta_1}) W_a^{\prime\theta_1\theta_2} - J_a^{\prime\theta_1\theta_2} \right] g_x(\bar{x}_{\theta_2}). \tag{5.112}$$

Here, in addition to the previous approximation $(\bar{x}_{j\theta_1}, \bar{y}_{j\theta_1}) \approx (\bar{x}_{\theta_1}, \bar{y}_{\theta_1})$ for all $m_j \geq 0$, we approximated $\bar{x}_{j\theta_2} \approx \bar{x}_{\theta_2}$ for $m_j < 0$. Usually $\bar{x}_{\theta_2} \neq \bar{x}_{\theta_1}$ since the fixed point should depend on the relative orientation between the bars and the orientation of the arrays.

Assuming $J_a^{\prime\theta_1\theta_2} \approx 0$ and $W_a^{\prime\theta_1\theta_2} \approx 0$ when $|\theta_1 - \theta_2| = \pi/2$, and noting that $\bar{x}_{\theta_1} \approx \bar{x}_{\theta_2}$ when $\theta_1 \approx \theta_2$,

$$\delta C \approx \begin{cases} 0 & \text{for } \theta_1 \approx \theta_2, \\ \sum_{a<0} \left[g_y'(\bar{y}_{\theta_1}) W_a^{\prime\theta_1} - J_a^{\prime\theta_1} \right] g_x(\bar{x}_{\theta_1}) > 0 & \text{for } \theta_1 \perp \theta_2, \\ \text{roughly increases} & \text{as } |\theta_1 - \theta_2| \text{ increases.} \end{cases} \tag{5.113}$$

Thus the border highlight diminishes as the orientation contrast approaches 0; this was seen in Fig. 5.23. Furthermore, even at a given contrast $|\theta_1 - \theta_2|$, the border enhancement δC depends on θ_1. For instance, with $|\theta_1 - \theta_2| = \pi/2$ and using the association field connections, the enhancement δC for border bars parallel to the border $\theta_1 = 0$ (which form a contour) is higher than that for border bars perpendicular to the border $\theta_1 = \pi/2$. This is because both the suppression $g_y'(\bar{y}_{\theta_1}) W_a^{\prime\theta_1} - J_a^{\prime\theta_1}$ between parallel contours ($\theta_1 = 0$ and $a \neq 0$) and the facilitation $J_0^{\prime\theta_1} - g_y'(\bar{y}_{\theta_1}) W_0^{\prime\theta_1}$ within a contour (Fig. 5.62 D) are much stronger than their counterparts for the vertical arrays of horizontal bars (Fig. 5.62 E). Thus the strength of the border highlight is predicted to be tuned to the relative orientation θ_1 between the border and the bars (Li 2000b). This explains the asymmetry in the responses in Fig. 5.22 B—the highlight of the vertical border is much stronger for the vertical texture bars at the border than for the horizontal ones.

The approximations $(\bar{x}_{i\theta_1}, \bar{y}_{i\theta_1}) \approx (\bar{x}_{\theta_1}, \bar{y}_{\theta_1})$ for $m_i \geq 0$ and $\bar{x}_{i\theta_2} \approx \bar{x}_{\theta_2}$ for $m_i < 0$, which were used to arrive at equation (5.113), clearly break down at the border. This is especially true at more salient borders like that in Fig. 5.22 B. This breakdown accentuates the tuning of the border highlight to θ_1.

As oft noted, iso-orientation suppression underlies the border highlight. By equation (5.105), its strength $\mathcal{I} - \mathcal{E}$ depends on input contrast through $g'_y(\bar{y})$. Since $g'_y(\bar{y})$ usually increases with increasing $\bar{y}$, the highlight is stronger at higher contrast. This is essentially the same as the dependence on input contrast of the contextual influence between two bars analyzed around equation (5.94). From just the perspective of dynamics, neural connections could be designed such that either of the following two situations holds: one is that iso-orientation suppression holds at all input contrasts; the other is that iso-orientation suppression holds only at sufficiently high input contrast and becomes iso-orientation facilitation at very low contrast (Li 1998a, Li 1999b). Psychophysically, texture segmentation does require an input contrast that is well above the texture detection threshold (Nothdurft 1994), suggesting that iso-orientation suppression diminishes at low input contrast. Computationally, facilitation certainly helps texture detection, which could be more important than segmentation at low input contrast.

For the parameters we have used here for the V1 model (Li 1998a, Li 1999b) (given in the appendix; see Section 5.9), contour facilitation ($F_{\text{contour}} > 0$) occurs at all contrasts, since no W connection links the contour segments. Connections different from the bowtie association field would have to be employed to model diminished contour enhancement at high contrast (Sceniak, Ringach, Hawken and Shapley 1999).

5.8.2.5 Translation invariance and pop-out

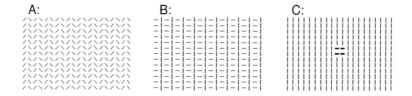

Fig. 5.63: Model responses to globally homogenous (A, B) and inhomogenous (C) input images, each composed of bars of equal input contrasts. A: The response to this globally homogenous (though locally inhomogenous) texture is uniform saliency. B: In this globally homogenous texture, the vertical bars are more salient than the horizontal ones; however, the whole texture has a translation invariant saliency distribution. C: The small figure pops out from the background; this is where translation invariance breaks down in the input, with the whole figure being its own boundary. Adapted with permission from Li, Z., Computational design and nonlinear dynamics of a recurrent network model of the primary visual cortex, *Neural Computation*, 13(8): 1749–1780, Fig. 8, copyright © 2001, MIT Press.

Consider the familiar case from above of two homogenous textures, each made of iso-oriented bars. There is an orientation contrast at the border between the two textures. This border is highlighted by higher responses to the border bars. However, if orientation contrasts are spatially homogenous within the texture itself, then their evoked responses will also be spatially homogenous. Figure 5.63 A shows an example in which the texture is made of alternating columns of bars at $\theta_1 = 45°$ and $\theta_2 = 135°$ in even and odd columns, respectively. Let $(\bar{x}_{\theta_1}, \bar{y}_{\theta_1})$ and $(\bar{x}_{\theta_2}, \bar{y}_{\theta_2})$ denote the states for the bars oriented at θ_1 and θ_2, respectively. The contextual suppression of a bar oriented at θ_1 is

$$C_{\text{complex-texture}} = \sum_{\text{even } a} \left[g'_y(\bar{y}_{\theta_1}) W'^{\theta_1}_a - J'^{\theta_1}_a \right] g_x(\bar{x}_{\theta_1})$$

$$+ \sum_{\text{odd } a} \left[g'_y(\bar{y}_{\theta_1}) W'^{\theta_1\theta_2}_a - J'^{\theta_1\theta_2}_a \right] g_x(\bar{x}_{\theta_2}).$$

Since $C_{\text{complex-texture}} \neq C^{\theta_1}_{\text{whole-texture}}$, the value of $\bar{x}_{\theta_1}$ is not the same as it would be in a simple texture of bars uniformly oriented at θ_1. This applies similarly to $\bar{x}_{\theta_2}$. Furthermore, for general θ_1 and θ_2, $\bar{x}_{\theta_1} \neq \bar{x}_{\theta_2}$. In Fig. 5.63 A, reflection symmetry leads to $\bar{x}_{\theta_1} = \bar{x}_{\theta_2}$, i.e., a uniform saliency within the whole texture. In Fig. 5.63 B, the vertical bars evoke higher responses than the horizontal ones because of contour facilitation. Nevertheless, no local patch within this complex texture is more salient than any other patch. This translation invariance in saliency is simply the result of the network preserving the translation invariance in the input (texture), as long as the translation symmetry is not spontaneously broken (see Section 5.8.1).

A special case of a texture boundary is when one small texture patch is embedded in a large and different texture. The small texture is small enough that the whole texture is its own boundary, and thus pops out from the background (Fig. 5.63 C). In general, orientation contrasts do not correspond to texture boundaries and thus do not necessarily pop out. Through contextual influences, the highlight at a texture border can alter responses to nearby locations up to a distance comparable to the length of the lateral connections. Hence, the response to a texture region is not homogenous unless this region is far enough away from the border. This was elaborated in Section 5.4.5.

5.8.2.6 Filling-in and leaking-out

Small fragments of a contour or homogenous texture can be missing in inputs due to input noise or to the visual scene itself. We define filling-in as the phenomenon of not noticing or not perceiving the missing visual input fragments as missing. It could be caused by one of the following two possible mechanisms. The first is that the neurons for the missing fragment are activated by contextual influences just as if they had actually received direct visual input themselves. This is a common assumption in models (Grossberg and Mingolla 1985). The second possibility is that, even though the missing fragments do not evoke any significant response in the neurons whose RFs cover their locations, the input locations bordering the missing fragments are not sufficiently salient or conspicuous to attract attention strongly. In other words, according to this explanation, filling-in arises if the "saliency of the hole," which we analyzed in Section 5.4.4.7, is insufficient. Consequently, the missing fragments are only noticeable by visual scrutiny. It is not yet clear from physiology (Kapadia et al. 1995) which mechanism is involved.

Consider these two mechanisms for the case of a single bar segment $i = (m_i = 0, n_i = 0)$ that is missing in a smooth contour—e.g., the horizontal line of Fig. 5.64 A. To excite the cell i to firing threshold, i.e., $x_i > x_{\text{th}}$ (such that $g_x(x_i) > 0$), contextual facilitation

$$\sum_{j=(m_j \neq 0, n_j = 0)} \left[J_{ij} - W_{ij} g'_y(\bar{y}_i) \right] g_x(\bar{x}_j)$$

should be strong enough, or, approximately,

$$F_{\text{contour}} + I_o = (\mathcal{E} - \mathcal{I}) g_x(\bar{x}) + I_o > x_{\text{th}}, \tag{5.114}$$

where I_o is the background input not caused by external visual input, F_{contour} and the effective net connections $\mathcal{E}$ and $\mathcal{I}$ are as defined in equations (5.100–5.103), and the approximation $(\bar{x}_j, \bar{y}_j) \approx (\bar{x}, \bar{y})$ is adopted as if responses to all the non-missing bars are unaffected by the missing fragment.

A: B:

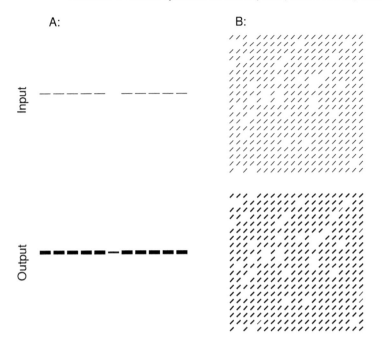

Fig. 5.64: Examples of filling-in; model responses $g_x(x_{i\theta})$ to inputs composed of bars of equal contrasts in each example. A: A line with a gap; the response to the gap is non-zero. B: A texture with missing bars; the responses to bars near the missing bars are not noticeably higher than the responses to other texture bars. Adapted with permission from Li, Z., Computational design and nonlinear dynamics of a recurrent network model of the primary visual cortex, *Neural Computation*, 13(8): 1749–1780, Fig. 9, copyright © 2001, MIT Press.

However, it is necessary that finite lines do not either grow longer or fatter as a result of contextual effects. For the first, we require that the neuron $i = (m_i = 0; n_i = 0)$ should not be excited above the threshold x_{th} if the left (or right) half $j = (m_j < 0, n_j = 0)$ of the horizontal contour is removed. Otherwise the contour will extend beyond its end or grow in length—this is referred to as leaking-out. To prevent leaking-out,

$$F_{\text{contour}}/2 + I_o < x_{\text{th}}, \tag{5.115}$$

since the contour facilitation to i is approximately $F_{\text{contour}}/2$, half of that F_{contour} in an infinitely long contour. The inequality (5.115) is satisfied for the line end in Fig. 5.61 B and should hold at any contour saliency $g_x(\bar{x})$. Not having leaking-out also means that large gaps in lines can not be filled in.

To the extent that segments within a smooth contour facilitate each other's firing, missing fragment i reduces the saliencies of the neighboring contour segments $j \approx i$. The missing segment and its vicinity are thus not easily noticed, even if the cell i for the missing segment does not fire.

To prevent contours growing fatter—e.g., the activation of $i = (m_i = 0, n_i = 1)$ along the side of an infinitely long horizontal contour such as that in Fig. 5.60 B, we require

$$\sum_{j \in \text{contour}} \left[J_{ij} - g'_y(\bar{y}_i) W_{ij} \right] g_x(\bar{x}) < x_{\text{th}} - I_o$$

for $i \notin$ contour. This condition is satisfied in Fig. 5.61 A.

Filling-in in an iso-orientation texture with missing fragments i (texture filling-in) can only arise from the second mechanism, i.e., to avoid conspicuousness near i. This is because i can not be excited to fire given that the net contextual influence within an iso-orientation texture is suppressive (which also means that textures do not suffer leaking-out around their borders). If i is not missing, its neighbor $k \approx i$ receives contextual suppression, as in equation (5.105) but omitting the index θ for the orientation of the bars for simplicity,

$$C_{\text{whole-texture}} = (\mathcal{I} - \mathcal{E})g_x(\bar{x}) \equiv \sum_{j \in \text{texture}} \left[g'_y(\bar{y})W_{kj} - J_{kj}\right] g_x(\bar{x}). \tag{5.116}$$

A missing i makes its neighbor k more salient by the removal of its contribution, which is approximately $\left[W_{ki}g'_y(\bar{y}) - J_{ki}\right] g_x(\bar{x})$, to the suppression. This contribution should be a negligible fraction of the total suppression, in order to ensure that the neighbors are not too conspicuous. Hence,

$$g'_y(\bar{y})W_{ki} - J_{ki} \ll (\mathcal{I} - \mathcal{E}) \equiv \sum_{j \in \text{texture}} \left[g'_y(\bar{y})W_{kj} - J_{kj}\right]. \tag{5.117}$$

This can be expected to hold when the lateral connections are extensive and reaching a large enough contextual area, i.e., when $W_{ki} \ll \sum_j W_{kj}$ and $J_{ki} \ll \sum_j J_{kj}$.

Note that there is an inevitable conflict between active filling-in by exciting the cells for a gap in a contour (equation (5.114)) and preventing leaking-out from contour ends (equation (5.115)). It is not difficult to build a model that achieves active filling-in. However, preventing the model from leaking-out and creating illusory contours that are not perceptually apparent implies a small range of choices for the connection strengths in J and W.

5.8.2.7 Hallucination prevention, and neural oscillations

To ensure that the model performs the desired computations analyzed in the previous sections, the mean or fixed points $(\bar{\mathbf{X}}, \bar{\mathbf{Y}})$ should correspond to the actual behavior of the model. Here we use bold face capitals to represent state vectors. Section 5.8.1 showed that an EI network exhibits oscillatory responses. It also showed that these oscillations can be exploited to prevent hallucinations (or spontaneous symmetry breaking) such that the temporally averaged model responses can correspond to the desired fixed points $(\bar{\mathbf{X}}, \bar{\mathbf{Y}})$. To ensure this correspondence, we now revisit the stability conditions in the general case, to examine constraints on J and W over and above the requirements for enhancing contours and texture borders (the inequalities (5.103), (5.105), (5.114), (5.115), and (5.117)).

To simplify the notation, we denote the deviation $(\mathbf{X} - \bar{\mathbf{X}}, \mathbf{Y} - \bar{\mathbf{Y}})$ from the fixed point $(\bar{\mathbf{X}}, \bar{\mathbf{Y}})$ just as $(\mathbf{X}, \mathbf{Y})$. A Taylor expansion of equations (5.8) and (5.9) around the fixed point gives the linear approximation

$$\begin{pmatrix} \dot{\mathbf{X}} \\ \dot{\mathbf{Y}} \end{pmatrix} = \begin{pmatrix} -1 + \mathbb{J} & -\mathbb{G}'_y \\ \mathbb{G}'_x + \mathbb{W} & -1 \end{pmatrix} \begin{pmatrix} \mathbf{X} \\ \mathbf{Y} \end{pmatrix}, \tag{5.118}$$

where $\mathbb{J}, \mathbb{W}, \mathbb{G}'_x$, and $\mathbb{G}'_y$ are matrices with elements $\mathbb{J}_{i\theta j\theta'} = J_{i\theta j\theta'}g'_x(\bar{x}_{j\theta'})$ for $i \neq j$, $\mathbb{J}_{i\theta,i\theta} = J_\theta g'_x(\bar{x}_{i\theta})$, $\mathbb{W}_{i\theta j\theta'} = W_{i\theta j\theta'}g'_x(\bar{x}_{j\theta'})$ for $i \neq j$, $\mathbb{W}_{i\theta,i\theta'} = 0$, $(\mathbb{G}'_x)_{i\theta j\theta'} = \delta_{ij}\delta_{\theta\theta'}g'_x(\bar{x}_{j\theta'})$, and $(\mathbb{G}'_y)_{i\theta j\theta'} = \delta_{ij}\psi(\theta - \theta')g'_y(\bar{y}_{j\theta'})$, with $\psi(0) = 0$. To focus on the output $\mathbf{X}$, eliminate variable $\mathbf{Y}$ to give

$$\ddot{\mathbf{X}} + (2 - \mathbb{J})\dot{\mathbf{X}} + \left[\mathbb{G}'_y(\mathbb{G}'_x + \mathbb{W}) + 1 - \mathbb{J}\right]\mathbf{X} = 0. \tag{5.119}$$

As inputs, we consider bars arranged in a translation-invariant fashion in a one- or two-dimensional array. For simplicity and approximation, we again omit bars outside the arrays and omit the index θ. This simplification and translation symmetry implies $(\bar{x}_i, \bar{y}_i) = (\bar{x}, \bar{y})$, $(\mathbb{G}'_y)_{ij} = \delta_{ij} g'_y(\bar{y})$, $(\mathbb{G}'_x)_{ij} = \delta_{ij} g'_x(\bar{x})$, $(\mathbb{G}'_y \mathbb{G}'_x)_{ij} = g'_x(\bar{x}) g'_y(\bar{y}) \delta_{ij}$, and $(\mathbb{G}'_y \mathbb{W})_{ij} = g'_y(\bar{y}) \mathbb{W}_{ij}$. Furthermore, $J(i-j) \equiv \mathbb{J}_{ij} = \mathbb{J}_{i+a,j+a}$ and $W(i-j) \equiv \mathbb{W}_{ij} = \mathbb{W}_{i+a,j+a}$ for any a. One can now perform a spatial Fourier transform to obtain

$$\ddot{\mathcal{X}}_k + (2 - \mathcal{J}_k)\dot{\mathcal{X}}_k + \{g'_y(\bar{y})\left[g'_x(\bar{x}) + \mathcal{W}_k\right] + 1 - \mathcal{J}_k\}\mathcal{X}_k = 0, \qquad (5.120)$$

where $\mathcal{X}_k$ is the Fourier component of $\mathbf{X}$ for frequency k such that $e^{ikn} = 1$ (note that, here, we use a non-italic font for "i" in a mathematical expression, such as e^{ikn}, to indicate that $i = \sqrt{-1}$ denotes an imaginary unit value and that "i" in a mathematical expression is not an index), where n is the number of spatial positions in visual space (in two-dimensional space, this becomes $e^{ik_x n_x + ik_y n_y} = 1$), $\mathcal{J}_k = \sum_a J(a)e^{-ika}$, and $\mathcal{W}_k = \sum_a W(a)e^{-ika}$. $\mathcal{X}_k$ evolves in time t as $\mathcal{X}_k(t) \propto e^{\gamma_k t}$ where

$$\gamma_k \equiv -1 + \mathcal{J}_k/2 \pm i\sqrt{g'_y(g'_x + \mathcal{W}_k) - \mathcal{J}_k^2/4}. \qquad (5.121)$$

When $Re(\gamma_k)$, the real part of γ_k, is negative for all k, any deviation $\mathbf{X}$ decays to zero, and hence no hallucination can occur. Otherwise, the mode k with the largest $Re(\gamma_k)$ will dominate the deviation $\mathbf{X}(t)$. If this mode has zero spatial frequency $k = 0$, then the dominant deviation is translation invariant and synchronized across space, and hence no spatially varying pattern can be hallucinated. Thus, the conditions to prevent hallucinations are

$$Re(\gamma_k) < 0 \qquad \text{for all } k, \qquad \text{or} \qquad Re(\gamma_k)|_{k=0} > Re(\gamma_k)|_{k\neq 0}. \qquad (5.122)$$

When $Re(\gamma_{k=0}) > 0$, the fixed point is not stable, and the homogenous deviation $\mathbf{X}$ is eventually confined by the threshold and saturating nonlinearity. All spatial units x_i oscillate synchronously over time when $g'_y(g'_x + \mathcal{W}_0) - \mathcal{J}_0^2/4 > 0$ or when there is no other fixed point which the system trajectory can approach.

Since $J(a) = J(-a) \geq 0$ and $W(a) = W(-a) \geq 0$, $\mathcal{J}_k$ and $\mathcal{W}_k$ are both real. They are largest, as $\mathcal{J}_0$ and $\mathcal{W}_0$, respectively, for the zero frequency $k = 0$ mode. Many simple forms of J and W make $\mathcal{J}_k$ and $\mathcal{W}_k$ decay with k. For example, $J(a) \propto e^{-a^2/2}$ gives $\mathcal{J}_k \propto e^{-k^2/2}$. However, the dominant mode is determined by the value of $Re(\gamma_k)$ and may be associated with $k \neq 0$. In principle, given a model interaction J and W and a translation-invariant input, whether it is arranged on a Manhattan grid or some other grid, $Re(\gamma_k)$ should be evaluated for all k to ensure appropriate behavior of the model (i.e., that inequalities (5.122) are satisfied). In practice, only a finite set of k modes needs to be examined, this is because there is a finite range of connections J and W, spatial locations in the image grid is discrete, and there is a rotational symmetry on this image grid.

Let us look at some examples using the bowtie connections shown in Fig. 5.15 B. For an isolated contour input like that in Fig. 5.60 B, $\mathcal{W}_{ij} = 0$. Then,

$$Re(\gamma_k) = Re\left(-1 + \mathcal{J}_k/2 \pm i\sqrt{g'_y g'_x - \mathcal{J}_k^2/4}\right)$$

increases with $\mathcal{J}_k$, whose maximum occurs at the translation-invariant mode $k = 0$, and $\mathcal{J}_0 = \sum_j \mathbb{J}_{ij}$. Then no hallucination can happen, though synchronous oscillations can occur when enough excitatory connections J link the units involved. For one-dimensional non-contour inputs like Fig. 5.60 CE, $\mathbb{J}_{ij} = 0$ for $i \neq j$; thus $\mathcal{J}_k = \mathbb{J}_{ii}$, and $\gamma_k = -1 + \mathbb{J}_{ii}/2 \pm i\sqrt{g'_y(g'_x + \mathcal{W}_k) - \mathbb{J}_{ii}^2/4}$. Hence $Re(\gamma_k) < -1 + \mathbb{J}_{ii} = -1 + J_0 g'_x(\bar{x}) < 0$ for all k, since $-1 + J_0 g'_x(\bar{x}) < 0$ is always satisfied (otherwise an isolated principal unit x, which follows equation $\dot{x} = -x + J_x g_x(x) + I$, is not well behaved). Hence there should be no hallucination or oscillation.

For two-dimensional texture inputs, frequency $k = (k_x, k_y)$ is a wave vector perpendicular to the peaks and troughs of the waves. When $k = (k_x, 0)$ is in the horizontal direction, $\mathcal{J}_k = g'(\bar{x}) \sum_a J'_a e^{-ik_x a}$ and $\mathcal{W}_k = g'(\bar{x}) \sum_a W'_a e^{-ik_x a}$, where J'_a and W'_a are the effective connections between two texture columns as defined in equation (5.104) (except for $J'_0 = J_o + \sum_{j, m_j = m_i} J_{ij}$). Hence, the texture can be analyzed as a one-dimensional array as above, substituting bar-to-bar connections with column-to-column connections. However, the column-to-column connections J'_a and W'_a are stronger, have a more complex Fourier spectrum ($\mathcal{J}_k, \mathcal{W}_k$), and depend on the orientation θ_1 of the texture bars. Again we use the bowtie connection pattern as an example. When $\theta_1 = 90°$ (horizontal bars), W'_b is weak between columns, i.e., $W'_b \approx \delta_{b0} W'_0$ and $\mathcal{W}_k \approx \mathcal{W}_0$. Then, $Re(\gamma^k)$ is largest when $\mathcal{J}_k$ is, at $k_x = 0$—a translation-invariant mode. Hence, illusory saliency waves (peaks and troughs) perpendicular to the texture bars are unlikely. Consider, however, vertical texture bars for the horizontal wave vector $k = (k_x, 0)$. The bowtie connection gives nontrivial J'_b and W'_b between vertical columns, or non-trivial dependencies of $\mathcal{J}_k$ and $\mathcal{W}_k$ on k. The dominant mode with the largest $Re(\gamma_k)$ is not guaranteed to be homogenous, and J and W must be designed carefully, or screened, to prevent hallucination.

Given a non-hallucinating system (i.e., when spontaneous symmetry breaking is prevented), and under simple or translation-invariant inputs, neural oscillations, if they occur, can only be homogenous, i.e., synchronous and identical among the units involved, with $k = 0$. Since $\gamma^0 = -1 + \mathcal{J}_0/2 \pm i \sqrt{g'_y(g'_x + \mathcal{W}_0) - \mathcal{J}_0^2/4}$, and $\mathcal{J}_k = \sum_j \mathbb{J}_{ij}$ for $k = 0$, the tendency to oscillate increases with increasing excitatory-to-excitatory links J_{ij} between units involved. Hence, this tendency is likely to be higher for two-dimensional texture inputs than for one-dimensional array inputs, and it is lowest for a single small bar input. This may explain why neural oscillations are observed in some but not all physiological experiments. Under the bowtie connections, a long contour is more likely to induce oscillations than a one-dimension input that does not form a contour (Li 2001). These predictions can be physiologically tested. Indeed, physiologically, grating stimuli are more likely to induce oscillations than bar stimuli (Molotchnikoff, Shumikhina and Moisan 1996).

The oscillation frequency for the model is $\sqrt{g'_y(g'_x + \mathcal{W}_0) - \mathcal{J}_0^2/4}$ in the linear approximation for a homogeneous one-dimensional or two-dimensional input. It increases with the total connection $\mathcal{W}_0$ from the pyramidal cells to the interneurons and decreases with the total connection $\mathcal{J}_0$ between pyramidal cells (even when considering nonlinearity when $Re(\gamma^0) > 0$). Adjusting these connection strengths can make the oscillation frequencies exhibited by the model resemble those observed physiologically, e.g., gamma oscillations.

5.8.3 Extensions and generalizations

Understanding neural circuit dynamics is essential to reveal the computational potential and limitations, and it has allowed an appropriate design of the V1 model (Li 1998a, Li 1999b). The analysis techniques presented here can be applied to other recurrent networks whose neural connections are translationally symmetric.

Many quantitatively different models that share the same qualitative architecture can satisfy the design principles described. My V1 model is one such, and interested readers can explore further comparisons between the behavior of that model and experimental data. Although the behavior of this model agrees reasonably well with experimental data, there must be better and quantitatively different models. In particular, connection patterns which are not bowtie like (unlike those in my model) could be more computationally flexible and could thus account for additional experimental data.

Additional or different computational goals might call for a more complex or different design; this might even be necessary to capture aspects of V1 that we have not yet modeled. For example, our model lacks an end-stopping mechanism for V1 neurons. Such a mechanism could highlight the ends of, or gaps in, a contour. By contrast, responses in our model to these features are reduced (relative to the rest of the contour) due to less contour facilitation (Li 1998a). Highlighting line ends can be desirable, especially for high input contrasts, when the gaps are clearly not due to input noise, since both the gaps and ends of contours can be behaviorally meaningful. Without end-stopping, our model is fundamentally limited in performing these computations. Our model also does not generate subjective contours like those evident in the Kanizsa triangle (see Fig. 6.6 B) or the Ehrenstein illusion (which could enable one to see a circle whose contour connects the interior line ends of bars in Fig. 5.60 E). However, these perceptions of illusory contours should be more related to decoding and, hence, not the same as saliency. Evidence (von der Heydt et al. 1984) suggests that area V2, rather than V1, is more likely to be responsible for these subjective contours; they are the focus of other models (Grossberg and Mingolla 1985, Grossberg and Raizada 2000).

5.8.3.1 Generalization of the model to other feature dimensions such as scale and color

The V1 model presented in this chapter omits for simplicity other input features such as color, scale, motion direction, eye of origin, and disparity. However, the same principle to detect and highlight deviations from input translation invariance should apply when these other dimensions are included. For example, it is straightforward to add at each spatial location i additional model neurons tuned to different colors or to different conjunctions of color and orientation. Iso-color suppression can be implemented analogously to iso-orientation suppression by having nearby pyramidal neurons tuned to the same or similar colors (or color-orientation conjunctions) suppress each other by disynaptic suppression. Such an augmented V1 model (Li 2002, Zhaoping and Snowden 2006) can explain feature pop-out by color, as expected, and can also explain interactions between color and orientation in texture segmentation.

It may also be desirable to generalize the notion of "translation invariance" to the case in which the input is not homogenous in the image plane but instead is generated from a homogenous flat-textured surface slanted in depth (Li 1999b, Li 2001). If so, V1 outputs to such images should be homogenous, preventing any visual location from being significantly more salient than any other. This would require multiscale image representations and recurrent interactions between cells tuned to different scales. More specifically, model neurons should be tuned to both orientation and scale; therefore, iso-feature suppression should be implemented between neurons tuned to the same scale, to make iso-scale suppression, and extended appropriately to between neurons tuned to neighboring scales.

5.9   Appendix: parameters in the V1 model

The following parameters have been used in equations (5.8) and (5.9) of the V1 model since its initial publication (Li 1998a) in 1998. They have been applied to representations of the visual space based on both hexagonal and Manhattan grids. The $K = 12$ preferred orientations of the model neurons are $\theta = m \cdot \pi/K$ for $m = 0, 1, 2, ..., K - 1$. The units of model time are the membrane time constant of the excitatory neurons, and $\alpha_x = \alpha_y = 1$.

$$g_x(x) = \begin{cases} 0 & \text{if } x < T_x = 1, \\ (x - T_x) & \text{if } T_x \le x \le T_x, \\ 1 & \text{if } x > T_x + 1; \end{cases} \tag{5.123}$$

and $\tag{5.124}$

$$g_y(y) = \begin{cases} 0 & \text{if } y < 0, \\ g_1 y & \text{if } 0 \le y \le L_y, \\ g_1 L_y + g_2(y - L_y) & \text{if } 0 < L_y \le y, \end{cases}$$

in which $T_x = 1$, $L_y = 1.2$, $g_1 = 0.21$, and $g_2 = 2.5$. The function $\psi(\theta) = 0$ except when $|\theta| = 0, \pi/K$, or $2\pi/K$, for which $\psi(\theta) = 1, 0.8$, or 0.7, respectively; $I_c = 1 + I_{c,\text{control}}$, with $I_{c,\text{control}} = 0$ for all simulations in this book. Li (1998a) discusses the case of $I_{c,\text{control}} \ne 0$ to model top-down feedback to V1; $I_o = 0.85 + I_{\text{normalization}}$. For each excitatory neuron $i\theta$,

$$I_{\text{normalization}} = -2.0 \left[\frac{\sum_{j \in S_i} \sum_{\theta'} g_x(x_{j\theta'})}{\sum_{j \in S_i} 1} \right]^2,$$

where S_i is a neighborhood of all grid points j no more than two grid points away from i along each of the axes (horizontal and vertical axes for the Manhattan grid, and hexagonal axes for the hexagonal grid). Noise input, I_{noise}, is random, with an average temporal width of 0.1 and an average height of 0.1, and is independent across different neurons.

The value $J_o = 0.8$. Connections $J_{i\theta, j\theta'}$ and $W_{i\theta, j\theta'}$ are determined as follows. Let $\Delta\theta \equiv \min(a, \pi - a)$ with $a \equiv |\theta - \theta'| < \pi$. Let the connecting line linking the centers of the two elements $i\theta$ and $j\theta'$ have length d in grid units, with θ_1 and θ_2 being the angles between the elements and the connecting line, such that $|\theta_1| \le |\theta_2| \le \pi/2$,

$$|\theta_1| \le |\theta_2| \le \pi/2$$

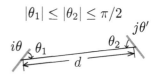

with $\theta_{1,2}$ being positive or negative, respectively, when the element bar rotates clockwise or anticlockwise toward the connecting line through an angle of no more than $\pi/2$. Let $\beta \equiv 2|\theta_1| + 2\sin(|\theta_1 + \theta_2|)$. Then $J_{i\theta, j\theta'}$ is zero except when the following three conditions are satisfied simultaneously: (1) $d > 0$, (2) $d \le 10$, and (3) either $\beta < \pi/2.69$, or $\beta < \pi/1.1$ while $|\theta_2| < \pi/5.9$. Given that these conditions are satisfied,

$$J_{i\theta, j\theta'} = 0.126 \cdot \exp\left[-(\beta/d)^2 - 2(\beta/d)^7 - d^2/90 \right].$$

Meanwhile, $W_{i\theta, j\theta'} = 0$ if any of the following expressions holds: $d = 0$, $[d/\cos(\beta/4)] \ge 10$, $\beta < \pi/1.1$, $|\theta_1| \le \pi/11.999$, $\Delta\theta \ge \pi/3$; otherwise,

$$W_{i\theta, j\theta'} = 0.141 \cdot \left\{ 1 - \exp\left[-0.4(\beta/d)^{1.5} \right] \right\} \exp\left\{ -[\Delta\theta/(\pi/4)]^{1.5} \right\}.$$

In equations (5.8) and (5.9), the external visual input value $I_{i\theta}$ to $x_{i\theta}$ is derived from the input contrast $\hat{I}_{i\gamma}$ of input bars at location i and oriented at γ which are within the orientation tuning width of neuron $i\theta$. In particular, $\hat{I}_{i\gamma}$ contributes $\hat{I}_{i\gamma}\phi(\theta - \gamma)$ to $I_{i\theta}$, as shown in equation (5.10). The function $\phi(x) = \exp[-|x|/(\pi/8)]$ for $|x| < \pi/6$, and $\phi(x) = 0$ otherwise.

6 Visual recognition as decoding

This chapter focuses on visual decoding, which is the last of the three stages of vision: encoding, selection, and decoding. Decoding is manifest partly in the recognition or discrimination of objects, e.g., recognizing a face or telling two objects apart, and partly by identifying the spatial configuration and movement of objects and environments so that they can be manipulated, have their orientations be discriminated, or be navigated, e.g., grasping a tool or walking up a staircase. The latter two operations can occur even without object recognition. Visual perception and discrimination are often identified with visual decoding.

Much less is known about the natural substrates of visual decoding than about visual encoding or bottom-up attentional selection. Since, in this book, we focus on natural vision, we will not use work on visual decoding by artificial or computer vision to fill the void. Instead, in this chapter, we aim for a perspective on visual decoding derived from experimental observations and computational principles. In particular,

1. we present examples in which visual perception and discrimination can be linked to neural response properties in the retina, V1, and extrastriate cortices;
2. we introduce Bayesian inference as a candidate framework for visual decoding, and use this framework to interpret examples of perception and discrimination (with or without a direct link to physiology);
3. we introduce some observed properties of visual recognition such as object invariance, perceptual ambiguity, and speed. We discuss their probable implications for the underlying algorithms, such as whether decoding is performed in a purely feedforward fashion or via a recurrent feedback network.

6.1 Definition of visual decoding

As discussed in Section 1.2, vision should lead to a description of visual scenes that is useful for the execution of tasks promoting the survival of the observer. The decoding stage is responsible for the bulk of this description and so should have a much smaller information content than the visual input does, before the drastic information reduction associated with visual selection. The description should also be dependent on the task, which largely determines the top-down part of the visual selection. Hence, if the task is to chase after a dog, the output description should be dominated by the direction in which the dog is moving but largely omit the information about the dog's fur or breed, or indeed about a butterfly also in the field of view. In such a case, the decoding problem is dominated by the sub-problem of estimating the direction of the dog's movement from the neural activities in the brain. In a different task which involves deciding which of two bananas is riper, the description produced by decoding might refer to the colors of the bananas, with the dominating sub-problem being estimation of the color of a surface patch, again based on neural activities.

Visual decoding must depend on the visual responses of one or more groups of neurons. However, we know very little about which groups are involved. Since we assume that decoding occurs after visual selection, we expect that the neural responses directly used for decoding are likely those in the extrastriate cortex, downstream along the visual pathway from V1,

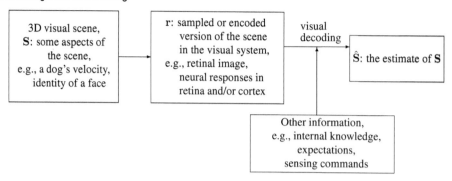

Fig. 6.1: Schematic of the definition of visual decoding in equation (6.1).

where at least part of the bottom-up selection occurs. Part of the V1 response to foveal inputs is also likely used for decoding, according to our discussion in Section 5.7, since selected spatial locations are placed at the fovea by gaze shifts. That much of V1 activities are not used for decoding is consistent with observation from Fig. 5.11 that human observers have great difficulty perceiving or discriminating the eye of origin of visual inputs, even though this information is available in the responses of monocular neurons in V1. However, the retina and the V1 responses are the *sources* of information at the eventual decoding stage after the attentional bottleneck, and so they, along with all other neural stages along the visual pathway up to the stage of decoded output description, might legitimately be regarded as parts of the implementation of visual selection and decoding algorithms.

Let S denote the aspect of the visual scene that is the target of visual decoding. Hence, for example, S might denote the movement direction of the dog, or the color of a surface patch. Let r denote the neural responses to visual input, or even input image pixel values, from which S is estimated. Then, visual decoding can be stated as follows (see Fig. 6.1).

Visual decoding estimates aspects S of a scene from the neural responses it evokes or the input image it forms, r. $\qquad$ (6.1)

Visual decoding can also be called *visual inference*, since aspects S of a scene are inferred from r.

In behavioral experiments, decoding is typically probed by having observers report their percepts or perceptual decisions. However, this does not mean that the observers do not decode aspects of the scene that they do not report. For example, even when observers are only asked to report the moving velocity of the dog, they may also have perceived, i.e., decoded, the breed of the dog. Hence, when we formulate or study a decoding problem focusing only on a certain aspect S of the scene, it should be understood that information in r about the other aspects of the scene may or may not be deleted by visual selection.

Let $\hat{S}$ denote the value of S of the visual scene that is estimated through decoding process. Veridical decoding gives $\hat{S} \approx S$, whereas a visual illusion occurs when $\hat{S}$ is very different from S. Since observers can only tell apart two similar, but not identical, S values up to a point, decoding has a finite resolution. The relationship between S and $\hat{S}$ should reflect the encoding properties of the visual scene in r, along with the mechanisms of decoding. Our way of probing decoding is to study this relationship through the medium of behavioral experiments.

6.2 Some notable observations about visual recognition

To make progress with understanding visual decoding, we first consider a range of critical experimental observations.

6.2.1 Recognition is after an initial selection or segmentation

Inattentional blindness (Simons and Chabris 1999), which is illustrated in Fig. 1.6, demonstrates that we do not perceive objects that are not selected by visual attention. This fact is also evident in many other observations. For example, Treisman (1980) noted that attention is necessary in order to perceive conjunctions of features correctly. If an image containing many green "X"s and red "O"s is shown too briefly, one often sees illusory red "X"s and green "O"s. These non-existent items are called illusory conjunctions, as if the shape of one visual item is mistakenly combined with the color of another item. Visual crowding (Levi 2008), demonstrated in Fig. 5.49, can be seen as another manifestation of the attentional bottleneck. Here, top-down attention cannot be fully effective since gaze is not directed at the peripheral object to be decoded, as directing gaze at attended object is mandatory under natural viewing conditions (Hoffman 1998). These observations are consistent with the proposal that decoding happens only after selection in the three-stage framework of vision (Fig. 1.5).

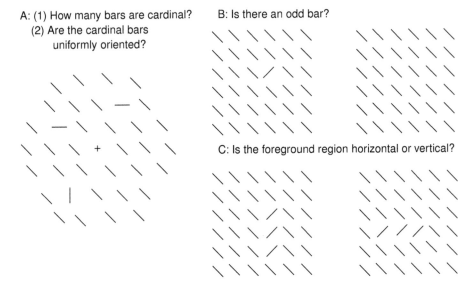

Fig. 6.2: "Where" and "what" visual processes in primates. A: Images such as this are viewed briefly (and are then replaced by a mask). In the "where" task, observers report the number (2, 3, or 4) of cardinal bars in the image; in the "what" task, observers report whether all the cardinal bars have the same orientation. The viewing duration needed for adequate performance is independent of the number (between 2–4) of cardinal bars present for the case of "where," but increases with this number for "what." The necessary viewing durations in the two tasks are similar when there are only two cardinal bars (Sagi and Julesz 1985). B: After a V2 lesion, monkeys can still report whether there is an odd bar in each of these two images. C: After V2 or V4 lesions, monkeys are less able to report whether the three foreground bars in each of such images form a vertical or horizontal array (Merigan et al. 1993, Merigan 1996). C is the same as Fig. 2.28 A.

Visual selection can be seen as a form of visual segmentation, which selects the image area for the visual object of interest. In space-based selection, selecting a location for an object also localizes the object, at least in the two-dimensional image space. Meanwhile, localizing and recognizing an object are different from each other. Indeed, in an image like Fig. 6.2 A, human observers can report (Sagi and Julesz 1985) whether there are 2, 3, or 4 cardinal bars among the oblique bars regardless of whether the cardinal bars are horizontal or vertical after a brief view of the image, and the viewing duration needed for such a report is independent of the number of cardinal bars present. However, if observers have to report whether all the cardinal bars have the same orientation, the viewing time needed increases with the number of cardinal bars present. This suggests that reporting the number of cardinal bars only requires multiple salient positions to be localized and that the (orientation) features that create the saliency need not be recognized. Thus, at least for images like Fig. 6.2 A, "localization" operates in a parallel process, whereas "recognition" is serial and perhaps comes after the "localization" process.

The processes of "localization" and "recognition" are what many call "where" and "what" vision. They are thought to be mainly associated with the dorsal and ventral visual streams, respectively, in the cortical visual pathway (Ungerleider and Mishkin 1982); see Fig. 2.3. If a scene contains multiple objects, knowing "where" the objects are enables one of them to be selected as the target of spatial attention. Hence, it is natural to expect that "where" should be processed before "what" can be completed. In a image like Fig. 6.2 A, it seems that the bottom-up saliency map in V1 can already serve as the answer to "where." Based on this answer, each conspicuous location can be subsequently selected (or segmented from the background), serially enter the attentional bottleneck, and be decoded to give rise to the answer to "what."

The separation between visual selection and decoding, or between segmentation and recognition, is also seen in brain lesion studies (Merigan et al. 1993, Merigan 1996). Monkeys with a lesion in V2 can still report whether there is an odd bar in a texture of uniformly oriented background bars; see Fig. 6.2 B. Here again, the task requires only the "where" operation—in principle, the monkey could simply monitor whether any V1 neuron is substantially more responsive than the other V1 neurons in response to the image, regardless of the preferred orientations of the various neurons concerned. However, a V2 (or V4) lesion severely impairs the animal's ability to report whether the three foreground bars in images like Fig. 6.2 C form a horizontal or vertical array. For this "what" task, the brain not only needs to locate the receptive fields of the most activated V1 neurons (which respond to the salient foreground bars) but must also identify or decode the orientation of an imaginary line linking these three foreground locations.

6.2.2 Object invariance

Figure 6.3 demonstrates that we can recognize the same object when it is viewed from different viewpoints and in different contexts. For example, we can recognize the same object whether it is in the left or right visual field (not far from fovea), demonstrating location invariance. We can also recognize the same object whether it is looked at from a closer or further distance, demonstrating scale invariance. Rotations and reflections of the image, background clutter, and partial object occlusion can also be tolerated typically (exceptions include a severe drop in the performance of face recognition when a face image is turned upside down). Retinal images arising from the same object under different situations are so different that object invariance is one of the biggest challenges for visual recognition.

Figure 6.4 revisits an observation introduced in Fig. 1.4, which suggests that invariance (at least rotation invariance) occurs after the attentional selection of an image location. Figures

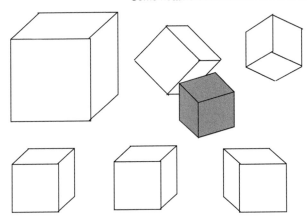

Fig. 6.3: We can recognize the same object under translation, scaling, image reflection, image rotation, three-dimensional object rotation, lighting changes, or when the object is partially occluded and in a busy background.

6.4 A and 6.4 B are identical, except for the orientations of their very salient orientation singletons. Similarly, Fig. 6.4 C and Fig. 6.4 D are identical, except for the same orientation singletons. Although the singletons in Fig. 6.4 C and Fig. 6.4 D are still likely the most salient locations in their images, each is far less salient than its counterpart in Fig. 6.4 A or Fig. 6.4 B, due to the non-uniform orientations of the additional cardinal bars. Observers can find the target in A and B equally fast, typically within a second. They can find the target in D in about 2–3 seconds but typically take at least a few more seconds to find the target in C. This is because the target bar in C is part of an "X" shape that is a rotated version of all the other "X" shapes in the image. Rotation invariance in shape recognition camouflages the target bar. In comparison, the target bar in D is part of an "X" which has a unique shape, being thinner than the other "X"s. Nevertheless, during the search for the orientation singleton, gaze locates the target bar in C just as fast as in D. This suggests that the visual selection that is reflected in the saccades precedes rotation invariance. This finding is consistent with the idea that rotation invariance is part of the visual decoding process for object shape recognition.

It is important to note that the dense placements of "X"s in Fig. 6.4 C and Fig. 6.4 D induces crowding, consistent with the fact that observers typically cannot recognize the "X" shape containing the target bar before their gaze reaches the target. This supports the idea that V1 can signal the high saliency of the target's location to guide gaze to it, without any requirement for recognition. It is also consistent with the observation that if the search image is suddenly removed when the observer's gaze arrives at the target, observers are typically unsure whether they had in fact seen the target at all. In many search trials, the observer's gaze reaches the target in Fig. 6.4 C, hesitates, and then abandons the target to search elsewhere, as seen in the gaze scan example in Fig. 1.4. Such gaze shifts to abandon the target are rare during search in Fig. 6.4 D.

6.2.3 Is decoding the shape of an object in the attentional spotlight a default routine?

The search task in Fig. 6.4 CD does not require the "X" to be recognized; locating and identifying the uniquely oriented oblique bar suffices. Nevertheless, recognizing the "X," which involves combining two oriented bars into a shape, appears to be a built-in default routine, even when observers are told ahead of time to ignore the cardinal bars. Thus, a search

A: Orientation singleton pops out

B: Same as A, except for the orientation of the singleton

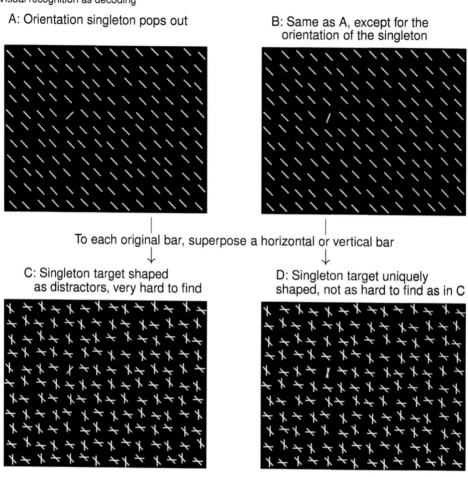

To each original bar, superpose a horizontal or vertical bar

C: Singleton target shaped as distractors, very hard to find

D: Singleton target uniquely shaped, not as hard to find as in C

Fig. 6.4: The rotation invariance of object recognition can interfere with visual search for a uniquely oriented bar. In each image, there is a uniquely tilted orientation singleton, which is likely to be the most salient bar. Images C and D are made from A and B by superposing a horizontal or vertical bar on each oblique bar, intersecting it in the middle, making the whole image contain "X"s. The orientation singleton is now camouflaged in C, because these "X"s are identical apart from a rotation. The orientation singleton in D remains easy to find, since it is part of a uniquely shaped, thinner "X" shape. However, in the task requiring the orientation singleton target to be found (Zhaoping and Guyader 2007), the observer's gaze take on average the same time (1–2 seconds) to locate the orientation singleton in both C and D. This suggests that rotational invariance, which causes the camouflage, occurs after the selection by saliency of the target's location by gaze. Adapted from *Current Biology*, 17(1), Li Zhaoping and Nathalie Guyader, Interference with Bottom-Up Feature Detection by Higher-Level Object Recognition, pp. 26–31, figure 1, copyright (2007), with permission from Elsevier.

task for a low-level image orientation feature that is processed by V1 is being interfered with by higher-level object shape recognition. The latter is likely to occur beyond V1, in areas that include the parietal cortex (as suggested by an experiment using transcranial magnetic stimulation (Oliveri, Zhaoping, Mangano, Turriziani, Smirni and Cipolotti 2010)). Recall from Fig. 5.32 that this interference, due to object invariance by reflection, is also present in searching for a letter "N" among its mirror images or the mirror image among letter "N"s, even though the task does not require recognizing the letter shapes.

However, observers can learn to avoid this interference from the object shape process by adopting a strategy in which they avoid looking at or scrutinizing the target during search. Indeed, they typically find that an efficient way to perform the task is to report the target's location as the most salient location sensed in peripheral vision (Zhaoping and Guyader 2007).

In the example of Fig. 6.4 CD, shape recognition was automatically engaged for objects admitted to the attentional spotlight. This might be a general default, perhaps because recognition normally facilitates visual performance. Figure 6.5 shows an example of a phenomenon known as the configural superiority effect. The RT to find a uniquely left-tilted bar among right-tilted bars can be reduced when these bars are parts of differently shaped objects. In Fig. 6.5 A, each bar can also be seen as an object, and all the four objects have the same shape (up to a rotation). However, in Fig. 6.5 B, the triangle containing the target bar has a unique shape. This eliminates interference and so makes the search faster.

One may notice that Fig. 6.5 A and Fig. 6.4 A are similar except for the set size, and so interference caused by object invariance should be present in both. Search is easier in the latter because the bottom-up saliency of the singleton target is stronger, since V1's iso-orientation suppression of the background that is responsible for the saliency is stronger in a denser texture, as shown in Section 5.4.4.6. (Indeed, gaze shifts more quickly to orientation singletons in denser background textures (Zhaoping and Frith 2011).) Stronger target saliency makes the task less vulnerable to interference from later object recognition. This explains the seemingly paradoxical observation that a search for a singleton bar among uniformly tilted background bars is faster when there are more background bars.

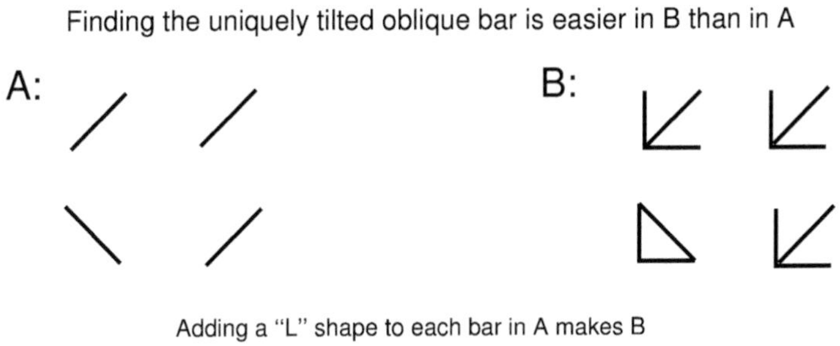

Finding the uniquely tilted oblique bar is easier in B than in A

Adding a "L" shape to each bar in A makes B

Fig. 6.5: Configural superiority effect (Pomerantz et al. 1977). A and B have the same oblique bars; B is made from A by adding an "L" shape to each texture grid location. Rotational invariance makes all bars identical in A. The uniquely left-tilted bar in B can be more easily found, since the triangle in which this bar is embedded has a unique shape.

6.2.4 Recognition by imagination or input synthesis

Figure 6.6 A demonstrates that some images are hard to parse (i.e., to identify the scene **S** that is responsible). Top-down knowledge can greatly help the analysis of this image, as a few hints can help the observer to recognize the foreground object. This suggests, although does not prove, that recognition is a process of imagination and verification—i.e., a process of generating an image from a hypothesized scene **S** (a process called synthesis, which requires a good model of image formation) and verifying whether this generated image resembles the

A: What is in this image?

B: Two triangles and three disks are quickly and vividly perceived

Fig. 6.6: Examples showing that the brain synthesizes visual inputs during object recognition. A: At the very first viewing, one typically cannot immediately recognize the contents of the scene in this image. However, once one recognizes, or is told, the contents of the scene (see the footnote at the bottom of this page for the scene content), it is difficult not to recognize the scene contents immediately the next time this image is viewed. B: The brain perceives a white occluding triangle (called the Kanizsa triangle, after the Italian psychologist) and fills in the occluded parts of three black discs and the outline black triangle. The white triangle may appear brighter than the background even though the actual luminances on the corresponding image locations are identical. Adapted with permission from Kanizsa, G., Subjective contours, *Scientific American*, 234: 48–52, copyright © 1976, Scientific American, INC.

input image on the retina. Recognition by imagination is formally called analysis by synthesis, i.e., analyzing the image by synthesizing it from a hypothesized scene.

Perceiving a white triangle in Fig. 6.6 B happens much more quickly than perceiving the contents in Fig. 6.6 A and can be done without any hints. The brain fills in the occluding object, i.e., a white triangle, as well as the occluded parts of other objects. The three black discs and one triangle with black outline are perceived as whole objects because of the occlusions. Perceiving occluded objects as being whole in such cases is called *amodal completion*. Meanwhile, seeing the occluding white triangle even though it is not drawn is called *modal completion*. This suggests that modal and amodal completion also involves synthesis. However, this synthesis may be qualitatively different, and certainly involve different neural mechanisms, from the synthesis involved in perceiving the duck in Fig. 6.6 A.[24]

---

[24] Figure 6.6 A contains a duck floating on water. The duck is seen roughly from its left; its head is very slightly below the center of the image and to the right, its beak points toward the lower right-hand corner of the image, and its body length is roughly half of the image length.

6.2.5   Visual perception can be ambiguous or unambiguous

A: A vase or two faces?   B: Which face of the cube is in front?   C: The nail illusion—
two nails side by side,
or one in front of the other?

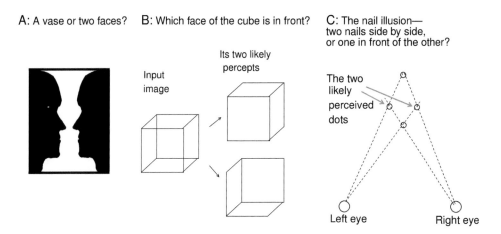

Fig. 6.7: Occasionally, visual perception can be ambiguous. A is called Rubin vase, named after Danish psychologist Edgar Rubin. B is the Necker cube and is named after Swiss crystallographer Louis Albert Necker. In C, the 3D visual scene could contain four dots, which are so aligned that they form an image of two dots in the left eye and another image of two dots in the right eye. The same two retinal images could also be formed from two dots, located either side by side (pointed to by red arrows) or one in front of the other. The percept is typically of two dots side by side, even if the scene really consists of one dot in front of the other. This is called the nail illusion since the dots could be nails or thin needles (depicted in a head-on view in this figure), too thin to provide perspective cues for disambiguating their 3D positions. In the field of stereo vision, when the scene contains two needles side by side, the other 3D locations are called *false matches*.

Visual perception is typically unambiguous, such that (human) observers perceive the same visual input in the same way over the course of a single extended view, or even multiple separate viewings. For example, we are usually certain about what objects are in front of our eyes in a bright environment. However, occasionally, perception can be ambiguous. For example, one can see one vase or two faces in the two-dimensional (2D) image Fig. 6.7 A, and these two percepts typically alternate slowly if viewing is prolonged. The percept depends on whether the white region is assigned as being in front of, or behind, the black region. The input image in Fig. 6.7 B can be seen as a cube viewed from one of the two viewpoints, as depicted. The Necker cube image could also be caused by an object which is not a cube, e.g., something which has a much smaller (or even zero) depth. However, it is almost impossible to see this. Figure 6.7 C shows another example called the nail illusion. It explicitly demonstrates that the same retinal images could arise from multiple possible scenes in a three-dimensional (3D) world. Given that two 2D images only incompletely constrain the full 3D world, it is not surprising that decoding or recognition, which aims to infer that underlying world (the inverse problem for vision discussed in Section 1.2), should be ambiguous. One might even be surprised that the perception is typically unambiguous—do we typically perceive only the most likely cause?

Binocular rivalry (Blake 2001) is another example of ambiguous perception. Figure 6.8 shows two pairs of images shown to the two eyes. The top pair typically evokes the percept of either the left image or the right image. As for the Necker cube, if this stimulus is viewed for

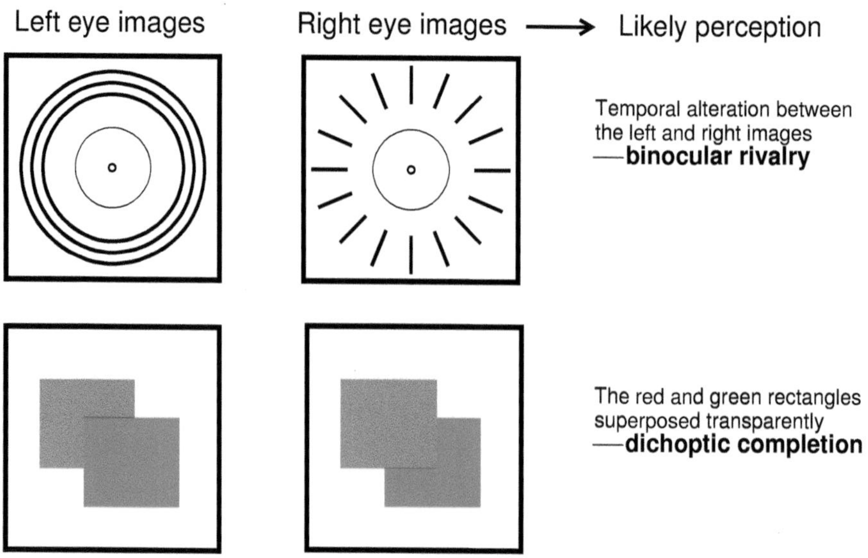

Left eye images   Right eye images ⟶ Likely perception

Temporal alteration between
the left and right images
—**binocular rivalry**

The red and green rectangles
superposed transparently
—**dichoptic completion**

Fig. 6.8: Binocular rivalry (Blake 2001) and dichoptic completion (Zhaoping and Meng 2011). In both pairs of images, the two eyes receive very different inputs at some spatially corresponding regions. When it is difficult to form a coherent hypothesis about the visual scene, rivalry ensues. In the bottom pair, each image alone evokes a perception of two rectangles, since the partially occluded (red or green) rectangle could be amodally completed. This synthesized rectangle actually exists in the real image presented to the other eye.

an extended time, the percept alternates from one possibility to the other, each persisting for a random duration with an average of around 1 second. It is also common to have patchwork rivalry, when the inputs to the two eyes determine different spatial parts of the percept. Figure 3.15 shows another kind of rivalry: the summation of the two monocular inputs rivals with the difference between the monocular inputs. These various forms of perceptual rivalry are consistent with the idea that the competition associated with perception is not so much between the monocular inputs in the dichoptic array, but rather the competition is between different hypotheses that the observer has about the kind of visual scene that could cause the inputs (Logothetis, Leopold and Sheinberg 1996, Dayan 1998).

The bottom pair of images in Fig. 6.8 are more closely related than is typical for binocular rivalry, with only some pixels in one monocular image contradicting those in the other. The percept arising from viewing them dichoptically is usually more stable over time than for typical binocular rivalry, and it involves two transparent, partly overlapping, rectangles. According to analysis by synthesis, this could arise if the inconsistency coming from amodal completion (for which the details from one monocular image have to be invented, or amodally completed, in the other monocular image) is less than the inconsistency coming from occlusion (for which the details from one monocular image have to overwrite the details in the other monocular image). This phenomenon is called *dichoptic completion* (Zhaoping and Meng 2011). It occurs when the unspecified but consistent details of the inferred objects (by amodal completion) from one monocular image appear in the other monocular image. Another example comes from showing the two eyes the two complementary views of a Necker cube; each view contains only three surfaces (like that in one of the two right images in Fig. 6.7 B) such that the two views contain all the six surfaces of the cube. This evokes the percept of a transparent cube comprising all six surfaces, even when the surfaces have different colors (Zhaoping and Meng 2011).

In the case of rivalry, observers often find that they can voluntarily control their percept, at least to some degree. This suggests that analysis by synthesis, if it indeed is an accurate description of visual decoding, is subject to both exogenous and endogenous influences.

6.2.6 Neural substrates for visual decoding

These various observations suggest that visual decoding involves at least the processes of (1) combining simple image features into surfaces or objects, (2) assigning a surface as being the occluder or being occluded, (3) achieving object invariance, and (4) completing the surfaces or boundaries of objects that are missing from the visual inputs. We can then ask about the order in which these processes occur, and about the neural mechanisms involved.

It has been argued that surface representations are built before objects are recognized (Nakayama, He and Shimojo 1995). Recall from Fig. 2.26 that neural responses in V2 contain information or hypotheses about the depth orders of object surfaces in the 3D world, or about surface boundaries, even when this information is ambiguous or not explicitly available in the 2D visual inputs. For example, in Fig. 6.7 A, this would suggest that one infers two faces or a vase by first deciding whether the border belongs to the white or black surface(s). Allocation of border ownership implies the assignment of foreground and background regions in the image. V2 neurons are sensitive to such border ownership signals (Zhou et al. 2000), such that the response of a V2 neuron tuned to the orientation of the border depends on whether the figure is to one side of the border or another. Sensitivity to border ownership (BOWN) is absent or weaker in area V1.

A network model of the V2 neural circuit (Zhaoping 2005a) suggests that V2's intracortical neural mechanisms alone (using lateral interactions between neurons) could feasibly achieve BOWN tuning. However, top-down attention can substantially influence the responses of BOWN tuned neurons (Qiu et al. 2007), with their activity depending on whether attention is directed to one surface or another. Inferring depth, as well as depth order, from images is also associated with V2 and extrastriate cortical areas further along the visual pathway (von der Heydt et al. 1984, Cumming and Parker 2000, Bakin et al. 2000, von der Heydt et al. 2000, Qiu and von der Heydt 2005, Janssen et al. 2003), and an intracortical circuit model of V2 has suggested its feasibility to infer depth (Zhaoping 2002). Given the close relationship between depth and figure-ground relationships, this is consistent with the idea that surface representations are achieved in extrastriate cortical areas, starting from V2.

Visual object recognition has been associated with the lateral occipital and temporal cortical regions (Desimone et al. 1984, Kobatake and Tanaka 1994, Rolls 2003, Tanaka 2003, Logothetis, Pauls and Poggio 1995, Kourtzi and Kanwisher 2001). Neurons tuned to image shapes resembling object components are found in these areas; see Fig. 2.25. Recall from Section 2.4.5 that the responses of many IT neurons do not change substantially in the face of moderate changes in position, scale, orientation of the objects in view, or the degree of background clutter. There are also cortical patches or regions within the temporal cortex that are specialized for recognizing faces (Freiwald et al. 2009, Kanwisher et al. 1997), scene (Epstein and Kanwisher 1998), and body parts (Downing et al. 2006).

Another way to probe cortical areas associated with visual decoding is to observe how neurons respond to images which evoke ambiguous percepts. If a neuron's response does not, or does, co-vary with perceptual changes given a fixed visual input, then this neuron may be upstream or downstream, respectively, of the computations or representation underlying perception. V1 responses are largely determined by visual inputs and are insensitive to perception (except perhaps the V1 representation corresponding to the foveal region), even for the highly disparate inputs giving rise to the marked perceptual switches of binocular rivalry. In contrast, covariations of perception (rather than visual inputs) with neural responses increases as one moves further downstream along the visual pathway (Leopold and Logothetis 1996) and can be dominant in V4. However, our knowledge about this remains incomplete. There are reports (Haynes, Deichmann and Rees 2005) that human perceptions during binocular rivalrous inputs can modulate fMRI responses in LGN. These modulations could be caused by cortical feedback or other modulatory factors, since LGN neurons are monocular and are primarily driven in a feedforward direction from the retina. Meanwhile, electrophysiologically, there are reports that effects on LGN neural responses from inputs to the eye that are not associated with their feedforward path are absent in alert fixating monkeys (Lehky and Maunsell 1996) but suppressive in anesthetized cats (Sengpiel, Blakemore and Harrad 1995).

Cortical areas in the dorsal stream of the visual pathway are more intimately involved in decoding positions, orientations, and movements of visual objects. We recall from Section 2.4.1 that lesioning cortical areas in the dorsal stream (and outside the ventral stream) impairs performance in visuospatial tasks while often sparing object recognition abilities. In particular, lesions in areas MT and MST selectively impair motion perception in monkeys (Newsome and Pare 1988, Rudolph and Pasternak 1999).

6.3 Visual decoding from neural responses

In the last section, we saw that lesioning some brain areas selectively impairs specific types of visual perception. However, it is unclear whether the neural activities in the brain regions concerned represent the inputs to, intermediate steps in, or outcomes of visual decoding routines. In this section, we present some examples in which neural responses in the retina, V1, and MT to visual inputs are used for decoding to obtain estimates of some aspect of the visual scene. Success in doing this does not imply that the brain normally performs this decoding based directly on these activities. However, it does provide some hints about neural processing, as follows.

Consider the case that we have a hypothetical decoding framework or algorithm which takes neural responses $\mathbf{r}$ to an input scene $\mathbf{S}$ to produces as its output an estimate $\hat{\mathbf{S}}$ of an aspect $\mathbf{S}$ of the scene. If $\hat{\mathbf{S}}$ agrees well with the subject's reported percept of the same scene, then it is possible that this framework or algorithm may be used in the brain, even though it does not usually reveal how this framework or algorithm might be implemented in neural hardware. This approach requires a good knowledge about how neural responses $\mathbf{r}$ depend on visual input, and particularly about how $\mathbf{r}$ depends on $\mathbf{S}$. This requirement is easier to satisfy for retinal and V1 neurons, since we have a better knowledge of their response properties, e.g., their input feature tuning curves. Although less is known in higher brain areas such as area MT, electrophysiological measurements can also help to ascertain how $\mathbf{r}$ depends on $\mathbf{S}$ in well-defined, restricted sets of circumstances.

We can also find the algorithm that is theoretically optimal (under certain constraints) at decoding from $\mathbf{r}$, in the sense that the perceptual performance such as visual discrimination is maximized. This is called an ideal observer analysis. If this optimal performance beats that of the subject, then this implies an inefficiency in the brain, such as the loss of available information in $\mathbf{r}$ due to the attentional bottleneck. By contrast, if the ideal algorithm is outperformed by the subject, then an additional source of information beyond that contained in the response $\mathbf{r}$ must be available.

6.3.1 Example: decoding motion direction from MT neural responses

This example is adapted from a pedagogical example in a textbook on theoretical neuroscience (Dayan and Abbott 2001). Consider a simplified visual field containing just moving and twinkling dots in a blank background on a two-dimensional plane. Furthermore, these dots have an overall direction of movement, either upwards or downwards, at a fixed speed, although many individual dots can move in random directions (uniformly distributed in 360 degrees) or twinkle in and out of existence. One aspect of this visual scene can be just about this overall direction of movement, and it thus ignores the details of each dot. This aspect S is now a scalar and is binary, with only two possible values: $S = S_1$ for up and $S = S_2$ for down. Let $\mathbf{r} = (r_1, r_2, ..., r_n)$ be the response pattern from n neurons in the visual cortical area MT such that r_i is the response of the ith neuron. A given S value can lead to many possible $\mathbf{r}$'s, since many other aspects of the input scene ignored by our variable S, as well as noise in the visual sampling and encoding stage, can also influence $\mathbf{r}$. Let $P(\mathbf{r}|S)$ be the conditional probability of each of the possible responses $\mathbf{r}$. This conditional probability is also called the *likelihood* of $\mathbf{r}$ given S. Given a response pattern $\mathbf{r}$, the likelihoods are $P(\mathbf{r}|S = S_1)$ and $P(\mathbf{r}|S = S_2)$ under scenes S_1 and S_2, respectively. If $P(\mathbf{r}|S_1) > P(\mathbf{r}|S_2) = 0$, a sensible decoding algorithm should give the decoding outcome as $\hat{S} = S_1$ for this response pattern $\mathbf{r}$, as long as there is no prior bias for either S_i. If $P(\mathbf{r}|S_1) \approx P(\mathbf{r}|S_2) > 0$, then the decoding outcome can be ambiguous.

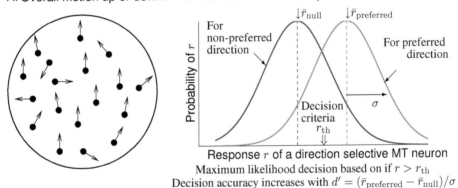

A: Overall motion up or down?   B: Two distributions of responses to the two directions

Maximum likelihood decision based on if $r > r_{\text{th}}$
Decision accuracy increases with $d' = (\bar{r}_{\text{preferred}} - \bar{r}_{\text{null}})/\sigma$

Fig. 6.9: Decoding motion direction from responses of a hypothetical, direction-selective MT neuron. A: Monkey subjects have to decide whether the overall direction of moving dots like those shown in the schematic is up or down (Britten et al. 1992). B: Probability distributions $P(r|S_1)$ and $P(r|S_2)$ of responses r from a hypothetical direction-tuned MT neuron, given the motion in the $S_1 = $ preferred and $S_2 = $ null direction of this neuron. When each $P(r|S_i)$ is a Gaussian distribution over r with a common standard deviation σ, the maximum-likelihood decision is S_1 or S_2 depending on the sign of $r - r_{\text{th}}$. The decision criteria r_{th} is where the two conditional probability curves cross, i.e., $P(r_{\text{th}}|S_1) = P(r_{\text{th}}|S_2)$. (Prior knowledge of motion direction or decision bias could shift this criterion; however, this no longer is the maximum likelihood choice.) When $\bar{r}_{\text{preferred}}$ and $\bar{r}_{\text{null}}$ are averages of r under $P(r|S_1)$ and $P(r|S_2)$, respectively, the accuracy of this decision rule increases with $d' = (\bar{r}_{\text{preferred}} - \bar{r}_{\text{null}})/\sigma$.

To be concrete, suppose that discrimination between S_1 and S_2 is based on the response of a single MT cell that prefers upward motion. Say that this neuron's response r (now a scalar) is on average $\bar{r}_{\text{preferred}} = 20$ spikes/second when the motion is upward, S_1, and $\bar{r}_{\text{null}} = 15$ spikes/second when the motion is downward, S_2 (the term "null" is often used to refer to the feature that is opposite to the preferred feature of a neuron). Due to response noise, the actual response r in any trial will deviate randomly from these average values. The two conditional probability distributions $P(r|S_1)$ and $P(r|S_2)$ can overlap when the noise is substantial, e.g., when the standard deviation σ of each distribution is $\sigma = 5$ spikes/second.

Let $P(S_1)$ and $P(S_2)$ be the probabilities that S_1 and S_2, respectively, occur a priori. For example, let $P(S_1) = P(S_2) = 0.5$, so that the two motion directions are equally probable. Consider a trial in which the neuron fires at $r = 17$ spikes/second. Then $P(r|S_2)$ is slightly larger than $P(r|S_1)$, when $\sigma = 5$ spike/second is the standard deviation for both $P(r|S_1)$ and $P(r|S_2)$. By Bayes theorem, the posterior probability $P(S_i|r)$ over the scene S_i causing response r is

$$P(S_i|r) = \frac{P(r|S_i)P(S_i)}{P(r|S_1)P(S_1) + P(r|S_2)P(S_2)} \tag{6.2}$$

$$= \frac{P(r|S_i)}{P(r|S_1) + P(r|S_2)}, \quad \text{when } P(S_i) = 0.5. \tag{6.3}$$

If the output of the decoding algorithm is $\hat{S} = S_i$ in response to r, then the probability that it is correct is $P(S_i|r)$ in the formula above. (As in Section 3.1, unless there is any confusion, we avoid notational clutter by using variable names inside the brackets of $P(.)$ to denote the variables concerned—so writing $P(\mathbf{S})$ and $P(\mathbf{r})$ for $P_{\mathbf{S}}(\mathbf{S})$ and $P_{\mathbf{r}}(\mathbf{r})$, respectively, which are over different spaces.)

6.3.1.1 Maximum-likelihood decoding

When $P(S_i)$ is independent of S_i, the decoding algorithm can minimize the probability of error by choosing the S_i for which the likelihood $P(r|S_i)$ is larger. This is called maximum-likelihood (ML) decoding, expressed in general as

$$\text{in maximum-likelihood decoding, the decoded scene } \hat{\mathbf{S}} = \text{argmax}_{\mathbf{S}} P(r|\mathbf{S}). \tag{6.4}$$

To use this decoding scheme, the brain would need to know the likelihood function $P(r|\mathbf{S})$. In our example of Fig. 6.9 B, for which $P(r|S_1)$ and $P(r|S_2)$ are both Gaussian distributions over r with the same standard deviation but different means, the ML decoding has

$$\begin{array}{c} \text{decoding outcome } \hat{S} \text{ is } S_1 \text{ or } S_2 \text{ if } r \text{ is larger or smaller than} \\ \text{a } \textit{decision criterion } r_{\text{th}} \text{ at which } P(r_{\text{th}}|S_1) = P(r_{\text{th}}|S_2). \end{array} \tag{6.5}$$

6.3.1.2 Comparing maximum-likelihood decoding from MT neurons with a monkey's perceptual decision

It has been possible to compare the ML decisions based on the responses of MT neurons with the perceptual decisions of subjects on the very same trials (Britten et al. 1992). Monkeys were presented with images of moving dots like that shown in Fig. 6.9 A. In each trial, a small (or zero) percentage (termed coherence) of the dots moved in a direction which was (with equal probability) either in the preferred ($S = S_1$) or null ($S = S_2$) direction of the recorded neuron. In each trial, the monkey had to report whether the overall moving direction was in the preferred or null direction. The difficulty of the task was adjusted by the proportion of the dots which moved in random directions. Estimates of the likelihoods $P(r|S_i)$ of the responses r for each direction S_i were compiled, allowing the ML choices to be estimated as in equation (6.5). The probability that these choices are correct is

$$P(\text{correct}) = P(S_1)P(r > r_{\text{th}}|S_1) + P(S_2)P(r \le r_{\text{th}}|S_2). \tag{6.6}$$

In Fig. 6.9, the likelihood of two input scenes S_1 and S_2 are modeled as Gaussian distributions over r with the same standard deviation σ but different means. In this case, $P(\text{correct})$ increases with the separation of the two curves, parameterized by $d' \equiv (\bar{r}_{\text{preferred}} - \bar{r}_{\text{null}})/\sigma$. In the experiment, it was found that this accuracy $P(\text{correct})$ based on each of many individual neurons[25] was comparable to, or sometimes better than, that of the monkey. This is illustrated in Fig. 6.10 A.

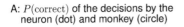

A: $P(\text{correct})$ of the decisions by the neuron (dot) and monkey (circle)

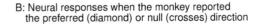

B: Neural responses when the monkey reported the preferred (diamond) or null (crosses) direction

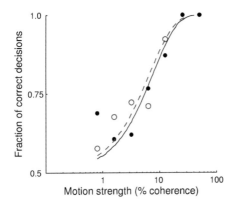

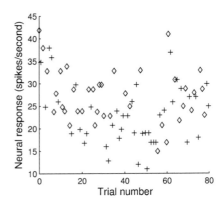

Fig. 6.10: Motion direction discrimination by neurons and monkeys (Shadlen et al. 1996). A: Performance of maximum-likelihood discrimination based on the responses of a direction-selective MT neuron (dots) and the psychophysical performance of the monkey (circles), as a function of the coherence of the moving dots (the percentage of dots moving in a particular non-random direction, S_1 or S_2). The solid and dashed curves are the best-fitting cumulative Weibull functions (defined as $F(x; k, \lambda) = 1 - \exp\left(-(x/\lambda)^k\right)$, a function with two parameters λ and k) to the neural and behavioral data. B: The responses r of the neuron in 80 trials; the diamonds and crosses denote the responses when the monkey reported the motion in the preferred and null directions, respectively, of the neuron. It is evident that the response of the neuron and the behavioral decision of the monkey is only weakly correlated. Data from Shadlen, M. N., Britten, K. H., Newsome, W. T., and Movshon, J. A., A computational analysis of the relationship between neuronal and behavioural responses to visual motion, *The Journal of Neuroscience*, 16(4): 1486–1510, 1996.

Of course, area MT has many directionally selective neurons. If their responses were uncorrelated given (i.e., conditional on) a stimulus S_i, then they could be pooled such that the overall d' of the ML decision would be much larger than that of a single neuron. Consider a simplistic example in which there are 100 neurons which resemble the single neuron being

---

[25] In the paper, the researchers actually used a different decision rule (Britten et al. 1992), based on this neuron and a hypothetical anti-neuron, notionally preferring the opposite direction of motion. The decision accuracy of this neuron-antineuron pair was calculated by randomly drawing responses r_1 and r_2 from $P(r_1|S_1)$ and $P(r_2|S_2)$, respectively, and deeming the decision correct or incorrect when $r_1 > r_2$ or $r_1 < r_2$, respectively.

measured, each of whose responses to input S_i is independently drawn at random from the same conditional probability distribution $P(r|S_i)$, with standard deviation σ. The distribution of the average response of these neurons will have a standard deviation that is 10 times smaller than σ. If the monkey uses the ML decoding from this average response, thereby utilizing information from all 100 neurons, it should have a much larger probability $P(\text{correct})$ of making a correct decision. This is contrary to the observation in Fig. 6.10 A, suggesting that the information present in MT neural responses is inefficiently used by decoding.

Is the low efficiency caused by correlations between responses from different neurons? Figure 6.10 B shows that, on a trial-by-trial basis, the responses of a single neuron were only modestly correlated with the perceptual decisions of the animal (Britten, Newsome, Shadlen, Celebrini and Movshon 1996). This low correlation cannot arise from a decoding algorithm based on thresholding the response from this neuron alone by a decision criterion r_{th} (as in equation (6.5)). (Hence, as expected, the monkey is unlikely to use a ML decoding scheme based on the response of a single neuron.) This low correlation also cannot arise from the ML decoding based on the average response of many highly correlated neurons, including the neuron in Fig. 6.10 B. These observations are consistent with the notion that the brain makes its decision by combining responses from many different neurons whose responses are not highly correlated. They are consistent with the notion that the information available in the population response pattern $\mathbf{r}$ is not efficiently utilized for this perceptual task. We will see more examples of the brain's low efficiency in decoding later in the chapter.

6.3.2 Example: discriminating two inputs based on photoreceptor responses

We saw above that when each input scene (or an aspect of the scene) S_i is equally likely, the ML decoding implements ideal observer analysis, since it minimizes the chance of error. (Here we are not concerned with minimizing other costs such as the cost associated with each error.) There are many more examples of this form of decoding.

Consider discriminating between two inputs, $S = A$ and $S = B$ (for instance, representing two different images), which lead to responses of the ith photoreceptor that are Poisson distributed (see Section 2.1.3) with mean values $\bar{r}_i(A)$ and $\bar{r}_i(B)$. Then, the likelihood that the whole set of responses $\mathbf{r} = (r_1, r_2, \ldots)$ is evoked by stimulus $\alpha = A$ or $\alpha = B$ is

$$P(\mathbf{r}|\alpha) = \Pi_i P(r_i|\alpha) = \Pi_i \left(\frac{[\bar{r}_i(\alpha)]^{r_i}}{r_i!} \exp\left[-\bar{r}_i(\alpha)\right] \right), \tag{6.7}$$

assuming that the noise corrupting the different receptors is independent.

The ML decoding based on response $\mathbf{r}$ reports that the input is A or B if $P(\mathbf{r}|A) > P(\mathbf{r}|B)$ or otherwise, i.e.,

$$\text{decide on } A \text{ or } B \text{ if the likelihood ratio} \frac{P(\mathbf{r}|A)}{P(\mathbf{r}|B)} > 1 \text{ or otherwise,} \tag{6.8}$$
$$\text{—the likelihood ratio test.}$$

When the noise corrupting the responses is independent, it is convenient to use the logarithm of the likelihoods or their ratio, making the test be whether the following expression

$$\ln\left[\frac{P(\mathbf{r}|A)}{P(\mathbf{r}|B)}\right] = \sum_i \ln \frac{\bar{r}_i(A)}{\bar{r}_i(B)} r_i - \sum_i [\bar{r}_i(A) - \bar{r}_i(B)] \tag{6.9}$$

is positive. The first term on the right-hand side is a linearly weighted sum of the responses

$$R \equiv \sum_i w_i r_i, \quad \text{where } w_i = \ln \frac{\bar{r}_i(A)}{\bar{r}_i(B)}. \tag{6.10}$$

The second term is constant, given A and B. Define

$$R_{\text{th}} \equiv \sum_i [\bar{r}_i(A) - \bar{r}_i(B)]. \tag{6.11}$$

Then, the ML decision is to report A when $R > R_{\text{th}}$, and B otherwise. Thus, the decision can be implemented by a feedforward network, as is illustrated schematically in Fig. 6.11. The output R of this feedforward network has the same role in this task as the stochastic response r of the directionally selective MT cell in the example in Section 6.3.1. Indeed, take the case that the input stimulus, A or B, excites many receptors with independent stochastic response r_i. Then R is the sum of many random variables, and so, by the law of large numbers, its distribution can be well approximated by a Gaussian distribution. For $\alpha = A$ or B, the mean and variance of this distribution is

$$\text{mean of } R(\alpha) = \sum_i w_i \bar{r}_i(\alpha) \quad \text{and variance of } R(\alpha) = \sum_i w_i^2 \bar{r}_i(\alpha), \tag{6.12}$$

since r_i is a Poisson random variable with a mean and variance both equal to $\bar{r}_i(\alpha)$.

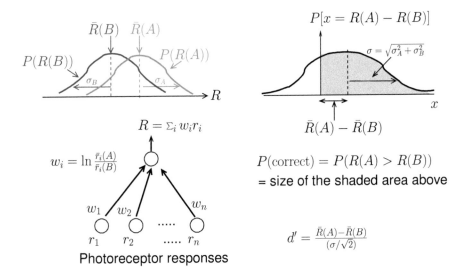

Fig. 6.11: ML decoding to discriminate between input images A and B from the responses $\mathbf{r} = (r_1, r_2, ..., r_n)$ of a population of photoreceptors. On average, A and B generate responses $\bar{r}_i(A)$ and $\bar{r}_i(B)$, respectively, in photoreceptor i, subject to independent Poisson noise. A task-specific feedforward network can generate a decision variable $R = \sum_i w_i r_i$ from the responses $\mathbf{r}$. The probability distributions $P(R)$ for input A and B are shown as red and blue curves, respectively. In a two-interval forced-choice (2IFC) paradigm, A and B are individually presented in two separate time intervals, leading to responses $R(1)$ and $R(2)$, respectively, from this task-specific network. The ML decision suggests reporting that A appeared in the first time interval if $R(1) - R(2) > 0$. The probability $P(\text{correct})$ that this is correct is shown on the right.

In one standard psychophysical paradigm, called a two-interval forced-choice (2IFC), the observers are shown stimuli A and B during separate time intervals, and have to decide which interval had A. The ML choice is to report that A came during the interval in which the R value was larger. The probability that this decision is correct is

$$P(\text{correct}) = \text{probability of } [R(A) > R(B)]. \tag{6.13}$$

Since both $R(A)$ and $R(B)$ are approximately Gaussian random variables, $R(A) - R(B)$ is also approximately Gaussian, with mean and variance

$$\sum_i [\bar{r}_i(A) - \bar{r}_i(B)] \ln \frac{\bar{r}_i(A)}{\bar{r}_i(B)} \quad \text{and} \quad \sum_i [\bar{r}_i(A) + \bar{r}_i(B)] \left[\ln \frac{\bar{r}_i(A)}{\bar{r}_i(B)} \right]^2. \tag{6.14}$$

Hence, from equation (6.13), the probability that the optimal ML decision is correct is roughly

$$P(\text{correct}) = \frac{1}{\sqrt{\pi}} \int_{-\infty}^{d'/2} dx \exp(-x^2), \quad \text{where}$$

$$d' = \frac{\sum_i [\bar{r}_i(A) - \bar{r}_i(B)] \ln \frac{\bar{r}_i(A)}{\bar{r}_i(B)}}{\sqrt{0.5 \sum_i [\bar{r}_i(A) + \bar{r}_i(B)] \left[\ln \frac{\bar{r}_i(A)}{\bar{r}_i(B)} \right]^2}}. \tag{6.15}$$

Here d' is the discriminability, which is commonly employed in signal detection theory. Achieving a fixed level of $P(\text{correct})$ (e.g., 75%) requires a particular d' value (e.g., $d' = 1$), which in turn, requires A to be sufficiently different from B. This difference, which is dependent on the required accuracy of discrimination, is sometimes called just-noticeable difference or the discrimination threshold. The inverse of this threshold is called the (discrimination) sensitivity.

Ideal observer analysis has been applied (Geisler 1989, Geisler 2003) to various aspects of photoreceptor responses. In one example, the threshold contrast was assessed for discriminating a pattern input A, a grating with 7.5 cycles, from a spatially uniform pattern B whose luminance is the same as the mean luminance in A, when both were viewed for 100 ms in central vision during separate time intervals (Banks, Geisler and Bennett 1987). From what is known about photoreceptors, one can calculate $\bar{r}_i(A)$ and $\bar{r}_i(B)$ and thereby estimate $P(\text{correct})$ and a resulting sensory threshold for discrimination. Figure 6.12 A shows that the sensitivities of the human and ideal observers to the grating agree well, but only after scaling down the ideal observer's sensitivity by a constant factor of about 20. This reflects the low efficiency of neural information processing in the use of information; see later discussions. Nevertheless, the fact that human and ideal observers have the same dependence of the sensitivity on the spatial frequency of the grating suggests a possible application of the ML decision rule in the brain.

One should note that the method here, in particular the likelihood function $P(\mathbf{r}|\alpha)$ in equation (6.7), is not applicable for discriminating two objects A and B, e.g., discriminating an apple from an table, in an object viewpoint invariant manner. This is because the responses $\mathbf{r}$ to the apple, for example, depend on factors such as the position and scale of the image of this apple. These factors are very different from the Poisson noise in light absorption, and hence they make $P(\mathbf{r}|\alpha)$ given α (the apple) very far from a Poisson distribution. The influence of the viewpoint variations on $P(\mathbf{r}|\alpha)$ is expected to overwhelm the influence of the Poisson noise in typical cases for object discrimination.

6.3.3 Example: discrimination by decoding the V1 neural responses

Although the quantification of the properties of V1 neural responses is less precise than that of the photoreceptors, efforts have been made to decode visual inputs from them. Geisler

A: Optimal contrast detection based on
photoreceptor responses

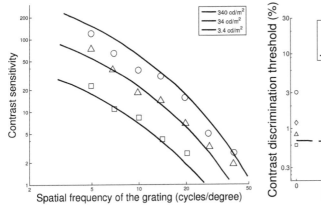

B: Optimal contrast discrimination
based on V1 neural responses

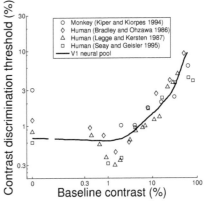

Fig. 6.12: Decoding from responses of photoreceptors or V1 neurons to perform contrast detection and discrimination. A: Contrast sensitivity to detect which of the two 100 ms intervals contained a grating (which has 7.5 cycles). The circles, triangles, and squares show data from a single human observer (MSB from Banks et al. (1987)) under three different intensities of background light levels, as indicated; the solid curves are sensitivities from the ML decoding performance after scaling down by a factor of 21.5. Data from Banks, M. S., Geisler, W. S., and Bennett, P. J., The physical limits of grating visibility, *Vision Research*, 27 (11): 1915–24, 1987. B: Contrast discrimination thresholds for primates (square, diamond, triangle, and circle symbols) and for an ideal observer (solid curve) decoding the responses from 71 V1 neurons in monkeys. The thresholds of the ideal observer (the solid curve) were shifted vertically to superpose on the primate behavioral thresholds. This was done by choosing an efficiency parameter $\epsilon = 0.5 \cdot 100/71 = 0.7042$ to scale down the d' value (by the transform $d'^2 \rightarrow \epsilon d'^2$) obtained from pooling the V1 neurons (using equations (6.18–6.19)). Data from Geisler, W. S., and Albrecht, D. G., Visual cortex neurons in monkeys and cats: detection, discrimination, and identification, *Visual Neuroscience*, 14: 897–919, 1997.

and Albrecht (1997) observed that the mean and variance of the responses of a V1 neuron in response to an input tend to be proportional to each other. That is, consider two inputs A and B, evoking mean responses $\bar{r}(A)$ and $\bar{r}(B)$ from a V1 neuron. Then the corresponding variances of the responses are $\sigma(A) = K\bar{r}(A)$ and $\sigma(B) = K\bar{r}(B)$, with a constant factor K specific for this neuron.

For example, Geisler and Albrecht (1997) found that the response r of a V1 neuron to an input contrast c can be reasonably well fitted phenomenologically as

$$\text{response mean} \equiv \bar{r}(c) = r_{\max}\frac{c^n}{c^n + c_{50}^n} + r_0, \quad \text{response variance} = \sigma^2(c) = K\bar{r}(c).$$
(6.16)

Hence, each neuron's responsivity can be characterized by five parameters: $r_{\max}, r_0, n, c_{50}^n$, and K.

When the average responses are large enough, the response r to each input is assumed to follow a Gaussian distribution, implying that the likelihood of a response r to input contrast c is

$$P(r|c) \propto \exp\left(-\frac{(r - \bar{r}(c))^2}{2\sigma^2(c)}\right).$$
(6.17)

Hence, according to Fig. 6.9, performance based on this neuron's response in a 2IFC task of discriminating between two input contrasts c_A and $c_B = c_A + \delta c$ should have a d' value of approximately

$$d' = \frac{|\bar{r}(c_A) - \bar{r}(c_B)|}{\sqrt{0.5\,[\sigma^2(c_A) + \sigma^2(c_B)]}}. \tag{6.18}$$

Given multiple (independent) such neurons, with response $\mathbf{r} = (r_1, r_2, ..., r_i, ...)$, the d' value associated with the optimal decision is approximately (Green and Swets 1988)

$$d' \approx \sqrt{\sum_i (d'_i)^2}, \text{ where } d'_i \text{ is the d prime value associated with the } i^{\text{th}} \text{ neuron.} \tag{6.19}$$

Using equations (6.18) and (6.19), Geisler and Albrecht (1997) calculated the contrast discrimination thresholds (i.e., $\delta c = |c_B - c_A|$ necessary to reach a criterion value of d') across a range of baseline contrasts c_A, pooling responses from 71 V1 neurons in monkey (in response to each stimulus for 200 ms); see Fig. 6.12 B. Taken as a function of the baseline contrast, these ideal thresholds are best matched with the thresholds by human or monkey observers after adjusting the d' value by a constant $\epsilon < 1$ in the formula $d' \approx \sqrt{\epsilon \sum_i (d'_i)^2}$ modified from equation (6.19). A similar such decoding scheme based on V1 neural responses was also applied to understand the quality of human discrimination of the spatial frequencies of gratings (Geisler and Albrecht 1997).

Note that $\epsilon < 1$ and that 71 neurons constitute a very small fraction of V1 neurons. Hence, the observation here again demonstrates a low efficiency in utilizing the information in neural responses by the brain for decoding.

6.3.4 Example: light wavelength discrimination by decoding from cone responses

The next task is more complex, since it involves discriminating whether two monochromatic light fields have different wavelengths, without knowing ahead of time what each might be. Here, in a design adopted by Pokorny and Smith (1970), one of the field is fixed for a given (standard) wavelength, and observers adjust the intensity of the other light field which has a slightly different wavelength. If the observer can find any intensity to make the two fields appear indistinguishable, then the difference between the two wavelengths is below the wavelength discrimination threshold. The smallest difference between the two wavelengths to make the two fields appear different regardless of the adjusted intensity is the discrimination threshold.

The central point of the experimental design by Pokorny and Smith is that two light fields can appear different based on their difference in wavelength, intensity, or both. For example, while one can be certain that two red-colored light fields appear different, it can be difficult to tell whether the difference in appearance is due to one field being redder than the other or due to a weaker input intensity of the first field. Assessing the just-noticeable difference in wavelength requires eliminating this confound. This is done here by allowing subjects to adjust intensity.

Since we have a good knowledge of photoreceptor properties, e.g., their selectivity to wavelength in response to light fields, it is possible to predict how an ideal observer would perform this task. These predictions can be compared with the performance of human observers in the behavioral data from Pokorny and Smith.

Each light field is a scene $\mathbf{S} = (\lambda, I)$, parameterized by wavelength λ and intensity I. Following the notation in the original literature, we use $a = S, M, L$ for cone types tuned to small, medium, and long wavelengths, respectively (rather than naming them as red, green, and blue or R,G, and B in the previous chapters). Some input photons are absorbed by the retinal media before they reach the cones. Let $\mathbf{r} = (r_S, r_M, r_L)$ denote the cone absorptions of the remaining photons. Two input fields, $\mathbf{S}$ and $\mathbf{S}'$, can be reliably discriminated only when the difference in the responses they evoke, $\mathbf{r}$ and $\mathbf{r}'$, is greater than the encoding noise (i.e., the spread of probable $\mathbf{r}$ or $\mathbf{r}'$ values) and so is unlikely caused only by response noise; see Fig. 6.13.

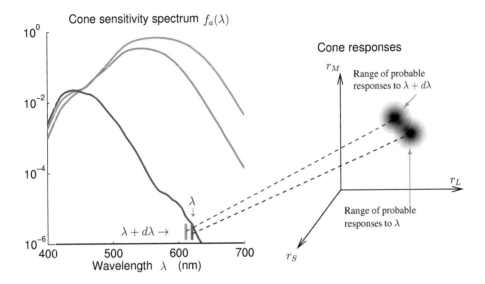

Fig. 6.13: Schematic of input sampling and the likelihoods of photoreceptor responses to monochromatic input. The left-hand graph shows the cone spectral sensitivities $f_a(\lambda)$ (with $f_a(\lambda) \propto n_a O_a \hat{f}_a(\lambda)$, where, for cone $a = S, M, L$, $\hat{f}_a(\lambda)$ is the Smith and Pokorny cone fundamental (Smith and Pokorny 1975), n_a is the cone density, with $n_L : n_M : n_S = 6 : 3 : 1$, O_a is the pre-receptor light transmission factor, with $O_L : O_M : O_S = 1 : 1 : 0.2$, and $\max_\lambda \sum_a f_a(\lambda) = 1$. A monochromatic input $\mathbf{S} = (\lambda, I)$ of wavelength λ and intensity I can evoke a range of probable responses $\mathbf{r} = (r_L, r_M, r_S)$ from the three cones, with a likelihood $P(\mathbf{r}|\mathbf{S} = (\lambda, I))$. If this range of probable responses overlaps substantially with that evoked by another monochromatic input $\mathbf{S}' = (\lambda + d\lambda, I')$ of a similar wavelength, it will be difficult to distinguish between $\mathbf{S}$ and $\mathbf{S}'$ perceptually. Adapted from Zhaoping, L., Geisler, W. S., and May, K. A., Human wavelength discrimination of monochromatic light explained by optimal wavelength decoding of light of unknown intensity. *PLoS One*, 6(5): e19248, Fig. 1, copyright © 2011, Zhaoping, L., Geisler, W. S., and May, K. A.

The mean absorption of each cone is $\bar{r}_a = I \cdot O_a \cdot \hat{f}_a(\lambda)$, where $O_a \leq 1$ (derived from optical density) models the attenuation of the input light by pre-receptor absorption, and $\hat{f}_a(\lambda)$ is the cone's wavelength sensitivity function or cone fundamental; see Section 2.2.5. The total response from n_a cones of type a exposed to the input is on average $\bar{r}_a = I \cdot n_a \cdot O_a \cdot \hat{f}_a(\lambda)$. To discriminate input wavelength, one may view the collection of n_a cones of type a as being equivalent to a single giant cone whose wavelength sensitivity (which, for simplicity, we write as $f_a(\lambda) \equiv n_a O_a \hat{f}_a(\lambda)$) is n_a times as sensitive as that of a single cone. In other words, the total absorptions of each cone type are sufficient statistics for decoding the wavelength. This holds as long as response noise is independent across different cones within a given cone type.

For convenience, we normalize the sensitivities so that $\max_\lambda \sum_a f_a(\lambda) = 1$, and absorb all the scale factors for the input intensity and the number of cones covered by the input fields into the parameter I.

Cone absorption is stochastic, following a Poisson distribution. Hence, the likelihoods are

$$P(r_a|\mathbf{S} = (\lambda, I)) = \frac{(\bar{r}_a)^{r_a}}{r_a!} \exp(-\bar{r}_a) = \frac{[I f_a(\lambda)]^{r_a}}{r_a!} \exp[-I f_a(\lambda)] . \qquad (6.20)$$

If the noise in different cone types is also independent, we have

$$P(\mathbf{r} = (r_S, r_M, r_L)|\mathbf{S}) = P(r_S|\mathbf{S}) P(r_M|\mathbf{S}) P(r_L|\mathbf{S})$$
$$= \left[\prod_a \frac{(I f_a(\lambda))^{r_a}}{r_a!} \right] \exp\left[-I \sum_a f_a(\lambda) \right] . \qquad (6.21)$$

The total log likelihood is therefore

$$\ln P(\mathbf{r}|\mathbf{S}) = \sum_a \{ r_a \ln[I f_a(\lambda)] - I f_a(\lambda) - \ln(r_a!) \} . \qquad (6.22)$$

According to the maximum-likelihood decoding scheme, the most probable $\hat{\mathbf{S}} = (\hat{\lambda}, \hat{I})$ is the solution to

$$\frac{\partial \ln P(\mathbf{r}|\mathbf{S})}{\partial I} = \frac{\sum_a r_a}{I} - \sum_a f_a(\lambda) = 0, \quad \text{and} \qquad (6.23)$$

$$\frac{\partial \ln P(\mathbf{r}|\mathbf{S})}{\partial \lambda} = \sum_a \frac{r_a f_a'(\lambda)}{f_a(\lambda)} - \sum_a I f_a'(\lambda) = 0. \qquad (6.24)$$

From the actual input $\mathbf{S} = (\lambda, I)$, this $\hat{\mathbf{S}}$ has a decoding error which is the two-dimensional vector:

$$\text{decoding error} \equiv \hat{\mathbf{S}} - \mathbf{S} \equiv (\hat{\lambda} - \lambda, \hat{I} - I). \qquad (6.25)$$

Different stochastic cone responses $\mathbf{r}$ give different decoded $\hat{\mathbf{S}}$ and thus different decoding errors. The covariance between the components of this decoding error characterize the degree and the structure of the uncertainty in the decoding. If $\ln P(\mathbf{r}|\mathbf{S})$ is a smooth, unimodal function of $\mathbf{S}$, this covariance is determined by

$$\left\langle \left(\hat{\mathbf{S}} - \mathbf{S} \right)_i \left(\hat{\mathbf{S}} - \mathbf{S} \right)_j \right\rangle = [I_F^{-1}(\mathbf{S})]_{ij}, \quad i, j = 1, 2, \qquad (6.26)$$

by the inverse of the 2×2 Fisher information matrix $I_F(\mathbf{S})$, which is defined as

$$I_F(\mathbf{S}) \equiv - \begin{pmatrix} \left\langle \frac{\partial^2 \ln P(\mathbf{r}|\mathbf{S})}{\partial \lambda^2} \right\rangle & \left\langle \frac{\partial^2 \ln P(\mathbf{r}|\mathbf{S})}{\partial \lambda \partial I} \right\rangle \\ \left\langle \frac{\partial^2 \ln P(\mathbf{r}|\mathbf{S})}{\partial \lambda \partial I} \right\rangle & \left\langle \frac{\partial^2 \ln P(\mathbf{r}|\mathbf{S})}{\partial I^2} \right\rangle \end{pmatrix}$$
$$= \sum_a \begin{pmatrix} I \frac{(f_a'(\lambda))^2}{f_a(\lambda)} & f_a'(\lambda) \\ f_a'(\lambda) & f_a(\lambda)/I \end{pmatrix} . \qquad (6.27)$$

In the above, $\langle ... \rangle$ denotes the ensemble average of the quantity inside the brackets. The second equality in the above equation is derived from calculating each component of the first matrix. For example, since $\int d\mathbf{r} P(\mathbf{r}|\mathbf{S}) r_a = I f_a(\lambda)$,

$$-\left\langle \frac{\partial^2 \ln P(\mathbf{r}|\mathbf{S})}{\partial I^2} \right\rangle = \int d\mathbf{r} P(\mathbf{r}|\mathbf{S}) \frac{\partial^2 \ln P(\mathbf{r}|\mathbf{S})}{\partial I^2} \tag{6.28}$$

$$= \int d\mathbf{r} P(\mathbf{r}|\mathbf{S}) \left(\sum_a r_a \right) / I^2 = \sum_a f_a(\lambda)/I. \tag{6.29}$$

Let the decoded value $\hat{\mathbf{S}}$ given input $\mathbf{S}$ follow a conditional probability distribution $P(\hat{\mathbf{S}}|\mathbf{S})$. For $\hat{\mathbf{S}} \approx \mathbf{S}$, $P(\hat{\mathbf{S}}|\mathbf{S})$ can thus be approximated as

$$P(\hat{\mathbf{S}}|\mathbf{S}) \approx P_{\max} \exp\left[-\frac{1}{2} \sum_{i,j=1}^{2} [I_F(\mathbf{S})]_{ij} \left(\hat{\mathbf{S}} - \mathbf{S} \right)_i \left(\hat{\mathbf{S}} - \mathbf{S} \right)_j \right], \tag{6.30}$$

where $P_{\max}$ is the value of $P(\hat{\mathbf{S}}|\mathbf{S})$ when $\hat{\mathbf{S}} = \mathbf{S}$. The function $P\left(\hat{\mathbf{S}}|\mathbf{S}\right)$ peaks at $\hat{\mathbf{S}} = \mathbf{S}$, as expected for an unbiased decoding scheme. Near $\hat{\mathbf{S}} = \mathbf{S}$, a contour of constant

$$P(\hat{\mathbf{S}}|\mathbf{S}) = P_\Delta \equiv P_{\max} \exp(-\Delta^2/2)$$

is an ellipse in the two-dimensional space $\hat{\mathbf{S}} = (\hat{\lambda}, \hat{I})$, centered at $\mathbf{S} = (\lambda, I)$ and defined by

$$[I_F(\mathbf{S})]_{11} (\hat{\lambda} - \lambda)^2 + 2 [I_F(\mathbf{S})]_{12} (\hat{\lambda} - \lambda)(\hat{I} - I) + [I_F(\mathbf{S})]_{22} (\hat{I} - I)^2 = \Delta^2, \tag{6.31}$$

as illustrated in Fig. 6.14 A. All $\hat{\mathbf{S}}$ inside or outside a given ellipse have $P(\hat{\mathbf{S}}|\mathbf{S})$ higher or lower, respectively, than the corresponding P_Δ. In other words, another input stimulus $\mathbf{S}'$, whose most likely, unbiased, decoding outcome is $\hat{\mathbf{S}}' = \mathbf{S}'$, should be less perceptually distinguishable from the standard reference stimulus $\mathbf{S} = (\lambda, I)$ when $\mathbf{S}'$ is inside the ellipse than when $\mathbf{S}'$ is outside the ellipse. Therefore, a threshold for a perceptually distinguishable difference between $\mathbf{S}$ and $\mathbf{S}'$ should correspond to a threshold value Δ_{th} of Δ, such that the ellipse defined by this $\Delta = \Delta_{\text{th}}$ separates the $\mathbf{S}'$ which are perceptually distinguishable from those which are indistinguishable from the standard $\mathbf{S}$. Let us call this threshold ellipse as the ellipse of (decoding) uncertainty.

In the behavioral experiment (Pokorny and Smith 1970), the wavelength discrimination threshold is the smallest wavelength difference $|\lambda' - \lambda|$ between the standard input $\mathbf{S} = (\lambda, I)$ and a test input $\mathbf{S}' = (\lambda', \hat{I}')$ fields to make them appear different regardless of how one adjusts the intensity I'. By the ML decoding, this threshold, $\sigma(\lambda)$, should be the absolute difference $|\hat{\lambda} - \lambda|$ when the line of the constant $\hat{\lambda}$ tangentially touches the uncertainty ellipse in equation (6.31); see Fig. 6.14 A. When $|\lambda' - \lambda| < \sigma(\lambda)$, then observer can find some I' value such that $\mathbf{S}' = (\lambda', I')$ is inside the uncertainty ellipse and be indistinguishable perceptually from $\mathbf{S} = (\lambda, I)$. This does not happen when $|\lambda' - \lambda| > \sigma(\lambda)$. The location of the tangential touch is where $d\hat{\lambda}/d\hat{I} = 0$ on the uncertainty ellipse. Let the ellipse be defined by $\Delta = \Delta_{\text{th}}$ in equation (6.31). Then, differentiating this equation with respect to $(\hat{\lambda}, \hat{I})$, we have[26]

$$[I_F(\mathbf{S})]_{11} (\hat{\lambda}-\lambda)d\hat{\lambda} + [I_F(\mathbf{S})]_{12} d\hat{\lambda}(\hat{I} - I) + [I_F(\mathbf{S})]_{12}(\hat{\lambda}-\lambda)d\hat{I} + [I_F(\mathbf{S})]_{22} (\hat{I} - I)d\hat{I} = 0.$$

Hence,

$$[I_F(\mathbf{S})]_{11} (\hat{\lambda}-\lambda)d\hat{\lambda}/d\hat{I} + [I_F(\mathbf{S})]_{12} (\hat{I}-I)d\lambda/d\hat{I} + [I_F(\mathbf{S})]_{12} (\hat{\lambda}-\lambda) + [I_F(\mathbf{S})]_{22}(\hat{I}-I) = 0.$$

When $d\hat{\lambda}/d\hat{I} = 0$, we have $\hat{I} - I = -[I_F(\mathbf{S})]_{12} (\hat{\lambda} - \lambda)/[I_F(\mathbf{S})]_{22}$. Substituting this into equation (6.31),

---

[26] I thank Mr. Bi Zedong for this simplification from the original proof in Zhaoping et al. (2011).

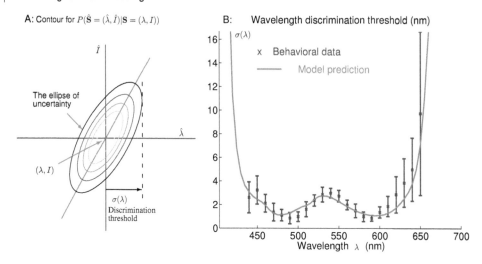

Fig. 6.14: Maximum-likelihood decoding and discrimination of monochromatic light (Zhaoping et al. 2011). A: Contour plots for the probability distribution of the inferred monochromatic light $\hat{\mathbf{S}} = (\hat{\lambda}, \hat{I})$ from cone responses $\mathbf{r}$ to input $\mathbf{S} = (\lambda, I)$. This distribution is approximately Gaussian around the actual input stimulus $\mathbf{S} = (\lambda, I)$, dropping to a certain fraction $\exp(-\Delta_{\text{th}}^2/2)$ of its peak value at the ellipse of uncertainty. Thus, input stimulus $\mathbf{S}' = (\lambda', I')$ outside this ellipse should be perceptually distinguishable from $\mathbf{S} = (\lambda, I)$ to a suitable degree. The threshold $\sigma(\lambda)$ is the smallest wavelength difference $\hat{\lambda} - \lambda$ such that input $\mathbf{S}' = (\lambda', I')$ can be distinguished from $\mathbf{S} = (\lambda, I)$ no matter how one adjusts its intensity I'. B: Comparison of the threshold $\sigma(\lambda)$ predicted from this model with those measured behaviorally by Pokorny and Smith (1970). Each data point and the associated error bar denotes the average and standard deviation of the measured thresholds from four observers.

$$\text{discrimination threshold } \sigma(\lambda) \equiv |\hat{\lambda} - \lambda| = \Delta_{\text{th}}\sqrt{[I_F(\mathbf{S})]_{22}/\det[I_F(\mathbf{S})]}$$

$$= \frac{\Delta_{\text{th}}}{\sqrt{I}}\left(\frac{\sum_a f_a(\lambda)}{\left[\sum_b \frac{(f_b'(\lambda))^2}{f_b(\lambda)}\right][\sum_c f_c(\lambda)] - [\sum_d f_d'(\lambda)]^2}\right)^{1/2}. \tag{6.32}$$

Since $f_a(\lambda)$ are known (see Fig. 6.13), only one free parameter Δ_{th}/I is needed to fit the whole form of the threshold $\sigma(\lambda)$ as a function of λ. Figure 6.14 B shows that the predicted and behaviorally measured thresholds match well when a single independent parameter $(\Delta_{\text{th}}/\sqrt{I})$ is adjusted to fit the data. The predicted thresholds are within the standard deviations of almost all experimentally measured thresholds, supporting the maximum-likelihood decoding scheme.

6.3.5 Perception, including illusion, of a visual feature value by neural population decoding

So far we have focused on discriminating between two visual inputs. The ML method can also be used to decode the value of a visual input S along a feature dimension, or in multiple feature dimensions, in which case $\mathbf{S}$ is a vector. For example, S might denote the orientation of a bar, or the disparity of an object. According to equation (6.4), given a response $\mathbf{r}$ from a population of neurons, and the likelihood function $P(\mathbf{r}|\mathbf{S})$ for input feature $\mathbf{S}$, the ML perception $\hat{\mathbf{S}}$ is $\hat{\mathbf{S}} = \text{argmax}_{\mathbf{S}} P(\mathbf{r}|\mathbf{S})$.

We illustrate this with a simple formulation adopted from Chapter 3 of the book *Theoretical Neuroscience* by Dayan and Abbott (2001). Here, the input feature is one-dimensional, and, for consistency with that book, we denote it as θ rather than S. The average response of the ith of a group of neurons to θ is called the feature-tuning curve of the neuron, given by

$$\bar{r}_i(\theta) = r_{\max} \exp\left(-\frac{(\theta - \theta_i)^2}{2\sigma^2}\right), \tag{6.33}$$

in which θ_i is the neuron's preferred feature, $r_{\max}$ is the average response to this preferred stimulus θ_i, and σ is the width of the feature-tuning curve. For simplicity, the neurons are homogenous and so share $r_{\max}$ and σ. Assuming that the neural response noise is Poisson and independent across neurons, a feature value θ evokes responses $\mathbf{r} = (r_1, r_2, ...)$ with likelihood

$$P(\mathbf{r}|\theta) = \Pi_i P(r_i|\theta) = \Pi_i \left(\frac{[\bar{r}_i(\theta)]^{r_i}}{r_i!} \exp\left[-\bar{r}_i(\theta)\right]\right). \tag{6.34}$$

The log likelihood is

$$\log P(\mathbf{r}|\theta) = \sum_i \left(r_i \log [\bar{r}_i(\theta)] - \bar{r}_i(\theta)\right) - \sum_i (r_i!). \tag{6.35}$$

Neural feature tuning, and average (blue) and single trial (red) responses to feature $\theta = 0$

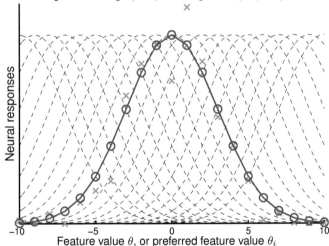

Fig. 6.15: Schematic illustration of population decoding for an input feature θ. Dashed black curves are the feature tuning curves $\bar{r}_i(\theta)$ as a function of the continuous value θ of a population of 21 simulated neurons $i = 1, 2, ..., 21$. The neurons have integer preferred values $\theta_i = \{-10, -9, ..., 10\}$ in arbitrary units. In blue are the average responses $\bar{r}_i(\theta = 0)$ versus θ_i from these neurons to an input feature $\theta = 0$. Red crosses show the noisy responses $r_i(\theta = 0)$ in an individual trial as a function of the discrete values θ_i. In this trial, decoding by the response-weighted average of θ_i (using equation (6.37)) would give $\hat{\theta} \approx 0$. This decoded value $\hat{\theta}$ is approximately the same as the preferred feature θ_i of the most activated neuron averaged across trials, i.e., the θ_i of the neuron with the highest $\bar{r}_i(\theta)$.

When the prior distribution of θ is $P(\theta) = $ constant, the most likely input feature θ given $\mathbf{r}$ is the maximum-likelihood solution, i.e., the θ value that makes $\frac{\partial \log P(\mathbf{r}|\theta)}{\partial \theta} = 0$. This gives

$$\frac{\partial \log P(\mathbf{r}|\theta)}{\partial \theta} = \sum_i \frac{r_i - \bar{r}_i(\theta)}{\bar{r}_i(\theta)} \frac{\partial \bar{r}_i(\theta)}{\partial \theta}$$

$$= \sum_i \frac{r_i - \bar{r}_i(\theta)}{\bar{r}_i(\theta)} r_{\max} \exp\left(-\frac{(\theta - \theta_i)^2}{2\sigma^2}\right)\left(\frac{-(\theta - \theta_i)}{\sigma^2}\right)$$

$$\propto \sum_i \frac{r_i - \bar{r}_i(\theta)}{\bar{r}_i(\theta)} \bar{r}_i(\theta)(\theta_i - \theta)$$

$$= \sum_i [r_i - \bar{r}_i(\theta)](\theta_i - \theta) = 0. \tag{6.36}$$

If the preferred features θ_i across the population of neurons are densely and evenly spaced along the θ dimension, such that $\theta_i = \delta\theta \cdot i + \text{constant}$ with $0 < \delta\theta \ll \sigma$, then $\sum_i \bar{r}_i(\theta)(\theta_i - \theta) \approx 0$ for almost any θ well within the range of maximum and minimum θ_i's. Hence, the solution to equation (6.36) is the maximum-likelihood value

$$\hat{\theta} = \frac{\sum_i r_i \theta_i}{\sum_j r_j} \text{—weighted average decoding.} \tag{6.37}$$

That is, the decoded value $\hat{\theta}$ is the weighted average of the preferred features θ_i of the responding neurons, with weights given by the firing rates. This is an intuitive method of decoding that has often been applied without justification from the maximum-likelihood decoding framework. However, this decoding formula is not necessarily applicable when the noise corrupting the responses is not Poisson, or in the inhomogenous case, when the feature dimension for θ is not evenly sampled by θ_i or with different feature-tuning widths for different neurons.

Let us return to the homogenous case when the θ_i densely sample the feature dimension. When we estimate the behaviorally perceived feature value averaged across trials, we can use the following approximation of the decoding algorithm in equation (6.37):

$$\hat{\theta} = \text{the preferred } \theta_i \text{ of the neuron with the highest average activation } \bar{r}_i(\theta). \tag{6.38}$$

This is illustrated in Fig. 6.15.

Discrimination threshold for θ can also be obtained like we did in Section 6.3.4 using Fisher information, which is now a scalar for the one-dimensional feature θ,

$$I_F(\theta) \equiv -\int d\mathbf{r} P(\mathbf{r}|\theta) \frac{\partial^2}{\partial \theta^2} \ln P(\mathbf{r}|\theta)$$

rather than the 2×2 matrix of equations (6.26) and (6.27). In many situations, it can be shown that (Dayan and Abbott 2001)

$$\text{discrimination threshold } \sigma(\theta) \propto \left[\int d\mathbf{r} \left(\hat{\theta} - \langle\hat{\theta}\rangle\right)^2 P(\mathbf{r}|\theta)\right]^{1/2} = [I_F(\theta)]^{-1/2}. \tag{6.39}$$

Given $P(\mathbf{r} = (r_1, r_2, ..., r_n)|\theta) = \Pi_i P(r_i|\theta)$,

$$I_F(\theta) = -\sum_{i=1}^{n} \int dr_i P(r_i|\theta) \frac{\partial^2}{\partial \theta^2} \ln[P(r_i|\theta)] \equiv \sum_i I_F(\theta; i). \tag{6.40}$$

Therefore, the Fisher information increases with the number n of the neurons tuned to sensory inputs θ, provided that they are independent (the same holds for the 2×2 Fisher information matrix in equation (6.27), in which the whole Fisher information matrix is the summation of the three separate Fisher information matrices for the three cones). Consequently, the sensory discrimination threshold is predicted to decrease as n grows. In particular, $\sigma(\theta) \propto \rho^{-1/2}$ in which ρ is the density of sampling neurons along θ (Dayan and Abbott 2001). From this derivation, one sees that this conclusion is general rather than specific to Poisson neurons or Gaussian feature tuning curves. This is analogous to $\sigma(\lambda) \propto I^{-1/2}$ relating the discrimination threshold for wavelength λ with the input intensity I in equation (6.32), since the intensity I was defined to scale with the number of cones activated by the input field. It is also analogous to the increases in discrimination sensitivity d' in equation (6.19) when more neurons independently encode the sensory signal.

6.3.5.1 Example: depth perception, Pulfrich effect, and its non-intuitive variations

Equation (6.37) (and its approximation equation (6.38)) specify the percept as the output of maximum-likelihood decoding. If the brain continues to decode in this way even when the visual input is altered so that the likelihood function $P(\mathbf{r}|\mathbf{S})$ is different, the perceived feature $\hat{\theta}$ can be erroneous, leading to illusions.

Here, we illustrate this with an example from depth perception. First, let us recall from equations (2.74–2.76) the disparity tuning of complex V1 cells. A complex cell is made from combining two simple cells. One simple cell has the respective spatial receptive fields for the inputs from the left and right eyes as

$$K_L(x) = \exp\left(-\frac{x^2}{2\sigma^2}\right)\cos(\hat{k}(x-x_L)), \quad K_R(x) = \exp\left(-\frac{x^2}{2\sigma^2}\right)\cos(\hat{k}(x-x_R)), \quad (6.41)$$

in which $\hat{k}$ is the preferred spatial frequency of the cell, $\hat{k}x_{L,R}$ are the phases of the receptive fields, and $x_L - x_R$ is the preferred disparity of this neuron; the other simple cell has the same property, in particular the same preferred disparity $x_L - x_R$, but its binocularly averaged phase $\hat{k}(x_L + x_R)/2$ differs from that of the first simple cell by $90°$. Let us denote this complex cell (made from these two simple cells) as the ith neuron in a population, and its preferred disparity is $D_i = x_L - x_R$. Let there be an input object whose image is $I(x)$ in the left eye and $I(x - D)$ in the right eye, so that this object has a input disparity D. The response evoked by this object in the complex cell (using the energy model) is approximately,[27] in the notation of this chapter,

$$\bar{r}_i \approx r_{\max}\cos^2\left[\frac{\hat{k}(D_i - D)}{2}\right] \propto \left|\mathcal{I}(\hat{k})\right|^2 \cos^2\left[\frac{\hat{k}(D_i - D)}{2}\right], \quad (6.42)$$

where $\mathcal{I}(k)$ is the Fourier transform of $I(x)$, and hence $\left|\mathcal{I}(\hat{k})\right|^2$ is the signal power of the input object in the spatial frequency $\hat{k}$ preferred by this complex cell.

Let this object evoke responses $\mathbf{r} = (r_1, r_2, ...)$ from a population of neurons with various preferred disparities D_i. Here the disparity tuning curve is a cosine squared curve, $\cos^2(.)$, rather than a Gaussian one (see equation (6.33)), and these neurons perhaps do not sample the disparity feature homogenously. Nevertheless, we assume that the feature sampling is sufficiently dense and that our approximation (in equation (6.38)) is still applicable, such that the disparity percept is estimated as

$$\hat{D} \approx D_i \text{ of the neuron with the highest } \bar{r}_i. \quad (6.43)$$

---

[27] For simplicity, we ignore the saturating nonlinearity, without affecting the main conclusions.

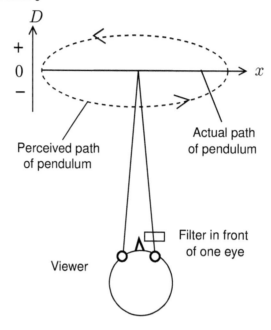

Fig. 6.16: The Pulfrich effect. A pendulum swinging back and forth horizontally in front of the viewer is viewed with a filter in front of the right eye. The perceived path of the pendulum follows an ellipse in depth, because the filter delays the responses of the photoreceptors in the right eye. The same qualitative depth illusion persists when the pendulum is only visible at discrete times (using flashed illumination). When the pendulum is replaced by a fronto-parallel display containing random dynamic noise, the perception contains multiple depths, with opposite depth signs for opposite moving directions and larger depth deviations from the screen for faster motions. Adapted from *Vision Research*, 37 (12), Ning Qian and Richard A. Andersen, A Physiological Model for Motion-Stereo Integration and a Unified Explanation of Pulfrich-like Phenomena, pp. 1683–98, figure 1, Copyright (1997), with permission from Elsevier.

Qian and Andersen (1997) showed how this population-decoding algorithm can be used to understand the Pulfrich effect, shown in Fig. 6.16, and its non-trivial variants. A pendulum swings horizontally in a frontoparallel plane in front of an observer. If a filter is placed in front of the viewer's right eye such that the photoreceptor responses in this eye are delayed by a time interval Δt, the pendulum is perceived to follow a elliptic trajectory extended in depth. The classical explanation is that if the pendulum moves at velocity V, the delay Δt leads to a spatial disparity of $\Delta x = V \Delta t$ between its positions imaged simultaneously on the two retinae. This disparity leads to the perceived depth.

To study the decoding, we must first endow each neuron with temporal as well as spatial tuning, including the time delay for the input from the right eye. The receptive fields (RFs) of the left- and right-eye simple cell components of the complex cell in equation (6.41) become

$$
\begin{aligned}
K_L(x,t) &= \exp\left(-\frac{x^2}{2\sigma^2} - \frac{t^2}{2\sigma_t^2}\right) \cos\left[\hat{k}(x - x_L) + \hat{\omega}t\right], \\
K_R(x,t) &= \exp\left(-\frac{x^2}{2\sigma^2} - \frac{(t-\Delta t)^2}{2\sigma_t^2}\right) \cos\left[\hat{k}(x - x_R) + \hat{\omega}(t - \Delta t)\right].
\end{aligned}
\tag{6.44}
$$

In the above, $\hat{\omega}$ is the preferred temporal frequency of the neuron, which is now direction selective (see equation (3.257)). The two RFs differ not only by spatial phase ($-\hat{k}x_L$ versus

$-\hat{k}x_R$), but also by a latency Δt, modeling the effect of the filter in front of the right eye. Let us assume that the effect of Δt on the exponential factor $\exp(...)$ in $K_{L,R}(x,t)$ is negligible (since the Gaussian envelope of the RFs is considerably larger than a typical on or off subregion in the RFs). Then, introducing Δt is equivalent to changing x_R to $x_R + \hat{\omega}\Delta t/\hat{k}$. Therefore, the $D_i = x_L - x_R$ in equation (6.42) for the expression of response r_i should be replaced by $x_L - x_R - \hat{\omega}\Delta t/\hat{k} = D_i - \hat{\omega}\Delta t/\hat{k}$. Hence, equation (6.42) becomes,

$$\bar{r}_i \approx r_{\max} \cos^2\left(\frac{\hat{k}(D_i - \hat{\omega}\Delta t/\hat{k} - D)}{2}\right). \tag{6.45}$$

Meanwhile, the input is now dynamic (rather than static), described by $I(x,t)$ as a function of both space and time, with signal power $|\mathcal{I}(k,\omega)|^2$ as a function of spatiotemporal frequency (k,ω). When the observer focuses on the plane of the pendulum, the input disparity $D = 0$. Extend $r_{\max} \propto |\mathcal{I}(\hat{k})|^2$ to $r_{\max} \propto |\mathcal{I}(\hat{k},\hat{\omega})|^2$, and use $D = 0$; the above equation becomes

$$\bar{r}_i \propto |\mathcal{I}(\hat{k},\hat{\omega})|^2 \cos^2\left(\frac{\hat{k}(D_i - \hat{\omega}\Delta t/\hat{k})}{2}\right). \tag{6.46}$$

Meanwhile, the pendulum, whose spatial image is $I(x)$ moving at velocity V, provides a spatiotemporal input pattern $I(x - Vt)$ to both eyes focusing on its depth plane. When $\mathcal{I}(k)$ is the spatial Fourier transform of $I(x)$, then the spatiotemporal Fourier transform $\mathcal{I}(k,\omega)$ of $I(x - Vt)$ is

$$\mathcal{I}(k,\omega) = \mathcal{I}(k)\delta(kV + \omega). \tag{6.47}$$

Hence, equation (6.46) becomes

$$\bar{r}_i \propto |\mathcal{I}(\hat{k})|^2 \delta(\hat{k}V + \hat{\omega}) \cos^2\left(\frac{\hat{k}(D_i - \hat{\omega}\Delta t/\hat{k})}{2}\right). \tag{6.48}$$

Hence, among this neural population, the highest responses are from neurons i with a preferred disparity $D_i = \hat{\omega}\Delta t/\hat{k} = -V\Delta t$. By the decoding algorithm in equation (6.38),

the perceived disparity of the Pulfrich pendulum is

$$\hat{D} \approx D_i = \frac{\hat{\omega}\Delta t}{\hat{k}} = -V\Delta t. \tag{6.49}$$

As the pendulum oscillates, its velocity V changes magnitude and direction. Each V activates a subpopulation of neurons whose preferred frequencies $(\hat{k},\hat{\omega})$ satisfy $V = -\hat{\omega}/\hat{k}$. Hence, the decoded or perceived disparity $\hat{D}$ oscillates with the pendulum, changing magnitude and sign between crossed and uncrossed disparities. Accordingly, the perceived 3D position of the pendulum follows an ellipse, as illustrated schematically in Fig. 6.16.

Understanding the Pulfrich effect through this decoding algorithm can be applied to understanding some variations of the original Pulfrich effect in which the classical intuitive $\Delta x = V\Delta t$ argument no longer applies. In one example, the oscillating pendulum is shown stroboscopically, such that it is visible only at discrete time steps (Morgan 1979). In such a case, the two eyes never see the pendulum at the same time. For example, if the pendulum is briefly illuminated at time $t = 0$ and at location $x = 0$, the left eye sees it at time $t = 0$ at location $x = 0$, whereas the right eye sees it at time $t = \Delta t$ but at the same location $x = 0$. There is no spatial disparity between the left eye and right eye views. Hence, the depth illusion can no longer be understood intuitively as it was in the original Pulfrich effect.

However, although stroboscopic presentation affects the overall response amplitudes, the neurons that were most strongly activated by the original pendulum are still likely to be the most strongly activated by the stroboscopic version. Hence, our decoding algorithm should give a qualitatively similar depth percept as that for the original pendulum. This is because our derivation of the results in equation (6.46) still holds, with the delay Δt still being equivalent to a disparity $D = \frac{\hat{\omega}\Delta t}{k}$ as far as the complex cell is concerned. The stroboscopic presentation, with illumination of the pendulum once each time interval T, modifies the input $I(x - Vt)$ and its Fourier transform $\mathcal{I}(k, \omega)$ such that

$$I(x - Vt) \quad \text{becomes} \quad \sum_{n=-\infty}^{\infty} I(x - Vt)\delta(t - nT),$$

$$\mathcal{I}(k, \omega) \quad \text{becomes} \quad \sum_{n=-\infty}^{\infty} \mathcal{I}(k, \omega - 2\pi n/T)$$

$$= \mathcal{I}(k, \omega) + \sum_{n \neq 0} \mathcal{I}(k, \omega - 2\pi n/T).$$

The new $\mathcal{I}(k, \omega)$ contains the original $\mathcal{I}(k, \omega)$ and its temporal frequency shifted versions $\mathcal{I}(k, \omega - 2\pi n/T)$ for $n \neq 0$. As long as the time interval T between successive illuminations is not longer than about 30 ms, the latter ($\mathcal{I}(k, \omega - 2\pi n/T)$ for $n \neq 0$) contains mainly temporal frequencies around or beyond 33 Hz ($\approx 1/30$ ms), which is too high to evoke responses in most neurons (which, for humans are generally sensitive to temporal frequency around 5–10 Hz). Hence, the stroboscopic and the non-stroboscopic inputs activate similar subpopulations of neurons, thereby evoking similar depth percepts.

In another variation of the original Pulfrich effect, the pendulum is replaced by spatiotemporal random noise on a frontoparallel display (Tyler 1974) containing no coherent motion. However, due to the filter in front of the right eye, each neuron responds according to equation (6.46), as if there is a input disparity $D = \hat{\omega}\Delta t/\hat{k}$ determined by its preferred $(\hat{k}, \hat{\omega})$. The response magnitude scales with $r_{\max}$, which increases with the input signal power in this frequency $(\hat{k}, \hat{\omega})$. If the noise has a broad spatiotemporal frequency spectrum, then it can excite almost any neuron, such that the population activity pattern no longer resembles a unimodal function of the preferred disparity D_i of the neurons (as in Fig. 6.15). Instead, the percept should contain multiple depths, invalidating our decoding scheme, which assumes that there is just a single depth. Nevertheless, at least qualitatively, we can expect the noise pattern to appear to contain depth locations both in front of and behind the display screen. Furthermore, noise in front of and behind the screen should appear to be moving in opposite directions, because cells preferring positive or negative disparities D_i are activated by spatiotemporal frequencies with negative or positive preferred velocity $-\hat{\omega}/\hat{k}$, respectively. Additionally, faster motions should evoke larger depth deviations from the $D = 0$ depth plane, since the preferred $|D_i|$ is proportional to the moving speed $|\hat{\omega}/\hat{k}|$.

It is notable that random dynamic noise, with no coherent motion (which means motion in different directions roughly cancel each other out) on the display screen, can evoke the perception of coherent motion when different motion directions appear to be separated in depth (here, by the temporal effect caused by the filter). We remind ourselves that MT neurons do not respond well to two groups of dots moving in opposite directions unless motion in different directions are separated in depth; see Fig. 2.27 B. However, it is premature to make conclusions about area MT's role in the Pulfrich effect just from this observation.

The classical Pulfrich effect and its less intuitive variants demonstrate that combining our knowledge of the underlying neural substrates with a population decoding framework can offer a powerful framework for understanding perception. They suggest the brain makes use of decoding algorithms similar to equations (6.37) and (6.38), for which the percept is the response-weighted average of the preferred features of the neurons or as the preferred feature of the most activated neurons. These algorithms are derived from the likelihood function $P(\mathbf{r}|\mathbf{S})$ assuming a normal visual environment, which lacks a filter obscuring one eye. Apparently, the brain continues to apply such algorithms when the environment is distorted, i.e., using the wrong likelihood functions. Sometimes, although all too rarely for the best illusions, it is possible to learn to correct the erroneous decoding.

6.3.6 Poisson-like neural noise and increasing perceptual performance for stronger visual inputs

The properties of maximum-likelihood decoding applied to wavelength discrimination (Section 6.3.4) are shared by many other ideal-observer analyses in visual perception. First, the discrimination threshold in equation (6.32) is inversely related to the square root of the input intensity parameter I, because the noise in the cone absorption is Poisson distributed. More explicitly, decompose the visual input $\mathbf{S}$ into two components: an overall intensity parameter I, and other feature parameters θ (we considered the wavelength λ; but this could more generally include other features such as orientation, color, motion directions, etc.), such that the average neural response is $\bar{r} = I \cdot f(\theta)$. This response scales with intensity I and is tuned to input features θ with tuning function $f(\theta)$. The Poisson noise in neural response r makes the likelihood function

$$P(r|\mathbf{S}) = \frac{(\bar{r})^r}{r!} \exp(-\bar{r}). \tag{6.50}$$

The response r has a variance

$$\sigma_r^2 \equiv \langle (r - \bar{r})^2 \rangle = \langle r^2 \rangle - (\bar{r})^2 = \bar{r}, \tag{6.51}$$

which is the same as the mean response. This means that the signal to noise ratio in the response is

$$\text{signal-to-noise in response} \equiv \frac{\bar{r}}{\sigma_r} = \sqrt{\bar{r}} \propto \sqrt{I}, \tag{6.52}$$

which increases with $\sqrt{I}$. This is the reason why the discrimination threshold decreases with increasing $\sqrt{I}$.

Many neurons in the brain respond with Poisson-like variability, such that their response variance is proportional to the average response (Geisler and Albrecht 1997, Dayan and Abbott 2001). The ratio between the variance and the average of a neuron's responses is called the *Fano factor*. It takes the value of 1 for Poisson responses. V1 neurons whose contrast response properties are given in equation (6.16) have Fano factor K; for the monkey V1 neurons described in Fig. 6.12 B, K is in the range 1.5–2. The signal-to-noise ratios for neurons that can be described this way tend to increase with the average response, and so, as in equation (6.32), lead to discrimination thresholds that decrease with increasing input intensity.

6.3.7 Low efficiency of sensory information utilization by the central visual system

However, even as input intensity increases, behavioral thresholds rarely decrease beyond a minimum value. Take the case of wavelength discrimination, and recall that the "input intensity" parameter I was taken to include multiple factors in order to sum the signals over space and time across all the cones of a particular type covered by the input field:

$$I \propto \text{input intensity (per unit area)} \times \text{input field size} \times \text{input viewing duration.} \quad (6.53)$$

The effective cone response $\bar{r}_a = I f_a(\lambda)$ should therefore scale with each factor—so, for example, a larger input field, a brighter intensity, or a longer viewing duration should all lead to stronger responses, and thus lower discrimination thresholds. Nevertheless, similar threshold values were observed by different research laboratories, using different intensities and sizes of the input light fields, and different or unconstrained viewing durations (Pokorny and Smith 1970). Thus, beyond some minimum values of the factors, the thresholds appear to be constrained instead by central neural processes that limit the efficiency with which sensory input information can be used. However, the fact that a single parameter (Δ_{th}/I) can quantitatively fit the whole curve of the threshold $\sigma(\lambda)$ as a function of λ implies that this efficiency is invariant to wavelength λ.

Similarly, for the example of contrast detection shown in Fig. 6.12 A, thresholds for human observers are about 20 times worse than for the ideal observer; however, this ratio between the performances of human and ideal observers is roughly invariant with the spatial frequency of the gratings. Furthermore, the thresholds for human observers do not substantially decrease further for gratings that are made bigger (beyond around 7.5 cycles) so that detection could be based on more spatial cycles (Banks et al. 1987).

For the contrast discrimination shown in Fig. 6.12 B, behavioral discrimination thresholds are comparable to those of an ideal observer that is restricted to use only a tiny group of less than 100 V1 monkey neurons! In fact, at some of the input contrast levels (along the horizontal axis of Fig. 6.12 B), the neuron that has the largest d_i' in equation (6.19) itself has a d' value that is comparable to that observed in behavior (Geisler and Albrecht 1997), although the most sensitive neuron is different at different input contrast. Figure 6.10 A showed another example in which an ideal observer using the responses of a single MT neuron could match the performance of the whole animal. Are these inefficiencies caused by correlations between variabilities in the responses from different neurons (Shadlen et al. 1996)? However, although observations suggest that there are some correlation between response noise in different neurons, such a correlation does not necessarily have a huge effect on decoding accuracy (Abbott and Dayan 1999, Averbeck, Latham and Pouget 2006).

6.3.8 Transduction and central inefficiencies in the framework of encoding, attentional selection, and decoding

To aid understanding, Barlow (1977) proposed separating into two components the brain's overall inefficiency in utilizing sensory information in the input: transduction inefficiency in the retina, and central inefficiency in post-retinal processing. The transduction inefficiency was attributed to suboptimal encoding of the photoreceptor signals (Pelli 1990). Specifically, visual input is first sampled by the photoreceptors, which introduces photon noise (such as the Poisson noise in equation (6.20)); subsequently, additional neural encoding noise is introduced. Adding a small amount of external noise to the visual input image (which emulates an increase in photon noise) typically does not affect visual performance. Thus, the additional encoding noise must dominate. The transduction inefficiency is attributed to this additional encoding noise beyond that of the photon noise. The power of this additional encoding noise can be quantified by that of the "equivalent internal encoding noise," which is the smallest external image noise needed to substantially impair perceptual performance. Meanwhile, central inefficiency is the inefficiency in processing the encoded visual input signal and the

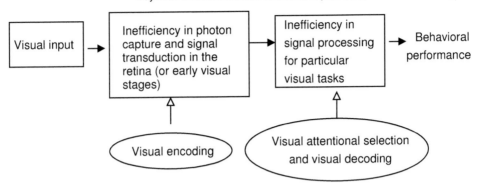

Fig. 6.17: Inefficiencies in visual processing can be separated into transduction and central components. These correspond to the encoding and the selection/decoding stages, respectively.

equivalent internal encoding noise by the central brain. The inefficiency introduced in this stage is attributed phenomenologically to the central noise (Barlow 1977).

   As oft noted in this book, we consider vision as having three stages: encoding, selection, and decoding; see Fig. 1.5. We place the transduction inefficiency at the stage of encoding, and central inefficiency at the stages of selection and decoding; see Fig. 6.17. According to this view, if decoding is assumed to be optimal or at least very efficient, the remaining cause of central inefficiency would have to be the attentional bottleneck. This observation may help us better understand the "central noise" that causes the central inefficiency.

6.4   Bayesian inference and the influence of prior belief in visual decoding

Behavioral observations suggest that the perception of scenes is influenced by the frequency with which they are encountered in experience or prior expectations based on the context. The left image in Fig. 2.31 B is an example of this—this image could be seen as either a scene illuminated from above, containing a central concave spherical dent among surrounding convex spherical bumps, or as another scene illuminated from below, containing a central convex bump among surrounding concave dents. The two scenes have an equal likelihood given this image. However, percept is typically the first scene.

   In this section, we introduce Bayesian inference, or Bayesian decoding, which captures the influence of prior expectation or experience on decoding or perception. In the case of Fig. 2.31 B, there is a strong prior presumption that lighting comes from above (according to our daily experience), and this biases the precept. Maximum-likelihood decoding is a special case of Bayesian decoding when there is no prior bias over the possible scenes. Most investigations of Bayesian decoding, especially of the influence of prior bias, have focused on behavioral rather than neural phenomena; we illustrate a few of these in this section. Meanwhile, investigations on Bayesian perceptual inference have motivated many theoretical and modeling studies aimed at probing the underlying neural mechanisms (Deneve, Latham and Pouget 2001, Ma, Beck, Latham and Pouget 2006, Fiser, Berkes, Orbán and Lengyel 2010).

6.4.1 The Bayesian framework

Bayesian inference is a comprehensive framework for understanding perception as statistical inference that can be traced back to Helmholtz's ideas of perception as inference. It is often formulated as follows (Dayan, Hinton, Neal and Zemel 1995, Kersten, Mamassian and Yuille 2004): let $P(\mathbf{S})$ denote the prior probability of a scene $\mathbf{S}$ in the visual world. Then, as we briefly mentioned in equation (6.2), the posterior probability $P(\mathbf{S}|\mathbf{r})$ of the scene $\mathbf{S}$ given neural response $\mathbf{r}$ should follow from Bayes' rule as

$$P(\mathbf{S}|\mathbf{r}) = \frac{P(\mathbf{r}|\mathbf{S})P(\mathbf{S})}{P(\mathbf{r})}, \tag{6.54}$$

where $P(\mathbf{r}) \equiv \int d\mathbf{S} P(\mathbf{r}|\mathbf{S})P(\mathbf{S})$ is the marginal probability of neural responses $\mathbf{r}$. A fully Bayesian view suggests that observers perceive according to this posterior probability distribution $P(\mathbf{S}|\mathbf{r})$, making scenes that have higher prior probabilities and explain the neural responses $\mathbf{r}$ well more likely percepts. This process of calculating and exploiting the posterior is called visual inference.

When the posterior distribution $P(\mathbf{S}|\mathbf{r})$ is very sharply peaked in the space of scenes $\mathbf{S}$ (given $\mathbf{r}$), the visual input should be unambiguously perceived as $\hat{\mathbf{S}} = \mathrm{argmax}_{\mathbf{S}} P(\mathbf{S}|\mathbf{r})$, which is called the maximum a posteriori or MAP value (to distinguish it from the maximum-likelihood or ML value $\hat{\mathbf{S}} = \mathrm{argmax}_{\mathbf{S}} P(\mathbf{r}|\mathbf{S})$). For the present, we consider MAP decoding; later in the section, we consider cases of ambiguous perception when the posterior distribution $P(\mathbf{S}|\mathbf{r})$ is multimodal and peaks around at least two qualitatively different scenes $\mathbf{S}$. We will not discuss Bayesian decision theory, in which the posterior distribution $P(\mathbf{S}|\mathbf{r})$ is used to make optimal choices. MAP is a particular sort of Bayesian decision.

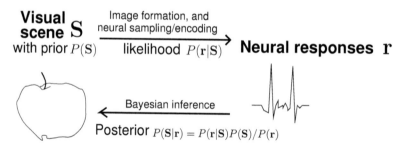

Visual scene S with prior $P(\mathbf{S})$ → Image formation, and neural sampling/encoding likelihood $P(\mathbf{r}|\mathbf{S})$ → **Neural responses r**

← Bayesian inference

Posterior $P(\mathbf{S}|\mathbf{r}) = P(\mathbf{r}|\mathbf{S})P(\mathbf{S})/P(\mathbf{r})$

Perception $\hat{\mathbf{S}}$ could be one of the following:
$\hat{\mathbf{S}} = \mathrm{argmax}_{\mathbf{S}} P(\mathbf{S}|\mathbf{r})$, the maximum a posteriori estimate of the scene $\mathbf{S}$,
$\hat{\mathbf{S}} = $ a sample from $P(\mathbf{S}|\mathbf{r})$, or average of samples,
the difference between $\hat{\mathbf{S}}$ and $\mathbf{S}$: the estimation error, leading to visual illusions.
The spread of $P(\mathbf{S}|\mathbf{r})$ in $\mathbf{S}$ characterizes perceptual uncertainty.

Fig. 6.18: A schematic of the Bayesian framework for visual perceptual inference.

MAP decoding is a ready replacement for the ML decoding of Section 6.3 when the prior $P(\mathbf{S})$ is not uniform across $\mathbf{S}$. For example, for the moving dots in Section 6.3.1, a prior bias for upward motion, i.e., $P(\mathbf{S} = \text{upward motion}) > P(\mathbf{S} = \text{downward motion})$, should make this direction the more likely percept given response $\mathbf{r}$. If the monkey has to report the perception, it should choose the MAP scene

$$\text{reported scene} = \hat{\mathbf{S}} = \mathrm{argmax}_{\mathbf{S}} P(\mathbf{S}|\mathbf{r}) = \mathrm{argmax}_{\mathbf{S}} P(\mathbf{r}|\mathbf{S})P(\mathbf{S}), \tag{6.55}$$

although the likelihood factor $P(\mathbf{r}|\mathbf{S})$ can still overwhelm the prior if it is strong enough.

Perception can be erroneous (i.e., illusory) if either prior $P(\mathbf{S})$ or likelihood $P(\mathbf{r}|\mathbf{S})$ does not reflect reality. This is exactly the case in the Pulfrich effect (in Section 6.3.5.1), which we argued arises from ML decoding (i.e., MAP decoding with a flat prior) when the likelihood $P(\mathbf{r}|\mathbf{S})$ is not adjusted in the light of the filter that is introduced in front of one of the two eyes.

6.4.2 Bayesian visual inference is highly complex unless the number and the dimensions of possible percepts are restricted

Inference is easy when there are only few possible percepts, or at least they vary in only a small number of dimensions. For example, to decide whether dots are moving up or down, whether a cube is below or above eye sight, or the color of a surface, it is only necessary to pick a single outcome from among a few alternatives. As long as the functional form of $P(\mathbf{S}|\mathbf{r}) \propto P(\mathbf{r}|\mathbf{S})P(\mathbf{S})$ is available, it is feasible to search in the space of $\mathbf{S}$ to find the MAP value within a reasonable amount of time.

However, general visual scenes are extremely complex. The variable $\mathbf{S}$ is high-dimensional, including all the objects in view as well as the lighting, etc. For example, given the visual input in Fig. 4.1 A, $\mathbf{S}$ must specify the number of objects in the view (one or two); along with the shape, surface color, location, orientation, and perhaps the identity of each object concerned. Some of these variables themselves are also high-dimensional. For example, object shape could include size, volume, angles between surfaces, and the length of each surface boundary; surface properties could include color, roughness, and specularity. That $\mathbf{S}$ is high-dimensional makes for a combinatorial explosion in the number of possible scenes. Given an $\mathbf{r}$, even if there is one $\mathbf{S}$ that makes $P(\mathbf{S}|\mathbf{r})$ much larger than for other scenes, it is far from clear how the brain could search the space of scenes to identify it, as required for MAP inference.

Computer or artificial vision algorithms, for which $\mathbf{r}$ typically refers to the output of a digital camera rather than the neural responses, face the same problem. To make recognition more tractable, they typically restrict dimensionality of $\mathbf{S}$. For example, computer algorithms can quickly detect faces in the scene and even identify them. They do this by restricting $\mathbf{S}$ to describe only the presence or absence of a face at an image location (which we can see as a form of selection) along with a reduced characterization of its features such as size, orientation, specific facial attributes, and lighting.

6.4.2.1 Bottom-up processes to aid complex inferences

Is the duck image $\mathbf{r}$ in Fig. 6.6 difficult to recognize because the posterior distribution $P(\mathbf{S}|\mathbf{r})$ contains too many possible scenes $\mathbf{S}$ embedded in a high-dimensional space? Human observers can quite easily recognize objects in typical photographs without hesitation or hints. Without any prior knowledge, the number of possible scenes $\mathbf{S}$ should be the same for different input images before they are seen. Are some images easy to recognize because good hints or guesses for candidate $\mathbf{S}$ come from the images $\mathbf{r}$ themselves, so that the brain need only search among a small number of alternative scenes?

Many proposals (MacKay 1956, Carpenter and Grossberg 1987, Kawato, Hayakawa and Inui 1993, Dayan et al. 1995, Yuille and Kersten 2006) along these lines have been made, suggesting that the brain uses bottom-up processes to generate from inputs $\mathbf{r}$ candidate guesses for the visual scenes $\mathbf{S}$. These are typically thought to involve heuristic rules of thumbs that are usually but not always true. Subsequently, top-down processes calculate the posterior probabilities for just these candidates by combining the prior $P(\mathbf{S})$ with a likelihood calculated by synthesizing $\mathbf{r}$ from $\mathbf{S}$, and comparing it with the bottom-up $\mathbf{r}$. Together, bottom-up and top-down processes verify and improve the candidate $\mathbf{S}$ which reaches perception. There are indeed extensive connections between different visual areas along the visual pathway, in both the upstream and downstream directions (Felleman and van Essen 1991). These connections could serve the bottom-up and top-down information flow during the likely iterative recognition processing. The fact that top-down processes generate **r** from **S** provides an opportunity for the bottom-up (called analysis) heuristics to be trained or tuned to become the statistical inverse of synthesis.

6.4.3 Behavioral evidence for Bayesian visual inference

To assess whether visual decoding follows Bayesian principles requires knowledge of neural response properties, described by $P(\mathbf{r}|\mathbf{S})$, and perceptual behavior (the perceived **S**, or at least the discrimination thresholds $\sigma(\mathbf{S})$) for a given cognitive task, so that we can compare the behavioral quantities with their counterparts predicted from the Bayesian framework using $P(\mathbf{r}|\mathbf{S})$ (and $P(\mathbf{S})$). (Furthermore, for this program to work, we need to ensure or assume that the neural responses **r** are actually the ones encoding the sensory input at the input stage to the decoding process, i.e., **r** is not an intermediate outcome in the implementation of decoding.) Unfortunately, most existing investigations of visual decoding have one but not the other. For example, from a reasonably good model of $P(\mathbf{r}|\mathbf{S})$ for the retinal ganglion cells, one could predict the MAP percept of a white noise input **S** from ganglion cell responses **r** (Paninski, Pillow and Lewi 2007). However, we do not know what animals actually perceive in white noise, except that not many details get through the attentional bottleneck. Similarly, although we can measure perceptual performance on recognizing and discriminating faces, not nearly enough is known about the neural responses involved to build up a reasonably complete $P(\mathbf{r}|\mathbf{S})$, which would be necessary to refute or support Bayesian inference.

There are cases, such as those presented in Section 6.3, for which the likelihood functions $P(\mathbf{r}|\mathbf{S})$ associated with retinal, V1, or MT neural responses and the percepts of input contrast, color, motion direction, or object depth are all known. However, for none of these examples is it clear whether prior expectations $P(\mathbf{S})$ affected the percept according to the Bayesian prescription.

Nevertheless, there are many purely behavioral investigations of the influence of priors $P(\mathbf{S})$. They have used 2D visual input images **r**, often the responses of camera pixel sensors, in place of the neural responses, and constructed $P(\mathbf{r}|\mathbf{S})$ based on models of the image formation process that links scene properties **S** with the images **r**, or from direct measurements of **r** and **S**. They have then investigated whether perceptual behavior is what would be expected from Bayesian inference based on this non-neural **r**, and in particular have considered the influence of priors on perception. This approach can indirectly probe whether the brain follows a Bayesian prescription if there is a one-to-one, or at least a reasonably tight, relationship between the visual input image **r** and the neural responses it evokes. Behavioral studies are much more flexible than neural ones; this has helped to identify and diagnose specific factors in visual inference algorithms, and to test, modify, or falsify hypotheses about Bayesian inference. We next present a few examples of such behavioral studies.

6.4.3.1 Example: perceived visual motion through a hole—the aperture problem

Weiss et al. (2002) showed that perception or illusions of motion of rigid bodies can be understood as Bayesian inference. They considered image $I(x, y, t)$ (as a function of pixel locations (x, y) and time t) caused by a rigid body moving in a blank background. Let the velocity of motion in the image plane be (v_x, v_y), with components v_x and v_y along the x and y directions.

The inference problem is to infer motion velocity (v_x, v_y) from $I(x, y, t)$. $\qquad$ (6.56)

In Bayesian terms:

$$\mathbf{r} \equiv I(x, y, t), \quad \text{and} \quad \mathbf{S} \equiv \mathbf{v} = (v_x, v_y). \tag{6.57}$$

If the motion is rigid, then the object does not change; so

$$I(x, y, t) = I(x + v_x \Delta t, y + v_y \Delta t, t + \Delta t) \tag{6.58}$$

for a small time step Δt. From the first term of a Taylor expansion of the equation, we can derive

$$D(x, y) \equiv \nabla_x I(x, y, t) v_x + \nabla_y I(x, y, t) v_y + \nabla_t I(x, y, t) = 0, \tag{6.59}$$

$$\text{in which} \quad (\nabla_x I, \nabla_y I, \nabla_t I) \equiv \left(\frac{\partial I(x, y, t)}{\partial x}, \frac{\partial I(x, y, t)}{\partial y}, \frac{\partial I(x, y, t)}{\partial t} \right). \tag{6.60}$$

In principle, given $\mathbf{r} = I(x, y, t)$, one can calculate $(\nabla_x I, \nabla_y I, \nabla_t I)$ and then find the $\mathbf{S} = (v_x, v_y)$ that satisfies equation (6.59) for all locations (x, y).

When the moving object is effectively one-dimensional, such as a straight and infinitely long line or edge, or an infinitely large grating, its static image $I(x, y)$ can be described by $I(x, y) = f(\cos(\theta)x + \sin(\theta)y)$ for some function $f(.)$, where the two-dimensional unit vector $\mathbf{n} \equiv (\cos(\theta), \sin(\theta))$ is perpendicular to the line or edge (or the grating). Let us say that the actual velocity of this one-dimensional object is $\check{\mathbf{S}} = \check{\mathbf{v}} \equiv (\check{v}_x, \check{v}_y)$; then

$$I(x, y, t) = f(\cos(\theta)(x - \check{v}_x t) + \sin(\theta)(y - \check{v}_y t)). \tag{6.61}$$

In such a case, using f' to denote the derivative of f,

$$\nabla_x I = \cos(\theta) \cdot f', \quad \nabla_y I = \sin(\theta) \cdot f', \quad \nabla_t I = - [\cos(\theta) \cdot \check{v}_x + \sin(\theta) \cdot \check{v}_y] f'. \tag{6.62}$$

Consequently, equation (6.59) becomes

$$D(x, y) = \{[\cos(\theta)v_x + \sin(\theta)v_y] - [\cos(\theta)\check{v}_x + \sin(\theta)\check{v}_y]\} \cdot f' = (\mathbf{n} \cdot \mathbf{v} - \mathbf{n} \cdot \check{\mathbf{v}}) \cdot f' = 0. \tag{6.63}$$

Thus, any velocity (v_x, v_y) that has the same component as the real velocity $\check{\mathbf{v}}$ along the direction $\mathbf{n}$ can account for the visual input $I(x, y, t)$. This includes the real velocity $\check{\mathbf{v}}$, but also an infinite set of other possibilities. This ambiguity comes from the one-dimensional structure in $I(x, y)$ and is called the *aperture problem*. This is illustrated in Fig. 6.19. Figure 6.19 A depicts a long moving white line seen through an aperture which occludes the ends of the line. The rigid body constraint $D(x, y) = 0$ does not need to be satisfied at the boundary of the aperture. Thus, this line can be treated as if one-dimensional, extending infinitely behind the aperture. Figure 6.19 B depicts the two-dimensional space of velocities. Here, the dashed line is parallel to the imaged line such that, according to equation (6.63), any velocity vector $\mathbf{v} = (v_x, v_y)$ that ends on this dashed line has the same projection as $\check{\mathbf{v}}$ along the direction $\mathbf{n}$. Thus, it can satisfy the rigid body constraint in equation (6.59) for the moving white line in Fig. 6.19 A.

Among these velocity vectors consistent with input $I(x, y, t)$, the one perpendicular to the line, shown in red in Fig. 6.19, has the smallest speed. This velocity is often the one perceived, even if the actual velocity is different. This is because there is a prior (Yuille and Grzywacz 1988, Stocker and Simoncelli 2006) favoring smaller speeds $|\mathbf{v}|$, arising from the fact that visual objects in our scenes tend to be not moving or moving only slowly.

In the Bayesian formulation, the rigid body equality in equation (6.59) can be seen as the extreme case of a likelihood function for the image $I(x, y, t)$. More generally, modest departures from the equality $D(x, y) = 0$ can be tolerated as coming from noise. Assuming a

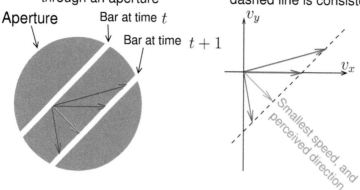

A: A rightward moving line seen through an aperture

B: Any velocity (v_x, v_y) ending on the dashed line is consistent with input in A

Fig. 6.19: The aperture problem. A: A long homogenous moving white line seen through an aperture. The line is longer than the aperture, so its velocity is ambiguous. The colored vectors mark four possible velocities consistent with the moving line. The red one has the smallest speed $|\mathbf{v}|$, since it is perpendicular to the moving line; all other possible velocities have the same velocity component parallel to the red one. B: Velocities in A depicted in the two-dimensional velocity space (v_x, v_y). All velocity vectors which start from the origin and end on the dashed line satisfy the constraint in equation (6.59) for the visual input in A. Even if the white line is moving horizontally, it is typically seen as moving perpendicular to itself (with the velocity depicted as the red vector), because that is the MAP value associated with the prior bias for slow speeds.

Gaussian distribution of this noise, which is independent at different image locations and has variance σ^2, the likelihood function can be modeled as

$$P\left(\mathbf{r} = I(x, y, t) | \mathbf{S} = (v_x, v_y)\right) \propto \exp\left(-\frac{1}{2\sigma^2} \int dx dy [D(x, y)]^2\right). \qquad (6.64)$$

Assuming that (v_x, v_y) also has an isotropic Gaussian prior favoring zero and smaller speeds with variance σ_p^2,

$$P\left(\mathbf{S} = (v_x, v_y)\right) \propto \exp\left(-\frac{v_x^2 + v_y^2}{2\sigma_p^2}\right), \qquad (6.65)$$

and the posterior distribution is

$$P\left(\mathbf{S} = (v_x, v_y) | \mathbf{r} = I(x, y, t)\right) \propto P\left(\mathbf{r} = I(x, y, t) | \mathbf{S} = (v_x, v_y)\right) P(\mathbf{S} = (v_x, v_y))$$
$$\propto \exp\left(-\frac{v_x^2 + v_y^2}{2\sigma_p^2} - \frac{1}{2\sigma^2} \int dx dy \left[\nabla_x I(x, y, t) v_x + \nabla_y I(x, y, t) v_y + \nabla_t I(x, y, t)\right]^2\right). \qquad (6.66)$$

Maximizing this posterior leads to a MAP value $\hat{\mathbf{v}} = (\hat{v}_x, \hat{v}_y)$ that balances the prior for small $|\mathbf{v}|$ with the rigid body constraint for small $|D(x, y)|$. For the image in Fig. 6.19, all velocity vectors ending on the dashed line in Fig. 6.19 B have the same (maximum) likelihood value since they all satisfy $D(x, y) = 0$. Among them, the velocity marked by the red vector is the one that maximizes the prior probability $P(\mathbf{S} = (v_x, v_y))$. This explains why lines seen through an aperture are seen as moving only perpendicular to themselves.

To see this mathematically, write velocity as $\mathbf{v} = (v_\perp, v_\parallel)$ (and analogously $\check{\mathbf{v}} = (\check{v}_\perp, \check{v}_\parallel)$), in terms of the two components that are perpendicular and parallel, respectively, to the line.

For simplicity, we use the noiseless $I(x, y, t)$ from equation (6.61), so that $I(x, y, t)$ is written in terms of $\check{\mathbf{v}}$, $\mathbf{n}$, and $f(.)$. Consequently, we have $D(x, y) = (v_\perp - \check{v}_\perp) f'$. Since $v_x^2 + v_y^2 = v_\perp^2 + v_\parallel^2$, the posterior for the aperture problem is

$$P\left(v_\perp, v_\parallel | I(x, y, t)\right) \propto \exp\left(-\frac{v_\perp^2 + v_\parallel^2}{2\sigma_p^2} - (v_\perp - \check{v}_\perp)^2 \frac{1}{2\sigma^2} \int dx dy (f')^2\right). \quad (6.67)$$

This posterior is maximized when the velocity $\mathbf{v} = (v_\perp, v_\parallel)$ is

$$\hat{v}_\parallel = 0, \quad \hat{v}_\perp = \left(1 - \frac{\sigma^2}{\sigma^2 + \sigma_p^2 \int dx dy (f')^2}\right) \check{v}_\perp. \quad (6.68)$$

This velocity is perpendicular to the moving line. This is why an infinitely long line seen through an aperture is perceived to move in the direction perpendicular to itself. Meanwhile, the perceived speed $|\hat{v}_\perp|$ is a fraction smaller than the actual velocity component perpendicular to the line. This fractional reduction in speed is caused by the prior: the reduction is greater for a stronger prior (i.e., smaller σ_p) or weaker image contrast power $\int dx dy (f')^2$, which itself depends on the length of the line. The perceived speed of longer lines, with $\sigma_p^2 \int dx dy (f')^2 \gg \sigma^2$, is close to the actual component of velocity perpendicular to the line; otherwise, the line can be seen as moving substantially slower than it really is. This speed illusion has been said (Weiss, Simoncelli and Adelson 2002) to cause drivers to speed up in fog, which reduces visual input contrast, and thus reduces the perceived speed, even without an aperture in this case.

6.4.3.2 Example: illusory perception of motion direction as a battle between prior and evidence (likelihood) in Bayesian perception

In the aperture problem, prior and likelihood are pitted against each other to determine the perceived velocity of motion. Figure 6.20 shows another example, simplified from the example by Weiss et al. (2002), who used a moving rhombus in which the directions of perceived and veridical motions deviated from each other even when the whole visual object was seen without a restrictive aperture. This perceptual distortion arises because the evidence from visual input is too weak to offset the influence of prior expectations. A moving bar that is high contrast is perceived as moving in its true direction (here, horizontally); but the same moving bar at a lower contrast can be seen to move toward the lower right, away from the actual direction of movement, and at a lower speed.

Analytically, the perceived direction should maximize the posterior probability, which, according to equation (6.66), can be written as

$$P\left(\mathbf{v} | I(x, y, t)\right) \propto P(\mathbf{S} = (v_x, v_y)) P(\mathbf{r} = I(x, y, t) | \mathbf{S} = (v_x, v_y))$$

$$\propto \exp\left(-\frac{v_x^2 + v_y^2}{2\sigma_p^2}\right) \exp\left(-\frac{1}{2\sigma^2} \int_{(x,y) \in R_1} dx dy [D(x, y)]^2\right)$$

$$\exp\left(-\frac{1}{2\sigma^2} \int_{(x,y) \in R_2} dx dy [D(x, y)]^2\right).$$

In the expression above, the first exponential factor comes from the prior. The second and third exponential factors come from the likelihood

$$P(\mathbf{r} = I(x, y, t) | \mathbf{S} = (v_x, v_y)) \propto \exp\left(-\frac{1}{2\sigma^2} \int dx dy [D(x, y)]^2\right).$$

The second factor comes from integrating the rigidity constraint over area $(x, y) \in R_1$, which is the whole image area except for the ends of the line. The third factor comes from

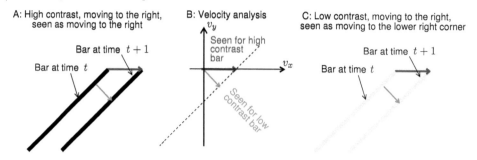

A: High contrast, moving to the right, seen as moving to the right

Bar at time $t+1$

Bar at time t

B: Velocity analysis

v_y

Seen for high contrast bar

v_x

Seen for low contrast bar

C: Low contrast, moving to the right, seen as moving to the lower right corner

Bar at time $t+1$

Bar at time t

Fig. 6.20: Weak image contrast leads to a bias toward the prior in the perceived motion. A and C contain moving bars which are identical except for the image contrast. B shows an analysis of the perceived velocity, as in Fig. 6.19 B. The blue velocity vectors in A, B, C are all identical; so are the red vectors (although strictly speaking, A and C are in image space, while B is in velocity space). If the ends of the long bars were obscured by an aperture, their velocities would be ambiguous (as in the aperture problem in Fig. 6.19), so that the blue and red velocity vectors would both satisfy the rigid body constraint. However, only the blue velocity vector satisfies the rigid body constraint at the bar ends. For sufficiently high image contrast, as in A, the evidence from the input tightly constrains the perceived velocity to satisfy the rigid body constraint imposed by the ends of the bar, thus to be consistent with the blue vector. When the image contrast is low, as in C, the prior is more influential, pulling the perceived velocity toward the red vector, which has a smaller speed than the veridical one.

integrating over the image area $(x, y) \in R_2$, which is just the ends of the line. Among the three exponential factors in $P(\mathbf{v}|I(x, y, t))$, the first two factors are exactly as in the aperture problem, without the line ends. They should thus favor the slow velocity perpendicular to the bar, which is indicated by the red vector in Fig. 6.20. The last exponential factor, for the line ends, peaks at the veridical velocity, indicated by the blue vector in Fig. 6.20. The force of the constraint associated with the likelihood scales with the square of the image contrast and is weak for low contrast input. When this is suitably weaker than the prior, the bar should appear to move toward the lower right and more slowly than it really does, and thus it appears to move at a non-veridical direction and speed.

6.4.3.3 Example: believing can be seeing, when the prior dictates perception against image contrast evidence

There are situations in which the image contrast is so weak that the prior dictates perception. Then one perceives objects that are not actually contained in the image, as if one could see ghosts. Figure 6.21 depicts three vertical arrays of bars; each array contains a gap between three bars above it and three bars below it. Observers are more likely to see a faint vertical bar in this gap when the gap is flanked by weak contrast vertical bars. However, this bias for seeing a faint vertical bar is absent when the flanking bars have high contrast or are horizontal.

For each array, focus only on the scene property at the "gap" location. This scene property is a scalar S with two possible values, $S = 1$ or $S = 0$, to denote, respectively, whether or not the gap contains a vertical bar. The response $\mathbf{r}$ is the image contrast c in the gap, i.e., we use c to describe $\mathbf{r}$. Here, the contextual inputs, i.e., the bars flanking the gap, can influence both the prior $P(S)$ and the likelihood $P(c|S)$ for image contrast c at the gap. For the prior: if the image location for the gap in each array is covered by an explicit occluder, and observers are asked whether they believe that a white vertical bar is likely to be present behind the occluder, they typically answer positively for arrays A and B but negatively for array C. This is because

Is there a faintly white vertical bar in each gap flanked by other white bars?

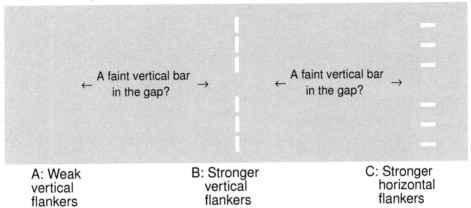

A: Weak
vertical
flankers

B: Stronger
vertical
flankers

C: Stronger
horizontal
flankers

Fig. 6.21: Contextual influences on visual filling-in of bars. A–C: Three very briefly presented vertical arrays of bars; each array has a central gap (with zero contrast). When observers are asked whether they see a very faintly white vertical bar in the gap (each time only one array is shown), they are more likely to answer positively for the gap in array A, and least likely for the gap in array C (Zhaoping and Jingling 2008).

observers believe that the vertical array of bars should continue behind the occluding cover from their experience of natural visual scenes (Geisler, Perry, Super and Gallogly 2001). This belief, in the absence of image evidence at the "gap," is the prior

$$P(S = 1) \text{ in array A} = P(S = 1) \text{ in array B} > P(S = 1) \text{ in array C.} \qquad (6.69)$$

Meanwhile, the two arrays on the left differ as to the likelihood $P(c|S)$ of image contrast c. Observers expect that, if a bar is present at the gap, it should be similar to those of the flankers, and in particular have a comparable contrast. This expectation likely arises from experience—the contrasts of images of objects should all be scaled up or down according to the overall level of illumination. Hence, the observer's conditional probability $P(c|S = 1)$ is likely to be a unimodal probability distribution of c centered around the contrast of the flanking bars. This probability distribution makes it highly likely that

$$P(c = 0|S = 1, \text{low contrast flankers}) > P(c = 0|S = 1, \text{high contrast flankers}).$$

Hence, the posterior for the tendency to fill in a ghost bar

$$P(S = 1|c \approx 0) \propto P(S = 1)P(c \approx 0|S = 1) \text{ is highest and lowest,}$$
$$\text{respectively, in the array A and array C in Fig. 6.21.}$$

Even when there is zero image contrast $c = 0$ in the gap, observers are likely to attribute this zero contrast to imaging noise rather than an absence of a vertical bar when the flankers are weak. However, they are unlikely to do so when the flankers are strong or, indeed, are horizontal.

To make the prior override the evidence provided by the image contrast, the image quality should be so poor that the likelihood function $P(c \approx 0|S)$ is insensitive to the value of S; in particular, $p(c = 0|S = 1) \approx p(c = 0|S = 0)$ such that $P(c = 0|S)$ has difficulty distinguishing between different scenes from a zero image contrast $c = 0$. This happens in the case of very weak image contrast. For example, in array A of Fig. 6.21, the flankers are

What is the foreground object in these two images, which differ only in contrast?

Fig. 6.22: Does lowering the contrast in a Mooney image make it easier to perceive the contents? See footnote 28 at the bottom of this page for the identity of the foreground object.

close to the threshold for visibility. This means that observers can legitimately expect that even when there is a bar in the scene, it could give rise to a near zero image contrast, i.e., the likelihood $P(c = 0|S = 1)$ is as substantial as $P(c = 0|S = 0)$. Poor image quality could also be arranged by reducing the viewing duration, such that observers only catch a brief glimpse of the scene.

One may extrapolate from this example to other cases of noisy input in which the Bayesian prior inspires filling-in of visual objects. For example, an impressionist painting encourages viewers to imagine objects that are not clearly depicted; a dimly lit house allows viewers to succumb to vivid imagination (e.g., of persons one wishes to see or avoid). Figure 6.22 shows two Mooney images made from a photograph by binarizing grayscale pixel values. If low contrast images, and/or image noise added onto Mooney images, encourage viewers to fill-in the missing details erased by the binarization from their priors, then the viewers should more readily see the foreground object in such images. Experimentally, one can test whether this prediction holds in typical images, or whether it only applies to certain image types like the two left ones in Fig. 6.21.

Unlike that in Fig. 6.22, the visual stimulus in Fig. 6.21 is suitable for physiological investigations into possible neural mechanisms of perceptual inference in the early visual cortex, since the bars are visual inputs preferred by neurons in V1 and V2. One can then investigate whether the cortical neurons respond according to the visual percept $\hat{S}$, rather than the visual input contrast c. Indeed, we previously discussed colinear facilitations in V1 (see Fig. 2.24 F), observing that a neuron that prefers vertical orientations with its receptive field at the gap between the vertical flankers will receive facilitatory inputs from other neurons responding to the flankers. This facilitation is stronger when the flankers have a high contrast (Nelson et al. 1985, Kapadia et al. 1995) and thus is opposite to the trend in perceptual filling-in. This suggests that V1 is unlikely to be the substrate for this perceptual filling-in. It would be interesting to investigate the relationship between the activities of V2 neurons and this percept.

---

[28] The foreground object in Fig. 6.22 is a teapot; its lid is near the middle of the image, and its spout points roughly to the top right-hand corner of the image.

6.4.3.4 Decoding by sampling

The second main tenet of Bayesian inference is the central role played by the full posterior distribution $P(\mathbf{S}|\mathbf{r})$. We have mostly considered cases in which this is well characterized by its peak—i.e., the MAP value. However, in the case of ambiguous perception, the distribution may be multi-modal, with a number of scenes $\mathbf{S}$ having similar posterior probabilities. In many ambiguous cases, the percept actually fluctuates over time. Is this fluctuation caused by temporal changes in neural responses $\mathbf{r}$, or by a different decoding scheme which expresses more than just one mode of the posterior—for instance, where the decoding output given $\mathbf{r}$ is stochastic and re-evaluated after some time even when the neural responses $\mathbf{r}$ are fixed? A natural example of the latter would be to assign the decoding outcome S_i according to the posterior probability $P(S_i|\mathbf{r})$, which could perhaps be represented by neural activities which fluctuate in time (Fiser et al. 2010). In this case, even less likely scenes S_i can be produced by decoding, i.e., be perceived, as long as $P(S_i|\mathbf{r})$ is non-zero.

Indeed, in reporting whether there is a bar in the "gap" of Fig. 6.21 A, although observers are more likely to report affirmatively, they do report negatively in a substantial percentage of trials (Zhaoping and Jingling 2008). Although this does not imply that observer's percept would fluctuate in time during a single trial (it is difficult to observe this with a very brief, e.g., 80 ms, presentation of the stimulus), observers typically feel uncertain about their reports, suggesting that competing percepts S_i have comparable posterior probabilities. Even if the neural response $\mathbf{r}$ is fixed, the perception could fluctuate over time if the decoding outcome is re-evaluated after some interval. This re-evaluation may be caused by a prior expectation that scene properties are not static and could potentially change over time (Bialek and DeWeese 1995). Some behavioral experiments have used ambiguous percepts to provide more substantial arguments in support of this notion of perception by sampling (Moreno-Bote, Knill and Pouget 2011, Battaglia, Kersten and Schrater 2011).

6.4.3.5 Other behavioral observations suggesting visual Bayesian inference in the brain

There are other examples of visual percepts which can be interpreted in Bayesian terms. For example, color constancy, the phenomenon that we see the same object surface color (e.g., orange) despite changes in the illumination of the scene, is argued as being (at least in many cases) a result of the Bayesian inference, estimating surface reflectance from visual input images which are influenced by both surface reflectance and the environmental illumination (Brainard and Freeman 1997, Maloney and Wandell 1986). Another example is the perception of surface slant in the 3D scene from 2D texture images (Knill 2003). Below are some simpler examples that can be more intuitively interpreted as Bayesian visual inference without lengthy mathematical analysis.

The effect of priors on lighting directions and surface convexity
Figure 6.23 A appears to contain a concave sphere on the left and a convex one on the right, both illuminated from above. In fact, the image could be made from illuminating from below a convex sphere on the left and a concave one on the right. This is a case of ambiguous perception. The image has the same likelihood regardless of whether the interpretation is the one or the other. However, observers are more likely to perceive the first interpretation, because there is a prior bias for lighting from above (Mamassian, Landy and Maloney 2002).

However, if observers have a prior bias for convex rather than concave surfaces, and if this prior bias is much stronger than the prior on the direction of the illumination, then the perceived surface would be convex regardless of the image shading (or perceived direction of illumination). This is shown in Fig. 6.23 B, in which the two hands appear bulging to some observers regardless of whether the illumination is perceived to come from above or below.

A: The prior bias for top lighting makes the left/right sphere appear concave/convex

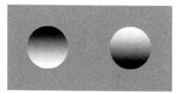

B: Handprints or bulging hands? The prior bias for convexity overcomes the top light bias

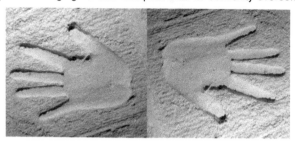

Fig. 6.23: The influence of prior biases for the direction of illumination and for convex shape in perception. A: A prior bias for illumination from above encourages one to perceive the left sphere as concave and the right sphere as convex. B: The photo of a handprint in sawdust and its 180° rotated version both appear to many observers to contain a bulging hand shape, insensitive to the inferred direction of illumination. (To many other observers, the left hand still appears as concave.) The prior bias for the convex shape of a hand can overcome the prior bias for the direction of the illumination.

Although both images are made from the same photo of a handprint in sawdust, the prior for the convex hand shape creates a illusion. An analogous, and more powerful, illusion is the hollow face illusion. Our prior bias for a convex facial surface, i.e., the nose sticks out, makes us perceive a convex facial surface when we look at a face mask from behind, even when we know that the surface is actually concave, and even when the convex perception implies an illumination from below. (Readers are encouraged to search the internet for videos of a rotating face mask to appreciate this illusion.)

The generic view

Figure 6.24 illustrates that the perceived object S by an image type r is also influenced by the likelihood $P(r|S)$. Here, we use a scalar S to denote an object, indicating that an object can take one of a set of discrete possible values, denoted as S_1, S_2, or S_3 in the figure. S_1 is a thin stick, S_2 is a rectangle surface, and S_3 is a cube. (The possibilities could be forced to be discrete by the experimenter, for example in a forced-choice discrimination task.) Similarly, the image type r can take one of many possible values, shown as r_1, r_2, r_3, r_4, or r_5 in the figure. The value r_1 is a point, r_2 is a line, r_3 is a parallelogram, r_4 contains two parallelograms sharing a side, and r_5 contains three parallelograms, each of which shares a side with each of the two other parallelograms. Note that here r_i does not denote a single image, but a set of images qualitatively similar to the one depicted.

If the image type r is rarely seen from an object S unless the object is viewed from a specific or a narrow range of viewing directions (more generally, including light directions), $P(r|S)$ is very small. This is because $P(r|S)$ is the outcome of integrating over the generic viewpoint variable x, such that

The generic image type r determines the
perceived object S by a higher likelihood $P(r|S)$

Objects   Image types

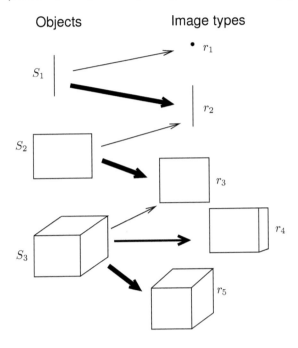

Fig. 6.24: Generic viewpoint—influence of the likelihood of the retinal image on the perceived object. A thicker arrow from an object S to an image type r means a higher likelihood $P(r|S)$. If an object S rarely gives a particular type of image r unless viewed in a particular narrow range of viewing directions, then S is not usually inferred or perceived from r. Adapted with permission from Nakayama, K. and Shimojo, S., Experiencing and perceiving visual surfaces, *Science*, 257 (5075): 1357–1363, Fig. 3, copyright © 1992, AAAS.

$$P(r|S) = \int dx P(r|S, x) P(x),$$

where $P(r|S, x)$ is the likelihood of image type r given the object S and viewpoint x, and $P(x)$ is the prior probability distribution over x (Freeman 1994). When image type r is seen only for a very small range of x, $P(r|S, x) \approx 0$ for almost all x.

Without a strong prior $P(S)$, observers perceive the S for which the likelihood $P(r|S)$ is larger. For example, a cube will not give rise to an image containing only a single parallelogram, unless the cube is viewed in a direction almost perpendicular to one of its faces. Hence, an object perceived in response to the image of a rectangle or parallelogram is more likely to be a rectangle surface than a cube. A likely view of the cube is an image containing three joint parallelograms in the lower right part of Fig. 6.24. This image is called a generic view of a cube. Meanwhile, a parallelogram is a generic view of a rectangle. Observers tend to report scenes whose generic view are consistent with the images shown.

Cue integration
Various examples of cue integration are also interpreted as supporting Bayesian inference. Popular examples include the combination of visual and haptic inputs (Ernst and Banks 2002) (albeit with some complexities (Battaglia et al. 2011)). However, perhaps the best known case is the McGurk effect (McGurk and MacDonald 1976), in which visual and auditory

inputs are integrated in the perception of speech. When viewing a film of a talking head, in which repeated utterances of the syllable "ba" had been dubbed on to lip movements for "ga," normal adults typically hear "da." According to a Bayesian interpretation, this is the result of inferring a single cause S, i.e., a speech utterance, from the combination of multiple signals $\mathbf{r} = (r_1, r_2, ..., r_i, ...)$, of which some are visual and some are auditory.

When multiple cues that are associated with the same stimulus are integrated, perceptual performance can be improved. To focus on the effect of combining different components of the $\mathbf{r}$, we consider a flat prior $P(\mathbf{S}) \propto 1$, whence MAP becomes ML. Consider decoding a one-dimensional scene described by θ. For example, one may decode the location θ of a briefly presented object from both the auditory and visual signals that it emits. When the two cues, one auditory, A, and the other visual, V, evoke independent responses from two separate populations of neurons (e.g., in the auditory and visual cortices) with Fisher information values $I_F(\theta; A)$ and $I_F(\theta; V)$, respectively, then, according to equation (6.40),

$$\text{total Fisher information } I_F(\theta) = I_F(\theta; A) + I_F(\theta; V)$$
is the sum of the contributions from groups (6.70)
of neurons that respond independently.

According to equation (6.39), the sensory discrimination threshold $\sigma(\theta) = [I_F(\theta)]^{-1/2}$. Hence, if the sensory discrimination thresholds associated with cue A or cue V alone are $\sigma(\theta; A)$ or $\sigma(\theta; V)$, combining the two cues gives a threshold that is predicted to be

$$\sigma(\theta) = \left[\sigma(\theta; A)^{-2} + \sigma(\theta; V)^{-2}\right]^{-1/2}$$
—Bayesian cue combination. (6.71)

Behavioral data on cue combination that are consistent with the above equation are sometimes used to support the claim that the brain performs Bayesian inference. Dorsal medial superior temporal (MSTd) neurons in monkeys have been observed to integrate visual and vestibular cues to discriminate heading direction in a way that is consistent with this (Gu, Angelaki and DeAngelis 2008).

6.4.3.6 Suboptimal, or suboptimal models of, visual inferences identified by the Bayesian formulation

It is also possible to use Bayesian analysis to identify when, and in what sense, visual inference is suboptimal. For example, equation (6.70) enables us to use the perceptual uncertainty associated with two separate cues to predict the perceptual uncertainty that should optimally be associated with their combination. If the actual uncertainty observed in the case of cue integration is more than predicted, some of the cue information must have been under-utilized. For instance, inefficient use of haptic information in judging the distance of an object has been observed (Battaglia et al. 2011). Human subjects can also be influenced by contextual information which encourages them to ignore or downplay some perceptual causes of their sensory information during inference (Knill 2003, Stocker and Simoncelli 2008), leading to inefficient or even biased perception.

Another way that Bayesian methods have been used is to test which sources of information might underlie visual inference. Consider the question as to whether human observers use a 3D model of the visual world to recognize objects from 2D images. If subjects perform more proficiently than a Bayesian ideal observer whose likelihood depends on only a 2D rather than a 3D model of the visual environment, then one can conclude that these subjects may be using at least some forms of 3D knowledge (Liu, Knill and Kersten 1995).

6.5 The initial visual recognition, feedforward mechanisms, and recurrent neural connections

As discussed in Section 6.4.2, decoding often requires searching for the probable scenes S in a large scene space for a visual input r, and this may require an algorithm that starts from an initial bottom-up guess of candidate scenes S, which are then refined through iterative processes. Let us examine some related observations in this section.

6.5.1 The fast speed of coarse initial recognition by the primate visual system

When human subjects are asked to report whether or not a photograph shown for just 20 ms contains an animal, event-related potentials (ERPs) in frontal areas in response to images from the two categories start to differ at 150 ms after the onset of the photograph on the display (Thorpe, Fize and Marlot 1996). The observers had never seen the photographs before and had no expectation of the kinds, sizes, poses, positions, or even number of the animals in the photographs. The median reaction time for monkeys performing a similar task was about 250 ms (Fabre-Thorpe, Richard and Thorpe 1998). Furthermore, results from studies in which multiple photographs were presented in quick succession have suggested that 100 ms viewing duration is sufficient to understand the scene (Potter 1976).

A related result is that by 125 ms after stimulus onset, it is possible to use an essentially linear classifier to categorize 77 objects into one of eight categories (Hung, Kreiman, Poggio and DiCarlo 2005) based on multiunit activities from more than 300 recording sites (from multiple animals) in monkey IT cortex. (The classifier is a cousin of the linear feedforward network in Fig. 6.11, with a threshold applied to the output value to distinguish between the categories. The feedforward weights should be set appropriately in the light of the discrimination task; and more than one output unit is typically needed to distinguish between more than two categories.) This form of object categorization is insensitive to the positions and sizes of the objects or to the degree of clutter in the background (Hung et al. 2005, Rust and DiCarlo 2010).

The earliest responses to visual inputs are seen about 40 ms after stimulus onset in V1, 70 ms in V4, and 100 ms in area IT (Nowak and Bullier 1997, Schmolesky et al. 1998). For at least the first stages of object recognition to have completed by 125 ms since stimulus onset leaves very little time for neural signals to feed back from area IT to lower visual cortical areas. These anatomical, physiological, and behavioral data are all consistent with the idea that much of the computation associated with initial coarse and simple object recognition employs essentially a feedforward algorithm. Feedback between hierarchical visual stages play a limited role in such recognitions. Such recognitions might correspond to the initial guesses of the scenes in the iterative decoding algorithm mentioned above.

6.5.2 Object detection and recognition by models of hierarchical feedforward networks

Although Section 6.3.2 showed that a feedforward network can classify photoreceptor responses into two input image categories, we argued that the method in that section would not be suitable for object categorization in a viewpoint insensitive manner since photoreceptor responses to objects are very sensitive to positions and sizes of objects. Since IT responses can be used by a feedforward network for object categorization, can a feedforward network transform photoreceptor responses to IT-like responses for this purpose?

Such feedforward networks have been demonstrated (Fukushima 1980, Riesenhuber and

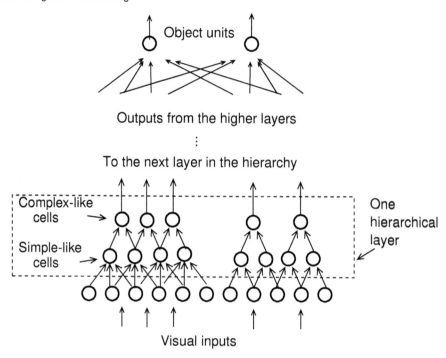

Fig. 6.25: Schematic of a hierarchical feedforward object recognition neural network model, based on an architecture proposed by Fukushima (1980) and others. Each circle denotes a neuron or a sensor; lines with arrows denote neural connections. Object units are output classifiers—one object unit for each object category.

Poggio 1999, Pinto, Barhomi, Cox and DiCarlo 2011); see Fig. 6.25. Coarsely, such models comprise two or more layers of neurons arranged in a feedforward hierarchy. Each layer is patterned after V1 (albeit without intracortical recurrent connections), with two groups of neurons. Let us call these two groups of neurons simple-like and complex-like, respectively. In each layer, there are multiple types of simple-like neurons; each type is characterized by a distinct shape of the neural receptive field. This receptive field transforms the outputs from the previous layer by a linear filtering followed by a local nonlinearity, such as that described in equations (2.42–2.43). These simple-like neurons serve to extract important visual features in their input, which is the original image for the first layer of the neurons. For example, the simple-like cells in the first layer can have receptive fields which are tuned to orientation and spatial frequency. Each complex-like neuron takes input from a group of simple-like neurons which are tuned to the same feature but to neighboring locations and scales of the feature. Its output is a highly nonlinear transform of its inputs, such as, for example, when the output is the maximum of the inputs. This output is less sensitive to the exact position and scale of the visual feature than the outputs from the simple-like neurons in the same layer. The outputs from the complex-like neurons are the outputs from each layer.

In terms of the raw visual inputs, the receptive fields of the simple-like neurons in the higher levels of the network hierarchy become progressively larger, tuned to more sophisticated features, and are less sensitive to the position and scale of input features. The receptive fields in the first layer of the network resemble those in V1, at least phenomenologically, while the last layer of the network is suggested to model the IT cortex. The outputs from the complex-like cells in the final layer (or layers) are used for object recognition, typically using a supervised learning method from machine learning such as backpropagation or support-vector classification (Bishop 2006), which we do not cover here.

6.5.3 Combining feedforward and feedback intercortical mechanisms, and recurrent intracortical mechanisms, for object inference

As we have extensively discussed, despite the popularity and power of feedforward networks, the majority of neural connections between different visual cortical areas are reciprocal (Felleman and Van Essen 1991, Markov et al. 2012), including feedback connections. Equally, behavioral evidence for recognition by imagination or input synthesis, demonstrated in Fig. 6.6, argues strongly for the role of feedback. The dynamic nature, and often slow emergence, of ambiguous perceptions, demonstrated in Fig. 6.7 and Fig. 6.8, suggest that the brain employs iterative and recurrent algorithms to make inferences about objects; these may also involve the extensive recurrent connections that are present within each hierarchical level.

As discussed earlier, the fast, initial recognition of the visual input $\mathbf{r}$ could be the candidate guess $\mathbf{S}'$ about the scene made by the bottom-up processes during decoding. The brain can then refine this guess $\mathbf{S}'$ to produce the ultimate percept through both bottom-up and top-down processes (MacKay 1956, Carpenter and Grossberg 1987, Kawato et al. 1993, Dayan et al. 1995, Rao and Ballard 1999, Yuille and Kersten 2006, Friston 2009). One popular notion is that feedback weights are used to generate the likely input $\mathbf{r}'$ caused by the initial guess $\mathbf{S}'$; then the discrepancy between $\mathbf{r}$ and $\mathbf{r}'$ is used in a second feedforward sweep to produce a refined guess $\mathbf{S}''$, and so forth, until convergence to the final percept.

In the monkey brain, intra-areal neural connections are by far the majority (about 80% of all connections according to recent data (Markov et al 2011)). They are typically recurrent, and this is certainly true of V1. The role of these heavily recurrent—intra-areal—connections in the overall computation has yet to be well specified or envisioned. Furthermore, although the majority of inter-areal connections link immediately neighboring brain areas to each other, a substantial fraction of the inter-areal connections link brain areas that are far apart (Felleman and Van Essen 1991, Markov et al. 2011, 2012). To understand visual decoding, we need to understand not only the complex dynamics in the recurrent neural circuits within and between the levels of visual hierarchy but also the representation of information across the levels.

7 Epilogue

7.1 Our ignorance of vision viewed from the perspective of vision as encoding, selection, and decoding

The organizing figure for the book, Fig. 1.5, views vision as being composed of three processing stages: encoding, selection, and decoding. Even from this, it is evident that the chapter on visual decoding is much shorter than those on encoding and selection. This is partly due to our (and particularly my) relative ignorance, but also as we have focused on those parts of the field with a richly explored link between the neural substrates and visual behavior.

Physiologically, more is known about the earlier than the later parts of the visual pathway. For example, the retina and V1 are far better studied than V4, area MT and area LIP, even though it is the later areas that are more closely associated with decoding. This ignorance about decoding is problematic for our overall understanding of vision. For example, Fig. 1.5 suggests that decoding processes influence selection via feedback—a proposal consistent with the massive feedback pathways in the brain, and with what we can understand of the mental introspection that occurs during ambiguous or difficult visual perception. Our ignorance about decoding is reflected in the near complete absence of coverage of top-down influences in visual selection. Worse, we have discussed decoding as if it is unitary, whereas it could itself be decomposed into multiple components or substages which may contain additional encoding and selection processes.

Nevertheless, our profound ignorance should not stop us from using this three-stage framework of encoding, selection, and decoding as a starting point for studying vision. This framework keeps in focus the nature of, and relationships between, some necessary sub-problems, particularly compared with alternatives such as dividing vision into low-level, mid-level, and high-level components. For example, given the extensive evidence that V1 is largely responsible for exogenous selection, it becomes necessary to understand extrastriate areas in light of the exogenous selection; this suggests that these areas are likely better understood in terms of post-selectional visual inference and endogenous selection. Thus, attempts to measure task-independent neural receptive fields sensitive to 2D visual images should be much less successful in extrastriate cortices than in V1 and the retina. This difficulty arises not merely from the greater complexity of the receptive fields in extrastriate areas for 2D images but likely also from the different or additional computation goals that these areas play. That is, they are more likely involved in decoding 3D scene properties than encoding 2D visual image properties. This viewpoint can hopefully shed light on many past observations about the extrastriate cortex (which are either not yet understood or are not mentioned in this book). Furthermore, the difference between selection and decoding, and the presence of feedback from higher to lower visual areas along the visual pathway, should encourage us to consider distinguishing central and peripheral V1 areas (and even V2 and higher visual areas with a retinotopic organization). Specifically, central and peripheral V1 areas are likely to be engaged in very different computations (see Fig. 5.50), emphasizing on decoding and selection, respectively. These considerations should motivate new investigations.

7.2   Computational vision

It is not uncommon to hear comments such as "most of vision is now known or finished, and new students entering brain sciences should better seek opportunities in other fields, such as audition." However, little of the vast volume of the experimental data on vision, which has been painstakingly collected over the centuries, has theories that adequately explain and organize them. This situation exposes a serious weakness of theoretical vision (more commonly called computational vision), and perhaps more importantly, it impedes progress in experimental vision. This weakness of computational vision is not due to a lack of talent—outstanding theoretical (and experimental) talents have been attracted, such as Issac Newton, Hermann von Helmholtz, and David Marr (2010). Meanwhile, the last half-century has seen an explosion of successful applications such as industrial robots and autonomous vehicles. Unfortunately, the notable success in computer vision has not led to substantial insights to biological vision.

What Issac Newton did not have was the vast amount of experimental data available today—and good data is essential for theoretical success. Hence a different comment for the new brain science students should be "vision has so much experimental data waiting to be understood, it is a theorist's gold mine."

Importantly, over the last 30 years, much data has emerged to suggest that visual processes are often very different from what we may intuitively expect. These include data on the separation between the visual pathways for "where" and "what" (Ungerleider and Mishkin 1982) or possibly between perception and action (Goodale and Milner 1992). The data also showed that one can act on visual objects without recognizing them or being aware of their presence. There are also data on inattentional blindness (Simons and Chabris 1999), which demonstrate that we can actually see much less than we intuitively believe we do.

It has been appreciated since the 1960s (Marr 2010) that vision is much more difficult than it appears to us to be, with our seemingly effortless daily visual experience. However, the more recent data made clear that the deceptive effortlessness is associated with the fact that much of the difficult processes of vision, such as preattentive or bottom-up visual selection and grouping, operate outside our awareness or voluntary control. (Here, I am using the words "awareness" and "consciousness" in their everyday, layman's, sense.) In particular, we can see in Fig. 2.2 that V1 and V2 are the largest cortical areas involved in vision, suggesting that they must perform a bulk of visual computations, and yet are furthest from the frontal part of the brain that is often associated with awareness. It is no wonder that understanding vision should be very difficult if so much is subconscious and thus dangerously incommensurable to our conscious brain—just like Earthlings trying to comprehend Martians! This drives home the importance of letting experimental data inspire and guide us, substantially more than has been common in the recent past. We need these data so that the conscious parts of our brain (which can understand college calculus) can comprehend the less conscious parts, by suitably "dumbing-down." It is for this reason that we have constantly emphasized the importance for theory and models to remain grounded by linking between physiology and behavior (Fig. 1.2).

Over recent years, methodological advances such as fMRI have helped link behavioral psychologists with neurobiologists. (Twenty years ago, and even today, many psychologists and neurobiologists had homes in different academic departments and attended separate academic conferences.) Furthermore, many theoretically minded individuals have conducted experiments, and many experimentalists have become better acquainted with computational ideas and methods. Consequently, experimentalists and theorists should be better able to interact with each other—toward the extent that Newton, the experimentalist, interacted with Newton, the theorist; except that there are many more of us! The resulting unified understanding of vision should not only provide insights into available data and link seemingly unrelated results, but it should also make falsifiable predictions and motivate new experimental investigations.

References

Abbott, L. and Dayan, P. (1999). The effect of correlated variability on the accuracy of a population code, *Neural Computation* **11**(1): 91–101.

Adelson, E. and Bergen, J. (1985). Spatiotemporal energy models for the perception of motion, *Journal of the Optical Society of America. A* **2**: 284–299.

Allman, J., Miezin, F. and McGuinness, E. (1985). Stimulus specific responses from beyond the classical receptive field: neurophysiological mechanisms for local-global comparisons in visual neurons, *Annual Review of Neuroscience* **8**: 407–30.

Andrews, T. J. and Coppola, D. M. (1999). Idiosyncratic characteristics of saccadic eye movements when viewing different visual environments, *Vision Research* **39**: 2947–2953.

Anstis, S. M. (1974). A chart demonstrating variations in acuity with retinal position, *Vision Research* **14**: 589–592.

Anzai, A., Bearse, M., Freeman, R. and Cai, D. (1995). Contrast coding by cells in the cat's striate cortex: monocular vs. binocular detection, *Visual Neuroscience* **12**: 77–93.

Anzai, A., Ohzawa, I. and Freeman, R. (1999). Neural mechanisms for encoding binocular disparity:receptive field position versus phase, *Journal of Neurophysiology* **82**: 874–890.

Arcizet, F., Mirpour, K. and Bisley, J. (2011). A pure salience response in posterior parietal cortex, *Cerebral Cortex* **21**(11): 2498–506.

Atick, J. (1992). Could information theory provide an ecological theory of sensory processing?, *Network: Computation in Neural Systems* **3**: 213–251.

Atick, J. and Redlich, A. (1990). Towards a theory of early visual processing, *Neural Computation* **2**: 308–320.

Atick, J., Li, Z. and Redlich, A. (1990). Color coding and its interaction with spatiotemporal processing in the retina, *Preprint IASSNS-HEP-90/75*, Institute for Advanced Study, Princeton, USA.

Atick, J., Li, Z. and Redlich, A. N. (1992). Understanding retinal color coding from first principles, *Neural Computation* **4**: 559–572.

Atick, J. J., Li, Z. and Redlich, A. N. (1993). What does post-adaptation color appearance reveal about cortical color representation?, *Vision Research* **33**(1): 123–9.

Averbeck, B. B., Latham, P. E. and Pouget, A. (2006). Neural correlations, population coding and computation, *Nature Reviews Neuroscience* **7**(5): 358–366.

Awh, E., Armstrong, K. and Moore, T. (2006). Visual and oculomotor selection: links, causes and implications for spatial attention, *Trends in Cognitive Sciences* **10**(3): 124–130.

Bahill, A., Adler, D. and Stark, L. (1975). Most naturally occurring human saccades have magnitudes of 15 degrees or less, *Investigative Ophthalmology* **14**: 468–469.

Bakin, J., Nakayama, K. and Gilbert, C. D. (2000). Visual responses in monkey areas V1 and V2 to three-dimensional surface configurations, *The Journal of Neuroscience* **20**: 8188–8198.

Banks, M., Geisler, W. and Bennett, P. (1987). The physical limits of grating visibility, *Vision Research* **27**(11): 1915–1924.

Barlow, H. (1961). Possible principles underlying the transformations of sensory messages, *in* W. A. Rosenblith (ed.), *Sensory Communication*, MIT Press, pp. 217–234.

Barlow, H. (1977). Retinal and central factors in human vision limited by noise, *in* H. Barlow and P. Fatt (eds), *Vertebrate Photoreception*, Academic Press, London, pp. 337–358.

Barlow, H. (1981). The Ferrier Lecture, 1980: Critical limiting factors in the design of the eye and visual cortex, *Proceedings of the Royal Society of London. Series B* **212**: 1–34.

Barlow, H. (1985). Cerebral cortex as model builder, *in* R. D. and V. G. Dobson (eds), *Models of the Visual Cortex*, John Wiley and Sons Ltd, Chichester, pp. 37–46.

Barlow, H., Fitzhugh, R. and Kuffler, S. (1957). Change of organization in the receptive fields of the cat's retina during dark adaptation, *The Journal of Physiology* **137**: 338–354.

Battaglia, P., Kersten, D. and Schrater, P. (2011). How haptic size sensations improve distance perception, *PLoS Computational Biology* **7**(6): e1002080.

Becker, W. (1991). Saccades, *in* R. Carpenter (ed.), *Eye Movements*, Macmillan, London, pp. 95–137.

Bell, A. and Sejnowski, T. (1997). The 'independent components' of natural scenes are edge filters, *Vision Research* **23**: 3327–38.

Ben Hamed, S., Duhamel, J., Bremmer, F. and Graf, W. (2001). Representation of the visual field in the lateral intraparietal area of macaque monkeys: a quantitative receptive field analysis, *Experimental Brain Research* **140**(2): 127–144.

Bergen, J. and Landy, M. (1991). Computational modeling of visual texture segregation, *in* M. Landy and J. Movshon (eds), *Computational Models of Visual Processing*, MIT Press, Cambridge, MA, USA, pp. 253–271.

Bethge, M. (2006). Factorial coding of natural images: How effective are linear model in removing higher-order dependencies?, *Journal of the Optical Society of America. A* **23**: 1253–1268.

Bialek, W. and DeWeese, M. (1995). Random switching and optimal processing in the perception of ambiguous signals, *Physical Review Letters* **74**(15): 3077–3080.

Bishop, C. M. (2006). *Pattern Recognition and Machine Learning*, Springer, New York.

Bislay, J. and Goldberg, M. (2011). Attention, intention, and priority in the parietal lobe, *Annual Review of Neuroscience* **33**: 1–21.

Blake, R. (2001). A primer on binocular rivalry, including current controversies, *Brain and Mind* **2**(1): 5–38.

Blatt, G., Andersen, R. and Stoner, G. (1990). Visual receptive field organization and cortico-cortical connections of the lateral intraparietal area (area LIP) in the macaque, *The Journal of Comparative Neurology* **299**(4): 421–445.

Born, R. and Bradley, D. (2005). Structure and function of visual area MT, *Annual Review of Neuroscience* **28**: 157–189.

Boussaoud, D., Desimone, R. and Ungerleider, L. (1991). Visual topography of area TEO in the macaque, *The Journal of Comparative Neurology* **306**(4): 554–575.

Box, G. and Draper, N. R. (1987). *Empirical Model Building and Response Surfaces*, John Wiley & Sons, New York.

Bradley, D., Qian, N. and Andersen, R. (1995). Integration of motion and stereopsis in middle temporal cortical area of macaques, *Nature* **373**(6515): 609–611.

Bradley, D., Chang, G. and Andersen, R. (1998). Encoding of three-dimensional structure-from-motion by primate area MT neurons, *Nature* **392**: 714–717.

Brainard, D. and Freeman, W. (1997). Bayesian color constancy, *Journal of the Optical Society of America. A* **14**(7): 1393–1411.

Braun, J., Niebur, E., Schuster, H. and Koch, C. (1994). Perceptual contour completion: A model based on local, anisotropic, fast-adapting interactions between oriented filters, *Society for Neuroscience Abstracts*, Vol. 20(1–2), p. 1665.

Bressloff, P., Cowan, J., Golubitsky, M., Thomas, P. and Wiener, M. (2002). What geometric visual hallucinations tell us about the visual cortex, *Neural Computation* **14**(3): 473–91.

Britten, K., Shadlen, M., Newsome, W. and Movshon, J. (1992). The analysis of visual motion: a comparison of neuronal and psychophysical performance, *The Journal of Neuroscience* **12**(12): 4745–4765.

Britten, K., Newsome, W., Shadlen, M., Celebrini, S. and Movshon, J. (1996). A relationship between behavioral choice and the visual responses of neurons in macaque MT, *Visual Neuroscience* **13**: 87–100.

Broadbent, D. (1958). *Perception and Communication*, Pergamon Press.

Bruce, C., Friedman, H., Kraus, M. and Stanton, G. (2004). The primate frontal eye field, *in* L. Chalupa and J. Werner (eds), *The Visual Neurosciences*, MIT Press, pp. 1429–1448.

Bruce, V., Green, P. and Georgeson, M. (2003). *Visual Perception, Physiology, Psychology, and Ecology*, 4 edn, Psychology Press, New York.

Buccigrossi, R. and Simoncelli, E. (1999). Image compression via joint statistical characterization in the wavelet domain, *IEEE Transactions on Image Processing* **8**(12): 1688–701.

Buchsbaum, G. and Gottschalk, A. (1983). Trichromacy, opponent colours coding and optimum colour information transmission in the retina, *Proceedings of the Royal Society of London. Series B* **220**(1218): 89–113.

Bullier, J. (2004). Communications between cortical areas of the visual system, *in* L. Chalupa and J. Werner (eds), *The Visual Neurosciences*, MIT Press, pp. 522–540.

Bullier, J. and Nowak, L. (1995). Parallel versus serial processing: new vistas on the distributed organization of the visual system, *Current Opinion Neurobiology* **5**(4): 497–503.

Carpenter, G. and Grossberg, S. (1987). Art 2: Self-organization of stable category recognition codes for analog input patterns, *Applied Optics* **26**(23): 4919–4930.

Cavanagh, P., Tyler, C. and Favreau, O. (1984). Perceived velocity of moving chromatic gratings, *Journal of the Optical Society of America. A* **1**(8): 893–899.

Cave, K. R. and Bichot, N. P. (1999). Visuospatial attention: Beyond a spotlight model, *Psychonomic Bulletin & Review* **6**(2): 204–223.

Chalupa, L. and Werner, J. (eds) (2004). *The Visual Neurosciences*, MIT Press.

Chelazzi, L., Miller, E., Duncan, J. and Desimone, R. (1993). A neural basis for visual search in inferior temporal cortex, *Nature* **363**(6427): 345–347.

Cohen, M. and Grossberg, S. (1983). Absolute stability of global pattern formation and parallel memory storage by competitive neural networks, *IEEE Transactions on Systems, Man, & Cybernetics* **13**: 815–26.

Colby, C. and Goldberg, M. (1999). Space and attention in parietal cortex, *Annual Review of Neuroscience* **22**(1): 319–349.

Corbetta, M. and Shulman, G. (2002). Control of goal-directed and stimulus-driven attention in the brain, *Nature Reviews Neuroscience* **3**: 201–15.

Cumming, B. G. and Parker, A. J. (2000). Local disparity not perceived depth is signaled by binocular neurons in cortical area V1 of the macaque, *The Journal of Neuroscience* **20**: 4758–4767.

Dan, Y., Atick, J. and Reid, R. (1996). Efficient coding of natural scenes in the lateral geniculate nucleus: experimental test of a computational theory, *The Journal of Neuroscience* **16**(10): 3351–62.

Daubechies, I. (1992). *Ten Lectures on Wavelets*, SIAM: Society for Industrial and Applied Mathematics.

Dayan, P. (1998). A hierarchical model of binocular rivalry, *Neural Computation* **10**(5): 1119–1135.

Dayan, P. and Abbott, L. (2001). *Theoretical Neuroscience: Computational and Mathematical Modeling of Neural Systems*, MIT Press.

Dayan, P., Hinton, G., Neal, R. and Zemel, R. (1995). The Helmholtz machine, *Neural Computation* **7**(5): 889–904.

De Valois, R., Albrecht, D. and Thorell, L. (1982). Spatial frequency selectivity of cells in macaque visual cortex, *Vision Research* **22**(5): 545–559.

DeAngelis, G. C., Freeman, R. D. and Ohzawa, I. (1994). Length and width tuning of neurons in the cat's primary visual cortex, *Journal of Neurophysiology* **71**: 347–374.

DeAngelis, G., Ohzawa, I. and Freeman, R. (1995). Receptive-field dynamics in the central visual pathways, *Trends in Neurosciences* **18**: 451–458.

DeAngelis, G., Ghose, G., Ohzawa, I. and Freeman, R. (1999). Functional micro-organization of primary visual cortex: receptive field analysis of nearby neurons, *The Journal of Neuroscience* **19**(9): 4046–4064.

Deneve, S., Latham, P. and Pouget, A. (2001). Efficient computation and cue integration with noisy population codes, *Nature Neuroscience* **4**: 826–831.

Desimone, R. and Duncan, J. (1995). Neural mechanisms of selective visual attention, *Annual Review of Neuroscience* **18**: 193–122.

Desimone, R. and Schein, S. (1987). Visual properties of neurons in area V4 of the macaque: sensitivity to stimulus form, *Journal of Neurophysiology* **57**(3): 835–868.

Desimone, R., Albright, T., Gross, C. and Bruce, C. (1984). Stimulus-selective properties of inferior temporal neurons in the macaque, *The Journal of Neuroscience* **4**(8): 2051–2062.

Deubel, H. and Schneider, W. (1996). Saccade target selection and object recognition: Evidence for a common attentional mechanism, *Vision Research* **36**(12): 1827–1837.

Deutsch, J. and Deutsch, D. (1963). Attention: Some theoretical considerations, *Psychological Review* **70**(1): 80–90.

Dong, D. and Atick, J. (1995). Temporal decorrelation: a theory of lagged and non-lagged responses in the lateral geniculate nucleus, *Network: Computation in Neural Systems* **6**: 159–178.

Douglas, R., Koch, C., Mahowald, M., Martin, K. and Suarez, H. (1995). Recurrent excitation in neocortical circuits, *Science* **269**(5226): 981–5.

Douglas, R. J. and Martin, K. A. (1990). Neocortex, *in* G. Shepherd (ed.), *Synaptic Organization of the Brain*, 3 edn, Oxford University Press, pp. 389–438.

Downing, P., Chan, A., Peelen, M., Dodds, C. and Kanwisher, N. (2006). Domain specificity in visual cortex, *Cerebral Cortex* **16**(10): 1453–1461.

Dubner, R. and Zeki, S. (1971). Response properties and receptive fields of cells in an anatomically defined region of the superior temporal sulcus in the monkey, *Brain Research* **35**(2): 528–32.

Duncan, J. (1980). The locus of interference in the perception of simultaneous stimuli, *Psychological Review* **87**(3): 272–300.

Duncan, J. and Humphreys, G. (1989). Visual search and stimulus similarity, *Psychological Review* **96**(3): 433–58.

Egeth, H. and Yantis, S. (1997). Visual attention: control, representation, and time course, *Annual Review of Pyschology* **48**: 269–97.

Einhäuser, W., Spain, M. and Perona, P. (2008). Objects predict fixations better than early saliency, *Journal of Vision* **8**(14): article 18.

Emerson, R., Bergen, J. and Adelson, E. (1992). Directionally selective complex cells and the computation of motion energy in cat visual cortex, *Vision Research* **32**(2): 203–218.

Enroth-Cugell, C. and Robson, J. (1966). The contrast sensitivity of retinal ganglion cells of the cat, *The Journal of Physiology* **187**: 517–552.

Epstein, R. and Kanwisher, N. (1998). A cortical representation of the local visual environment, *Nature* **392**(6676): 598–601.

Ermentrout, G. and Cowan, J. (1979). A mathematical theory of visual hallucination patterns, *Biological Cybernetics* **34**(3): 137–150.

Ernst, M. and Banks, M. (2002). Humans integrate visual and haptic information in a statistically optimal fashion, *Nature* **415**(6870): 429–433.

Fabre-Thorpe, M., Richard, G. and Thorpe, S. (1998). Rapid categorization of natural images by rhesus monkeys, *Neuroreport* **9**(2): 303–8.

Felleman, D. and van Essen, D. (1991). Distributed hierarchical processing in the primate cerebral cortex, *Cerebral Cortex* **1**(1): 1–47.

Field, D. (1987). Relations between the statistics of natural images and the response properties of cortical cells, *Journal of the Optical Society of America. A* **4**(12): 2379–94.

Field, D. (1989). What the statistics of natural images tell us about visual coding, *Proceedings of SPIE, Human Vision, Visual Processing, and Digital Display*, Vol. 1077, pp. 269–276.

Field, D., Hayes, A. and Hess, R. (1993). Contour integration by the human visual system: Evidence for a local "association field", *Vision Research* **33**(2): 173–193.

Findlay, J. and Gilchrist, I. (2003). *Active Vision, the Psychology of Looking and Seeing*, Oxford University Press, UK.

Finlay, B., Schiller, P. and Volman, S. (1976). Quantitative studies of single-cell properties in monkey striate cortex. iv. corticotectal cells, *Journal of Neurophysiology* **39**(6): 1352–1361.

Fiorani, M., Gattass, R., Rosa, M. and Sousa, A. (1989). Visual area MT in the cebus monkey: location, visuotopic organization, and variability, *The Journal of Comparative Neurology* **287**(1): 98–118.

Fiser, J., Berkes, P., Orbán, G. and Lengyel, M. (2010). Statistically optimal perception and learning: from behavior to neural representations, *Trends in Cognitive Sciences* **14**(3): 119–130.

Freeman, W. and Adelson, E. (1991). The design and use of steerable filters, *IEEE Transactions on Pattern Analysis and Machine Intelligence* **13**(9): 891–906.

Freeman, W. T. (1994). The generic viewpoint assumption in a framework for visual perception, *Nature* **368**(6471): 542–545.

Freiwald, W. and Tsao, D. (2010). Functional compartmentalization and viewpoint generalization within the macaque face-processing system, *Science* **330**(6005): 845–851.

Freiwald, W., Tsao, D. and Livingstone, M. (2009). A face feature space in the macaque temporal lobe, *Nature Neuroscience* **12**: 1187–1196.

Friedman, H. S., Zhou, H. and Von der Heydt, R. (2003). The coding of uniform colour figures in monkey visual cortex, *The Journal of Physiology* **548**(2): 593–613.

Friston, K. (2009). The free-energy principle: a rough guide to the brain?, *Trends in Cognitive Sciences* **13**(7): 293–301.

Frith, U. (1974). A curious effect with reverse letters explained by a theory of schema, *Perception and Psychophysics* **16**(1): 113–116.

Fukushima, K. (1980). Neocognitron: A self-organizing neural network model for a mechanism of pattern recognition unaffected by shift in position, *Biological Cybernetics* **36**(4): 193–202.

Gallant, J. L., Van Essen, D. C. and Nothdurft, H. C. (1995). Two-dimensional and three-dimensional texture processing in visual cortex of the macaque monkey, *in* T. Papathomas, C. Chubb, A. Gorea and E. Kowler (eds), *Early Vision and Beyond*, MIT Press, pp. 89–98.

Galletti, C., Battaglini, P. and Fattori, P. (1995). Eye position influence on the parieto-occipital area PO(V6) of the macaque monkey, *The European Journal of Neuroscience* **7**(12): 2486–2501.

Galletti, C., Fattori, P., Gamberini, M. and Kutz, D. (1999). The cortical visual area V6: brain location and visual topography, *The European Journal of Neuroscience* **11**(11): 3922–3936.

Garrigan, P., Ratliff, C., Klein, J., Sterling, P., Brainard, D. and Balasubramanian, V. (2010). Design of a trichromatic cone array, *PLoS Computational Biology* **6**(2): e1000677.

Gattass, R., Sousa, A. and Rosa, M. (1987). Visual topography of V1 in the cebus monkey, *The Journal of Comparative Neurology* **259**(4): 529–548.

Gattass, R., Sousa, A. and Gross, C. (1988). Visuotopic organization and extent of V3 and V4 of the macaque, *The Journal of Neuroscience* **8**(6): 1831–1845.

Gegenfurtner, K. R., Kiper, D. C. and Fenstemaker, S. B. (1996). Processing of color, form, and motion in macaque area V2, *Visual Neuroscience* **13**: 161–172.

Geisler, W. (1989). Sequential ideal-observer analysis of visual discriminations, *Psychological Review* **96**(2): 267–314.

Geisler, W. (2003). Ideal observer analysis, *in* L. Chalupa and J. Werner (eds), *The Visual Neurosciences*, MIT Press, Cambridge, MA, USA, pp. 826–837.

Geisler, W. and Albrecht, D. (1997). Visual cortex neurons in monkeys and cats: detection, discrimination and identification, *Visual Neuroscience* **14**: 897–919.

Geisler, W., Perry, J., Super, B. and Gallogly, D. (2001). Edge co-occurrence in natural images predicts contour grouping performance, *Vision Research* **41**(6): 711–724.

Gilbert, C. and Wiesel, T. (1983). Clustered intrinsic connections in cat visual cortex, *The Journal of Neuroscience* **3**(5): 1116–33.

Gilbert, C. and Wiesel, T. (1990). The influence of contextual stimuli on the orientation selectivity of cells in primary visual cortex of the cat, *Vision Research* **30**(11): 1689–1701.

Goodale, M. and Milner, A. (1992). Separate visual pathways for perception and action, *Trends in Neurosciences* **15**(1): 20–25.

Goodall, M. (1960). Performance of stochastic net, *Nature* **185**: 557–558.

Gottlieb, J., Kusunoki, M. and Goldberg, M. (1998). The representation of visual salience in monkey parietal cortex, *Nature* **391**(6666): 481–4.

Green, D. and Swets, J. (1988). *Signal Detection Theory and Psychophysics*, Peninsula Publishing, Los Altos, California, USA.

Grossberg, S. and Mingolla, E. (1985). Neural dynamics of perceptual grouping: textures, boundaries, and emergent segmentations, *Perception & Psychophysics* **38**(2): 141–71.

Grossberg, S. and Raizada, R. (2000). Contrast-sensitive perceptual grouping and object-based attention in the laminar circuits of primary visual cortex, *Vision Research* **40**(10-12): 1413–1432.

Gu, Y., Angelaki, D. and DeAngelis, G. (2008). Neural correlates of multisensory cue integration in macaque MSTd, *Nature Neuroscience* **11**(10): 1201–1210.

Haenny, P. and Schiller, P. (1988). State dependent activity in monkey visual cortex. i. single cell activity in V1 and V4 on visual tasks, *Experimental Brain Research* **69**(2): 225–244.

Hamilton, D., Albrecht, D. and Geisler, W. (1989). Visual cortical receptive fields in monkey and cat: spatial and temporal phase transfer function, *Vision Research* **29**(10): 1285–308.

Haynes, J.-D., Deichmann, R. and Rees, G. (2005). Eye-specific effects of binocular rivalry in the human lateral geniculate nucleus, *Nature* **438**(7067): 496–499.

He, Z. J. and Nakayama, K. (1995). Visual attention to surfaces in three-dimensional space, *Proceedings of the National Academy of Sciences of the USA* **92**: 11155–11159.

Hebb, D. (1949). *The Organization of Behavior: A Neuropsychological Theory*, Wiley, New York.

Heeger, D. (1992). Normalization of cell responses in cat striate cortex, *Visual Neuroscience* **9**(02): 181–197.

Hirsch, J. A. and Gilbert, C. D. (1991). Synaptic physiology of horizontal connections in the cat's visual cortex, *The Journal of Neuroscience* **11**(6): 1800–9.

Hoffman, J. (1998). Visual attention and eye movements, *in* H. Pashler (ed.), *Attention*, Psychology Press, pp. 119–153.

Holub, R. and Morton-Gibson, M. (1981). Response of visual cortical neurons of the cat to moving sinusoidal gratings: response-contrast functions and spatiotemporal interactions, *Journal of Neurophysiology* **46**(6): 1244–59.

Hopfield, J. (1984). Neurons with graded response have collective computational properties like those of two-state neurons, *Proceedings of the National Academy of Sciences of the USA* **81**: 3088–3092.

Horwitz, G. and Albright, T. (2005). Paucity of chromatic linear motion detectors in macaque V1, *Journal of Vision* **5**(6): article 4.

Hubel, D. and Wiesel, T. (1968). Receptive fields and functional architecture of monkey striate cortex, *The Journal of Physiology* **195**(1): 215–43.

Hubel, D. H. and Wiesel, T. N. (1965). Binocular interaction in striate cortex of kittens reared with artificial squint, *Journal of Neurophysiology* **28**: 1041–1059.

Hubel, D. H., Wiesel, T. N. and LeVay, S. (1977). Plasticity of ocular dominance columns in monkey striate cortex, *Philosophical Transactions of the Royal Society of London, Series B* **278**: 377–409.

Hung, C., Kreiman, G., Poggio, T. and DiCarlo, J. (2005). Fast readout of object identity from macaque inferior temporal cortex, *Science* **310**(5749): 863–866.

Hupé, J., James, A., Girard, P. and Bullier, J. (2001). Response modulations by static texture surround in area V1 of the macaque monkey do not depend on feedback connections from V2, *Journal of Neurophysiology* **85**(1): 146–163.

Isa, T. and Yoshida, M. (2009). Saccade control after V1 lesion revisited, *Current Opinion in Neurobiology* **19**(6): 608–614.

Itti, L. and Baldi, P. (2006). Bayesian surprise attracts human attention, *Advances in Neural Information Processing Systems*, Vol. 19, MIT Press, Cambridge, MA, USA, pp. 1–8.

Itti, L. and Koch, C. (2000). A saliency-based search mechanism for overt and covert shifts of visual attention, *Vision Research* **40**(10-12): 1489–506.

Itti, L. and Koch, C. (2001). Computational modelling of visual attention, *Nature Reviews Neuroscience* **2**(3): 194–203.

Janssen, P., Vogels, R., Liu, Y. and Orban, G. (2003). At least at the level of inferior temporal cortex, the stereo correspondence problem is solved, *Neuron* **37**(4): 693–701.

Jeffreys, D. and Axford, J. (1972). Source locations of pattern-specific components of human visual evoked potentials. i. component of striate cortical origin, *Experimental Brain Research* **16**(1): 1–21.

Jones, H., Grieve, K., Wang, W. and Sillito, A. (2001). Surround suppression in primate V1, *Journal of Neurophysiology* **86**(4): 2011–28.

Jonides, J. (1981). Voluntary versus automatic control over the mind's eye's movement, *in* J. B. Long and A. D. Baddeley (eds), *Attention and Performance IX*, Lawrence Erlbaum Associates Inc, Hillsdale, NJ, USA, pp. 187–203.

Joo, S. J., Boynton, G. M. and Murray, S. O. (2012). Long-range, pattern-dependent contextual effects in early human visual cortex, *Current Biology* **22**(9): 781–786.

Julesz, B. (1981). Textons, the elements of texture perception, and their interactions, *Nature* **290**(5802): 91–7.

Kadir, T. and Brady, M. (2001). Saliency, scale, and image description, *International Journal of Computer Vision* **45**(2): 83–105.

Kanwisher, N., McDermott, J. and Chun, M. (1997). The fusiform face area: a module in human extrastriate cortex specialized for face perception, *The Journal of Neuroscience* **17**(11): 4302–4311.

Kapadia, M., Ito, M., Gilbert, C. and Westheimer, G. (1995). Improvement in visual sensitivity by changes in local context: parallel studies in human observers and in V1 of alert monkeys, *Neuron* **15**(4): 843–56.

Kaplan, E., Marcus, S. and So, Y. (1979). Effects of dark adaptation on spatial and temporal properties of receptive fields in cat lateral geniculate nucleus, *The Journal of Physiology* **294**: 561–80.

Kawato, M., Hayakawa, H. and Inui, T. (1993). A forward-inverse optics model of reciprocal connections between visual cortical areas, *Network: Computation in Neural Systems* **4**(4): 415–422.

Kelly, D. H. (1962). Information capacity of a single retinal channel, *IRE Transactions on Information Theory* **8**(3): 221–226.

Kersten, D. (1987). Predictability and redundancy of natural images, *Journal of the Optical Society of America. A* **4**(12): 2395–400.

Kersten, D., Mamassian, P. and Yuille, A. (2004). Object perception as Bayesian inference, *Annual Review of Psychology* **55**: 271–304.

Knierim, J. and Van Essen, D. (1992). Neuronal responses to static texture patterns in area V1 of the alert macaque monkey, *Journal of Neurophysiology* **67**(4): 961–80.

Knill, D. (2003). Mixture models and the probabilistic structure of depth cues, *Vision Research* **43**(7): 831–854.

Kobatake, E. and Tanaka, K. (1994). Neuronal selectivities to complex object features in the ventral visual pathway of the macaque cerebral cortex, *Journal of Neurophysiology* **71**(3): 856–867.

Koch, C. and Ullman, S. (1985). Shifts in selective visual attention: towards the underlying neural circuitry, *Human Neurobiology* **4**(4): 219–27.

Koene, A. and Zhaoping, L. (2007). Feature-specific interactions in salience from combined feature contrasts: Evidence for a bottom-up saliency map in V1, *Journal of Vision* **7**(7): article 6.

Kourtzi, Z. and Kanwisher, N. (2001). Representation of perceived object shape by the human lateral occipital complex, *Science* **293**(5534): 1506–1509.

Kowler, E., Anderson, E., Dosher, B. and Blaser, E. (1995). The role of attention in the programming of saccades, *Vision Research* **35**(13): 1897–1916.

Krummenacher, J., Müller, H. and Heller, D. (2001). Visual search for dimensionally redundant pop-out targets: Evidence for parallel-coactive processing of dimensions, *Perception & Psychophysics* **63**(5): 901–917.

Kustov, A. and Robinson, D. (1996). Shared neural control of attentional shifts and eye movements, *Nature* **384**: 74–77.

Lamme, V. (1995). The neurophysiology of figure-ground segregation in primary visual cortex, *The Journal of Neuroscience* **15**(2): 1605–1615.

Lamme, V., Rodriguez-Rodriguez, V. and Spekreijse, H. (1999). Separate processing dynamics for texture elements, boundaries and surfaces in primary visual cortex of the macaque monkey, *Cerebral Cortex* **9**(4): 406–413.

Laughlin, S. (2001). Energy as a constraint on the coding and processing of sensory information, *Current Opinion in Neurobiology* **11**(4): 475–480.

Laughlin, S. B. (1981). A simple coding procedure enhances a neuron's information capacity, *Zeitschrift für Naturforschung. Section C* **36**: 910–2.

Laughlin, S., Howard, J. and Blakeslee, B. (1987). Synaptic limitations to contrast coding in the retina of the blowfly calliphora, *Proceedings of the Royal Society of London. Series B* **231**(1265): 437–67.

Lee, B., Pokorny, J., Smith, V., Martin, P. and Valberg, A. (1990). Luminance and chromatic modulation sensitivity of macaque ganglion cells and human observers, *Journal of the Optical Society of America. A* **7**(12): 2223–2236.

Lee, T. (1996). Image representation using 2D gabor wavelets, *IEEE Transactions on Pattern Analysis and Machine Intelligence* **18**: 959–971.

Lee, T., Mumford, D., Romero, R. and Lamme, V. (1998). The role of the primary visual cortex in higher level vision, *Vision Research* **38**(15-16): 2429–2454.

Lee, T., Yang, C., Romero, R. and Mumford, D. (2002). Neural activity in early visual cortex reflects behavioral experience and higher-order perceptual saliency, *Nature Neuroscience* **5**(6): 589–597.

Legge, G. and Yuanchao, G. (1989). Stereopsis and contrast, *Vision Research* **29**(8): 989–1004.

Lehky, S. R. and Maunsell, J. H. (1996). No binocular rivalry in the LGN of alert macaque monkeys, *Vision Research* **36**(9): 1225–1234.

Lennie, P. (1998). Single units and visual cortical organization, *Perception* **27**: 889–935.

Lennie, P. (2003a). The cost of cortical computation, *Current Biology* **13**(6): 493–7.

Lennie, P. (2003b). The physiology of color vision, *in* S. Shevell (ed.), *The Science of Color*, 2 edn, Optical Society of America, pp. 217–242.

Lennie, P., Krauskopf, J. and Sclar, G. (1990). Chromatic mechanisms in striate cortex of macaque, *The Journal of Neuroscience* **10**(2): 649–669.

Leopold, D. and Logothetis, N. (1996). Activity changes in early visual cortex reflect monkeys' percepts during binocular rivalry, *Nature* **379**(6565): 549–553.

Levi, D. (2008). Crowding—an essential bottleneck for object recognition: a mini-review, *Vision Research* **48**: 635–654.

Levy, W. and Baxter, R. (1996). Energy efficient neural codes, *Neural Computation* **8**(3): 531–43.

Lewis, A. and Zhaoping, L. (2005). Saliency from natural scene statistics, *Program No. 821.11. Abstract Viewer/Itinerary Planner, Online*, Annual Meeting, Society for Neuroscience, Washington, DC, USA.

Lewis, A. and Zhaoping, L. (2006). Are cone sensitivities determined by natural color statistics?, *Journal of Vision* **6**(3): article 8.

Lewis, A., Garcia, R. and Zhaoping, L. (2003). The distribution of visual objects on the retina: connecting eye movements and cone distributions, *Journal of Vision* **3**(11): article 21.

Li, C. and Li, W. (1994). Extensive integration field beyond the classical receptive field of cat's striate cortical neurons—classification and tuning properties, *Vision Research* **34**(18): 2337–55.

Li, N., Cox, D., Zoccolan, D. and DiCarlo, J. (2009). What response properties do individual neurons need to underlie position and clutter "invariant" object recognition?, *Journal of Neurophysiology* **102**(1): 360–376.

Li, W., Piëch, V. and Gilbert, C. (2004). Perceptual learning and top-down influences in primary visual cortex, *Nature Neuroscience* **7**: 651–657.

Li, W., Piëch, V. and Gilbert, C. (2006). Contour saliency in primary visual cortex, *Neuron* **50**(6): 951–962.

Li, Z. (1992). Different retinal ganglion cells have different functional goals, *International Journal of Neural Systems* **3**(3): 237–248.

Li, Z. (1995). Understanding ocular dominance development from binocular input statistics, *in* J. Bower (ed.), *The Neurobiology of Computation*, Kluwer Academic Publishers, pp. 397–402.

Li, Z. (1996). A theory of the visual motion coding in the primary visual cortex, *Neural Computation* **8**(4): 705–30.

Li, Z. (1997). Primary cortical dynamics for visual grouping, *in* K. Wong, I. King and D. Yeung (eds), *Theoretical Aspects of Neural Computation: A Multidisciplineary Perspective (proceeding from International Workshop (TANC'97), in May, 1997, Hong Kong)*, Springer-Verlag, Hong Kong, pp. 155–164.

Li, Z. (1998a). A neural model of contour integration in the primary visual cortex, *Neural Computation* **10**(4): 903–40.

Li, Z. (1998b). Visual segmentation without classification: A proposed function for primary visual cortex, *Perception* **27**: ECVP Abstract supplement, page 45. Proceedings of ECVP, 1998, Oxford, England.

Li, Z. (1999a). Contextual influences in V1 as a basis for pop out and asymmetry in visual search, *Proceedings of the National Academy of Sciences of the USA* **96**(18): 10530–10535.

Li, Z. (1999b). Visual segmentation by contextual influences via intra-cortical interactions in primary visual cortex, *Network: Computation in Neural Systems* **10**(2): 187–212.

Li, Z. (2000a). Can V1 mechanisms account for figure-ground and medial axis effects?, *in* S. Solla, T. Leen and K.-R. Muller (eds), *Advances in Neural Information Processing Systems*, Vol. 12, MIT Press, Cambridge, MA, USA, pp. 136–142.

Li, Z. (2000b). Pre-attentive segmentation in the primary visual cortex, *Spatial Vision* **13**(1): 25–50.

Li, Z. (2001). Computational design and nonlinear dynamics of a recurrent network model of the primary visual cortex, *Neural Computation* **13**(8): 1749–1780.

Li, Z. (2002). A saliency map in primary visual cortex, *Trends in Cognitive Sciences* **6**(1): 9–16.

Li, Z. and Atick, J. J. (1994a). Efficient stereo coding in the multiscale representation, *Network: Computation in Neural Systems* **5**(2): 157–174.

Li, Z. and Atick, J. J. (1994b). Towards a theory of striate cortex, *Neural Computation* **6**: 127–146.

Li, Z. and Dayan, P. (1999). Computational differences between asymmetrical and symmetrical networks, *Network: Computation in Neural Systems* **10**(1): 59–77.

Li, Z. and Hopfield, J. (1989). Modeling the olfactory bulb and its neural oscillatory processings, *Biological Cybernetics* **61**(5): 379–392.

Linsker, R. (1988). Self-organization in a perceptual network, *Computer* **2193**: 105–117.

Linsker, R. (1990). Perceptual neural organization: some approaches based on network models and information theory, *Annual Review of Neuroscience* **13**: 257–81.

Liu, Z., Knill, D. and Kersten, D. (1995). Object classification for human and ideal observers, *Vision Research* **35**(4): 549–568.

Livingstone, M. (1996). Ocular dominance columns in new world monkeys, *The Journal of Neuroscience* **16**(6): 2086–2096.

Livingstone, M. and Hubel, D. (1984). Anatomy and physiology of a color system in the primate visual cortex, *The Journal of Neuroscience* **4**(1): 309–56.

Logothetis, N., Pauls, J. and Poggio, T. (1995). Shape representation in the inferior temporal cortex of monkeys, *Current Biology* **5**(5): 552–563.

Logothetis, N. K., Leopold, D. A. and Sheinberg, D. L. (1996). What is rivalling during binocular rivalry?, *Nature* **380**: 621–624.

Luck, S., Chelazzi, L., Hillyard, S. and Desimone, R. (1997). Neural mechanisms of spatial selective attention in areas V1, V2, and V4 of macaque visual cortex, *Journal of Neurophysiology* **77**(1): 24–42.

Lyu, S. (2011). Dependency reduction with divisive normalization: Justification and effectiveness, *Neural Computation* **23**(11): 2942–2973.

Lyu, S. and Simoncelli, E. P. (2009). Nonlinear extraction of independent components of natural images using radial Gaussianization, *Neural Computation* **21**(6): 1485–1519.

Ma, W. J., Beck, J. M., Latham, P. E. and Pouget, A. (2006). Bayesian inference with probabilistic population codes, *Nature Neuroscience* **9**(11): 1432–1438.

MacKay, D. (1956). Towards an information flow model of human behavior, *British Journal of Psychology* **47**(1): 30–43.

Maloney, L. and Wandell, B. (1986). Color constancy: a method for recovering surface spectral reflectance, *Journal of the Optical Society of America. A* **3**(1): 29–33.

Mamassian, P., Landy, M. S. and Maloney, L. T. (2002). Bayesian modelling of visual perception, *in* R. Rao,

B. Olshausen and M. Lewicki (eds), *Probabilistic Models of the Brain: Perception and Neural Function*, MIT Press, Cambridge, MA, USA, pp. 13–36.

Mannan, S., Kennard, C. and Husain, M. (2009). The role of visual salience in directing eye movements in visual object agnosia, *Current Biology* **19**(6): R247–8.

Marr, D. (2010). *VISION, a computational investigation into the human representation and processing of visual information*, MIT Press.

Martínez-Trujillo, J. and Treue, S. (2004). Feature-based attention increases the selectivity of population responses in primate visual cortex, *Current Biology* **14**(9): 744–751.

May, K., Zhaoping, L. and Hibbard, P. (2012). Perceived direction of motion determined by adaptation to static binocular images, *Current Biology* **22**: 28–32.

Mazer, J. and Gallant, J. (2003). Goal-related activity in V4 during free viewing visual search: Evidence for a ventral stream visual salience map, *Neuron* **40**(6): 1241–1250.

McCollough, C. (1965). Color adaptation of edge-detectors in the human visual system, *Science* **149**(3688): 1115–1116.

McGurk, H. and MacDonald, J. (1976). Hearing lips and seeing voices, *Nature* **264**: 746–748.

Melloni, L., Van Leeuwen, S., Alink, A. and Müller, N. G. (2012). Interaction between bottom-up saliency and top-down control: how saliency maps are created in the human brain, *Cerebral Cortex* **22**(12): 2943–2952.

Merigan, W. (1996). Basic visual capacities and shape discrimination after lesions of extrastriate area V4 in macaques, *Visual Neuroscience* **13**: 51–60.

Merigan, W., Nealey, T. and Maunsell, J. (1993). Visual effects of lesions of cortical area V2 in macaques, *The Journal of Neuroscience* **13**(7): 3180–3191.

Michael, C. R. (1978). Color vision mechanisms in monkey striate cortex: simple cells with dual opponent-color receptive fields, *Journal of Neurophysiology* **41**(5): 1233–1249.

Mitchison, G. (1992). Axonal trees and cortical architecture, *Trends in Neurosciences* **15**(4): 122–6.

Mohler, C., Goldberg, M. and Wurtz, R. (1973). Visual receptive fields of frontal eye field neurons, *Brain Research* **61**: 385–389.

Molotchnikoff, S., Shumikhina, S. and Moisan, L. (1996). Stimulus-dependent oscillations in the cat visual cortex: differences between bar and grating stimuli, *Brain Research* **731**(1-2): 91–100.

Moran, J. and Desimone, R. (1985). Selective attention gates visual processing in the extrastriate cortex, *Science* **229**(4715): 782–4.

Moreno-Bote, R., Knill, D. and Pouget, A. (2011). Bayesian sampling in visual perception, *Proceedings of the National Academy of Sciences of the USA* **108**(30): 12491–12496.

Morgan, M. (1979). Perception of continuity in stroboscopic motion: a temporal frequency analysis, *Vision Research* **19**(5): 491–500.

Motter, B. (1993). Focal attention produces spatially selective processing in visual cortical areas V1, V2, and V4 in the presence of competing stimuli, *Journal of Neurophysiology* **70**(3): 909–919.

Motter, B. and Belky, E. (1998). The zone of focal attention during active visual search, *Vision Research* **38**(7): 1007–22.

Movshon, J., Adelson, E., Gizzi, M. and Newsome, W. (1985). The analysis of moving visual patterns, *Pattern Recognition Mechanisms* **54**: 117–151.

Müller, H. J. and Rabbitt, P. M. (1989). Reflexive and voluntary orienting of visual attention: time course of activation and resistance to interruption, *Journal of Experimental Psychology: Human Perception and Performance* **15**(2): 315–330.

Nadal, J. and Parga, N. (1993). Information processing by a perceptron in an unsupervised learning task, *Network: Computation in Neural Systems* **4**(3): 295–312.

Nadal, J. and Parga, N. (1994). Nonlinear neurons in the low-noise limit: a factorial code maximizes information transfer, *Network: Computation in Neural Systems* **5**: 565–581.

Nakamura, K. and Colby, C. (2002). Updating of the visual representation in monkey striate and extrastriate cortex during saccades, *Proceedings of the National Academy of Sciences of the USA* **99**(6): 4026–4031.

Nakayama, K. and Mackeben, M. (1989). Sustained and transient components of focal visual attention, *Vision Research* **29**(11): 631–47.

Nakayama, K. and Silverman, G. (1986). Serial and parallel processing of visual feature conjunctions, *Nature* **320**(6059): 264–5.

Nakayama, K., He, Z. and Shimojo, S. (1995). Visual surface representation: A critical link between lower-level and higher-level vision, *in* S. M. Kosslyn and D. N. Osherson (eds), *An Invitation to Cognitive Science: Visual Cognition*, Vol. 2, MIT Press, Cambridge, MA, USA, pp. 1–70.

Nelson, J., and Frost, B. (1985). Intracortical facilitation among co-oriented, co-axially aligned simple cells in cat striate cortex, *Experimental Brain Research* **61**(1): 54–61.

Newsome, W. T. and Pare, E. B. (1988). A selective impairment of motion perception following lesions of the middle temporal visual area (MT), *The Journal of Neuroscience* **8**(6): 2201–2211.

Nirenberg, S., Carcieri, S. M., Jacobs, A. L. and Latham, P. E. (2001). Retinal ganglion cells act largely as independent encoders, *Nature* **411**: 698–701.

Nothdurft, H. (1994). Common properties of visual segmentation, *in* G. R. Bock and J. A. Goode (eds), *Higher-order Processing in the Visual System, Ciba Foundation Symposium 184*, Wiley & Sons, pp. 245–268.

Nothdurft, H. (2000). Salience from feature contrast: variations with texture density, *Vision Research* **40**(23): 3181–200.

Nothdurft, H. C. (1985). Sensitivity for structure gradient in texture discrimination tasks, *Vision Research* **25**: 1957–68.

Nothdurft, H. C. (1991). Texture segmentation and pop-out from orientation contrast, *Vision Research* **31**(6): 1073–8.

Nothdurft, H., Gallant, J. and Van Essen, D. (1999). Response modulation by texture surround in primate area V1: correlates of "popout" under anesthesia, *Visual Neuroscience* **16**: 15–34.

Nowak, L. and Bullier, J. (1997). The timing of information transfer in the visual system, *in* K. Rockland, J. Kaas and A. Peters (eds), *Cerebral Cortex: Extrastriate Cortex in Primate*, New York, Plenum Publishing Corporation, pp. 205–242.

Ohzawa, I., DeAngelis, G. and Freeman, R. (1990). Stereoscopic depth discrimination in the visual cortex: neurons ideally suited as disparity detectors, *Science* **249**(4972): 1037–1041.

Oja, E. (1982). A simplified neuron model as a principal component analyzer, *Journal of Mathematical Biology* **15**: 267–273.

Oliveri, M., Zhaoping, L., Mangano, G., Turriziani, P., Smirni, D. and Cipolotti, L. (2010). Facilitation of bottom-up feature detection following rTMS-interference of the right parietal cortex, *Neuropsychologia* **48**: 1003–1010.

Olshausen, B. and Field, D. (1997). Sparse coding with an overcomplete basis set: a strategy employed by V1?, *Vision Research* **37**: 3311–3325.

Palmer, S. (1999). *Vision Science: Photons to Phenomenology*, MIT Press.

Paninski, L., Pillow, J. and Lewi, J. (2007). Statistical models for neural encoding, decoding, and optimal stimulus design, *Progress in Brain Research* **165**: 493–507.

Pelli, D. (1990). The quantum efficiency in vision, *in* C. Blakemore (ed.), *Vision: Coding and Efficiency*, Cambridge University Press, Cambridge, pp. 3–24.

Petrov, Y. and Zhaoping, L. (2003). Local correlations, information redundancy, and sufficient pixel depth in natural images, *Journal of the Optical Society of America. A* **20**(1): 56–66.

Petrov, Y., Carandini, M. and McKee, S. (2005). Two distinct mechanisms of suppression in human vision, *The Journal of Neuroscience* **25**(38): 8704–8707.

Pinto, N., Barhomi, Y., Cox, D. and DiCarlo, J. (2011). Comparing state-of-the-art visual features on invariant object recognition tasks, *Applications of Computer Vision (WACV), 2011 IEEE Workshop on*, IEEE, pp. 463–470.

Pokorny, J. and Smith, V. (1970). Wavelength discrimination in the presence of added chromatic fields, *Journal of the Optical Society of America. A* **69**: 562–9.

Polat, U. and Sagi, D. (1993). Lateral interactions between spatial channels: suppression and facilitation revealed by lateral masking experiments, *Vision Research* **33**(7): 993–999.

Pollen, D. and Ronner, S. (1981). Phase relationships between adjacent simple cells in the visual cortex, *Science* **212**(4501): 1409–1411.

Pomerantz, J., Sager, L. and Stoever, R. (1977). Perception of wholes and of their component parts: Some configural superiority effects, *Journal of Experimental Psychology: Human Perception and Performance* **3**(3): 422–435.

Popple, A. (2003). Context effects on texture border localization bias, *Vision Research* **43**(7): 739–43.

Posner, M. I. (1980). Orienting of attention, *Quarterly Journal of Experimental Psychology* **32**(1): 3–25.

Potter, M. (1976). Short-term conceptual memory for pictures, *Journal of Experimental Psychology: Human Learning and Memory* **2**(5): 509–522.

Puchalla, J., Schneidman, E., Harris, R. and Berry, M. (2005). Redundancy in the population code of the retina, *Neuron* **46**(3): 493–504.

Qian, N. (1994). Computing stereo disparity and motion with known binocular cell properties, *Neural Computation* **6**(3): 390–404.

Qian, N. (1997). Binocular disparity and the perception of depth, *Neuron* **18**(3): 359–368.

Qian, N. and Andersen, R. (1994). Transparent motion perception as detection of unbalanced motion signals. ii. physiology, *The Journal of Neuroscience* **14**(12): 7367–7380.

Qian, N. and Andersen, R. (1997). A physiological model for motion-stereo integration and a unified explanation of pulfrich-like phenomena, *Vision Research* **37**(12): 1683–1698.

Qian, N., Andersen, R. and Adelson, E. (1994). Transparent motion perception as detection of unbalanced motion signals. i. psychophysics, *The Journal of Neuroscience* **14**(12): 7357–7366.

Qiu, F. and von der Heydt, R. (2005). Figure and ground in the visual cortex: V2 combines stereoscopic cues with gestalt rules, *Neuron* **47**(1): 155–66.

Qiu, F. and von der Heydt, R. (2007). Neural representation of transparent overlay, *Nature Neuroscience* **10**: 283–4.

Qiu, F., Sugihara, T. and von der Heydt, R. (2007). Figure-ground mechanisms provide structure for selective attention, *Nature Neuroscience* **10**: 1492–9.

Ramsden, B. M., Hung, C. P. and Roe, A. W. (2001). Real and illusory contour processing in area V1 of the primate: a cortical balancing act, *Cerebral Cortex* **11**(7): 648–665.

Rao, R. and Ballard, D. (1999). Predictive coding in the visual cortex: a functional interpretation of some extra-classical receptive field effects, *Nature Neuroscience* **2**: 79–87.

Reynolds, J., Chelazzi, L. and Desimone, R. (1999). Competitive mechanisms subserve attention in macaque areas V2 and V4, *The Journal of Neuroscience* **19**(5): 1736–1753.

Reynolds, J., Pasternak, T. and Desimone, R. (2000). Attention increases sensitivity of V4 neurons, *Neuron* **26**(3): 703–714.

Riesenhuber, M. and Poggio, T. (1999). Hierarchical models of object recognition in cortex, *Nature Neuroscience* **2**: 1019–1025.

Robinson, D. and Petersen, S. (1992). The pulvinar and visual salience, *Trends in Neurosciences* **15**(4): 127–132.

Rockland, K. and Lund, J. (1983). Intrinsic laminar lattice connections in primate visual cortex, *The Journal of Comparative Neurology* **216**(3): 303–18.

Rolls, E. T. (2003). Invariant object and face recognition, *in* L. M. Chalupa and J. S. Werner (eds), *The Visual Neurosciences*, Vol. 2, MIT Press, Cambridge, MA, USA, pp. 1165–1178.

Rosa, M., Sousa, A. and Gattass, R. (1988). Representation of the visual field in the second visual area in the cebus monkey, *The Journal of Comparative Neurology* **275**(3): 326–345.

Rossi, A., Desimone, R. and Ungerleider, L. (2001). Contextual modulation in primary visual cortex of macaques, *The Journal of Neuroscience* **21**(5): 1698–1709.

Rubenstein, B. and Sagi, D. (1990). Spatial variability as a limiting factor in texture-discrimination tasks: implications for performance asymmetries, *Journal of the Optical Society of America. A* **7**(9): 1632–43.

Ruderman, D. and Bialek, W. (1994). Statistics of natural images: Scaling in the woods, *Physical Review Letters* **73**(6): 814–817.

Ruderman, D., Cronin, T. and Chiao, C.-C. (1998). Statistics of cone responses to natural images: implications for visual coding, *Journal of the Optical Society of America. A* **15**(8): 2036–45.

Rudolph, K. and Pasternak, T. (1999). Transient and permanent deficits in motion perception after lesions of cortical areas MT and MST in the macaque monkey, *Cerebral Cortex* **9**(1): 90–100.

Rust, N. and DiCarlo, J. (2010). Selectivity and tolerance ("invariance") both increase as visual information propagates from cortical area V4 to IT, *The Journal of Neuroscience* **30**(39): 12978–12995.

Sagi, D. and Julesz, B. (1985). "Where" and "what" in vision, *Science* **228**(4704): 1217–1219.

Salinas, E. and Abbott, L. (2000). Do simple cells in primary visual cortex form a tight frame?, *Neural Computation* **12**(2): 313–35.

Sceniak, M., Ringach, D., Hawken, M. and Shapley, R. (1999). Contrast's effect on spatial summation by macaque V1 neurons, *Nature Neuroscience* **2**(8): 733–9.

Schall, J. (2004). Selection of targets for saccadic eye movements, *in* L. Chalupa and J. Werner (eds), *The Visual Neurosciences*, MIT Press, pp. 1369–1390.

Schiller, P. (1993). The effects of V4 and middle temporal (MT) area lesions on visual performance in the rhesus monkey, *Visual Neuroscience* **10**: 717–746.

Schiller, P. (1998). The neural control of visually guided eye movements, *in* J. E. Richards (ed.), *Cognitive Neuro-science of Attention, a Developmental Perspective*, Lawrence Erlbaum Associates, Inc., Mahwah, New Jersey, USA, pp. 3–50.

Schiller, P. and Lee, K. (1991). The role of the primate extrastriate area V4 in vision, *Science* **251**(4998): 1251–1253.

Schiller, P. and Malpeli, J. (1977). Properties and tectal projections of monkey retinal ganglion cells, *Journal of Neurophysiology* **40**: 428–445.

Schiller, P. and Tehovnik, E. (2005). Neural mechanisms underlying target selection with saccadic eye movements, *Progress in Brain Research* **149**: 157–171.

Schiller, P., Stryker, M., Cynader, M. and Berman, N. (1974). Response characteristics of single cells in the monkey superior colliculus following ablation or cooling of visual cortex, *Journal of Neurophysiology* **37**: 181–184.

Schira, M., Tyler, C., Breakspear, M. and Spehar, B. (2009). The foveal confluence in human visual cortex, *The Journal of Neuroscience* **29**(28): 9050–9058.

Schmolesky, M., Wang, Y., Hanes, D., Thompson, K., Leutgeb, S., Schall, J. and Leventhal, A. (1998). Signal timing across the macaque visual system, *Journal of Neurophysiology* **79**: 3272–3278.

Scholte, H., Jolij, J., Fahrenfort, J. and Lamme, V. (2008). Feedforward and recurrent processing in scene segmenta-tion: Electroencephalography and functional magnetic resonance imaging, *Journal of Cognitive Neuroscience* **20**(11): 2097–2109.

Schwartz, E. L. (1977). Spatial mapping in the primate sensory projection: analytic structure and relevance to perception, *Biological Cybernetics* **25**(4): 181–194.

Schwartz, O. and Simoncelli, E. (2001). Natural signal statistics and sensory gain control, *Nature Neuroscience* **4**(8): 819–25.

Sengpiel, F., Blakemore, C. and Harrad, R. (1995). Interocular suppression in the primary visual cortex: a possible neural basis of binocular rivalry, *Vision Research* **35**(2): 179–195.

Sengpiel, F., Baddeley, R., Freeman, T., Harrad, R. and Blakemore, C. (1998). Different mechanisms underlie three inhibitory phenomena in cat area 17, *Vision Research* **38**(14): 2067–2080.

Shadlen, M. and Carney, T. (1986). Mechanisms of human motion perception revealed by a new cyclopean illusion,

Science **232**(4746): 95–97.

Shadlen, M., Britten, K., Newsome, W. and Movshon, J. (1996). A computational analysis of the relationship between neuronal and behavioral responses to visual motion, *The Journal of Neuroscience* **16**(4): 1486–1510.

Shannon, C. and Weaver, W. (1949). *The Mathematical Theory of Communication*, University of Illinois Press, Urbana IL, USA.

Shapley, R. and Perry, V. (1986). Cat and monkey retinal ganglion cells and their visual functional roles, *Trends in Neurosciences* **9**: 229–235.

Shepherd, G. (1990). *The Synaptic Organization of the Brain*, 3 edn, Oxford University Press, Oxford.

Sherman, S. and Guillery, R. (2004). The visual relays in the thalamus, *in* L. Chalupa and J. Werner (eds), *The Visual Neurosciences*, MIT Press, pp. 565–591.

Sherman, S. and Koch, C. (1986). The control of retinogeniculate transmission in the mammalian lateral geniculate nucleus, *Experimental Brain Research* **63**(1): 1–20.

Shipp, S. (2004). The brain circuitry of attention, *Trends in Cognitive Sciences* **8**(5): 223–230.

Sillito, A., Grieve, K., Jones, H., Cudeiro, J. and Davis, J. (1995). Visual cortical mechanisms detecting focal orientation discontinuities, *Nature* **378**: 492–496.

Simoncelli, E. and Olshausen, B. (2001). Natural image statistics and neural representation, *Annual Review of Neuroscience* **24**: 1193–216.

Simons, D. and Chabris, C. (1999). Gorillas in our midst: sustained inattentional blindness for dynamic events, *Perception* **28**: 1059–1074.

Smith, V. C. and Pokorny, J. (1975). Spectral sensitivity of the foveal cone photopigments between 400 and 500 nm, *Vision Research* **15**: 161–171.

Somers, D., Todorov, E., Siapas, A., Toth, L., Kim, D. and Sur, M. (1998). A local circuit approach to understanding integration of long-range inputs in primary visual cortex, *Cerebral Cortex* **8**(3): 204–217.

Sommer, M. and Wurtz, R. (2004). What the brain stem tells the frontal cortex. ii. role of the SC-MD-FEF pathway in corollary discharge, *Journal of Neurophysiology* **91**(3): 1403–1423.

Sommer, M. and Wurtz, R. (2006). Influence of the thalamus on spatial visual processing in frontal cortex, *Nature* **444**(7117): 374–377.

Srinivasan, M., Laughlin, S. and Dubs, A. (1982). Predictive coding: a fresh view of inhibition in the retina, *Proceedings of the Royal Society of London. Series B* **216**(1205): 427–59.

Stemmler, M., Usher, M. and Niebur, E. (1995). Lateral interactions in primary visual cortex: a model bridging physiology and psychophysics, *Science* **269**(5232): 1877–1880.

Stocker, A. A. and Simoncelli, E. P. (2006). Noise characteristics and prior expectations in human visual speed perception, *Nature Neuroscience* **9**(4): 578–585.

Stocker, A. A. and Simoncelli, E. P. (2008). A Bayesian model of conditioned perception, *in* J. C. Platt, D. Koller, Y. Singer and S. Roweis (eds), *Advances in Neural Information Processing Systems*, Vol. 20, MIT Press, Cambridge, MA, USA, pp. 1409–1416.

Stoner, G. and Albright, T. (1992). Neural correlates of perceptual motion coherence, *Nature* **358**(6385): 412–414.

Stryker, M. (1986). The role of neural activity in rearranging connections in the central visual system, *in* R. Ruben, T. Van De Water and E. Rubel (eds), *The Biology of Change in Otolaryngology*, Elsevier Science Amsterdam, pp. 211–224.

Sziklai, G. (1956). Some studies in the speed of visual perception, *IRE Transactions on Information Theory* **2**(3): 125–8.

Tanaka, K. (1996). Inferotemporal cortex and object vision, *Annual Review of Neuroscience* **19**(1): 109–139.

Tanaka, K. (2003). Inferotemporal response properties, *in* L. M. Chalupa and J. S. Werner (eds), *The Visual Neurosciences*, Vol. 2, MIT Press, Cambridge, MA, USA, pp. 1151–1164.

Tehovnik, E., Slocum, W. and Schiller, P. (2003). Saccadic eye movements evoked by microstimulation of striate cortex, *The European Journal of Neuroscience* **17**(4): 870–8.

Theeuwes, J. (1992). Perceptual selectivity for color and form, *Perception & Psychophysics* **51**(6): 599–606.

Thomas, O., Cumming, B. and Parker, A. (2002). A specialization for relative disparity in V2, *Nature Neuroscience* **5**(5): 472–478.

Thompson, K. and Bichot, N. (2005). A visual salience map in the primate frontal eye field, *Progress in Brain Research* **147**: 249–262.

Thorpe, S., Fize, D. and Marlot, C. (1996). Speed of processing in the human visual system, *Nature* **381**(6582): 520–522.

Treisman, A. (1985). Preattentive processing in vision, *Computer Vision, Graphics, and Image Processing* **31**(2): 156–177.

Treisman, A. and Gormican, S. (1988). Feature analysis in early vision: evidence from search asymmetries, *Psychological Review* **95**(1): 15–48.

Treisman, A. M. and Gelade, G. (1980). A feature-integration theory of attention, *Cognitive Psychology* **12**(1): 97–136.

Treue, S. and Martínez-Trujillo, J. (1999). Feature-based attention influences motion processing gain in macaque visual cortex, *Nature* **399**(9): 575–579.

Treue, S. and Maunsell, J. (1999). Effects of attention on the processing of motion in macaque middle temporal and medial superior temporal visual cortical areas, *The Journal of Neuroscience* **19**(17): 7591–7602.

Troy, J. and Lee, B. (1994). Steady discharges of macaque retinal ganglion cells, *Visual Neuroscience* **11**(1): 111–8.

Troy, J. and Robson, J. (1992). Steady discharges of X and Y retinal ganglion cells of cat under photopic illuminance, *Visual Neuroscience* **9**(6): 535–53.

Tsotsos, J. (1990). Analyzing vision at the complexity level, *Behavioral and Brain Sciences* **13**(3): 423–445.

Tyler, C. (1974). Stereopsis in dynamic visual noise, *Nature* **250**: 781–782.

Umeno, M. and Goldberg, M. (1997). Spatial processing in the monkey frontal eye field. i. predictive visual responses, *Journal of Neurophysiology* **78**(3): 1373–1383.

Ungerleider, L. and Pasternak, T. (2004). Ventral and dorsal cortical processing streams, *in* L. Chalupa and J. S. Werner (eds), *The Visual Neurosciences*, Vol. 1, MIT Press, Cambridge, Chapter 34, pp. 541–562.

Ungerleider, L. G. and Mishkin, M. (1982). Two cortical visual systems, *in* D. Ingle, M. A. Goodale and R. W. Mansfield (eds), *Analysis of Visual Behavior*, MIT Press, Cambridge, MA, USA, pp. 549–586.

van Essen, D. and Anderson, C. (1995). Information processing strategies and pathways in the primate visual system, *in* S. Zornetzer, J. Davis, C. Lau and T. McKenna (eds), *An Introduction to Neural and Electronic Networks*, 2 edn, Academic Press, Florida, USA, pp. 45–76.

van Hateren, J. (1992). A theory of maximizing sensory information, *Biological Cybernetics* **68**(1): 23–9.

van Hateren, J. and Ruderman, D. (1998). Independent component analysis of natural image sequences yields spatio-temporal filters similar to simple cells in primary visual cortex, *Proceedings. Biological Sciences/ The Royal Society* **265**(1412): 2315–20.

von der Heydt, R., Peterhans, E. and Baumgartner, G. (1984). Illusory contours and cortical neuron responses, *Science* **224**: 1260–2.

von der Heydt, R., Zhou, H. and Friedman, H. S. (2000). Representation of stereoscopic edges in monkey visual cortex, *Vision Research* **40**: 1955–1967.

Vos, J. J. and Walraven, P. L. (1971). On the derivation of the foveal receptor primaries, *Vision Research* **11**: 799–818.

Wachtler, T., Sejnowski, T. and Albright, T. (2003). Representation of color stimuli in awake macaque primary visual cortex, *Neuron* **37**(4): 681–91.

Walker, M., Fitzgibbon, E. and Goldberg, M. (1995). Neurons in the monkey superior colliculus predict the visual result of impending saccadic eye movements, *Journal of Neurophysiology* **73**(5): 1988–2003.

Wandell, B. A. (1995). *Foundations of Vision*, Sinauer Associates Inc.

Watanabe, M., Tanaka, H., Uka, T. and Fujita, I. (2002). Disparity-selective neurons in area V4 of macaque monkeys, *Journal of Neurophysiology* **87**(4): 1960–1973.

Watson, A. (1986). Temporal sensitivity, *in* K. Boff, L. Kaufman and J. Thomas (eds), *Handbook of Perception and Human Performance*, Vol. 1, Wiley, New York, Chapter 6.

Watson, A. and Ahumada, A. (1985). Model of human visual-motion sensing, *Journal of the Optical Society of America. A* **2**: 322–342.

Webster, M. and Mollon, J. (1991). Changes in colour appearance following post-receptoral adaptation, *Nature* **349**(6306): 235–238.

Weiss, Y., Simoncelli, E. and Adelson, E. (2002). Motion illusions as optimal percepts, *Nature Neuroscience* **5**(6): 598–604.

White, E. (1989). *Cortical Circuits*, Birkhauser.

Wiesel, T. (1959). Recording inhibition and excitation in the cat's retinal ganglion cells with intracellular electrodes, *Nature* **183**: 264–265.

Williford, T. and Maunsell, J. (2006). Effects of spatial attention on contrast response functions in macaque area V4, *Journal of Neurophysiology* **96**(1): 40–54.

Wolfe, J. (2001). Asymmetries in visual search: an introduction, *Perception & Psychophysics* **63**(3): 381–9.

Wolfe, J. and Franzel, S. (1988). Binocularity and visual search, *Perception & Psychophysics* **44**(1): 81–93.

Wolfe, J., Cave, K. and Franzel, S. L. (1989). Guided search: an alternative to the feature integration model for visual search, *Journal of Experimental Psychology: Human Perception and Performance* **15**: 419–433.

Wolfe, J. M. (1998). Visual search, a review, *in* H. Pashler (ed.), *Attention*, Psychology Press Ltd., Hove, East Sussex, UK, pp. 13–74.

Wolfson, S. and Landy, M. (1995). Discrimination of orientation-defined texture edges, *Vision Research* **35**(20): 2863–77.

Wurtz, R., Goldberg, M. and Robinson, D. (1982). Brain mechanisms of visual attention, *Scientific American* **246**(6): 124–135.

Yen, S.-C. and Finkel, L. (1998). Extraction of perceptually salient contours by striate cortical networks, *Vision Research* **38**(5): 719–41.

Yuille, A. and Grzywacz, N. (1988). A computational theory for the perception of coherent visual motion, *Nature* **333**(6168): 71–74.

Yuille, A. and Kersten, D. (2006). Vision as Bayesian inference: analysis by synthesis?, *Trends in Cognitive Sciences* **10**(7): 301–308.

Zhang, K. (1996). Representation of spatial orientation by the intrinsic dynamics of the head-direction cell ensemble:

a theory, *The Journal of Neuroscience* **16**(6): 2112–2126.

Zhang, X., Zhaoping, L., Zhou, T. and Fang, F. (2011). Neural activities in V1 create a bottom-up saliency map, *Neuron* **73**: 183–192.

Zhaoping, L. (2002). Pre-attentive segmentation and correspondence in stereo, *Philosophical Transactions of the Royal Society of London. Series B: Biological Sciences* **357**(1428): 1877–1883.

Zhaoping, L. (2003). V1 mechanisms and some figure-ground and border effects, *Journal of Physiology, Paris* **97**(4-6): 503–515.

Zhaoping, L. (2004). V1 mechanisms explain filling-in phenomena in texture perception and visual search, *Journal of Vision* **4**(8): article 689.

Zhaoping, L. (2005a). Border ownership from intracortical interactions in visual area V2, *Neuron* **47**(1): 143–153.

Zhaoping, L. (2005b). The primary visual cortex creates a bottom-up saliency map, *in* L. Itti, G. Rees and J. Tsotsos (eds), *Neurobiology of Attention*, Elsevier, Chapter 93, pp. 570–575.

Zhaoping, L. (2006a). Overcomplete representation for fast attentional selection by bottom up saliency in the primary visual cortex, *Perception* **35**: ECVP Abstract Supplement, page 233. Presented at European Conference on Visual Perception, August, 2006, St Petersburg, Russia.

Zhaoping, L. (2006b). Theoretical understanding of the early visual processes by data compression and data selection, *Network: Computation in Neural Systems* **17**(4): 301–334.

Zhaoping, L. (2008). Attention capture by eye of origin singletons even without awareness—a hallmark of a bottom-up saliency map in the primary visual cortex, *Journal of Vision* **8**(5): article 1.

Zhaoping, L. (2011). A saliency map in cortex: Implications and inference from the representation of visual space, *Perception* **40**: ECVP Abstract Supplement, page 162. Presented at European Conference on Visual Perception, August, 2011, Toulouse, France.

Zhaoping, L. (2012). Gaze capture by eye-of-origin singletons: Interdependence with awareness, *Journal of Vision* **12**(2): article 17.

Zhaoping, L. (2013a). Different perceptual decoding architectures for the central and peripheral vision revealed by dichoptic motion stimuli, *Perception* **42**: ECVP Abstract Supplement, page 21.

Zhaoping, L. (2013b). A theory of the primary visual cortex (V1): Predictions, experimental tests, and implications for future research, *Perception* **42**: ECVP Abstract Supplement, page 84.

Zhaoping, L. and Anzai,, A. (in preparation). A chart demonstrating variations in acuity with retinal position.

Zhaoping, L. and Frith, U. (2011). A clash of bottom-up and top-down processes in visual search: the reversed letter effect revisited, *Journal of Experimental Psychology: Human Perception and Performance* **37**(4): 997–1006.

Zhaoping, L. and Guyader, N. (2007). Interference with bottom-up feature detection by higher-level object recognition, *Current Biology* **17**(1): 26–31.

Zhaoping, L. and Jingling, L. (2008). Filling-in and suppression of visual perception from context: A Bayesian account of perceptual biases by contextual influences, *PLoS Computational Biology* **4**(2): e14.

Zhaoping, L. and May, K. (2007). Psychophysical tests of the hypothesis of a bottom-up saliency map in primary visual cortex, *PLoS Computational Biology* **3**(4): e62.

Zhaoping, L. and Meng, G. (2011). Dichoptic completion, rather than binocular rivalry or binocular summation, *i-Perception* **2**(6): 611–614.

Zhaoping, L. and Snowden, R. (2006). A theory of a saliency map in primary visual cortex (V1) tested by psychophysics of color-orientation interference in texture segmentation, *Visual Cognition* **14**(4-8): 911–933.

Zhaoping, L. and Zhe, L. (2012a). Properties of V1 neurons tuned to conjunctions of visual features: application of the V1 saliency hypothesis to visual search behavior, *PLoS One* **7**(6): e36223.

Zhaoping, L. and Zhe, L. (2012b). V1 saliency theory makes quantitative, zero parameter, prediction of reaction times in visual search of feature singletons, *Journal of Vision* **12**(9): 1160–1160.

Zhaoping, L., Guyader, N. and Lewis, A. (2009). Relative contributions of 2D and 3D cues in a texture segmentation task, implications for the roles of striate and extrastriate cortex in attentional selection, *Journal of Vision* **9**(11): article 20.

Zhaoping, L., Geisler, W. and May, K. (2011). Human wavelength discrimination of monochromatic light explained by optimal wavelength decoding of light of unknown intensity, *PLoS One* **6**(5): e19248.

Zhou, H., Friedman, H. and von der Heydt, R. (2000). Coding of border ownership in monkey visual cortex, *The Journal of Neuroscience* **20**(17): 6594–6611.

Zhou, L. and Zhaoping, L. (2010). The salience of absence: when a hole is more than the sum of its parts, *Journal of Vision* **10**(7): article 1281.

Zigmond, M., Bloom, F. E., Landis, S. C., Roberts, J. L. and Squire, L. R. (1999). *Fundamental Neuroscience*, Academic Press, New York.

Zucker, S. W., Dobbins, A. and Iverson, L. (1989). Two stages of curve detection suggest two styles of visual computation, *Neural Computation* **1**: 68–81.

Index